K	stress concentration factor
k	spring constant, constant
L	length
M	moment, bending moment
M_p	plastic moment
m	mass, moment caused by virtual unit load
N	number of revolutions per minute
n	number, ratio of moduli of elasticity
P	force, concentrated load
p	pressure intensity
Q	first or statical moment of area A_{fghj} around neutral axis
q	distributed load intensity, shear flow
R	reaction, radius
S	elastic section-modulus ($S = I/c$)
r	radius, radius of gyration
T	torque, temperature
t	thickness, width, tangential deviation
U	strain energy, work
u	internal force caused by virtual unit load, axial or radial displacement
V	shearing force (often vertical), volume
v	transverse deflection of beam, velocity
W	total weight, work
w	weight or load per unit of length
y	distance from neutral axis
Z	plastic section modulus
psi	pounds per square inch
rad	radian
rpm	revolutions per minute
S	S-shape (standard) steel beam
s	second
ult	ultimate
W	W-shape (wide flange) steel beam
yp	yield point

GREEK LETTER SYMBOLS

α	(alpha)	linear coefficient of thermal expansion, general angle
γ	(gamma)	shearing strain, weight per unit volume
Δ	(delta)	total deformation or deflection, change of any designated function
ε	(epsilon)	normal strain
θ	(theta)	slope angle for elastic curve, angle of inclination of line on body
κ	(kappa)	curvature
λ	(lambda)	eigenvalue in column buckling problems
ν	(nu)	Poisson's ratio
ρ	(rho)	radius, radius of curvature
σ	(sigma)	tensile or compressive stress (i.e., normal stress)
τ	(tau)	shearing stress
ϕ	(phi)	total angle of twist, general angle

Mechanics
of Materials SECOND EDITION

Mechanics
of Materials SECOND EDITION

E. P. POPOV
Professor of Civil Engineering
University of California, Berkeley

Text in Collaboration with:

S. Nagarajan
Associate Research Scientist
Lockheed Missiles & Space Company
Sunnyvale, California

Problems with Assistance of

Z.A. Lu
Associate, T. Y. Lin International, Inc.
San Francisco, California

PRENTICE-HALL, INC., Englewood Cliffs, New Jersey

Library of Congress Cataloging in Publication Data

POPOV, EGOR PAUL.
　　Mechanics of materials.

　Includes index.
　　1. Strength of materials. I. Nagarajan, Sam-
murthy, joint author. II. Lu, Zung An,
joint author. III. Title.
TA405.P68 1976　　620.1'12　　75-34466
ISBN　0-13-5713560-0

Civil Engineering and Engineering Mechanics Series
N. M. Newmark and W. J. Hall, Editors

Printed in the United States of America

10　9　8　7　6

Prentice-Hall International, Inc., *London*
Prentice-Hall of Australia Pty. Limited, *Sydney*
Prentice-Hall of Canada, Ltd., *Toronto*
Prentice-Hall of India Private Limited, *New Delhi*
Prentice-Hall of Japan, Inc., *Tokyo*
Prentice-Hall of Southeast Asia Pte. Ltd., *Singapore*

Contents

Abbreviations and Symbols: See Inside Front Cover

8. Analysis of Plane Stress and Strain   235

Appendix to Chapter 8
Transformation of Moments of Inertia of Areas to Different Axes   266

9. Combined Stresses—Pressure Vessels—Failure Theories   275

*14. Structural Connections 501

15. The Energy Methods 525

*16. Thick-Walled Cylinders 557

Preface to
the Second Edition

In the mid-sixties strong pressures developed to revise the first edition of this book. At that time this was prompted mainly by the need for a text with a more mathematically rigorous approach, which was favored in some of our engineering schools. In revising and re-arranging the original text, some of the sections were entirely re-written and some new material added. This changed the character of the book, and it was re-titled *Introduction to Mechanics of Solids* (Prentice-Hall, Inc., 1968). The new book had an excellent reception. Nevertheless, demand for the original text continued. It is remarkable that an unrevised engineering text, with numerous others available, was being reprinted twenty-three years after it had made its first appearance. Therefore it was decided to up-date the *original* text wherever necessary, but basically leave the treatment very similar to what it was before. This implied a practical orientation of the text, a gradual development of the subject, and adherance to the widely used beam sign convention for shears and moments. The coming change to the Systeme Internationale system of units provided the immediate impetus for this work.

In this edition the SI system of units is used side-by-side with the English. It is hoped that this approach will prove useful, both to the student and the teacher, for the necessary transition to the new units. If this subject were pure Newtonian mechanics, a strong temptation for consistency may have suggested the use of the SI system of units only. But in a design-oriented subject such as this is, and for a text which definitely emphasizes practical applications, this does not seem possible now. The data available on the mechanical properties of materials, as well as the commercially available sizes of angles, bolts, steel beams, timber, etc., in the United States, are still being given in the English system of units. It will take some time before such information becomes generally available in SI units; in the meantime, confusing as this may be, an engineering student will have to become acquainted with the dual system of units.

This book is designed for use in an undergraduate course in Strength or Mechanics of Materials. Fundamental principles of the subject are emphasized throughout. Applications are selected from the various fields of engineering.

It is assumed that the reader has completed a course in Statics. However, those topics which are particularly important in Mechanics of Materials are reviewed where introduced.

The various articles are arranged in a logical sequence that has proved very effective in the author's teaching experience. However, some instructors may wish to combine the study of Chapter 1 with Chapter 4, treating this part of the course as a review of Statics. Some may find it advantageous to proceed to the articles in Chapter 10 on the construction of shear and moment diagrams right after the study of Chapter 4. For this purpose Articles 10-5 through 10-8 are recommended. Other instructors may find it desirable to introduce combined stresses by assigning for simultaneous reading the early articles in Chapter 8 with those in Chapter 9.

The book contains more material than can be covered in a one-quarter or a one-semester course. To assist instructors in selecting material for such courses, articles of an advanced or specialized nature are preceded by an asterisk for possible omission. Moreover, with very few exceptions, each chapter is written so as to introduce *gradually* the more complex material. Thus, study of a particular topic may be terminated where desired. Chapters 14 and 16 may be entirely omitted. On the other hand, Chapters 10, 11, 12, 13, and 14 can form a basis for an introductory design course in structural steel, whereas Chapter 16 may be covered with more mathematically inclined classes.

More advanced topics are interspersed throughout the book wherever justified for logical development of the subject. This treatment has two desirable effects. First, the more inquisitive reader is presented with the elaborate treatment he prefers. Second, the book can serve as a reference work after it has served its purpose as a text.

Among the more advanced topics treated in this text are the generalized Hooke's law, stress concentrations, inelastic torsion of shafts, plastic analysis of beams, curved bars, shear center, Mohr's circles of stress and strain, strain rosettes, a description of the photoelastic method of stress analysis, force and displacement methods of indeterminate analysis, virtual work method for deflection of beams and trusses, and analysis of thick-walled cylinders. The book includes an extensive practical treatment of concentrically and eccentrically loaded columns, as well as self-contained treatment of structural connections, including those with high-strength bolts, and welding.

Numerous illustrative examples are given to show not only how to set up a problem, but to explain the limitations of the solution. A large number of problems for solution appear at the end of each chapter. These are presented to parallel the text discussion and are arranged approximately in order of difficulty. The longer or more difficult problems are identified with an asterisk. Answers are given to many problems. More than a third of all the problems were used in examinations. In many instances the data given are selected so as to simplify numerical solution for the reader. Some problems are academic, designed to emphasize the principles studied. And to maintain student interest, realistic problems are interspersed throughout the text. Many problem solutions require the use of free-body diagrams to bolster the student's knowledge of Statics and to make the course in Mechanics of Materials truly continuous with the one in Statics.

The development of this book was strongly influenced by the author's colleagues, his students, and the numerous books published both here and abroad. The privilege of studying under S. Timoshenko and T. von Karman remains memorable. Special gratitude is due, however, as in the first edition, to the author's colleagues in the Division of Structural Engineering and Structural Mechanics in the Department of Civil Engineering at the University of California, Berkeley. Of this group the author wishes to especially thank Professors H. D. Eberhart, K. S. Pister, and A. C. Scordelis for stimulating discussions, for constructive criticism, and their generous assistance with the problems for solution. Professor R. W. Clough kindly provided the photograph of a photoelastic experiment and several problems for solution. Other present and former members of the staff, including colleagues from the Mechanical Engineering Department, provided much valuable material for the problems for solution, among these it is a pleasure to acknowledge Professors F. Baron, J. Bouwkamp, B. Bresler, C. L. Monismith, J. Penzien, D. Pirtz, M. Polivka, C. W. Radcliffe, R. A. Seban, C. F. Scheffey, E. L. Wilson and the late C. T. Wiskocil.

In preparing the second edition of the book the author is greatly indebted to two of his former students who made the appearance of this book possible. Dr. S. Nagarajan reviewed the entire text, revised example problems where appropriate into the SI units, and prepared drafts on the new material. Dr. Z. A. Lu assisted with the assembly of the problems for solution. In an effort to reduce possible errors, all problems offered for solution by the student were worked through mainly by J. K. Watt, and some by P. Hashimoto.

The Prentice-Hall staff was most cooperative in bringing out this new edition of the book. Lastly, the author again is deeply indebted to his wife, Irene, for her continual help with the manuscript.

E. P. POPOV
Berkeley, California

Mechanics
of Materials SECOND EDITION

1 Stress — Axial Loads

1-1. INTRODUCTION

In all engineering construction the component parts of a structure must be assigned definite physical sizes. Such parts must be properly proportioned to resist the actual or probable forces that may be imposed upon them. Thus, the walls of a pressure vessel must be of adequate strength to withstand the internal pressure; the floors of a building must be sufficiently strong for their intended purpose; the shaft of a machine must be of adequate size to carry the required torque; a wing of an airplane must safely withstand the aerodynamic loads which may come upon it in flight or landing. Likewise, the parts of a composite structure must be rigid enough so as not to deflect or "sag" excessively when in operation under the imposed loads. A floor of a building may be strong enough but yet may deflect excessively, which in some instances may cause misalignment of manufacturing equipment, or in other cases result in the cracking of a plaster ceiling attached underneath. Also a member may be so thin or slender that, upon being subjected to compressive loading, it will collapse through buckling; i.e., the initial configuration of a member may become unstable. Ability to determine the maximum load that a slender column can carry before buckling occurs, or determination of the safe level of vacuum that can be maintained by a vessel is of great practical importance.

In engineering practice, such requirements must be met with minimum expenditure of a given material. Aside from cost, at times—as in the design of satellites—the feasibility and success of the whole mission may depend on the weight of a package. The subject of *mechanics of materials*, or the *strength of materials*, as it has been traditionally called in the past, involves analytical methods for determining the **strength, stiffness** (deformation characteristics), and **stability** of the various load-carrying members. Alternately, the subject may be termed the *mechanics of solid deformable bodies*.

Mechanics of materials is a fairly old subject, generally dated from the work of Galileo in the early part of the seventeenth century. Prior to his investigations into the behavior of solid bodies under loads, constructors

1

followed precedents and empirical rules. Galileo was the first to attempt to explain the behavior of some of the members under load on a rational basis. He studied members in tension and compression, and notably beams used in the construction of hulls of ships for the Italian navy. Of course much progress has been made since that time, but it must be noted in passing that much is owed in the development of this subject to the French investigators, among whom a group of outstanding men such as Coulomb, Poisson, Navier, St. Venant, and Cauchy, who worked at the break of the nineteenth century, has left an indelible impression on this subject.

The subject of mechanics of materials cuts broadly across all branches of the engineering profession with remarkably many applications. Its methods are needed by designers of offshore structures; by civil engineers in the design of bridges and buildings; by mining engineers and architectural engineers, each of whom is interested in structures; by nuclear engineers in the design of reactor components; by mechanical and chemical engineers, who rely upon the methods of this subject for the design of machinery and pressure vessels; by metallurgists, who need the fundamental concepts of this subject in order to understand how to improve existing materials further; finally, by electrical engineers, who need the methods of this subject because of the importance of the mechanical engineering phases of many portions of electrical equipment. Mechanics of materials has characteristic methods all its own. It is a definite discipline and one of the most fundamental subjects of an engineering curriculum, standing alongside such other basic subjects as fluid mechanics, thermodynamics, and basic electricity.

The behavior of a member subjected to forces depends not only on the fundamental laws of Newtonian mechanics that govern the equilibrium of the forces but also on the *physical characteristics* of the materials of which the member is fabricated. The necessary information regarding the latter comes from the laboratory where materials are subjected to the action of accurately known forces and the behavior of test specimens is observed with particular regard to such phenomena as the occurrence of breaks, deformations, etc. Determination of such phenomena is a vital part of the subject, but this branch of the subject is left to other books.* Here the end results of such investigations are of interest, and this course is concerned with the analytical or mathematical part of the subject in contradistinction to experimentation. For the above reasons, it is seen that mechanics of materials is a blended science of experiment and Newtonian postulates of analytical mechanics. From the latter is borrowed the branch of the science called *statics*, a subject with which the reader of this book is presumed to be familiar, and on which the subject of this book primarily depends.

This text will be limited to the simpler topics of the subject. In spite of the relative simplicity of the methods employed here, however, the resulting

---

*See H. E. Davis, G. E. Troxell, and C. T. Wiskocil, *Testing and Inspection of Engineering Materials* (2nd ed.), New York: McGraw-Hill, 1955. See also L. H. Van Vlack, *Material Science for Engineers*, New York: Addison-Wesley, 1970.

techniques are unusually useful as they do apply to a vast number of technically important problems.

The subject matter can be mastered best by solving numerous problems. The number of formulas necessary for the analysis and design of structural and machine members by the methods of mechanics of materials is remarkably small; however, throughout this study the student must develop an ability to *visualize* a problem and the nature of the quantities being computed. *Complete, carefully drawn diagrammatic sketches of problems to be solved will pay large dividends in a quicker and more complete mastery of this subject.*

1-2. METHOD OF SECTIONS

One of the main problems of mechanics of materials is the investigation of the internal resistance of a body, that is, *the nature of forces set up within a body to balance the effect of the externally applied forces.* For this purpose, a uniform method of approach is employed. A complete diagrammatic sketch of the member to be investigated is prepared, on which *all* of the external forces acting on a body are shown at their respective points of application. Such a sketch is called a *free-body* diagram. All forces acting on a body, including the reactive forces caused by the supports and the weight* of the body itself, are considered external forces. Moreover, since a stable body at rest is in equilibrium, the forces acting on it satisfy the equations of static equilibrium. Thus, if the forces acting on a body such as shown in Fig. 1-1(a) satisfy the equations of static equilibrium and are all shown acting on it, the sketch represents a free-body diagram. Next, since a determination of the internal forces caused by the external ones is one of the principal concerns of this subject, an arbitrary section is passed through the body, completely separating it into two parts. The result of such a process can be seen in Figs. 1-1(b) and (c) where an arbitrary plane $ABCD$ separates the original solid body of Fig. 1-1(a) into two *distinct* parts. This process will be referred to as the *method of sections.*

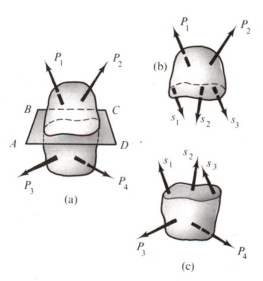

Fig. 1-1. Sectioning of a body

Then, if the body as a whole is in equilibrium, *any part* of it must also be in equilibrium. For such parts of a body, however, some of the forces necessary to maintain equilibrium must act at the cut section. These consi-

---

*Strictly speaking, the weight of the body, or more generally, the inertial forces due to acceleration, etc., are "body forces," and act throughout the body in a manner associated with the units of volume of the body. However, in most instances, these body forces can be considered as external loads.

derations lead to the following fundamental conclusion:

the externally applied forces to one side of an arbitrary cut must be balanced by the internal forces developed at the cut, or briefly, the external forces are balanced by the internal forces. Later it will be seen that the cutting planes will be oriented in a particular direction to fit special requirements. However, the above concept will be relied upon as a first step in solving *all* problems where internal forces are being investigated.

In discussing the method of sections, it is significant to note that some bodies, although not in static equilibrium, may be in dynamic equilibrium. These problems can be reduced to problems of static equilibrium. First, the acceleration of the part in question is computed, then it is multiplied by the mass of the body, giving a force $F = ma$. If the force so computed is applied to the body at its mass center in a direction opposite to the acceleration, the dynamic problem is reduced to one of statics. This is the so-called *d'Alembert principle*. With this point of view, all bodies can be thought of as being instantaneously in a state of static equilibrium. Hence for any body, whether in static or dynamic equilibrium, a free-body diagram can be prepared on which the necessary forces to maintain the body as a whole in equilibrium can be shown. From then on the problem is the same as discussed above.

1-3. STRESS

In general, the internal forces acting on infinitesimal areas of a cut may be of varying magnitudes and directions, as is shown diagrammatically in Figs. 1-1(b) and (c). These internal forces are vectorial in nature and maintain in equilibrium the externally applied forces. In mechanics of materials it is particularly significant to determine the *intensity* of these forces on the various portions of the cut, as resistance to deformation and the capacity of materials to resist forces depend on these intensities. In general, these intensities of force acting on infinitesimal areas of the cut vary from point to point, and, in general, they are inclined with respect to the plane of the cut. In engineering practice it is customary to resolve this intensity of force perpendicular and parallel to the section investigated. Such resolution of the intensity of a force on an infinitesimal area is shown in Fig. 1-2. The intensity of the force *perpendicular* or *normal to the section* is called the *normal stress* at a point. In this book it will be designated by the Greek letter σ (sigma). As a particular stress generally holds true only at a point, it is defined mathematically as

$$\sigma = \lim_{\Delta A \to 0} \frac{\Delta F}{\Delta A}$$

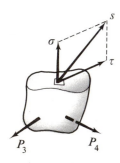

Fig. 1-2. The normal and shearing components of stress

where F is a force acting *normal* to the cut, while A is the corresponding area. It is customary to refer to the normal stresses that cause traction or tension on the surface of a cut as *tensile* stresses. On the other hand, those that are pushing against the cut are *compressive* stresses.

The other component of the intensity of force acts *parallel to the plane of the elementary area*, as in Fig. 1-2. This component of the intensity of force is called the *shearing stress*. It will be designated by the Greek letter τ(tau). Mathematically it is defined as

$$\tau = \lim_{\Delta A \to 0} \frac{\Delta V}{\Delta A}$$

where A represents area, and V is the component of the force parallel to the cut. It should be noted that these definitions of stresses at a point involve the concept of letting $\Delta A \to 0$ and may be questionable from a strictly atomic view of matter. However, the homogeneous model implied by these equations has been a good approximation to inhomogenous matter on the macroscopic level. Therefore, this so-called phenomenological approach is used.

The student should form a clear mental picture of the stresses called normal and those called shearing. To repeat, normal stresses result from force components perpendicular to the plane of the cut, while shearing stresses result from components parallel to the plane of the cut.

It is seen from the above definitions of normal and shearing stresses that, since they represent the intensity of force on an area, stresses* are measured in units of force divided by units of area. Since a force is a vector and an area is a scalar, their ratio, which represents the component of stress in a given direction, is a vectorial quantity.†

It should be noted that *stresses multiplied by the respective areas on which they act give forces, and it is the sum of these forces at an imaginary cut that keeps a body in equilibrium.*

In the English system, the usual units for stress are pounds per square inch, abbreviated in this text as "psi." In many cases it will be found convenient to use as a unit of force the coined word "kip," meaning kilo-pound or 1,000 lb. The stress in kips per square inch is abbreviated as "ksi." It should be noted that the unit pound referred to here implies a pound-force, not a pound-mass. Such ambiguities are avoided in the modernized version of the metric system referred to as the International System of Units or SI units.‡ SI units are being increasingly adopted and will be used in this text along with the conventional English system in order to facilitate a smooth transition. The base units in the SI are *meter* (m) for length, *kilogram* (kg) for mass, and *second* (s) for time. The derived unit for area is a *square meter* (m²), and for acceleration a *meter per second squared* (m/s²). The unit of force is defined as a unit mass subjected to a unit acceleration, i.e., *kilogram-meter per second squared* (kg·m/s²), and is designated a *newton* (N). The unit of stress is the *newton per square meter* (N/m²), also designated a *pascal* (Pa). Multiple and submultiple prefixes representing steps of 1 000 are recommen-

*In some books the term "unit stress" is used to indicate stress per unit of area. However, in this text the word "stress" is used for this concept.

†For further details see Art. 8-2.

‡From the French, Systéme International d'Unités.

ded. For example, force can be shown in *millinewtons* (1 mN = 0.001 N), *newtons*, or *kilonewtons* (1 kN = 1 000 N), length in *millimeters* (1 mm = 0.001 m), *meters*, or *kilometers* (1 km = 1 000 m), stresses in *kilopascals* (1 kPa = 10^3 Pa), *megapascals* (1 MPa = 10^6 Pa), *gigapascals* (1 GPa = 10^9 Pa), etc.*

The stress expressed numerically in units of N/m² may appear to be unusually small to those familiar with the English system of units. This is because the force of one newton is small in relation to a pound-force, and a square meter is associated with a much larger area than one square inch. Therefore, to some it may be more acceptable to think in terms of a force of one newton acting on one square millimeter. Since the notation N/mm² is not recommended, however, one can simply employ its equivalent, the megapascal (MPa).

If in addition to a plane such as *ABCD* in Fig. 1-1(a) another plane an infinitesimal distance away and parallel to the first were passed through the body, a thin element of the body would be isolated. Then, if an additional two pairs of planes were passed normal to the first pair, an elementary cube of infinitesimal dimensions would be isolated from the body. Such a cube is shown in Fig. 1-3. Here, for identification purposes, the process of resolution of stresses into components has been carried further than discussed above. At each surface the shearing stress τ has been resolved into two components parallel to a particular set of axes. The subscripts of the σ's designate the direction of the normal stress along a particular axis, while the stress itself acts on a plane perpendicular to the same axis. The first subscripts of the τ's associate the shearing stress with a plane that is perpendicular to a given axis, while the second designate the direction of the shearing stress.

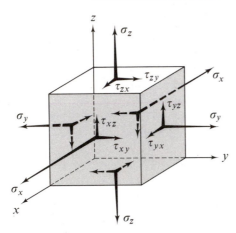

Fig. 1-3. The most general state of stress acting on an element

An infinitesimal cube, as shown in Fig. 1-3, could be used as the basis for an exact formulation of the problem in mechanics of materials. However, the methods for the study of the behavior of such a cube (which involve the writing of an equation for its equilibrium and making certain that such a cube, after deformations caused in it by the action of forces will be geometrically compatible with the adjoining infinitesimal cubes) are beyond the scope of this course. They are in the realm of the mathematical theory of elasticity. The procedures used in this text do not resort to the generality implied in Fig. 1-3. The methods used here will be much simpler.

*A detailed discussion of SI units, including conversion factors, rules for SI style, and usage can be found in a comprehensive guide published by the American Society for Testing and Materials under the designation ASTM E-380-1974. For convenience, a short table of conversion factors is included on the inside of the back cover.

1-4. AXIAL LOAD; NORMAL STRESS

In many practical situations, if the direction of the imaginary plane cutting a member is judiciously selected, the stresses that act on the cut will be found both particularly significant and simple to determine. One such important case occurs in a straight *axially* loaded rod in tension, *provided a plane is passed perpendicular to the axis of the rod*. The tensile stress acting on such a cut is the *maximum* stress, as any other cut not perpendicular to the axis of the rod provides a larger surface for resisting the applied force. The maximum stress is the most significant one, as it tends to cause the failure of the material.*

To obtain an algebraic expression for this maximum stress, consider the case illustrated in Fig. 1-4(a). If the rod is assumed weightless, two equal and opposite forces P are necessary, one at each end to maintain equilibrium. Then, as stated in Art. 1-2, since the body as a whole is in equilibrium, any part of it is also in equilibrium. A part of the rod to either side of the cut x-x is in equilibrium. At the cut, where the cross-sectional area of the rod is A, a force equivalent to P, as shown in Figs. 1-4(b) and (c), must be developed. Whereupon, from the definition of stress, the normal stress, or the stress that acts perpendicularly to the cut, is

$$\sigma = \frac{P}{A} \quad \text{or} \quad \frac{force}{area} \quad \left[\frac{\text{N}}{\text{m}^2}\right] \quad \text{or} \quad \left[\frac{\text{lb}}{\text{in.}^2}\right] \quad (1\text{-}1)$$

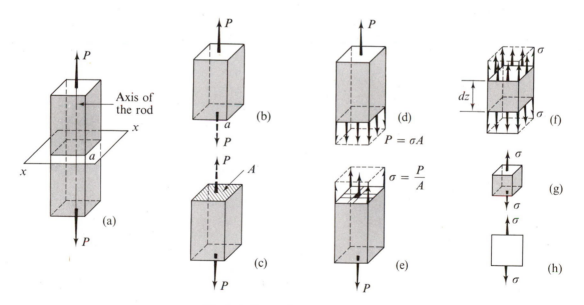

Fig. 1-4. Successive steps in the analysis of a body for stress

*Some materials exhibit a far greater relative strength to normal stresses than to shearing stresses. For such materials, failure takes place on an oblique plane. This will be discussed in Chapter 9.

This *normal* stress is *uniformly* distributed over the cross-sectional area A.* The nature of the quantity computed by Eq. 1-1 may be seen graphically in Figs. 1-4(d) and (e). *In general, the force P is a resultant of a number of forces to one side of the cut or another.*

If an additional cut is made parallel to the plane *x-x* in Fig. 1-4(a), the isolated section of the rod could be represented as in Fig. 1-4(f), and upon further "cutting," an infinitesimal cube as in Fig. 1-4(g) results. The only kind of stresses that appear here are the normal stresses on the two surfaces of the cube. Such a *state of stress* on an element is referred to as *uniaxial stress.* In practice, isometric views of a cube as shown in Fig. 1-4(g) are seldom employed; the diagrams are simplified to look like those of Fig. 1-4(h). Nevertheless, the student must never lose sight of the three-dimensional aspect of the problem at hand.

At a cut, the system of tensile stresses computed by Eq. 1-1 provides an equilibrant to the externally applied force. When these normal stresses are multiplied by the corresponding infinitesimal areas and then summed over the whole area of a cut, the summation is equal to the applied force P. Thus the system of stresses is *statically equivalent* to the force P. Moreover, the resultant of this sum must act through the *centroid* of a section. Conversely, to have a uniform stress distribution in a rod, the applied axial force must act through the centroid of the cross-sectional area investigated. For example, in the machine part shown in Fig. 1-5(a) the stresses cannot be obtained from Eq. 1-1 alone. Here, at a cut such as *A-A*, a statically equivalent system of forces developed within the material must consist not only of the force P but also of a bending moment M that must maintain the externally applied force in equilibrium. This causes nonuniform stress distribution in the member. This will be treated in Chapter 7.

In accepting Eq. 1-1, it must be kept in mind that the material's behavior is *idealized*. Each and every particle of a body is assumed to contribute equally to the resistance of the force. A perfect homogeneity of the material is implied by such an assumption. Real materials, such as metals, consist of a great many grains, while wood is fibrous. In real materials, some particles will contribute more to the resistance of a force than others. Stresses as shown in Figs. 1-4(d) and (e) actually do not exist. The diagram of true stress distribution varies in each particular case and is a highly irregular, jagged affair. However, on the average, or statistically speaking, computations based on Eq. 1-1 are

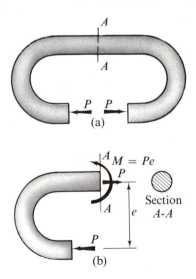

Fig. 1-5. A member with a nonuniform stress distribution at Section *A-A*

---

*Equation 1-1 strictly applies only if the cross-sectional area is constant along the rod. For a discussion of situations where an abrupt discontinuity in the cross-sectional area occurs, see Art. 2-11.

correct, and hence the computed stress represents a highly significant quantity.

Similar reasoning applies to compression members. The maximum normal or compressive stress can also be obtained by passing a section perpendicular to the axis of a member and applying Eq. 1-1. The stress so obtained will be of uniform intensity as long as the resultant of the applied forces coincides with the *centroid* of the area at the cut. However, one must exercise additional care when compression members are investigated. These may be so slender that they may not behave in the fashion considered. For example, an ordinary yardstick under a rather small axial compression force has a tendency to buckle sideways and collapse. The consideration of such *instability* of compression members is deferred until Chapter 13. *Equation 1-1 is applicable only for axially loaded compression members that are rather chunky*, i.e., to short blocks. As will be shown in Chapter 13, a block whose *least* dimension is approximately one-tenth of its length may usually be considered a short block. For example, a 2 in. by 4 in. wooden piece may be 20 in. long and still be considered a short block.

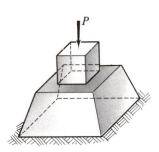

Fig. 1-6. Bearing stresses occur between the block and pier

Situations often arise where one body is supported by another. If the resultant of the applied forces coincides with the centroid of the contact area between the two bodies, the intensity of force, or stress, between the two bodies can again be determined from Eq. 1-1. It is customary to refer to this normal stress as a *bearing stress*. Figure 1-6, where a short block bears on a concrete pier and the latter bears on the soil, illustrates such a stress. The bearing stresses are obtained by dividing the applied force P by the corresponding area of contact.

1-5. AVERAGE SHEARING STRESS

Another situation that frequently arises in practice is shown in Figs. 1-7(a), (c), and (e). In all of these cases the forces are transmitted from one part of a body to the other by causing stresses in the plane parallel to the applied force. To obtain stresses in such instances, cutting planes as A-A are selected and free-body diagrams* as shown in Figs. 1-7(b), (d), and (f) are used. The forces are transmitted through the respective cut areas. Hence, *assuming* that the stresses that *act in the plane of these cuts are uniformly distributed*, one obtains a relation for stress

$$\tau = \frac{P}{A} \quad \text{or} \quad \frac{force}{area} \quad \left[\frac{N}{m^2}\right] \quad \text{or} \quad \left[\frac{lb}{in.^2}\right] \tag{1-2}$$

*A small unbalance in moment equal to Pe exists in the first two cases shown in Fig. 1-7, but, being small, is commonly ignored.

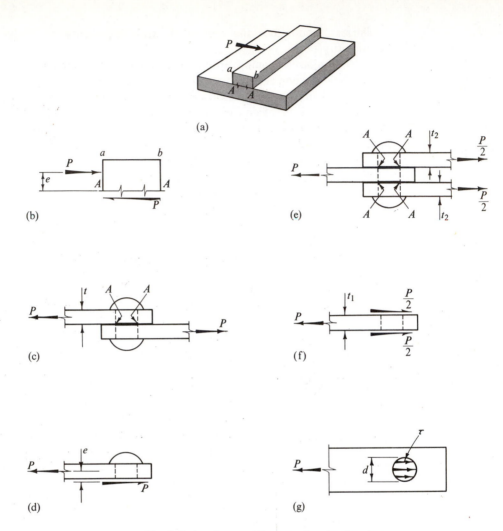

Fig. 1-7. Loading conditions causing shearing stresses

where τ by definition is the *shearing* stress, P is the total force acting across and parallel to the cut, often called *shear*, and A is the cross-sectional area of the cut member. For reasons to be discussed later, unlike normal stress, the shearing stress given by Eq. 1-2 is only *approximately* true. For the cases shown, the shearing stresses actually are distributed in a nonuniform fashion across the area of the cut. The quantity given by Eq. 1-2 represents an *average* shearing stress.

The shearing stress, as computed by Eq. 1-2, is shown diagrammatically in Fig. 1-7(g). Note that for the case shown in Fig. 1-7(e) there are *two planes* of the rivet that resist the force. Such a rivet is referred to as being in *double* shear.

In cases such as those in Figs. 1-7(c) and (e), as the force P is applied, a highly irregular pressure develops between a rivet or a bolt and the plates.

The *average* nominal intensity of this pressure is obtained by dividing the force transmitted by the projected area of the rivet onto the plate. This is referred to as the *bearing stress*. The bearing stress in Fig. 1-7(c) is $\sigma_b = P/(td)$, where t is the thickness of the plate and d is the diameter of the rivet. For the case in Fig. 1-7(e) the bearing stresses for the middle plate and the outer plates are $\sigma_1 = P/(t_1d)$ and $\sigma_2 = P/(2t_2d)$, respectively.

1-6. PROBLEMS IN NORMAL AND SHEARING STRESS

Once P and A are determined in a given problem, Eqs. 1-1 and 1-2 are easy to apply. These equations have a clear physical meaning. Moreover, it seems reasonably clear that the desired magnitudes of stresses are the *maximum stresses*, as they are the greatest imposition on the strength of a material. The greatest stresses occur at a cut or section of *minimum* cross-sectional area and the greatest axial force. Such sections are called *critical sections*. The critical section for the particular arrangement being analyzed can usually be found by inspection. However, to determine the force P that acts through a member is usually a more difficult task. In the majority of problems treated in this text the latter information is obtained from statics.

For the equilibrium of a body in space, the equations of statics require the fulfillment of the following conditions:

$$\left. \begin{array}{ll} \sum F_x = 0 & \sum M_x = 0 \\ \sum F_y = 0 & \sum M_y = 0 \\ \sum F_z = 0 & \sum M_z = 0 \end{array} \right\} \qquad (1\text{-}3)$$

The first column of Eq. 1-3 states that the sum of *all* forces acting on a body in any (x, y, z) direction must be zero. The second column notes that the summation of moments of *all* forces around *any* axis parallel to any (x, y, z) direction must also be zero for equilibrium. In a *planar* problem, i.e., all members and forces lie in a single plane such as the x-y plane, relations $\sum F_z = 0$, $\sum M_x = 0$, and $\sum M_y = 0$, while still valid, are trivial.

These equations of statics are directly applicable to deformable solid bodies. The deformations tolerated in engineering structures are usually negligible in comparison with the over-all dimensions of structures. Therefore, *for the purposes of obtaining the forces in members, the initial undeformed dimensions of members are used in computations.*

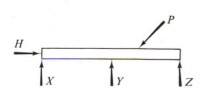

Fig. 1-8. A statically indeterminate beam

There are problems where equations of statics are not sufficient to determine the forces *in*, or those acting *on*, the member. For example, the reactions for a straight beam, shown in Fig. 1-8, supported vertically at three points, cannot be determined from statics alone. In this planar problem there are four unknown reaction components, while only three *independent* equations of statics are available. Such problems are termed *statically indeterminate*. The consideration of statically indeterminate problems is postponed until Chapter 11. For the present, and in the succeeding nine chapters of this text, *all structures and members considered will be statically determinate*, i.e., all of the external forces acting on such bodies can be determined by Eqs. 1-3. There is no dearth of statically determinate problems that are practically significant.

Equations 1-3 should already be familiar to the student. However, several examples where they are applied will now be given, the professional techniques for their use being stressed. These examples will serve as an informal review of some of the principles of statics and will show applications of Eqs. 1-1 and 1-2.

EXAMPLE 1-1

The beam *BE* in Fig. 1-9(a) is used for hoisting machinery. It is anchored by two bolts at *B*, and at *C* it rests on a parapet wall. The essential details are given in the figure. Note that the bolts are threaded as shown in Fig. 1-9(d) with $d = 16$ mm at the root of the threads. If this arrangement is used to lift equipment of 10 kN, determine the stress in the bolts *BD* and the bearing stress at *C*. Assume that the weight of the beam is negligible in comparison with the loads handled.

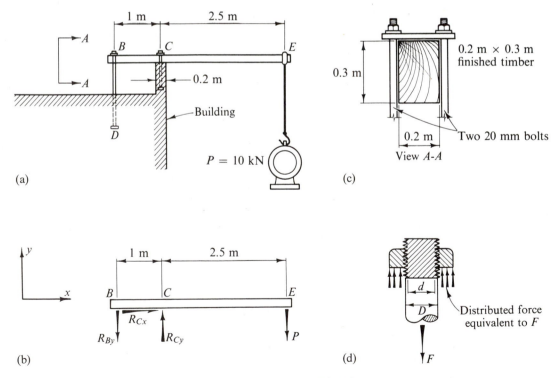

Fig. 1-9

SOLUTION

To solve this problem, the actual situation is idealized, and a free-body diagram is made on which all known and unknown forces are indicated. This is shown in Fig. 1-9(b). The vertical reactions at B and C are unknown. They are indicated respectively as R_{By} and R_{Cy}, where the first subscript identifies the location and the second the line of action of the unknown force. As the long bolts BD are not effective in resisting the horizontal force, only an unknown horizontal reaction at C is assumed and marked as R_{Cx}. The applied known force P is shown in its proper location. After a free-body diagram is prepared, the equations of statics are applied and solved for the unknown forces.

$$\Sigma F_x = 0 \qquad\qquad\qquad\qquad\qquad\qquad\qquad R_{Cx} = 0$$
$$\Sigma M_B = 0 \,\circlearrowleft +, \qquad 10(2.5 + 1) - R_{Cy}(1) = 0, \qquad R_{Cy} = 35\text{ kN} \uparrow$$
$$\Sigma M_C = 0 \,\circlearrowright +, \qquad 10(2.5) - R_{By}(1) = 0, \qquad R_{By} = 25\text{ kN} \downarrow$$
$$Check: \quad \Sigma F_y = 0 \uparrow +, \qquad\qquad\qquad -25 + 35 - 10 = 0$$

These steps complete and check the work of determining the forces. The various areas of the material that resist these forces are determined next, and Eq. 1-1 is applied.

Cross-sectional area of one 20 mm bolt: $A = \pi(0.02/2)^2 = 0.000\ 314\text{ m}^2$ This is not the minimum area of a bolt; threads reduce it.

The cross-sectional area of one 20 mm bolt at the root of the threads is

$$A_{net} = \pi(0.016/2)^2 = 0.000\ 201\text{ m}^2$$

Maximum normal tensile stress* in each of the two bolts BD:

$$\sigma_{max} = \frac{R_{By}}{2A} = \frac{25}{2(0.000\ 201)} = 62\ 000\text{ kN/m}^2 = 62 \times 10^6\text{ N/m}^2 = 62\text{ MPa}$$

Tensile stress in the shank of the bolts BD:

$$\sigma = \frac{25}{2(0.000\ 314)} = 39\ 800\text{ kN/m}^2 = 39.8 \times 10^6\text{ N/m}^2 = 39.8\text{ MPa}$$

Contact area at C:

$$A = 0.2 \times 0.2 = 0.04\text{ m}^2$$

Bearing stress at C:

$$\sigma_b = \frac{R_{Cy}}{A} = \frac{35}{0.04} = 875\text{ kN/m}^2 = 0.875 \times 10^6\text{ N/m}^2 = 0.875\text{ MPa}$$

EXAMPLE 1-2

The concrete pier shown in Fig. 1-10(a) is loaded at the top with a uniformly distributed load of 20 kN/m². Investigate the state of stress at a level of 1 m above the base. Concrete weighs approximately 25 kN/m³.

*See also discussion on stress concentrations, Art. 2-11.

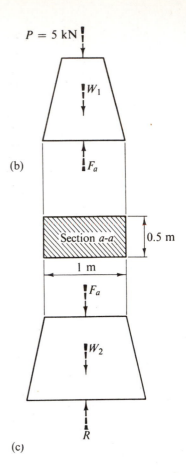

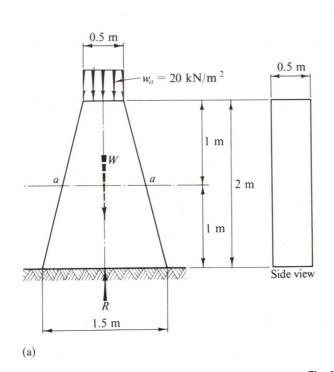

(a)

(b)

(c)

Fig. 1-10

SOLUTION

In this problem the weight of the structure itself is appreciable and must be included in the calculations.

Weight of the whole pier:

$$W = [(0.5 + 1.5)/2](0.5)(2)(25) = 25 \text{ kN}$$

Total applied force:

$$P = 20(0.5)(0.5) = 5 \text{ kN}$$

From $\sum F_y = 0$, reaction at base:

$$R = W + P = 30 \text{ kN}$$

These forces are shown schematically in the diagrams as concentrated forces acting through their respective centroids. Then, to determine the stress at the desired level, the body is cut into two separate parts. A free-body diagram for either part is sufficient to solve the problem. For comparison the problem is solved both ways.

Using the upper part of the pier as a free body, Fig. 1-10(b), weight of the pier above the cut:

$$W_1 = (0.5 + 1)(0.5)(1)(25/2) = 9.4 \, \text{kN}$$

From $\sum F_y = 0$, force at the cut:

$$F_a = P + W_1 = 14.4 \, \text{kN}.$$

Hence, using Eq. 1-1, the normal stress at the level a-a is

$$\sigma_a = \frac{F_a}{A} = \frac{14.4}{(0.5)(1)} = 28.8 \, \text{kN/m}^2$$

This stress is compressive as F_a acts on the cut.

Using the lower part of the pier as a free body, Fig. 1-10(c), weight of the pier below the cut:

$$W_2 = (1 + 1.5)(0.5)(1)(25/2) = 15.6 \, \text{kN}$$

From $\sum F_y = 0$, force at the cut:

$$F_a = R - W_2 = 14.4 \, \text{kN}$$

The remainder of the problem is the same as before. The pier considered here has a vertical axis of symmetry, making the application of Eq. 1-1 possible.*

EXAMPLE 1-3

A bracket of negligible weight shown in Fig. 1-11(a) is loaded with a force P of 3 kips. For interconnection purposes the bar ends are clevised (forked). Pertinent dimensions are shown in the figure. Find the normal stresses in the members AB and BC and the bearing and shearing stresses for the pin C. All pins are 0.375 in. in diameter.

SOLUTION

First an idealized free-body diagram consisting of the two bars pinned at the ends is prepared, Fig. 1-11(b). As there are no intermediate forces acting on the bars and the applied force acts through the joint at B, the forces in the bars are directed along the lines AB and BC, and the bars AB and BC are loaded axially. The magnitudes of the forces are unknown and are labeled F_A and F_C in the diagram.† These forces can be determined graphically by completing a triangle of forces F_A, F_C, and P. These forces may also be

---

*Strictly speaking the solution obtained is not exact, as the sides of the pier are sloping. If the included angle between these sides is large, this solution is altogether inadequate. For further details see S. Timoshenko and J. N. Goodier, *Theory of Elasticity* (3rd ed.), New York: McGraw-Hill, 1970, p. 139.

†In frameworks it is convenient to assume all unknown forces are tensile. A negative answer in the solution then indicates that the bar is in compression.

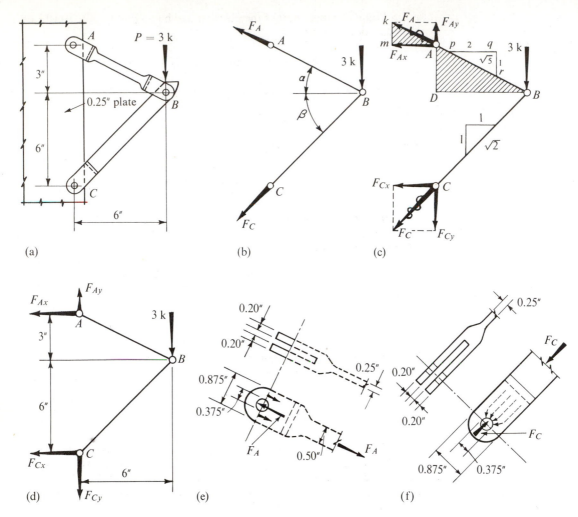

Fig. 1-11

found analytically from two simultaneous equations $\sum F_y = 0$ and $\sum F_x = 0$, written in terms of the unknowns F_A and F_C, a known force P, and two known angles α and β. Both these procedures are possible. However, in this course it will usually be found advantageous to proceed in a different way. Instead of treating forces F_A and F_C directly, their components are used; and instead of $\sum F = 0$, $\sum M = 0$ becomes the main tool.

Any force can be resolved into components. For example, F_A can be resolved into F_{Ax} and F_{Ay} as in Fig. 1-11(c). Conversely, if any one of the components of a directed force is known, the force itself can be determined. This follows from similarity of dimension and force triangles. In Fig. 1-11(c) the triangles Akm and BAD are similar triangles (both are shaded in the diagram). Hence, if F_{Ax} is known,

$$F_A = (AB/DB)F_{Ax}$$

Similarly, $F_{Ay} = (AD/DB)F_{Ax}$. However, note further that AB/DB or AD/DB are ratios, hence relative dimensions of members can be used. Such relative dimensions are shown by a little triangle on the member AB and again on BC. In the problem at hand

$$F_A = (\sqrt{5}/2)F_{Ax} \quad \text{and} \quad F_{Ay} = F_{Ax}/2$$

Adopting the above procedure of resolving forces, a revised free-body diagram, Fig. 1-11(d), is prepared. Two components of force are necessary at the pin joints. After the forces are determined by statics, Eq. 1-1 is applied several times, thinking in terms of a free body of an individual member:

$$\sum M_C = 0 \circlearrowright +, \qquad +F_{Ax}(3 + 6) - 3(6) = 0, \qquad F_{Ax} = +2 \text{ kips}$$
$$F_{Ay} = F_{Ax}/2 = 2/2 = 1 \text{ kip,}$$
$$F_A = 2(\sqrt{5}/2) = +2.23 \text{ kips}$$
$$\sum M_A = 0 \circlearrowright +, \qquad +3(6) + F_{Cx}(9) = 0,$$
$$F_{Cx} = -2 \text{ kips} \quad \text{(compression)}$$
$$F_{Cy} = F_{Cx} = -2 \text{ kips,}$$
$$F_C = \sqrt{2}(-2) = -2.83 \text{ kips}$$

Check: $\quad \sum F_x = 0, \qquad F_{Ax} + F_{Cx} = 2 - 2 = 0$
$$\sum F_y = 0, \qquad F_{Ay} - F_{Cy} - P = 1 - (-2) - 3 = 0$$

Stress in main bar AB:

$$\sigma_{AB} = \frac{F_A}{A} = \frac{2.23}{(0.25)(0.50)} = 17.8 \text{ ksi} \quad \text{(tension)}$$

Stress in clevis of bar AB, Fig. 1-11(e):

$$(\sigma_{AB})_{\text{clevis}} = \frac{F_A}{A_{\text{net}}} = \frac{2.23}{2(0.20)(0.875 - 0.375)} = 11.2 \text{ ksi} \quad \text{(tension)}$$

Stress in main bar BC:

$$\sigma_{BC} = \frac{F_C}{A} = \frac{2.83}{(0.875)(0.25)} = 12.9 \text{ ksi} \quad \text{(compression)}$$

In the compression member the net section at the clevis need not be investigated; see Fig. 1-11(f) for the transfer of forces. The bearing stress at the pin is more critical. Bearing between pin C and clevis:

$$\sigma_b = \frac{F_C}{A_{\text{bearing}}} = \frac{2.83}{(0.375)(0.20)2} = 18.8 \text{ ksi}$$

Bearing between the pin C and the main plate:

$$\sigma_b = \frac{F_C}{A} = \frac{2.83}{(0.375)(0.25)} = 30.2 \text{ ksi}$$

Double shear in the pin C:

$$\tau = \frac{F_C}{A} = \frac{2.83}{2\pi(0.375/2)^2} = 12.9 \text{ ksi}$$

For a complete analysis of this bracket, other pins should be investigated. However, it can be seen by inspection that the other pins in this case are stressed either the same amount as computed above or less.

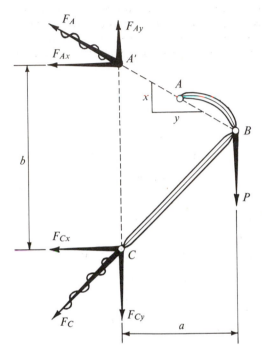

Fig. 1-12

The advantages of the method used in the above example for finding forces in members should now be apparent. It can also be applied with success in a problem such as the one shown in Fig. 1-12. The force F_A transmitted by the curved member AB acts through points A and B, since the forces applied at A and B must be collinear. By resolving this force at A', the same procedure can be followed. Wavy lines through F_A and F_C indicate that these forces are replaced by the two components shown. Alternatively, the force F_A can be resolved at A, and since $F_{Ay} = (x/y)F_{Ax}$, the application of $\sum M_C = 0$ yields F_{Ax}.

In frames where the applied forces do not act through a joint, proceed as above as far as possible. Then isolate an individual member, and using its free-body diagram, complete the determination of forces. If inclined forces are acting on the structure, resolve them into convenient components.

1-7. ALLOWABLE STRESSES; FACTOR OF SAFETY

The determination of stresses would be altogether meaningless were it not for the fact that physical testing of materials in a laboratory provides information regarding a material's resistance to stress. In a laboratory, specimens of known material, manufacturing process, and heat treatment are carefully prepared to desired dimensions. Then these specimens are subjected to successively increasing known forces. In the most widely used test, a round rod is subjected to tension and the specimen is loaded until it finally ruptures. The force necessary to cause rupture is called the *ultimate* load. By dividing this ultimate load by the *original* cross-sectional area of the specimen, the *ultimate strength* (stress) of a material is obtained. Figure 1-13 shows a testing machine used for this purpose. Figure 1-14 is a photograph of a tension-test specimen. The tensile test is used most widely. However, compression, bending, torsion, and shearing tests are also employed. Table 1 of the Appendix

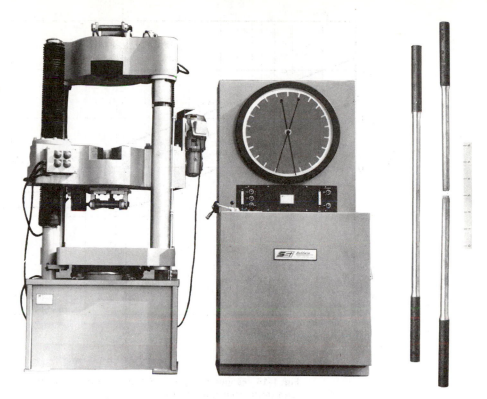

Fig. 1-13. Universal testing machine (Courtesy SATEC Systems).

Fig. 1-14. A typical tension test specimen of mild steel: (a) before fracture, (b) after fracture.

gives ultimate strengths and other physical properties for a few materials.

For the design of members the stress level called the *allowable stress* is set considerably lower than the ultimate strength found in the so-called "static" test mentioned above. This is necessary for several reasons. The exact magnitudes of the forces that may act upon the designed structure are seldom accurately known. Materials are not entirely uniform. Some of the materials stretch unpermissible amounts prior to an actual break, so to hold down these deformations, stresses must be kept low.* Some materials seriously corrode. Some materials flow plastically under a sustained load, a phenomenon called *creep*. With a lapse of time, this can cause large deformations that cannot be tolerated.

For applications where a force comes on and off the structure a number of times, the materials cannot withstand the ultimate stress of a static test. In such cases the "ultimate strength" depends on the number of times the force is applied as the material works at a particular stress level. Figure 1-15 shows the results of tests† on a number of the same kind of specimens at

*See Chapter 2 for more details.
†Zambrow, J. L., and Fontana, M. G., "Mechanical Properties, including Fatigue, of Aircraft Alloys at Very Low Temperatures," *Trans. ASM*, 1949, vol. 41, p. 498.

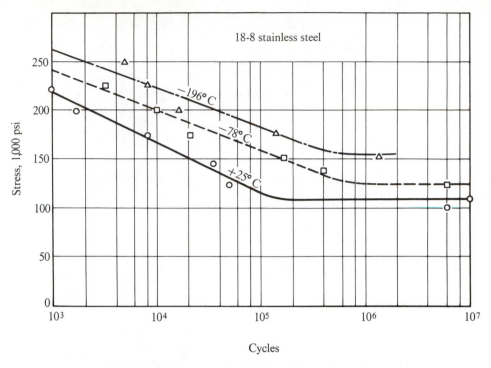

Fig. 1-15. Fatigue strength of 18-8 stainless steel at various temperatures (reciprocating beam test).

different stresses. Experimental points indicate the number of cycles required to break the specimen at a particular stress under the application of a fluctuating load. Such tests are called "fatigue tests," and the corresponding curves are termed *S-N* (stress-number) diagrams. As can be seen from Fig. 1-15, at smaller stresses the material can withstand an ever-increasing number of cycles of load application. For some materials, notably steels, the *S-N* curve for low stresses becomes essentially horizontal. This means that at a low stress an infinitely large number of reversals of stress can take place before the material fractures. The limiting stress at which this occurs is called the *endurance limit* of the material. This limit, being dependent on stress, is measured in pounds per square inch or newtons per square meter.

Some care must be exercised in interpreting *S-N* diagrams, particularly with regard to the range of the applied stress. In some tests, complete reversal (tension to compression) of stress is made; in others the applied load is varied in a different manner, such as tension to no load and back to tension. The major part of fatigue testing is done on specimens in bending.

In some cases another item deserves attention. As materials are manufactured, they are often rolled, peened, and hammered. In castings, materials cool unevenly. These processes set up high internal stresses, which are called *residual stresses*. In all cases treated in this text the materials are assumed to be entirely free of such stresses.

The aforementioned facts, coupled with the impossibility of determining stresses accurately in complicated structures and machines, necessitate a substantial reduction of stress compared to the ultimate strength of a material in a static test. For example, ordinary steel will withstand an ultimate stress in tension of 60,000 psi and more. However, it deforms rather suddenly and severely at the stress level of about 36,000 psi, and it is customary in the United States to use an allowable stress of around 22,000 psi for structural work. This allowable stress is even further reduced to about 12,000 psi for parts that are subjected to alternating loads because of the fatigue characteristics of the material. *Fatigue properties of materials are of utmost importance in mechanical equipment.* Many failures in machine parts can be traced to disregard of this important consideration. (Also see Art. 2-11.)

Large companies, as well as city, state, and federal authorities, prescribe or recommend* allowable stresses for different materials, depending on the application. Often such stresses are called the allowable *fiber*† stresses.

Since according to Eq. 1-1, stress times area is equal to a force, the allowable and ultimate stresses may be converted into the allowable and ultimate forces or "loads" that a member can resist. Also a significant ratio may be formed:

$$\frac{ultimate\ load\ for\ a\ member}{allowable\ load\ for\ a\ member}$$

This ratio is called a *factor of safety* and must always be greater than unity. Although not commonly used, perhaps a better term for this ratio is a *factor of ignorance*.

This factor is identical with the ratio of ultimate to allowable stress for tension members. For more complexly stressed members, the former definition is implied, although the ratio of stresses is actually used. As will become apparent from subsequent reading, the two are not synonymous since the stresses do not necessarily vary linearly with load.

In the aircraft industry the term factor of safety is replaced by another, defined as

$$\frac{ultimate\ load}{design\ load} - 1$$

and is known as the *margin of safety*. In normal usage this also reverts to

$$\frac{ultimate\ stress}{maximum\ stress\ caused\ by\ the\ design\ load} - 1$$

*For example, see the American Institute of Steel Construction *Manual*, Building Construction Code of any large city, ANC-5 *Strength of Aircraft Elements* issued by the Army-Navy Civil Committee on Aircraft Design Criteria, etc.

†The adjective *fiber* in the above sense is used for two reasons. Many original experiments were made on wood, which is fibrous in character. Also, in several derivations that follow, the concept of a continuous filament or fiber in a member is a convenient device for visualizing its action.

An alternative approach is to determine the ultimate collapse load* of a structure and then divide by a suitably chosen *load factor* to obtain the allowable or working load. Conversely, once the working load has been determined, members are proportioned such that the ultimate load of the structure is equal to the working load multiplied by the load factor. The two concepts, allowable stress design and ultimate load design, lead to the same results for a bar in simple tension or compression as well as for more complicated structures where failure is defined by an elastic criterion. However, significantly different designs may be obtained in most cases where inelastic material behavior is taken into account and the failure criterion is excessive plastic deformation.

1-8. DESIGN OF AXIALLY LOADED MEMBERS AND PINS

The design of members for axial forces is rather simple. From Eq. 1-1 the required area of a member is

$$A = \frac{P}{\sigma_{\text{allow}}} \tag{1-1a}$$

In all statically determinate problems the axial force P is determined from statics, and the intended use of the material sets the allowable stress. For tension members, the area A so computed is the required *net* cross-sectional area of a member. For short compression blocks, Eq. 1-1a is also applicable; however, *for slender members, do not attempt to use the above equation prior to study of the chapter on columns.*

The simplicity of Eq. 1-1a is unrelated to its importance. A large number of problems requiring its use occur in practice. The following problems illustrate some applications of Eq. 1-1a as well as provide additional review in statics.

EXAMPLE 1-4

Reduce the size of bar *AB* in Example 1-3 by using a better material such as chrome-vanadium steel. The ultimate strength of this steel is approximately 120,000 psi. Use a factor of safety of $2\frac{1}{2}$.

SOLUTION

$\sigma_{\text{allow}} = 120/2.5 = 48$ ksi. From Example 1-3 the force in the bar *AB*: $F_A = +2.23$ kips. Required area: $A_{\text{net}} = 2.23/48 = 0.0464$ in.^{2}. Adopt: 0.20-in. by 0.25-in. bar. This provides an area of $(0.20)(0.25) = 0.050$ in.^{2}, which is slightly in excess of the required area. Many other proportions of the bar are possible.

With the cross-sectional area selected, the actual or working stress is somewhat below the allowable stress: $\sigma_{\text{actual}} = 2.23/(0.050) = 44.6$ ksi.

*See Art. 12-10 for further details.

The actual factor of safety is $120/(44.6) = 2.69$, and the actual margin of safety is 1.69.

In a complete redesign, clevis and pins should also be reviewed and, if possible, decreased in dimensions.

EXAMPLE 1-5

Select members FC and CB in the truss of Fig. 1-16(a) to carry an inclined force P of 650 kN. Set the allowable tensile stress at 140 000 kN/m².

SOLUTION

If all members of the truss were to be designed, forces in all members would have to be found. In practice this is now done by employing computer programs developed on the basis of matrix structural analysis* or by directly

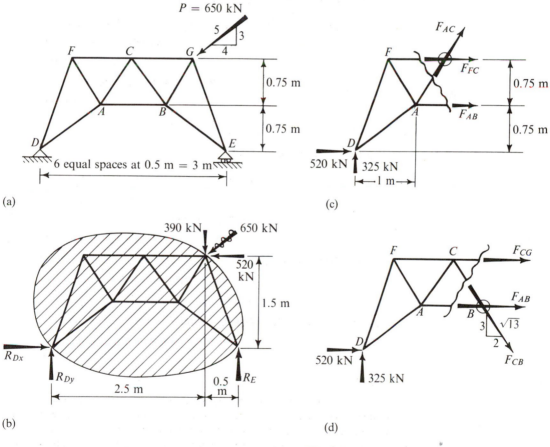

(a)

(b)

(c)

(d)

Fig. 1-16

*For example, see J. M. Gere, and W. Weaver, Jr., *Analysis of Framed Structures*, New York: Van Nostrand Reinhold, 1965. See also J. L. Meek, *Matrix Structural Analysis*, New York: McGraw-Hill, 1971.

analyzing the truss by the method of joints. However, if only a few members are to be designed or checked, the method of sections illustrated here is quicker.

It is generally understood that a planar truss such as shown in the figure is stable in the direction perpendicular to the plane of the paper. Practically this is accomplished by introducing braces at right angles to the plane of the truss. In this example the design of compression members is avoided as this will be treated in the chapter on columns.

To determine the forces in the members to be designed, the reactions for the whole structure are computed first. This is done by completely disregarding the interior framing. Only reaction and force components definitely located at their points of application are indicated on a free-body diagram of the whole structure, Fig. 1-16(b). After the reactions are determined, free-body diagrams of a part of the structure are used to determine the forces in the members considered, Figs. 1-16(c) and (d).

Using free body in Fig. 1-16(b):

$$\sum F_x = 0 \qquad R_{Dx} - 520 = 0, \qquad\qquad R_{Dx} = 520 \text{ kN}$$

$$\sum M_E = 0 \circlearrowleft +, \qquad + R_{Dy}(3) - 390(0.5) - 520(1.5) = 0$$

$$R_{Dy} = 325 \text{ kN}$$

$$\sum M_D = 0 \circlearrowleft +, \qquad + R_E(3) + 520(1.5) - 390(2.5) = 0$$

$$R_E = 65 \text{ kN}$$

$$Check: \quad \sum F_y = 0, \qquad + 325 - 390 + 65 = 0$$

Using free body in Fig. 1-16(c):

$$\sum M_A = 0 \circlearrowleft +, \qquad + F_{FC}(0.75) + 325(1) - 520(0.75) = 0$$

$$F_{FC} = +86.7 \text{ kN}$$

$$A_{FC} = F_{FC}/\sigma_{\text{allow}} = 86.7/(140\,000) = 0.000\,620 \text{ m}^2 = 620 \text{ mm}^2$$

$$(\text{use } 12.5 \text{ mm} \times 50 \text{ mm bar})$$

Using free body in Fig. 1-16(d):

$$\sum F_y = 0, \qquad -(F_{CB})_y + 325 = 0, \qquad (F_{CB})_y = +325 \text{ kN}$$

$$F_{CB} = \sqrt{13}(F_{CB})_y/3 = +391 \text{ kN}$$

$$A_{CB} = F_{CB}/\sigma_{\text{allow}} = 391/(140\,000) = 0.002\,790 \text{ m}^2 = 2\,790 \text{ mm}^2$$

$$(\text{use two bars } 30 \text{ mm} \times 50 \text{ mm})$$

EXAMPLE 1-6

Consider the idealized dynamic system shown in Fig. 1-17. The shaft AB rotates at a constant frequency of 10 Hz.* A light rod CD is attached to this shaft at point C, and at the end of this rod a weight of 50 N is fastened. In describing a complete circle the weight at D spins on a "frictionless" plane. Select the size of the rod CD so that the stress in it will not exceed 70 000 kN/m². In calculations, neglect the weight of the rod.

*Hz (abbreviation for hertz) or cycles per second is the SI unit for frequency.

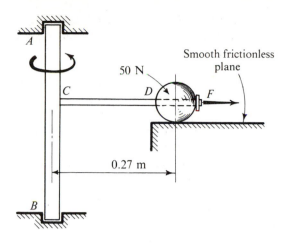

Fig. 1-17

SOLUTION

The acceleration of gravity g is 9.81 m/s². The angular velocity ω is 20π radians* per second. For the given motion the body W is accelerated toward the center of rotation with an acceleration of $\omega^2 R$, where R is the distance CD. By multiplying this acceleration a by the mass m of the body, force F is obtained. This force acts in the opposite direction to that of the acceleration (d'Alembert principle); see Fig. 1-17.

$$F = ma = \frac{W}{g}\omega^2 R = \frac{50}{9.81}(20\pi)^2(0.27) = 5\,430 \text{ N} = 5.43 \text{ kN}$$

$$A_{\text{net}} = \frac{F}{\sigma_{\text{allow}}} = \frac{5.43}{70\,000} = 0.000\,077 \text{ m}^2 = 77 \text{ mm}^2$$

A 10 mm round rod provides the required cross-sectional area. The additional pull at C caused by the mass of the rod, not considered above, is

$$F_1 = \int_0^R (m_1\,dr)\omega^2 r$$

where m_1 is the mass of the rod per unit length and $(m_1 dr)$ is its infinitesimal mass at a variable distance r from the vertical rod AB. The total pull at C caused by the rod and the weight W at the end is $F + F_1$.

1-9. BASIC APPROACH

The method of attack for problems in mechanics of materials follows along remarkably uniform lines. Now, by way of a bird's-eye view of the subject, the typical procedure will be outlined. It has already been used, and

*2π radians correspond to one cycle or complete revolution of the shaft.

the reader will recognize the same method of approach in other problems that follow. At times it is obscured by intermediate steps, but in the final analysis it is always applied.

1. From a particular arrangement of parts, a single member is isolated. Such a member is indicated on a diagram with *all* the forces and reactions acting on it. *This is a free body of the whole member.*
2. The reactions are determined by the application of the equations of statics. In indeterminate problems, statics is supplemented by additional considerations.
3. At a point where the magnitude of the stress is wanted, a section *perpendicular* to the axis of the body is passed, and a portion of the body, to either one side of the section or the other, is *completely* removed.
4. At the section investigated, the system of internal forces necessary to keep the isolated part of the member in *equilibrium* is determined. In general, this system of forces consists of an axial force, a shear, a bending moment, and a torque.* These quantities are found by treating a *part* of the member as a free body.
5. With the system of forces at the section *properly resolved,* the formulas of mechanics of materials enable one to determine the stresses at the section considered.
6. If the magnitude of the maximum stress at a section is known, one can provide proper material for such a section; or, conversely, if the physical properties of a material are known, one can select a member of adequate size.
7. In certain other problems, a further study of a member at a section enables one to predict the deformation of the structure as a whole and hence, if necessary, to design members that do not deflect or "sag" excessively.

Very few basic formulas are used in mechanics of materials. These will be learned by their repeated application. *However, visualization of the nature of the quantities being computed is essential.* Free-body diagrams help visualization immensely.

PROBLEMS FOR SOLUTION

1-1. If an axial tensile force of 110 kips is applied to a member made of a W 8 × 31 section, what will the tensile stress be? What will the stress be if the member is a C 12 × 20.7 section? For designation and cross-sectional areas of these members see Tables 4 and 5 in the Appendix. *Ans.* 12-1 ksi, 18.1 ksi.

1-2. Revise the data in Example 1-1 to read as follows: the distance BC is 3 ft, the distance CE is 8 ft, the thickness of the parapet wall at C is 8 in., the weight being lifted is 1 ton (2,000 lb), the actual timber size is 7.5 in. by 11.5 in. (8 in. by 12 in. nominal, see Table 10 in the Appendix), the bolts are $\frac{3}{4}$ in. in diameter, and the cross-sec-

---

*A complete appreciation of these terms will result only after a study of Chapters 3 and 4.

tional area at the root of the threads is 0.302 in.² Solve for the same quantities as done in the above example. *Ans.* $\sigma_{max} = 8{,}800$ psi, $\sigma_b = 122$ psi.

1-3 and 1-4. Short cast iron members have the cross-sectional dimensions shown in the figures. If they are subjected to axial compressive forces of 45 kN each, find the points of application for these forces to cause no bending, and determine the normal stresses. All dimensions are in mm.

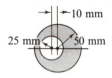

PROB. 1 – 3

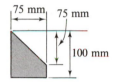

PROB. 1 – 4

1-5. A gear transmitting a torque of 4,000 in.-lb to a $2\frac{3}{16}$ in. shaft is keyed to it as shown in the figure. The $\frac{1}{2}$ in. square key is 2 in. long. Determine the shearing stress in the key. *Ans.* 3,650 psi.

PROB. 1 – 5

1-6. Two 10 mm thick steel plates are fastened together as shown in the figure, by means of two 20 mm bolts that fit tightly into the holes. If the joint transmits a tensile force of 45 kN, determine

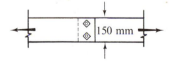

PROB. 1 – 6

(a) the average normal stress in the plates at a section where no holes occur; (b) the average normal stress at the critical section; (c) the average shearing stress in the bolts; and (d) the average bearing stress between the bolts and the plates.

1-7. In Example 1-2, find the stress 0.5 m above the base. Show the result on an infinitesimal element.

1-8. Determine the bearing stresses caused by the applied force at A, B, and C for the structure shown in the figure. *Ans.* $\sigma_A = -100$ psi, $\sigma_B = -167$ psi.

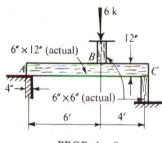

PROB. 1 – 8

1-9. A lever mechanism used to lift panels of a portable army bridge is shown in the figure. Calculate the shearing stress in pin A caused by a load of 2.5 kN. *Ans.* 20.3 MPa.

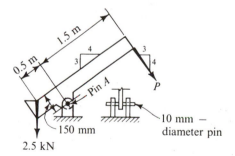

PROB. 1 – 9

1-10. Calculate the shearing stress in pin A of the bulldozer if the total forces acting on the blade are as shown in the figure. Note that there is a $1\frac{1}{2}$ in. diameter pin on each side of the bulldozer. Each pin is in single shear. *Ans.* $\tau = 2.83$ ksi.

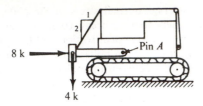

PROB. 1 – 10

1-11. A steel bar 1 in. in diameter is loaded in double shear until failure; the ultimate load is found to be 100,000 lb. If the allowable stress is to be based on a safety factor of 4, what must the diameter of a pin designed for an allowable load of 6,000 lb in single shear be? *Ans.* 0.69 in.

1-12. A 150 mm × 150 mm wooden post delivers a force of 50 kN to a concrete footing, as shown in Fig. 1-6. (a) Find the bearing stress of the wood on the concrete. (b) If the allowable pressure on the soil is 100 kN/m², determine in plan view the required dimensions of a square footing. Neglect the weight of the footing.

1-13. An arrangement of three rods is used to suspend a 50 kN weight as shown in the figure. The rods AB and BD are 20 mm in diameter, the rod BC is 13 mm in diameter. Find the stresses in the rods. *Ans.* $\sigma_{AB} = 151$ MPa.

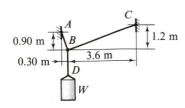

PROB. 1 – 13

1-14. A rod of variable cross section built in at one end is subjected to three axial forces as shown in the figure. Find the maximum normal stress. *Ans.* 22.5 ksi.

PROB. 1 – 14

1-15. Rework the preceding problem, assuming that the (axial) end force, instead of being 40 k, is to be such as to cause the same maximum normal stresses in the two sizes of the rod. The 20 k and the 70 k axial forces remain applied, and the maximum normal stress for the smaller part of the rod may be either between these two forces, or nearer to the free end. Investigate both conditions. *Ans.* 25 ksi.

1-16. A short column is made up of two standard steel pipes, one on top of the other as shown in the figure. If the allowable stress in compression is 15 ksi, (a) what is the allowable axial load P_1 if the axial load $P_2 = 50$ kips; (b) what is the allowable load P_1 if the load $P_2 = 15$ kips? Neglect the weight of the pipes. *Ans.* (a) 14.5 k, (b) 33.45 k.

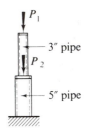

PROB. 1 – 16

1-17. Rework the above problem, assuming that the direction of the force P_1 is reversed, i.e., P_1 becomes a tensile force. Assume that the allowable tensile stress is also 15 ksi.

1-18. For the structure shown in the figure, calculate the size of the bolt and area of the bearing plates required if the allowable stresses are 18,000 psi in tension and 500 psi in bearing. Neglect the weight of the beams. *Ans.* 1.25 in., 30 in.²

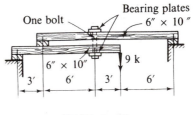

PROB. 1 – 18

1-19. Rework Example 1-5 after revising the data as follows: the total height of the truss is 60 in., the total width is 120 in., the applied force *P* is 150 kips. Let the allowable tensile stress be 20,000 psi. *Ans.* 1 in.², 4.51 in.²

1-20. A 30 kN weight is supported by means of a pulley as shown in the figure. The pulley is supported by the frame *ABC*. Find the required cross-sectional areas for members *AC* and *BC* if the allowable stress in tension is 140 000 kN/m² and in compression, determined by the method of Chapter 13, is 96 000 kN/m². *Ans.* 306 mm².

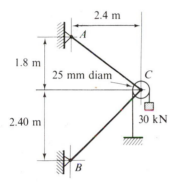

PROB. 1 – 20

1-21. A force of 500 kN is applied at joint *B* to a system of two pin-joined bars as shown in the figure. Determine the required cross-sectional area of the bar *BC* if the allowable stresses are 100 MPa in tension and 70 MPa in compression.

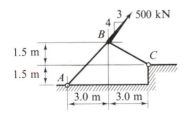

PROB. 1 – 21

1-22. Find the stress in the mast of the derrick shown in the figure. All members are in the same vertical plane and are joined by pins. The mast is made from an 8 in. standard steel pipe weighing 28.55 lb/ft. Neglect the weight of the members. *Ans.* −446 psi.

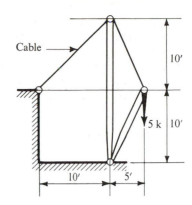

PROB. 1 – 22

1-23. Find the required cross-sectional areas for all tension members in Example 1-5. The allowable stress is 140 000 kN/m² = 140 MPa.

1-24. A signboard 4.5 m by 6.0 m in area is supported by two frames as shown in the figure. All members are actually 50 mm by 100 mm in cross section. Calculate the stress in each member due to a horizontal wind load of 960 N/m² on the sign. Assume all joints to be connected by pins and that one-quarter of the total wind force acts at *B* and at *C*. Neglect the possibility of buckling of the compression members. Neglect the weight of the structure. *Ans.* σ_{AD} = 1.08 MPa.

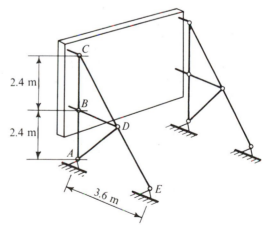

PROB. 1 – 24

1-25. What distances, *a* and *b*, are required beyond the notches in the horizontal member of the truss shown? All members are nominally 8 in. by 8 in. in cross section (see Table 10, Appendix for actual size). Assume the ultimate strength of wood in shear parallel to the grain to be 500 psi. Use a factor of safety of 5. (This detail is not recommended.) *Ans. a = 10.7 in.*

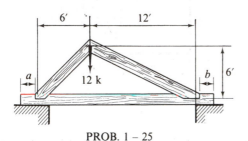

PROB. 1 – 25

1-26. What is the required diameter of the pin *B* for the bell crank mechanism shown in the figure if an applied force of 60 kN at *A* is resisted by a force *P* at *C*? The allowable shearing stress is 100 MPa. *Ans.* 16.4 mm.

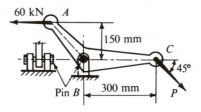

PROB. 1 – 26

1-27. What is the shearing stress in the bolt *A* caused by the applied load shown in the figure? The bolt is 6 mm in diameter, and it acts in double shear. All dimensions are in mm.

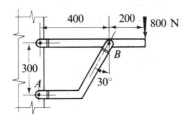

PROB. 1 – 27

1-28. A control pedal for actuating a spring mechanism is shown in the figure. Calculate the

shearing stress in pins *A* and *B* due to a force *P* when it causes a stress of 10,000 psi in the rod *AB*. Both pins are in double shear. *Ans.* 2,380 psi, 2,240 psi.

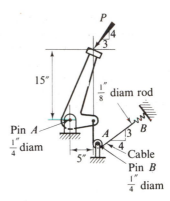

PROB. 1 – 28

1-29. · A beam with a force of 100 kips at one end is supported by a strutted cable as shown in the figure. Find the horizontal and vertical components of the reactions at *A*, *B*, and *D*. If the allowable tensile stress is 20,000 psi and the allowable compressive stress is 10,000 psi, what is the required cross-sectional area of members *AC*, *BC*, and *CE*? (*Hint:* Isolate the beam *DF* first.) *Ans.* $A_{AC} = A_{CE} = 10$ in.^{2} and $A_{BC} = 5.66$ in.^{2}.

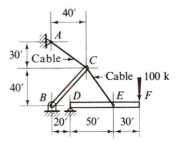

PROB. 1 – 29

1-30. A tower used for a highline is shown in the figure. If it is subjected to a horizontal force of 540 kN and the allowable stresses are 100 MPa in compression and 140 MPa in tension, what is the required cross-sectional area of each member? All members are pin-connected. All dimensions are in meters. *Ans.* $A_{AD} = 3640$ mm.^{2}

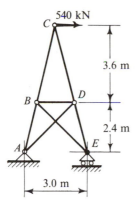

540 kN

C

3.6 m

B

D

2.4 m

A

E

3.0 m

PROB. 1 – 30

1-31. To support a load, $P = 180$ kN, determine the necessary diameter for the rods AB and AC for the tripod shown in the figure. Neglect the weight of the structure and assume that the joints are pin-connected. No allowance need be made for threads. The allowable tensile stress is 125 MPa. All dimensions are in meters.

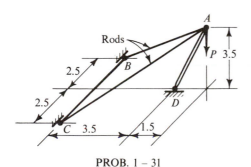

Rods

2.5

2.5

C 3.5

B

P 3.5

A

D

1.5

PROB. 1 – 31

1-32. A 10 lb weight moves in a horizontal circle at the end of a 5 ft wire with such an angular velocity that the wire makes an angle of 30° with the vertical. What is the proper diameter for the wire if the allowable tensile stress for high strength steel is 40 ksi? *Ans.* 0.0192 in.

1-33. A pin-connected frame for supporting

a force P is shown in the figure. The stress σ in both members AB and BC is to be the same. Determine the angle α necessary to achieve the minimum weight of construction. Members AB and BC have a constant cross section. *Ans.* $\cos^2 \alpha = \frac{1}{3}$ or $\alpha \approx 55°$.

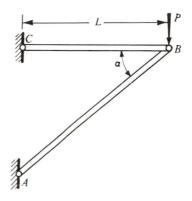

L

P

C

B

α

A

PROB. 1 – 33

1-34. A joint for transmitting a tensile force is to be made by means of a pin as shown in the figure. If the diameter of the rods being connected is D, what should the diameter d of the pin be? Assume that the allowable shearing stress in the pin is one-half the maximum tensile stress in the rods. (In Art. 9-18 it will be shown that this ratio for the allowable stresses is an excellent assumption for many materials.) *Ans.* $D = d$.

D

d

PROB. 1 – 34

2 Strain — Hooke's Law — Axial Load Problems

2-1. INTRODUCTION

This chapter will be devoted to the further examination of some of the physical properties of the materials of construction. An investigation of the nature of the deformations that take place in a stressed body will be the primary objective. These deformations will be related to the magnitudes of the stresses that cause them. Also, some additional limitations that must be imposed on Eqs. 1-1 and 1-2 of the previous chapter are pointed out.

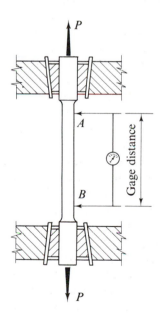

Fig. 2-1. Diagram of a tension specimen in a testing machine

2-2. STRAIN

In Art. 1-7 it was stated that information regarding the physical properties of materials comes from the laboratory. In the particularly common tension test, not only the ultimate strength, but other properties are usually observed, especially those pertaining to the study of deformation as a function of the applied force. Thus, while a specimen is being subjected to an increasing force *P*, as shown in Fig. 2-1, a change in length between two points, as *A* and *B*, on the specimen is observed. Initially, two such points can be selected an arbitrary distance apart. Thus, depending on the test, either a 2 in. or an 8 in. distance is commonly used. This initial distance between the two points is called a *gage distance*.* In an experiment it is the change in length of this distance that is noted. With the same load and a longer gage distance, a larger deformation is observed, or vice versa. Therefore it is more fundamental to refer to the observed elon-

*Now electric resistance strain gages that are bonded directly to the specimen are extensively used. Their gage lengths vary widely from about 4 mm to 150 mm depending on application. See Art. 8-13 for further details.

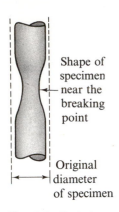

Shape of
specimen
near the
breaking
point

Original
diameter
of specimen

Fig. 2-2. Typical contraction of a specimen of mild steel in tension near the breaking point

gation *per unit of length of the gage.* If Δ is the total elongation in a given *original* gage length L, the elongation per unit of length, ε (epsilon), is

$$\varepsilon = \frac{\Delta}{L} \tag{2-1}$$

This elongation per unit of length is termed *strain.** It is a dimensionless quantity, but it is customary to refer to it as having the dimensions of inches per inch or meters per meter. Sometimes strain is given in percent. The quantity ε is a *very small* one, except for a few materials such as rubber. If the strain is known, the total deformation of an axially loaded bar is εL. This relationship holds for any gage length until some local deformation takes place on an appreciable scale. The latter effect, exemplifying the behavior of a mild steel rod near a breaking point, is shown in Fig. 2-2. This phenomenon is referred to as "necking." Brittle materials do not exhibit this at usual temperatures, although they too contract transversely a little in a tension test and expand in a compression test.

2-3. STRESS-STRAIN DIAGRAM

It is apparent from this discussion that for general purposes the deformations of a rod in tension or compression are most conveniently expressed in terms of strain. Similarly, stress rather than force is the more significant parameter in the study of materials, since the effect on a material of an applied force P depends primarily on the cross-sectional area of the member. As a consequence, in the study of the properties of materials, it is customary to plot diagrams on which a relationship between *stress* and *strain* for a particular test is reported. Such diagrams establish a relationship between stress and strain, and for most practical purposes are assumed to be independent of the size of the specimen or its gage length. For these *stress-strain diagrams,* it is customary to use the ordinate scale for stresses and the abscissa for strains. Stresses are usually computed on the basis of the *original* area of a specimen, although, as mentioned earlier, some transverse contraction or expansion of a material always takes place. If the stress is computed by dividing the applied force by the corresponding *actual* area of a specimen at the same instant, the so-called *true* stress is obtained. A plot of true stress vs. strain is called a *true* stress-strain diagram. Such diagrams are seldom used in practice.

Experimentally determined stress-strain diagrams differ considerably for different materials. Even for the same material they differ, depending on the temperature at which the test was conducted, the speed of the test, and several other variables.† However, broadly speaking, two types of diagrams

*The term *unit* strain is sometimes used.

†For more details, see Davis, H. E., G. E. Troxell, and C. T. Wiskocil, *Testing and Inspection of Engineering Materials*, (2nd .ed.) New York: McGraw-Hill, 1955.

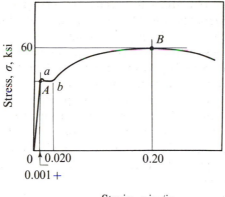

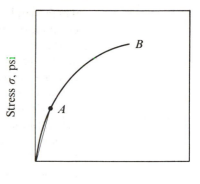

Fig. 2-3. Stress-strain diagram for mild steel

Fig. 2-4. Stress-strain diagram for a brittle material

can be recognized. One type is shown in Fig. 2-3, which is for mild steel, a *ductile* material widely used in construction. The other type is shown in Fig. 2-4. Such diverse materials as tool steel, concrete, copper, etc., have curves of this variety, although the extreme value of strain that these materials can withstand *differs drastically*. The "steepness" of these curves varies considerably. Numerically speaking, each material has its own curve. The terminal point on a stress-strain diagram represents the complete failure (rupture) of a specimen. Materials capable of withstanding large strains are referred to as *ductile materials*. The converse applies to *brittle materials*.

2-4. HOOKE'S LAW

Fortunately, one feature of stress-strain diagrams is applicable with sufficient accuracy to nearly all materials. It is a fact that for a *certain distance* from the origin the experimental values of stress vs. strain lie essentially on a straight line. This holds true almost without reservations for glass. It is true for mild steel up to some point, as *A* in Fig. 2-3. It holds nearly true up to very close to the failure point for many high-grade alloy steels. On the other hand, the straight part of the curve hardly exists in concrete, annealed copper, or cast iron. Nevertheless, for all practical purposes, up to some such point as *A* (also in Fig. 2-4), *the relationship between stress and strain may be said to be linear for all materials*. This sweeping idealization and generalization applicable to all materials became known as *Hooke's law*.* Symbolically, this law can be expressed by the equation

$$\sigma = E\varepsilon \qquad \text{or} \qquad E = \frac{\sigma}{\varepsilon} \qquad (2\text{-}2)$$

*Actually Robert Hooke, an English scientist, worked with springs and not with rods. In 1676 he announced an anagram "c e i i i n o s s s t t u v," which in Latin is *Ut Tensio sic Vis* (the force varies as the stretch).

which simply means that stress is directly proportional to strain where the constant of proportionality is E. This constant E is called the *elastic modulus*, modulus of elasticity, or Young's modulus.* As ε is dimensionless, E has the units of stress in this relation. In the English system of units it is usually measured in pounds per square inch, while in the SI units it is measured in newtons per square meter (or pascals).

Graphically E is interpreted as the slope of a straight line from the origin to the rather vague point A on a stress-strain diagram. The stress corresponding to the latter point is termed the *proportional limit* of the material. Physically the elastic modulus represents the stiffness of the material to an imposed load. *The value of the elastic modulus is a definite property of a material.* From experiments it is known that ε is *always a very small quantity*, hence E must be a large one. Its approximate values are tabulated for a few materials in Table 1 of the Appendix. For most steels, E is between 29 and 30×10^6 psi, or 200 and 207×10^9 N/m².

It follows from the foregoing discussion that *Hooke's law applies only up to the proportional limit of the material.* This is highly significant as in most of the subsequent treatment the derived formulas are based on this law. Clearly then, such formulas will be limited to the material's behavior in the lower range of stresses.

Some materials, notably single crystals, possess different elastic moduli in different directions with reference to their crystallographic planes. Such materials, having different physical properties in different directions, are termed nonisotropic. A consideration of such materials is *excluded* from this text. The vast majority of engineering materials consist of a large number of *randomly* oriented crystals. Because of this random orientation of crystals, properties of materials become essentially alike in any direction.† Such materials are called *isotropic. Throughout this text complete homogeneity* (sameness) *and isotropy of materials is assumed.*

2-5. FURTHER REMARKS ON STRESS-STRAIN DIAGRAMS

In addition to the proportional limit defined in Art. 2-4, several other interesting points can be observed on the stress-strain diagrams. For instance, the highest points (B in Figs. 2-3 and 2-4) correspond to the *ultimate* strength of a material. *Stress* associated with the remarkably long plateau *ab* in Fig. 2-3 is termed the *yield point* of a material. As will be brought out later, this remarkable property of mild steel, in common with other *ductile* materials, is significant in stress analysis. For the present, note that at an essentially constant stress, strains 15 to 20 times those that take place up to the proportional limit occur during yielding. At the yield point a large amount of

---

*Young's modulus is so called in honor of Thomas Young, an English scientist. His *Lectures on Natural Philosophy*, published in 1807, contain a definition of the modulus of elasticity.

†Rolling operations produce preferential orientation of crystalline grains in some materials.

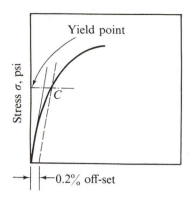

Yield point

Stress σ, psi

C

$\longrightarrow$ $\longleftarrow$ 0.2% off-set

Strain, ε, in./in.

Fig. 2-5. Off-set method of determining the yield point of a material

deformation takes place at a constant stress. The yielding phenomenon is absent in most materials, particularly in those that behave in a brittle fashion.

A study of stress-strain diagrams shows that the yield point is so near the proportional limit that for most purposes the two may be taken as one. However, it is much easier to locate the former. For materials that do not possess a well-defined yield point, one is actually "invented" by the use of the so-called "offset method." This is illustrated in Fig. 2-5 where a line offset an *arbitrary* amount of 0.2% of strain is drawn parallel to the straight-line portion of the initial stress-strain diagram.* Point C is then taken as the *yield point* of the material at 0.2% offset.

Finally, the technical definition of the *elasticity* of a material should be made. In such usage it means that a material is able to regain *completely* its original dimensions upon removal of the applied forces. At the beginning of loading, if a small force is applied to a body, the body deforms a certain small amount. If such a force is removed, the body returns to its initial size and shape. With increasing magnitude of force this continues to take place while the material behaves elastically. However, eventually a stress is reached that causes permanent deformation, or set, in the material. The corresponding stress level is called the *elastic limit* of the material. Practically speaking, the *elastic limit* corresponds closely to the proportional limit of the material.

For the majority of materials, stress-strain diagrams obtained for short compression blocks are reasonably close to those found in tension. However, there are some notable exceptions. For example, cast iron and concrete are very weak in tension but not in compression. For these materials the diagrams differ considerably, depending on the sense of the applied force.

2-6. DEFLECTION OF AXIALLY LOADED RODS

Equations 1-1, 2-1, and 2-2, plus a known elastic modulus for a given material, are sufficient to determine the deformations of axially loaded rods. However, the usual calculations *apply only within the elastic range of a material's behavior* inasmuch as Hooke's law (Eq. 2-2) is used. To formulate this problem in general terms, consider the *axially* loaded bar shown in Fig. 2-6(a). In this bar the cross-sectional area varies along the length, and forces of various magnitudes are applied at several points. Now suppose that in this problem the change in length of the bar *between two points A and B* caused

---

*For decreasing loads the stress-strain diagram is parallel to the straight-line portion of the initial stress-strain diagram. For further details see Fig. 15-2 and the accompanying text.

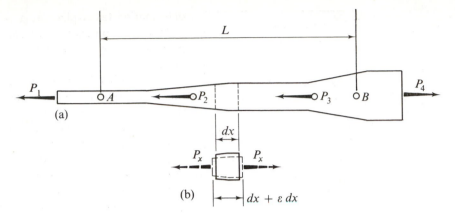

Fig. 2-6. An axially loader bar

by the applied forces is sought. The quantity wanted is the sum (or accumulation) of the deformations that take place in infinitesimal lengths of the rod. Hence the amount of deformation that takes place in an arbitrary element of length dx is first formulated, then the sum or integral of this effect over the given length gives the quantity sought.

An arbitrary element cut out from the bar is shown in Fig. 2-6(b). From free-body considerations, this element is subjected to a pull P_x which, in general, is a variable quantity. The infinitesimal deformation $d\Delta$ that takes place in this element upon application of the forces is equal to strain ε multiplied by the length dx. From Eq. 2-2, strain is equal to the stress σ_x divided by the elastic modulus E. However, in general, σ_x is a variable quantity obtained by dividing the variable force P_x by the *corresponding* area A_x. Hence, since $\varepsilon = \sigma_x/E$ and $\sigma_x = P_x/A_x$,

$$d\Delta = \varepsilon\,dx = \frac{\sigma_x}{E}\,dx = \frac{P_x\,dx}{A_x E}$$

Since the contribution of individual elements is now known, the total deformation between any two given points on a bar is simply their sum,* that is,

$$\Delta = \int_A^B d\Delta = \int_A^B \frac{P_x\,dx}{A_x E} \qquad (2\text{-}3)$$

Three examples will now be solved to show applications of the above equations.

---

*The limits of integration as stated represent the range of integration. Actually, they must be expressed in terms of the values of the variable. This nonrigorous usage of the limits will occasionally be employed in this text.

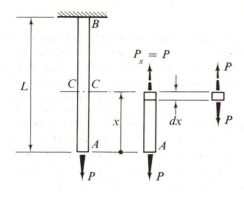

(a)   (b)   (c)

Fig. 2-7

EXAMPLE 2-1

Consider the rod AB of constant cross-sectional area A and of length L shown in Fig. 2-7(a). Determine the relative displacement of the end A with respect to B when a force P is applied, i.e., find the *deflection* of the free end, caused by the application of a concentrated force P. The elastic modulus of the material is E.

SOLUTION

In this problem the rod may be treated as being weightless as only *the effect of P on the deflection* is investigated. Hence, no matter where a cut C-C is made through the rod, $P_x = P$, Fig. 2-7(b). The infinitesimal elements, Fig. 2-7(c), are everywhere the same, subjected to a *constant* pull P. Likewise, A_x everywhere has a constant value A. Applying Eq. 2-3,

$$\Delta = \int_A^B \frac{P_x\,dx}{A_x E} = \frac{P}{AE}\int_0^L dx = \frac{P}{AE}\,x\Big|_0^L = \frac{PL}{AE} \tag{2-4}$$

It is seen from Eq. 2-4 that the deflection of the rod is directly proportional to the applied force and the length, and is inversely proportional to A and E.

This equation will be referred to in subsequent work.

EXAMPLE 2-2

Determine the relative displacement of points A and D of the steel rod of *variable* cross-sectional area shown in Fig. 2-8(a) when it is subjected to the four concentrated forces $P_1, P_2, P_3,$ and P_4. Let $E = 200 \times 10^6$ kN/m².

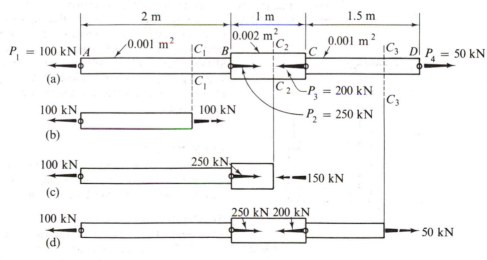

Fig. 2-8

SOLUTION

In attacking such a problem, a check must first be made to ascertain that the body as a whole is in equilibrium, i.e., $\sum F_x = 0$. Here, by inspection it can be seen that such is the case. Next, the variation of P_x along the length of the bar must be studied. This may be done conveniently with the aid of sketches as shown in Figs. 2-8(b), (c), and (d), which show that *no matter where* a section C_1-C_1 is taken between points A and B, the force in the rod is $P_x = +100$ kN. Similarly, between B and C, $P_x = -150$ kN, and between C and D, $P_x = +50$ kN. The variation of A_x is shown in Fig. 2-8(a). Both P_x and A_x, mathematically speaking, are not continuous functions along the rod. Both have "jumps" or *sudden* changes in their values. Hence, in integrating, the limits of integration must be "broken." Thus, from Eq. 2-3,

$$\Delta = \int_A^D \frac{P_x\,dx}{A_x E} = \int_A^B \frac{P_{AB}\,dx}{A_{AB}E} + \int_B^C \frac{P_{BC}\,dx}{A_{BC}E} + \int_C^D \frac{P_{CD}\,dx}{A_{CD}E}$$

In the last three integrals the respective P_x and A_x are constants between the limits shown. The subscripts of P and A denote the range of applicability of the function; thus P_{AB} applies in the interval AB, etc. These integrals revert to the solution of the previous example, i.e., Eq. 2-4. Applying it and substituting numerical values,

$$\Delta = \sum \frac{PL}{AE} = +\frac{(100)(2)}{(0.001)(200 \times 10^6)} - \frac{(150)(1)}{(0.002)(200 \times 10^6)}$$
$$+ \frac{(50)(1.5)}{(0.001)(200 \times 10^6)}$$

$$= +0.001 - 0.000\,375 + 0.000\,375 = +0.001 \text{ m} = +1 \text{ mm}$$

The operation performed means that the individual deformations of the three "separate" rods have been added, or *superposed*. Each one of these "rods" is subjected to a constant force. The positive sign of the answer indicates that the rod *elongates*, as a positive sign is associated with tensile forces. The equality of the absolute values of the deformations in lengths BC and CD is purely accidental. Note that in spite of the relatively large stresses present in the rod, the value of Δ is small.

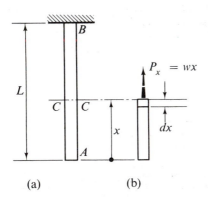

(a)   (b)

Fig. 2-9

EXAMPLE 2-3

Find the deflection, caused by its own weight, of the free end A of the rod AB having a constant cross-sectional area A and weighing w N/m, Fig. 2-9(a).

SOLUTION

Here again Eq. 2-3 must be applied. However, in this case P_x is a variable quantity. It is conveniently expressed as wx if the origin of x is taken at A. Hence

$$\Delta = \int_A^B \frac{P_x\,dx}{A_x E} = \frac{1}{AE} \int_0^L wx\,dx = \frac{w}{AE}\left.\frac{x^2}{2}\right|_0^L = \frac{(wL)L}{2AE} = \frac{WL}{2AE}$$

where wL is the *total* weight of the rod, which is designated by capital W. Compare this expression with Eq. 2-4.

If a concentrated force P, in *addition* to the bar's own weight, were acting on the bar AB at the end A, the total deflection due to the *two* causes would be obtained by *superposition* (direct addition) as

$$\Delta = \frac{PL}{AE} + \frac{WL}{2AE} = \frac{[P + (W/2)]L}{AE}$$

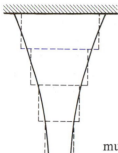

Fig. 2-10

In problems where the area of a rod is variable, a proper *function* for it must be substituted into Eq. 2-3 to determine deflections. In practice, it is sometimes sufficiently accurate to analyze such problems by approximating the shape of a rod by a *finite* number of elements as shown in Fig. 2-10. The deflections for each one of these elements are added to obtain the total deflection.

2-7. POISSON'S RATIO

In addition to the deformation of materials in the direction of the applied force, another remarkable property can be observed in all solid materials, namely, that at right angles to the applied force, a certain amount of lateral (transverse) expansion or contraction takes place. This phenomenon is illustrated in Figs. 2-11(a) and (b), where the deformations are *greatly exaggerated*. For clarity this physical fact may be restated thus: if a solid body is subjected to an axial tension, it contracts laterally; on the other hand, if it is compressed, the material "squashes out" sideways. With this in mind, directions of lateral deformations are easily determined, depending on the sense of the applied force. Mathematically, a plus sign is usually assigned to an *increment* of lateral dimension, and vice versa.

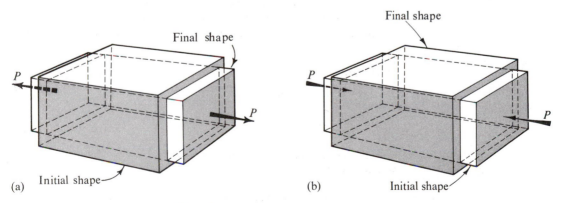

Fig. 2-11. Lateral contraction and expansion of solid bodies subjected to axial forces (Poisson effect)

For a general theory, it is preferable to refer to these lateral deformations on the basis of deformation per *unit* of length of the transverse dimension. Thus the lateral deformations on a *relative* basis can be expressed in inches per inch or m/m. These relative unit lateral deformations are termed *lateral strains*. Moreover, it is known from experiments that lateral strains bear a *constant* relationship to the longitudinal or axial strains caused by an axial force, provided a material remains *elastic* and is homogeneous and isotropic. This constant is a definite property of a material, just like the elastic modulus E, and is called *Poisson's ratio*.* It will be denoted by v (nu) and is defined as follows:

$$v = \left| \frac{lateral\ strain}{axial\ strain} \right| = - \frac{lateral\ strain}{axial\ strain} \qquad (2\text{-}5)$$

where the strains are caused by uniaxial stress only. The second, alternative form of Eq. 2-5 is true because the lateral and axial strains are always of opposite sign for uniaxial stress, thus invariably giving a positive value of v. The value of v fluctuates for different materials over a relatively narrow range. Generally it is in the neighborhood of 0.25 to 0.35. In extreme cases values as low as 0.1 (some concretes) and as high as 0.5 (rubber) occur. The latter value is the *largest possible*. It is normally attained by materials during plastic flow and signifies constancy of volume.† In this text, Poisson's ratio will be used only when materials behave elastically.

In conclusion, note that the Poisson effect exhibited by materials causes *no additional stresses* other than those considered earlier *unless the transverse deformation is inhibited or prevented*. The same is found to be true with regard to thermal expansion or contraction of materials. This topic will be treated in the chapter on statically indeterminate structures.

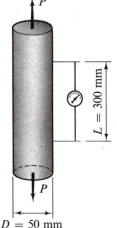

P

$L = 300$ mm

P

$D = 50$ mm

Fig. 2-12

EXAMPLE 2-4

Consider a carefully conducted test where an aluminum bar of 50 mm diameter is stressed in a testing machine as shown in Fig. 2-12. At a certain instant the applied force P is 100 kN, while the measured elongation of the rod is 0.219 mm in a 300 mm gage length, and the diameter's dimension is decreased by 0.012 15 mm. Calculate the two physical constants v and E of the material.

SOLUTION

Transverse or lateral strain:

$$\varepsilon_t = \frac{\Delta_t}{D} = - \frac{0.000\ 012\ 15}{0.050} = -0.000\ 243 \text{ m/m}.$$

In this case the lateral strain ε_t is negative, since the diameter of the bar *decreases* by Δ_t,

*Named after S. D. Poisson, a French scientist who formulated this concept in 1828.

†A. Nadai, *Theory of Flow and Fracture of Solids*, vol. 1., New York: McGraw-Hill, 1950.

$$\text{Axial strain:} \quad \varepsilon_a = \frac{\Delta}{L} = \frac{0.000\,219}{0.3} = 0.000\,73 \text{ m/m.}$$

$$\text{Poisson's ratio:} \quad \nu = -\frac{\varepsilon_t}{\varepsilon_a} = -\frac{(-0.000\,243)}{0.000\,73} = 0.333$$

Next, as the area of the rod $A = \frac{1}{4}\pi(0.050)^2 = 1.96 \times 10^{-3}$ m², from Eq. 2-4,

$$E = \frac{PL}{A\Delta} = \frac{(100)(0.3)}{1.96 \times 10^{-3}(0.000\,219)} = 70 \times 10^6 \text{ kN/m}^2$$

In practice, when a study of physical quantities, as E and ν, is being made, it is best to work with the corresponding stress-strain diagram to be sure that the quantities determined are associated with the elastic range of the material. Also note that it makes no difference whether the initial or the final lengths are used in computing strains, since the deformations are very small.

*2-8. GENERALIZED HOOKE'S LAW

In the previous article, Poisson's ratio was defined as the ratio of lateral strain to the axial strain for an axially loaded member. This applies only to a uni-axial state of stress on an element. Now a more general state of stress acting upon an *isotropic* body will be considered, and equations relating stress to deformation will be developed.

A block whose sides are a, b, c respectively is acted upon by tensile stresses *uniformly* distributed on all faces as shown in Fig. 2-13(a).* This diagram approaches the generality of Fig. 1-3. However, for the present, shearing stresses have been deleted, as it is a known experimental fact that the strains caused by normal stresses are *independent of small shearing deformations*. The normal stresses are designated by σ's with appropriate subscripts referring to the directions in which the stresses act.

For the moment, attention will be directed to the change in length of the block in the x-direction. To find this change, use is made of the *principle of superposition*, which is based upon the premise that the resultant stress or strain in a system due to several forces *is the algebraic sum of their effects when separately applied*. This assumption is true only if each effect is directly and linearly related to the force causing it. It is only approximately true when the deflections or deformations due to one force cause an abnormal change in the effect of another force. Fortunately the magnitudes of deflections are relatively small in most engineering structures. Hence, by proceeding on the basis of the above principle, the separate effects shown in Figs. 2-13(b), (c), and (d) can be summed. The stress in the x-direction causes a positive strain $\varepsilon_x' = \sigma_x/E$. Each of the positive stresses in the y- and z-directions causes a negative strain *in the x-direction* as a result of Poisson's effect. These strains are $\varepsilon_x'' = -\nu\sigma_y/E$ and $\varepsilon_x''' = -\nu\sigma_z/E$, respectively. The strains in the y- and

---

*This figure is adapted from G. Dreyer, *Festigkeitslehre und Elastizitätslehre*, Leipzig: Jänecke, p. 151.

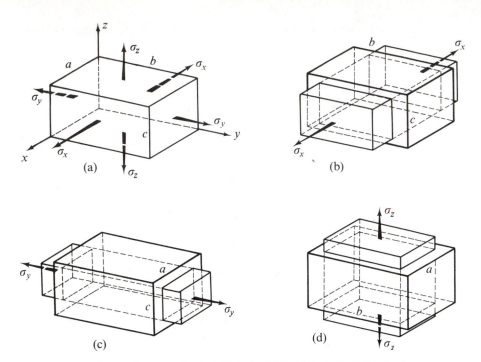

Fig. 2-13. An element subjected to normal stresses acting in the directions of coordinate axes

z-directions can also be obtained in a similar manner, and the net axial strains in the three coordinate directions are

$$\varepsilon_x = +\frac{\sigma_x}{E} - v\frac{\sigma_y}{E} - v\frac{\sigma_z}{E}$$

$$\varepsilon_y = -v\frac{\sigma_x}{E} + \frac{\sigma_y}{E} - v\frac{\sigma_z}{E}$$

$$\varepsilon_z = -v\frac{\sigma_x}{E} - v\frac{\sigma_y}{E} + \frac{\sigma_z}{E}$$

$$(2\text{-}6)$$

The application of Eq. 2-6 is limited to isotropic materials in the elastic range. *If a particular stress is compressive, the sign of the corresponding term changes.* The reader should verify this statement by *physical* reasoning, visualizing the Poisson effect with reference to Fig. 2-11. Also, it should be noted particularly that the above expression, which is known as the *generalized Hooke's law*, gives the *deformation per unit of length* or strain in a body. The strain given by Eq. 2-6 *must be multiplied by the dimension* of an element or member *in the corresponding direction* to obtain the total deformation in that direction.* The *total* extensional deformation in the x-direction is

$$\Delta_x = \varepsilon_x L_x \qquad (2\text{-}7)$$

*The stresses must remain constant in the interval considered.

where L_x is the dimension in the x-direction, as the a dimension in Fig. 2-13(a). Similar relations exist for Δ_y and Δ_z.

EXAMPLE 2-5

A 50 mm cube of steel is subjected to a uniform pressure of 200 000 kN/m² acting on all faces. Determine the change in dimension between two parallel faces of the cube. Let $E = 200 \times 10^6$ kN/m² and $\nu = 0.25$.

SOLUTION

Using Eq. 2-6 and noting that pressure is a compressive stress,

$$\varepsilon_x = \frac{(-200\ 000)}{(200)10^6} - \left(\frac{1}{4}\right)\frac{(-200\ 000)}{(200)10^6} - \left(\frac{1}{4}\right)\frac{(-200\ 000)}{(200)10^6}$$

$$= -5 \times 10^{-4}\ \text{m/m}.$$

$$\Delta_x = \varepsilon_x L_x = -(5)10^{-4} \times 0.050 = -0.000\ 025\ \text{m} \quad \text{(contraction)}$$

In this case $\Delta_x = \Delta_y = \Delta_z$.

2-9. SHEARING STRESSES ON MUTUALLY PERPENDICULAR PLANES

Article 2-8 dealt with a general case of deformations caused by normal stresses. Now the effect of shearing stresses on deformation will be considered. This requires some preparatory remarks. First, return to Fig. 1-3 and simplify it to the case shown in Fig. 2-14(a). In this figure only the τ_{zy} and τ_{yz} shearing stresses are shown. As before, the first subscript of τ associates the shearing stress with a plane *perpendicular* to a given axis and the second specifies its *direction* relative to another axis. The dimensions of the *infinitesimal* element considered are $(dx)(dy)(dz)$. For any such element, as with normal stresses, the shearing stresses on *parallel* planes are numerically equal. This follows directly from the equilibrium of an element. Thus, multiplying

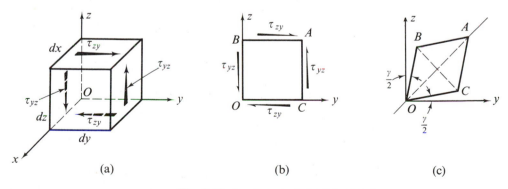

Fig. 2-14. An element of a body in pure shear

stresses by their respective areas and applying $\sum F_z = 0$,

$$(\tau_{yz})_{\text{left-hand face}} \, dx \, dz = (\tau_{yz})_{\text{right-hand face}} \, dx \, dz$$

Hence, the shearing stress τ_{yz} on the left- and the right-hand face of an infinitesimal element is numerically the same, but is *opposite in direction*. Similar reasoning applies to τ_{zy}.

Although the three-dimensional aspect of the problem should not be forgotten, it is customary, for the sake of convenience, to deal with a plane representation of the problem as shown in Fig. 2-14(b). Using this representation and summing the moments of *forces* about axis O,

$$\sum M_O = 0 \circlearrowleft +, \quad +(\tau_{zy})(dy\,dx)(dz) - (\tau_{yz})(dx\,dz)(dy) = 0$$

where the parenthetical expressions correspond respectively to stress, area, and moment arm. Simplifying,

$$\tau_{zy} = \tau_{yz} \tag{2-8}$$

Similarly it can be shown that $\tau_{xz} = \tau_{zx}$, and $\tau_{yx} = \tau_{xy}$. Hence the subscripts for the shearing stresses are commutative, i.e., their order may be interchanged.

The implication of Eq. 2-8 is very significant. The fact that subscripts are commutative signifies that *shearing stresses on mutually perpendicular planes of an infinitesimal element are numerically equal.* (Note that the mutually perpendicular planes referred to contain shearing stresses that act only toward or away from the intersection of such planes.) Moreover, it is possible to have an element in equilibrium only when *shearing stresses occur on four sides of an element simultaneously.* That is, in any stressed body where shearing stresses exist, *two* pairs of such stresses act on the mutually perpendicular planes. Hence $\sum M_O = 0$ is not satisfied by a single pair of shearing stresses.

In the subsequent work, situations where more than two pairs of shearing stresses act on an element simultaneously will seldom occur. Hence the subscripts used above to identify the planes and directions of the shearing stresses become superfluous. Shearing stresses will normally be designated by τ without any subscripts. However, one must remember that shearing stresses always occur in *two pairs.* Moreover, on diagrams, as in Fig. 2-14(b), *the arrowheads of the shearing stresses must meet at diametrically opposite corners* of an element to satisfy the equilibrium conditions for the element.

2-10. HOOKE'S LAW FOR SHEARING STRESS AND STRAIN

In the above article it was shown that in *an element* of a body the shearing stresses must occur in two pairs acting on mutually perpendicular planes. When *only* these stresses occur, the element is said to be in *pure shear.* Such a system of stresses distorts an element of an elastic body in the fashion shown in Fig. 2-14(c). Of course such a distortion is true only for a perfectly homogeneous, isotropic body having equal properties in all directions. The diagonals OA and BC are axes of symmetry for a distorted element.

If attention is confined to the study of *small* deformations, and further, if the behavior of an element is considered only in its elastic range, it is again found experimentally that there is a *linear* relationship between the shearing stress and the angle γ (gamma) shown in Fig. 2-14(c). Hence, if γ is defined as the *shearing strain*, mathematically the extension of Hooke's law for shearing stress and strain is

$$\tau = G\gamma \tag{2-9}$$

where G is a constant of proportionality called *the shearing modulus of elasticity* or the modulus of rigidity. Like E, G is a constant for a given material. It is measured in the same units as E, while γ is measured in *radians*, a dimensionless quantity. (The shearing strain γ can be stated in percent, in the same way as ε). The expressions for the three different sets of shearing strains can be stated as follows:

$$\left. \begin{array}{l} \gamma_{xy} = \tau_{xy}/G \\ \gamma_{yz} = \tau_{yz}/G \\ \gamma_{zx} = \tau_{zx}/G \end{array} \right\} \tag{2-9a}$$

For convenience, Fig. 2-14(c) is redrawn with a different set of axes so that the complete angle* γ appears on only one side of the distorted element, Fig. 2-15. Note that the shearing strains considered, numerically given by γ, are *always small*. It is sufficiently accurate to assume that tan γ, sin γ, or γ in *radian* measure are numerically equal. Likewise, the linear dimensions of a distorted element do not change appreciably. For example, in Fig. 2-15, $OB \cos \gamma \approx OB$.

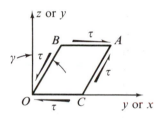

Fig. 2-15. Distortion due to pure shear

The best arrangement available for direct experimental verification of Eq. 2-9 is a *thin* tube subjected to a twist or torque. As will be explained in the next chapter, in this arrangement there is *uniform* shearing stress throughout the walls of the tube. From such experiments it is known that the appearance of τ-γ diagrams is similar to that of the σ-ε diagrams of a tension test for the same material. Similar points for the elastic limit in *shear*, yield point, and ultimate shearing stress can be obtained. However, *for the same material, the numerical values of these points for the shearing stresses are generally much lower than* (approximately one-half) *the corresponding values for the normal stress.*

In Art. 8-14 it will be shown that the three elastic constants E, ν, and G are not independent of each other for isotropic materials. In fact,

$$G = \frac{E}{2(1 + \nu)} \tag{2-10}$$

*Shearing strain is independent of the individual angles made with the coordinate axes.

Thus, for example, in the tension experiment described in Example 2-4,

$$G = (70)10^6/2(1 + 0.333) = 26.2 \times 10^6 \text{ kN/m}^2.$$

*2-11. STRESS CONCENTRATIONS

The first fundamental formulas of stress analysis, Eqs. 1-1 and 1-2, were discussed in Chapter 1, and from the preceding articles of this chapter it is seen that stresses are accompanied by deformations. If such deformations take place at the same uniform rate in *adjoining* elements, no additional stresses other than those given by the above equations exist in isotropic materials. However, if the *uniformity* of the cross-sectional area of an axially loaded member is interrupted, or if the applied force is actually applied over a very small area, a perturbation in stresses takes place. This is caused by the fact that the adjoining elements must be physically *continuous* in a deformed state. They must stretch or contract equal amounts at the adjoining sides of all particles. These deformations result from extensional and shearing deformations involving the properties of materials E, G, and ν and the applied forces. Methods of obtaining this disturbed stress distribution are beyond the scope of this text. Such problems are treated in the *mathematical theory of elasticity*. Even by those advanced methods only the simpler cases have been solved. The mathematical difficulties become too great for many practically significant problems. For the group of problems that are not tractable mathematically, approximate numerical procedures formulated on the basis of finite elements or finite difference equations are now widely used for the solution of complex problems. Digital computers are indispensable in such work. Alternatively, special experimental techniques (mainly photoelasticity, briefly discussed in Art. 9-4) have been developed to determine the actual stress distribution.

In this text it seems significant to examine qualitatively the results of more advanced investigations. For example, in Fig. 2-16(a) a short block is shown loaded by a concentrated force P. This problem could be solved by using Eq. 1-1, i.e., $\sigma = P/A$. But is this answer really correct? Reasoning in a qualitative way, it is apparent that the strains must be maximum in the *vicinity* of the applied force, hence the corresponding stresses must also be maximum. That indeed is the answer given by the *theory of elasticity*.* The end results for normal stress distribution at various sections are shown in the adjoining stress distribution diagrams, Figs. 2-16(b), (c), and (d). For present purposes, physical intuition is sufficient to justify these results. Note particularly the high peak of the normal stress at a section close to the applied force.† Also note how rapidly this peak smoothes out to a nearly uniform

---

*S. Timoshenko, and J. N. Goodier, *Theory of Elasticity* (3rd ed.), New York: McGraw-Hill, 1970, p. 60. Fig. 2-16 is adopted from this source.

†In a purely elastic material the stress is infinite right under a "concentrated" force.

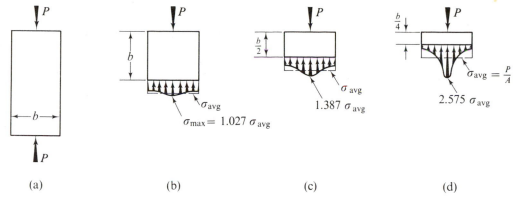

Fig. 2-16. Stress distribution near a concentrated force

stress distribution at a section below the top equal to the width of the bar. This illustrates the famed *Saint Venant's principle* of rapid dissipation of *localized* stresses. This principle asserts that the effect of forces or stresses applied over a small area may be treated as a statically equivalent system which, at a distance approximately equal to the width or thickness of a body, causes stress distribution that follows a simple law. Hence Eq. 1-1 is nearly true at a distance equal to the width of the member from the point of application of a concentrated force. Note also that at every level where the stress is investigated accurately, the *average* stress is still correctly given by Eq. 1-1. This follows, since the equations of statics *must always be satisfied*. No matter how irregular the nature of the stress distribution at a given section through a member, an integral (or sum) of $\sigma \, dA$ over the whole area *must be equal to the applied force.*

Because of the great difficulty encountered in solving for the above-mentioned "peak" or *local* stresses, a convenient scheme has been developed in practice. This scheme consists simply of calculating the stress by the elementary equations (as Eqs. 1-1 or 1-2) and then multiplying the stress so computed by a number called the *stress-concentration factor*. In this text this number will be designated by K. The values of the stress-concentration factor depend *only* on the *geometrical proportions* of the member. These factors are available in technical literature in various tables and graphs* as a function of the geometrical parameters of members. Using this scheme, Eq. 1-1 may now be rewritten as

$$\sigma_{max} = K \frac{P}{A} \qquad (2\text{-}11)$$

where K is the stress-concentration factor. From Fig. 2-16(d), at a depth below the top equal to one-quarter of the width of the member, $K = 2.575$. Hence $\sigma_{max} = 2.575 \, \sigma_{av}$.

*R. J. Roark and W. C. Young, *Formulas for Stress and Strain* (5th ed.). New York: McGraw-Hill, 1975.

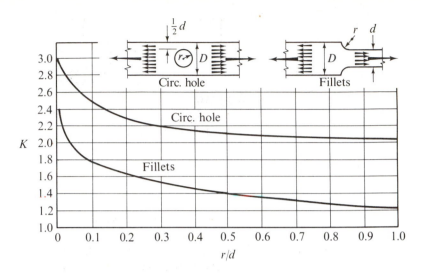

Fig. 2-17. Stress-concentration factors for flat bars in tension

Two other particularly significant stress-concentration factors for *flat* axially loaded members are shown in Fig. 2-17.* The corresponding factors that may be read from this graph represent a *ratio* of the peak stress of the actual stress in the *net* or small section of the member as shown in Fig. 2-18 to the average stress in the net section given by Eq. 1-1. A considerable stress concentration also occurs at the root of threads. This depends to a large degree upon the sharpness of the cut. For ordinary thread, the stress-concentration factor is in the neighborhood of $2\frac{1}{2}$. The application of Eq. 2-11 presents no difficulties, provided proper graphs or tables of K are available.

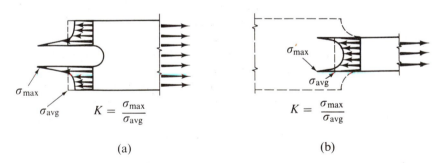

Fig. 2-18. Meaning of the stress-concentration factor K

EXAMPLE 2-6

Find the maximum stress in the member AB in the forked end A in Example 1-3.

*M. M. Frocht, "Factors of Stress Concentration Photoelastically Determined." *Trans., ASME*, 1935, vol. 57, p. A-67.

SOLUTION

Geometrical proportions:

$$\frac{\text{radius of the hole}}{\text{net width}} = \frac{3/16}{1/2} = 0.375$$

From Fig. 2-17:* $K \approx 2.15$ for $r/d = 0.375$.
Average stress from Example 1-3: $\sigma_{av} = P/A_{net} = 11.2$ ksi
Maximum stress, Eq. 2-11: $\sigma_{max} = K\sigma_{av} = (2.15)(11.2) = 24.1$ ksi

This answer indicates that actually a large local increase in stress occurs at this hole, a fact that may be highly significant.

In considering stress-concentration factors in design, it must be remembered that their theoretical or photoelastic determination is based on the use of Hooke's law. If members are *gradually* stressed beyond the proportional limit of the material, these factors lose their significance. For example, consider a flat bar of *mild steel*, of the proportions shown in Fig. 2-19, that is subjected to a gradually increasing force P. The stress distribution will be geometrically similar to that shown in Fig. 2-18 until σ_{max} reaches the yield point of the material. However, with a further increase in the applied force, σ_{max} *remains* the same, as a great deal of deformation can take place while the material yields. Therefore the stress at A remains virtually "frozen" at the same value. Nevertheless, for equilibrium, stresses acting over the net area must be high enough to resist the increased P.

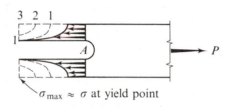

$\sigma_{max} \approx \sigma$ at yield point

Fig. 2-19. Behavior of a flat bar of mild steel when stressed beyond the yield point

As a result of this, the stress distribution begins to look something like that shown by line 1-1 in Fig. 2-19; then as 1-2, and finally as 1-3. Hence, for *ductile* materials prior to rupture, the local stress concentration is practically wiped out, and a nearly uniform, distribution of stress across the net section occurs prior to necking.

The above argument is not quite as true for materials less ductile than mild steel. Nevertheless, the tendency is in that direction unless the material is unusually brittle, like glass or some alloy steels. The argument presented applies to situations where the force is gradually applied or is static in character. *It is not applicable for fluctuating loads as found in some machine parts.* There the working stress level that is actually reached *locally* determines the fatigue behavior of the member. The maximum permissible stress is set from an *S-N* diagram (Art. 1-7). *Failure of most machine parts can be traced to progressive cracking that originates at points of high stress.* In machine design, then, stress concentrations are of paramount importance, although some machine designers feel that the theoretical concentrations are somewhat high.

---

*Strictly speaking the stress concentration depends on the condition of the hole, whether it is empty, or filled with a bolt or pin.

Apparently some tendency is present to smooth out the stress peaks even in members subjected to cyclic loads.

From the above discussion and accompanying charts it should be apparent why a trained machine designer tries to "streamline" the junctures and transitions of elements that make up a structure.

PROBLEMS FOR SOLUTION

2-1. A standard steel specimen of $\frac{1}{2}$ in. diameter elongated 0.0087 in. in an 8 in. gage length when it was subjected to a tensile force of 6,250 lb. If the specimen was known to be in the elastic range, what is the elastic modulus of the steel? *Ans:* 29.3×10^6 psi.

2-2. A steel rod 10 m long used in a control mechanism must transmit a tensile force of 5 kN without stretching more than 3 mm, nor exceeding an allowable stress of 150 MN/m². What must the diameter of the rod be? Give the answer to the nearest millimeter. $E = 210\,000$ MN/m².

2-3. A solid cylinder of 50 mm diameter and 900 mm length is subjected to a tensile force of 120 kN. One part of this cylinder, L_1 long, is of steel; the other part, fastened to steel, is aluminum and is L_2 long. (a) Determine the lengths L_1 and L_2 so that the two materials elongate an equal amount. (b) What is the total elongation of the cylinder? $E_{St} = 210\,000$ MN/m² $= 210$ GPa; $E_{Al} = 70\,000$ MN/m² $= 70$ GPa. *Ans:* (b) 0.39mm.

2-4. Rework Example 2-2 if $P_1 = 40$ k, $P_2 = 100$ k, $P_3 = 80$ k, and $P_4 = 20$ k. Let $A_{AB} = 2$ in.², $A_{BC} = 4$ in., $A_{CD} = 2$ in., $L_{AB} = 4$ ft., $L_{BC} = 2$ ft., $L_{CD} = 3$ ft., and $E = 30 \times 10^6$ psi. *Ans:* 0.032 in.

2-5. Revise the data in Example 2-2 to read as follows: $P_1 = 10$ kips; $P_3 = 100$ kips; $P_4 = 30$ kips. Then find (a) the force P_2 necessary for equilibrium and (b) the total elongation of the rod AD. The cross-sectional area of the rod from A to B is 1 in.², from B to C is 4 in.², and from C to D is 2 in.². *Ans:* 0.020 in.

2-6. In Example 2-2, what two additional (equal and opposite) forces applied at the ends will bring the total deformation back to zero?

2-7. A $\frac{1}{4}$ in. by 3 in. plate, hanging vertically, consists of an aluminum portion 6 ft long fastened to a steel portion 8 ft long. At the lower end a load of 6,000 lb is suspended. Neglecting the weight of the plate, calculate the deflection of the lower end. $E_{St} = 30 \times 10^6$ psi; $E_{Al} = 10 \times 10^6$ psi. *Ans:* 0.083 in.

2-8. A round steel bar having a cross section of 0.5 in.² is attached at the top and is subjected to three axial forces, as shown in the figure. Find the deflection of the free end caused by these forces. Plot the axial force and the axial deflection diagrams. Let $E = 30 \times 10^6$ psi. *Ans:* $\Delta_{min} = 0$, $\Delta_{max} = 0.040$ in.

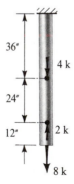

PROB. 2 – 8

2-9. A bar of steel and a bar of aluminum have the dimensions shown in the figure. Calculate the magnitude of the force P that will cause the total length of the two bars to decrease 0.010 in. Assume that the normal stress distribution over all cross sections of both bars is uniform and that the bars are prevented from buckling sideways. Plot the axial deflection diagram. Let $E_{St} = 30 \times 10^6$ psi, and $E_{Al} = 10 \times 10^6$ psi. *Ans:* 51.6 k.

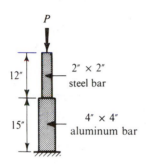

PROB. 2 – 9

2-10. In one of the California oil fields, a very long steel drill pipe got stuck in hard clay (see figure). It was necessary to determine at what depth this occurred. The engineer on the job ordered the pipe subjected to a large upward tensile force. As a result of this operation the pipe came up elastically 2 ft. At the same time the pipe elongated 0.0014 in. in an 8-in. gage length. Approximately where was the pipe stuck? Assume that the cross-sectional area of the pipe was constant and that the media surrounding the pipe hindered elastic deformation of the pipe in a static test very little. *Ans:* 11,400 ft.

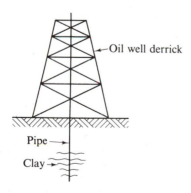

PROB. 2 – 10

2-11. A wall bracket is constructed as shown in the figure. All joints may be considered pin-connected. The steel rod AB has a cross-sectional area of 5 mm². The member BC is a rigid beam. If a 1.00 m diameter frictionless drum weighing 4 500 N is placed in the position shown, what will the elongation of the rod AB be? $E = 200$ GN/m².

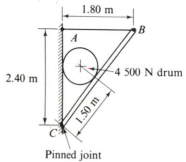

PROB. 2 – 11

2-12. For the truss shown in the figure, determine the total elongation of the member BC due to the application of the force $P = 450$ kN. The member BC is made from steel and is 60 mm² in cross-sectional area. $E = 210\,000$ MN/m².

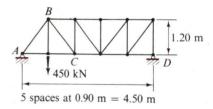

PROB. 2 – 12

2-13. If the deformation of any one member in Prob. 2-12 cannot exceed 0.1% of its length, which member requires the largest cross-sectional area and what is this area? *Ans:* 21.4 $\times 10^2$ mm².

2-14. If in Example 2-3 the rod is a 1 in. square aluminum bar weighing 1.17 lb/ft, what should its length be for the free end to elongate 0.250 in. under its own weight? $E = 10 \times 10^6$ psi.

2-15. What will the deflection of the free end of the rod in Example 2-3 be if, instead of Hooke's law, the stress-strain relationship is $\sigma = E\varepsilon^n$, where n is a number dependent on the properties of the material?

2-16. The tapered steel bar shown in the figure is cut out from a steel plate 25 mm thick and is welded at the top to a rigid structure. Find the deflection of the end A caused by the force of 40 kN applied at B. Consider the origin of the coordinate axes at the point of intersection of the sloping lines. $E = 210$ GN/m². *Ans:* 0.093 mm.

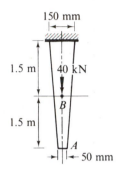

PROB. 2 – 16

2-17. Rework Prob. 2-16 taking the origin at point B.

2-18. Two bars are to be cut from a 1 in. thick metal plate so that both bars have a constant thickness of 1 in. Bar A is to have a constant width of 2 in. throughout its entire length. Bar B is to be 3 in. wide at the top and 1 in. wide at the bottom. Each bar is to be subjected to the same load P. Determine the ratio L_A/L_B so that both bars will stretch the same amount. Neglect the weight of the bar. *Ans:* 1.10.

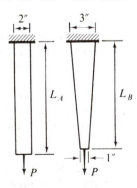

PROB. 2 – 18

2-19. The load P applied to the pin-connected frame shown stretches cable CD 2.5 mm. The area of the cable is 150 mm² and $E = 210\,000$ MN/m². Determine P. *Ans:* 39.4 kN.

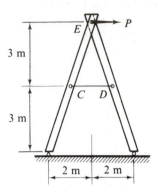

PROB. 2 – 19

2-20. Two wires are connected to a rigid bar as shown in the figure, and a weight of 9 000 N is applied. The cross-sectional area of the left wire is 60 mm², and its $E = 200\,000$ MN/m²; the corresponding quantities for the right wire are 120 mm², and $E = 70\,000$ MN/m². Calculate the vertical displacement of the weight.

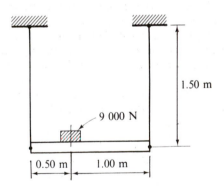

PROB. 2 – 20

2-21. The dimensions of a frustum of a right circular cone supported at the large end on a rigid base are shown in the figure. Determine the deflection of the top due to the weight of the body. The unit weight of material is γ; the elastic modulus is E. (*Hint:* Consider the origin of the coordinate axes at the vertex of the extended cone.) *Ans:* $1.0 \times 10^5 \gamma/E$ mm.

25 mm

600 mm

75 mm

PROB. 2 – 21

2-22. Find the total elongation Δ of a slender elastic bar of constant cross-sectional area A, such as shown in the figure, if it is rotated in a horizontal plane with an angular velocity of ω radians per second. The unit weight of the material is γ. Neglect the small amount of extra material by the pin. (*Hint:* First find the stress at a section a distance r from the pin by integrating the effect of the inertial forces between r and L. See Example 1-6.) *Ans: $2\gamma\omega^2 L^3/3gE$.*

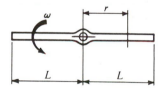

ω

r

L L

PROB. 2 – 22

2-23. A cast brass rod 2.25 in. in diameter and 6 in. long is compressed axially by a uniformly distributed force of 45,000 lb. Determine the increase in diameter caused by the applied force. $E = 12.5 \times 10^6$ psi; $\nu = 0.30$.

2-24. A piece of 50 mm by 250 mm by 10 mm steel plate is subjected to uniformly distributed stresses along its edges (see figure). (a) If $P_x = 100$ kN and $P_y = 200$ kN, what change in thickness occurs due to the application of these forces? (b) To cause the same change in thickness as in (a) by P_x alone, what must be its magnitude? Let $E = 200\,000$ MN/m² and $\nu = 0.25$.

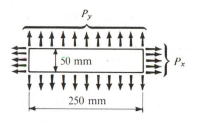

P_y

50 mm

P_x

250 mm

PROB. 2 – 24

2-25. A rectangular steel block, such as shown in Fig. 2-13(a), has the following dimensions: $a = 50$ mm, $b = 75$ mm, and $c = 100$ mm. The faces of this block are subjected to uniformly distributed forces of 180 kN (tension) in the x-direction, 200 kN (tension) in the y-direction, and 240 kN (compression) in the z-direction. Determine the magnitude of a single system of forces acting only in the y-direction that would cause the same deformation in the y-direction as the initial forces. Let $\nu = \frac{1}{4}$. *Ans: 250 kN.*

2-26. A 6 mm by 75 mm plate 600 mm long has a circular hole of 25 mm diameter located in its center. Find the axial tensile force that may be applied to this plate in the longitudinal direction without exceeding an allowable stress of 220 MPa.

2-27. Determine the extent by which a machined flat tensile bar used in a mechanical application is weakened by having an enlarged section as shown in the figure. Since the bar is to be loaded cyclically, consider stress concentrations. *Ans: About 57%.*

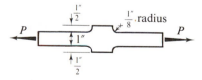

P $\frac{1}{2}''$ $\frac{1''}{8}$ radius P

$1''$

$\frac{1''}{2}$

PROB. 2 – 27

2-28. A long slot is cut out from a 1 in. by 6 in. steel bar 10 ft long as shown in the figure. (a) Find the maximum stress if an axial force $P = 50$ kips is applied to the bar. Assume that the upper curve in Fig. 2-17 is applicable. (b) For the same case, determine the total elongation of the rod. Neglect local effects of stress concentrations and assume that the reduced cross-sectional area extends for 2 ft. (c) Estimate the elongation of the same rod if $P = 160$ kips. Assume that steel yields 0.020 in. per inch at a stress of 40 ksi. (d) On removal of the load in (c), what is the residual deflection? Let $E = 30 \times 10^6$ psi. *Ans:* (a) 28.7 ksi; (b) 0.0367 in.; (c) 0.56 in.; (d) 0.448 in.

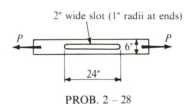

PROB. 2 – 28

2-29. The bar shown in the figure is cut from a 1 in. thick piece of steel. At the changes in section, approximate stress concentration factors are as indicated. A force P is applied, producing a total change of length in the bar of 0.016 in.

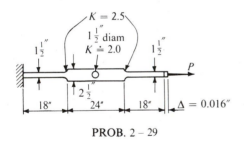

PROB. 2 – 29

Determine the maximum stress in the bar caused by this force. Neglect the effect of the hole and the stress concentrations on the axial deformation. Let $E = 30 \times 10^6$ psi. *Ans:* $\sigma_{max} = 28,600$ psi.

*2-30.   A uniform timber pile, which has been driven to a depth L in clay, carries an applied load of F at the top. This load is resisted entirely by friction f along the pile, which varies in the parabolic manner shown in the figure. (a) Determine the total shortening of the pile in terms of F, L, A, and E. (b) If $P = 420$ kN, $L = 12$ m, $A = 64\,000$ mm^2, and $E = 10\,000$ MN/m^2 = 10 GPa, how much does such a pile shorten? (*Hint:* From the equilibrium requirement, first determine the constant k.)   *Ans:* (a) $FL/(4AE)$.

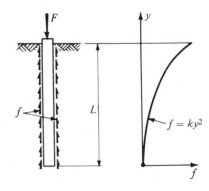

PROB. 2 – 30

3 Torsion

3-1. INTRODUCTION

The first two chapters, besides giving general concepts of the subject of mechanics of materials, investigated in detail the behavior of axially loaded rods. By the application of the method of sections and by the assumption of equal strains in longitudinal fibers, a formula for stress in an axially loaded rod was developed. Then an expression was established for obtaining the axial deformation of members. In this chapter, similar relations for statically determinate members subjected only to torque about their longitudinal axes will be determined. Thus *the investigation will be confined to the effect of a single type of action, i.e, of a torque causing twist or torsion in a member.* Members subjected simultaneously to torque and bending, frequently occurring in practice, will be treated in Chapter 10. Statically indeterminate cases are discussed in Chapter 12.

A major part of this chapter is devoted to the treatment of members with circular, or tubular, cross-sectional areas. Noncircular sections are discussed only briefly. In practice, members that transmit torque, such as shafts of motors, torque tubes of power equipment, etc., are predominantly circular or tubular in cross section. Thus, although mainly special cases of the torsion problem will be treated, the majority of important applications fall within the scope of the formulas developed.

Shaft couplings are considered briefly at the end of the chapter, since their analysis is related to the method of analysis used for circular shafts.

3-2. APPLICATION OF METHOD OF SECTIONS

In analyzing members subjected to torque, the basic approach outlined in Art. 1-9 is followed. First, the system as a whole is examined for equilibrium, and then the method of sections is applied by passing a *cutting plane perpendicular to the axis of the member. Everything* to either side of a cut is then removed, and the *internal or resisting torque* necessary to maintain equilibrium of the isolated part is determined. For finding this internal torque in *statically determinate* members, *only one* equation of statics, $\sum M_x = 0$, where the x-axis is directed along the member, is required. By applying this equation to *an isolated part* of a shaft, it may be found that the externally applied torques* are *balanced* by the internal resisting torque developed in the material. Hence *the external and internal torques are numerically equal*, but act in opposite directions.

In this chapter, shafts will be assumed "weightless" or supported at frequent enough intervals to make the effect of bending negligible. Axial forces that may also act simultaneously on the member are excluded for the present.

EXAMPLE 3-1

Find the internal torque at section K-K for the shaft shown in Fig. 3-1(a) and acted upon by the three torques indicated.

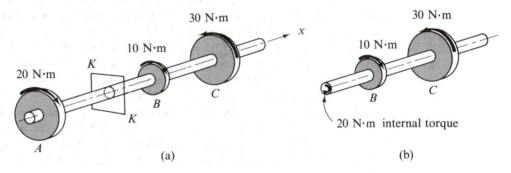

Fig. 3-1

SOLUTION

The 30 N·m torque at C is balanced by the two torques of 20 N·m and 10 N·m at A and B, respectively. Therefore, the body as a whole is in equilibrium. Next, by passing a cutting plane K-K perpendicular to the axis of the rod *anywhere* between A and B, a free body of a part of the shaft, shown in Fig. 3-1(b), is obtained. Whereupon, from $\sum M_x = 0$, or

externally applied torque = internal torque

the conclusion is reached that the internal or resisting torque developed in the shaft between A and B is 20 N·m. Similar considerations lead to the conclusion that the internal torque resisted by the shaft between B and C is 30 N·m.

It may be seen intuitively that for a member of constant cross-sectional area the maximum internal torque causes the maximum stress and imposes

---

*If *two* planes are used to isolate a portion of a body, the internal torque at one end of the *isolated body* must be treated as an external torque when the other section is considered.

the most severe condition on the material. Hence, in investigating a torsion member, several sections may have to be examined to determine the largest internal torque. A section where the largest internal torque is developed is the *critical section*. In Example 3-1 the critical section is anywhere between points B and C. If the torsion member varies in size, it is more difficult to decide where the material is critically stressed. Several sections may have to be investigated and *stresses computed* to determine the critical section. These situations are analogous to the case of an axially loaded rod, and means must be developed to determine stresses as a function of the internal torque and the size of the member. In the next several articles the necessary formulas are derived.

3-3. BASIC ASSUMPTIONS

To establish a relation between the internal torque and the stresses it sets up in members with *circular and tubular cross sections*, it is necessary to make several assumptions, the validity of which will be justified further on. These, in addition to homogeneity of material, are as follows:

1. A plane section of material perpendicular to the axis of a circular member remains *plane* after the torques are applied, i.e., no *warpage* or distortion of parallel planes normal to the axis of a member takes place.*

2. In a circular member subjected to torque, *shearing strains*, γ, *vary linearly from the central axis*. This assumption is illustrated in Fig. 3-2 and means that an imaginary plane such as AO_1O_3C moves to $A'O_1O_3C$ when the torque is applied. Alternatively, if an imaginary radius O_3C is considered fixed in direction, similar radii initially at O_2B and O_1A rotate to the respective new positions O_2B' and O_1A'. These radii *remain* straight.

 It must be emphasized that these assumptions *hold only for circular and tubular members*. For this class of members these assumptions work so well that they *apply beyond the limit of the elastic behavior of a material*. These assumptions will be used again in Art. 3-8 where stress distribution beyond the

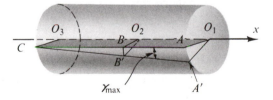

Fig. 3-2. Variation of strain in a circular member subjected to torque.

---

*Actually it is also implied that parallel planes perpendicular to the axis *remain a constant distance apart*. This is not true if deformations are large. However, since the usual deformations are very small, stresses not considered here are negligible. For details see S. Timoshenko, *Strength of Materials* (3rd. ed.), Part II, Advanced Theory and Problems. New York: Van Nostrand, 1956 Chapter VI.

proportional limit is discussed. However, if attention is confined to the *elastic* case, Hooke's law applies.

3. Thus it follows that shearing stress is proportional to shearing strain.

In the interior of a member it is difficult to justify the first two assumptions directly. However, after deriving stress and deformation formulas based on them, unquestionable agreement is found between measured and computed quantities. Moreover, their validity may be rigorously demonstrated by the methods of the theory of elasticity based on the generalized Hooke's law, and by the requirements of strain compatibility.

3-4. THE TORSION FORMULA

In the elastic case, on the basis of the above assumptions, since stress is proportional to strain, and the latter varies linearly from the center, *stresses vary linearly from the central axis of a circular member.* The stresses

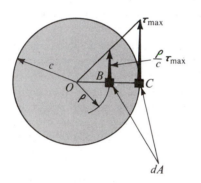

induced by the assumed distortions are *shearing* stresses and lie in the plane parallel to the section taken normal to the axis of a rod. The variation of shearing stress is illustrated in Fig. 3-3. Unlike the case of an axially loaded rod, this stress is *not* of uniform intensity. The maximum shearing stress occurs at points most remote from the center O and is designated τ_{max}. These points, such as point C in Fig. 3-3, lie at the periphery of a section at a distance c from the center. While, by virtue of a linear stress variation, at *any* arbitrary point at a distance ρ from O, the shearing stress is $(\rho/c)\tau_{max}$.

Fig. 3-3. Variation of stress in a circular member in the elastic range.

Once the stress distribution at a section is established, the resistance to torque in terms of stress can be expressed. The resistance to the torque so developed must be *equivalent* to the internal torque. Hence an equality can be formulated thus:

$$\int_A \underbrace{\underbrace{\frac{\rho}{c}\tau_{max}}_{\text{(stress)}} \, \underbrace{dA}_{\text{(area)}}}_{\text{(force)}} \, \underbrace{\rho}_{\text{(arm)}} = T$$

$$\underbrace{}_{\text{(torque)}}$$

where the integral sums up all torques developed on the cut by the infinitesimal forces acting at a distance ρ from a member's axis, O in Fig. 3-3, over the whole area A of the cross section, and where T is the resisting torque. At any given section, τ_{max} and c are constant; hence the above relation can be written as

$$\frac{\tau_{max}}{c} \int_A \rho^2 \, dA = T \tag{3-1}$$

However, $\int_A \rho^2 \, dA$, *the polar moment of inertia* of a cross-sectional area, is also a constant for a particular cross-sectional area. It will be designated by J in this text. For a circular section, $dA = 2\pi\rho \, d\rho$, where $2\pi\rho$ is the circumference of an annulus* with a radius ρ of width $d\rho$. Hence

$$J = \int_A \rho^2 \, dA = \int_0^c 2\pi\rho^3 \, d\rho = 2\pi \left.\frac{\rho^4}{4}\right|_0^c = \frac{\pi c^4}{2} = \frac{\pi d^4}{32} \tag{3-2}$$

where d is the diameter of a solid circular shaft. If c or d is measured in meters, J has the units of m^4; if in inches, the units become in.^{4}.

By using the symbol J for the polar moment of inertia of a circular area, Eq. 3-1 may be written more compactly as

$$\tau_{max} = \frac{Tc}{J} \tag{3-3}$$

This equation is the well-known *torsion formula*† for circular shafts that expresses the maximum shearing stress in terms of the resisting torque and the dimensions of a member. In applying this formula the internal torque T can be expressed in newton-meters (N·m)‡ or inch-pounds, c in meters or inches, and J in m^4 or in.^{4} Such usage makes the units of the torsional shearing stress,

$$\frac{[\text{N·m}][\text{m}]}{[\text{m}^4]} = \left[\frac{\text{N}}{\text{m}^2}\right]$$

or *pascals* in SI units, or

$$\frac{[\text{in.-lb}][\text{in.}]}{[\text{in.}^4]} = [\text{lb per in.}^2]$$

or *psi* in English units.

A more general relation than Eq. 3-3 for a shearing stress, τ, at *any* point a distance ρ from the center of a section is

$$\tau = \frac{\rho}{c}\tau_{max} = \frac{T\rho}{J} \tag{3-3a}$$

*An annulus is an area contained between two concentric circles.

†It was developed by Coulomb, a French engineer, in about 1775 in connection with his work on electric instruments. His name has been immortalized by its use for a practical unit of quantity in electricity.

‡Alternatively one (N·m) is equal to one joule (J). However, in this text the symbol J is used only for the polar moment of inertia of a section.

Equations 3-3 and 3-3(a) *are applicable* with equal rigor *to circular tubes*, since the same assumptions as used in the above derivation apply. It is necessary, however, to modify J. For a tube, as may be seen from Fig. 3-4, the limits of integration for Eq. 3-2 extend from b to c. Hence for a *circular tube*,

$$J = \int_A \rho^2 \, dA = \int_b^c 2\pi\rho^3 \, d\rho = \frac{\pi c^4}{2} - \frac{\pi b^4}{2} \qquad (3\text{-}4)$$

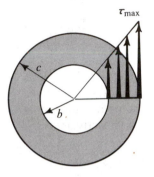

or stated otherwise: J for a circular tube equals J for a solid shaft, using the outer diameter, minus J for a solid shaft, using the inner diameter.

For *thin* tubes, if b is nearly equal to c, and $c - b = t$, the thickness of the tube, J reduces to a simple approximate expression:

$$J \approx 2\pi c^3 t \qquad (3\text{-}4a)$$

which is sufficiently accurate in many applications.

The concepts used in deriving the torsion formula for circular members are summarized as follows:

Fig. 3-4. Variation of stress in a hollow circular member in the elastic range.

1. *Equilibrium requirements* are used to determine the internal or resisting torque.
2. *Deformation* is assumed so that shearing strain varies linearly from the axis of the shaft.
3. *Material properties* in the form of Hooke's law are used to relate the assumed strain variation to stress.

Only item 3 must be modified suitably in treating the inelastic behavior of circular shafts subjected to the action of torques.

3-5. REMARKS ON THE TORSION FORMULA

So far the shearing stresses as given by Eqs. 3-3 and 3-3(a) have been thought of as acting only in the plane of a cut perpendicular to the axis of the shaft. There indeed they are acting to form a couple resisting the externally applied torques. However, to understand the problem further, an infinitesimal cylindrical element,* shown in Fig. 3-5(b), is isolated from the member of Fig. 3-5(a).

The shearing stresses acting in the planes perpendicular to the axis of the rod are known from Eq. 3-3(a). *Their directions coincide with the direction*

---

*Two planes perpendicular to the axis of the rod, two planes through the axis, and two surfaces at different radii are used to isolate this element. Properties of such an element are expressible mathematically in cylindrical coordinates.

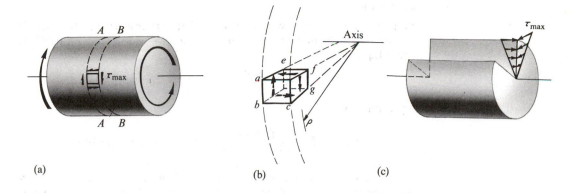

(a)

(b)   (c)

Fig. 3-5. Existence of shearing stresses on mutually perpendicular planes in a shaft subjected to torque.

of the internal resisting torque. (This should be clearly visualized by the reader.) On adjoining parallel planes of a disk-like element these stresses act in opposite directions. However, these shearing stresses acting in the plane of the cuts taken normal to the axis of a rod *cannot exist alone*, as was shown in Art. 2-9. Numerically equal shearing stresses must act on the axial planes (such as the planes *aef* and *bcg* in Fig. 3-5(b)) to fulfill the requirements of static equilibrium for an element.*

Shearing stresses acting in the axial planes follow the same variation in intensity as do the shearing stresses in the planes perpendicular to the axis of the rod. This variation of shearing stresses on the mutually perpendicular planes is shown in Fig. 3-5(c), where a portion of the shaft has been removed for the purposes of illustration.

In isotropic materials it makes little difference in which direction the shearing stresses act. However, not all materials used in construction are isotropic. For example, wood exhibits drastically different properties of strength in different directions. The shearing strength of wood on planes parallel to the grain is much less than on planes perpendicular to the grain. Hence, although equal intensities of shearing stress exist on mutually perpendicular planes, wooden shafts of inadequate size fail longitudinally along axial planes. Wooden shafts are occasionally used in the process industries.

EXAMPLE 3-2

Find the maximum torsional shearing stress in the shaft *AC* shown in Fig. 3-1(a). Assume the shaft from *A* to *C* to be of 10 mm diameter.

*Note that maximum shearing stresses, as shown diagrammatically in Fig. 3-5(a), actually act on planes perpendicular to the axis of the rod and on planes passing through the axis of the rod. The representation shown is purely schematic. The free *surface* of a shaft is *free* of all stresses.

SOLUTION

From Example 3-1 the maximum internal torque resisted by this shaft is known to be 30 N·m. Hence $T = 30$ N·m, and $c = d/2 = 5$ mm $= 0.005$ m. From Eq. 3-2

$$J = \frac{\pi d^4}{32} = \frac{\pi (0.01)^4}{32} = 9.82 \times 10^{-10} \text{ m}^4$$

From Eq. 3-3

$$\tau_{max} = \frac{Tc}{J} = \frac{(30)(0.005)}{9.82 \times 10^{-10}} = 153 \times 10^6 \text{ N/m}^2 \text{ (or Pa).}$$

This maximum shearing stress at 5 mm from the axis of the rod acts in the plane of a cut perpendicular to the axis of the rod *and* along the longitudinal planes passing through the axis of the rod (Fig. 3-5(c)).

EXAMPLE 3-3

Consider a long tube of 20 mm outside diameter, d_o, and of 16 mm inside diameter, d_i, twisted about its longitudinal axis with a torque T of 40 N·m. Determine the shearing stresses at the outside and the inside of the tube, Fig. 3-6.

Fig. 3-6

SOLUTION

From Eq. 3-4

$$J = \frac{\pi(c^4 - b^4)}{2} = \frac{\pi(d_o^4 - d_i^4)}{32} = \frac{\pi(0.02^4 - 0.016^4)}{32}$$

$$= 9.27 \times 10^{-9} \text{ m}^4$$

From Eq. 3-3

$$\tau_{max} = \frac{Tc}{J} = \frac{(40)(0.01)}{9.27 \times 10^{-9}} = 43.1 \times 10^6 \text{ N/m}^2$$

From Eq. 3-3(a)

$$\tau_{inside} = \frac{T\rho}{J} = \frac{(40)(0.008)}{9.27 \times 10^{-9}} = 34.5 \times 10^6 \text{ N/m}^2$$

Note that, for a tube, less material is required to transmit a given torque at the same stress than for a solid shaft, since no material operates at a low stress. By making the wall thickness small and the diameter large, nearly uniform shearing stress τ is obtained in the wall. This fact makes thin tubes suitable for experiments where a uniform "field" of pure shearing stress is wanted (Art. 2-10). To avoid local crimping or buckling, however, the wall thickness cannot be excessively thin.

3-6. DESIGN OF CIRCULAR MEMBERS IN TORSION

In designing members for strength, allowable shearing stresses must be selected. These depend on the information available from experiments and on the intended application. Accurate information on the capacity of materials to resist shearing stresses comes from tests on thin-walled tubes. Solid shafting is employed in routine tests. Moreover, as torsion members are so often used in power equipment, many fatigue experiments are done. Characteristically, the shearing stress that a material can withstand is lower than the normal stress. The ASME (American Society of Mechanical Engineers) code of recommended practice for transmission shafting gives an allowable value in shearing stress of 8,000 psi for unspecified steel and 0.3 of yield, or 0.18 of ultimate, shearing strength, whichever is smaller.* In practical designs, suddenly applied and shock loads warrant special considerations.

Once the torque to be transmitted by the shaft is known and the maximum shearing stress is selected, the proportions of the member become fixed. Thus, from Eq. 3-3,

$$\frac{J}{c} = \frac{T}{\tau_{\max}} \tag{3-5}$$

where J/c is the *parameter* on which the elastic strength of a shaft depends. For an axially loaded rod such a parameter is the cross-sectional area of a member. For a *solid* shaft, $J/c = \pi c^3/2$, where c is the outside radius. By using this expression and Eq. 3-5, the required radius of a shaft can be determined. For a *hollow* shaft, a number of tubes can provide the same numerical value of J/c, so the problem has an infinite number of possible solutions.

Members subjected to torque are very widely used as rotating shafts for transmitting power. For future reference, a formula will be established for the conversion of horsepower, the conventional unit used in the industry, into torque acting through the shaft. By definition, 1 hp does the work of 745.7 N·m/s. One N·m/s is conveniently referred to as a watt (W) in the SI units. Thus 1 hp can be converted into 745.7 W. Likewise, it will be recalled from dynamics that power is equal to torque multiplied by the angle, measured in radians, through which the shaft rotates per unit of time. For a shaft rotating with a frequency of f Hz, the angle is $2\pi f$ rad/s. Hence, if a

---

*Recommendations for other materials may be found in machine design books. For example, see J. E. Shigley, *Mechanical Engineering Design* (2nd ed.), New York: McGraw-Hill, 1972, or R. C. Juvinal, *Stress, Strain, and Strength*, New York: McGraw-Hill, 1967.

shaft were transmitting a constant torque T measured in N·m, it would do $2\pi f T$ N·m of work per second. Equating this to the horsepower supplied

$$\text{hp}(745.7)\,[\text{N}\cdot\text{m/s}] = 2\pi f T\,[\text{N}\cdot\text{m/s}]$$

or

$$T = \frac{119\,\text{hp}}{f} \qquad [\text{N}\cdot\text{m}] \qquad (3\text{-}6)$$

where f is the frequency in hertz of the shaft transmitting the horsepower hp. This equation converts the horsepower delivered to the shaft into a constant torque acting through it as the power is applied.

In the English system, 1 hp does work of 550 ft-lb per second, or 550(12)60 in.-lb per minute. If the shaft rotates at N rpm (revolutions per minute), an equation similar to Eq. 3-6 can be obtained as

$$T = \frac{63{,}000\,\text{hp}}{N} \qquad [\text{in.-lb}] \qquad (3\text{-}6a)$$

EXAMPLE 3-4

Select a solid shaft for a 10 hp motor operating at 30 Hz. The maximum shearing stress is limited to 55 000 kN/m².

SOLUTION

From Eq. 3-6

$$T = \frac{119\,\text{hp}}{f} = \frac{(119)(10)}{30} = 39.7\ \text{N}\cdot\text{m}$$

From Eq. 3-5

$$\frac{J}{c} = \frac{T}{\tau_{\max}} = \frac{39.7}{55 \times 10^6} = 0.722 \times 10^{-6}\ \text{m}^3$$

$$\frac{J}{c} = \frac{\pi c^3}{2} \qquad \text{or} \qquad c^3 = \frac{2}{\pi}\frac{J}{c} = \frac{2(722 \times 10^{-9})}{\pi} = 460 \times 10^{-9}\ \text{m}^3$$

Hence, $c = 0.007\,72$ m or $d = 2c = 0.015\,4$ m $= 15.4$ mm.

For practical purposes a 16 mm shaft would probably be selected.

EXAMPLE 3-5

Select solid shafts to transmit 200 hp each without exceeding a shearing stress of 10,000 psi. One of these shafts operates at 20 rpm and the other at 20,000 rpm.

SOLUTION

Subscript 1 applies to the low-speed shaft; 2 to the high-speed shaft.
From Eq. 3-6

$$T_1 = \frac{(\text{hp})(63{,}000)}{N_1} = \frac{200(63{,}000)}{20} = 630{,}000\ \text{in.-lb}$$

Similarly
$$T_2 = 630 \text{ in.-lb}$$

From Eq. 3-5

$$\frac{J_1}{c} = \frac{T_1}{\tau_{max}} = \frac{630,000}{10,000} = 63 \text{ in.}^3$$

$$\frac{J_1}{c} = \frac{\pi d_1^3}{16} \quad \text{or} \quad d_1^3 = \frac{16}{\pi}(63) = 321 \text{ in.}^3$$

and
$$d_1 = 6.85 \text{ in.}$$

Similarly
$$d_2 = 0.685 \text{ in.}$$

This example illustrates the reason for the modern tendency to use high-speed machines in mechanical equipment. The difference in size of the two shafts is striking. Further saving in the weight of the material can be effected by making use of hollow tubes.

3-7. ANGLE OF TWIST OF CIRCULAR MEMBERS

So far in this chapter, methods of determining stresses in solid and hollow circular shafts subjected to torque have been discussed. Now attention will be directed to the method of finding the angle of twist for shafts subjected to torsional loading. The interest in this problem is at least three-fold. First, it is important to predict the twist of a shaft per se since at times it is not sufficient to design it only to be strong enough: it also must not deform excessively. Then, magnitudes of angular rotations of shafts are needed in the torsional vibration analysis of machinery, although this subject is not treated here. Finally, the angular twist of members is needed in dealing with statically indeterminate torsional problems, which are discussed in Chapter 12.

According to Assumption 1 stated in Art. 3-3, planes perpendicular to the axis of a circular rod do not warp. The elements of a shaft undergo deformation of the type shown in Fig. 3-7(b). The shaded element is shown in its undistorted form in Fig. 3-7(a). From such a shaft a typical element of length dx is shown isolated in Fig. 3-8.

In the element shown a line or "fiber" such as AB is parallel initially to the axis of the shaft. After the torque is applied, it assumes a new position AD. At the same time, by virtue of Assumption 2, Art. 3-3, radius OB remains straight and rotates through a small angle $d\phi$ to a new position OD.

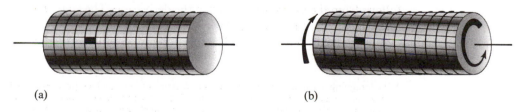

(a)   (b)

Fig. 3-7. Circular shaft (a) before (b) after torque is applied.

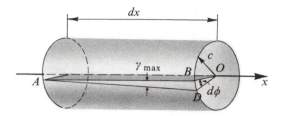

Fig. 3-8. An element of a circular shaft subjected to torque.

Denoting the small angle DAB by $\gamma_{\max}$, from geometry one has

$$\text{arc } BD = \gamma_{\max} \, dx \qquad \text{or} \qquad \text{arc } BD = d\phi \, c$$

where both angles are small and are measured in radians. Hence

$$\gamma_{\max} \, dx = d\phi \, c \qquad (3\text{-}7)$$

$\gamma_{\max}$ applies only in the zone of an infinitesimal "tube" of uniform maximum shearing stress $\tau_{\max}$. Limiting attention to linearly elastic response makes Hooke's law applicable. Therefore, according to Eq. 2-9, the angle $\gamma_{\max}$ is proportional to $\tau_{\max}$, i.e., $\gamma_{\max} = \tau_{\max}/G$. Moreover, by Eq. 3-3, $\tau_{\max} = Tc/J$. Hence $\gamma_{\max} = Tc/(JG)$.* Substituting the latter expression into Eq. 3-7 and canceling c,

$$\frac{d\phi}{dx} = \frac{T}{JG} \qquad \text{or} \qquad d\phi = \frac{T \, dx}{JG}$$

This is the relative angle of twist of two adjoining sections an infinitesimal distance dx apart. To find the total angle of twist ϕ between any two sections A and B on a shaft, the rotations of all elements must be summed. Hence, the general expression for the angle of twist at any section for a shaft of a linearly elastic material is

$$\phi = \int_A^B d\phi = \int_A^B \frac{T_x \, dx}{J_x G} \qquad (3\text{-}8)$$

The internal torque T_x and the polar moment of inertia J_x may vary along the length of a shaft. The direction of the angle of twist ϕ coincides with the direction of the applied torque T.

Equation 3-8 is valid for both solid and hollow circular shafts, which follows from the assumptions used in the derivation. The angle ϕ is measured in radians. Note the great similarity of this relation to Eq. 2-3 for the deformation of axially loaded rods. The following two examples illustrate applications of Eq. 3-8.

---

*The foregoing argument can be carried out in terms of any γ, which progressively becomes smaller as the axis of the rod is approached. The only difference in derivation consists in taking an arc corresponding to BD an arbitrary distance ρ from the center and using $T\rho/J$ instead of Tc/J for τ.

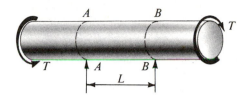

EXAMPLE 3-6

Find the relative rotation of section *B-B* with respect to section *A-A* of the solid shaft shown in Fig. 3-9 when a constant torque *T* is being transmitted through it. The polar moment of inertia of the cross-sectional area *J* is constant.

Fig. 3-9   SOLUTION

In this case $T_x = T$ and $J_x = J$; hence from Eq. 3-8

$$\phi = \int_A^B \frac{T_x \, dx}{J_x G} = \int_0^L \frac{T \, dx}{JG} = \frac{T}{JG} \int_0^L dx = \frac{TL}{JG} \qquad (3\text{-}9)$$

Equation 3-9 is an important relation. It can be used in the design of shafts for stiffness, i.e., for limiting the amount of twist that may take place in their length. For such an application *T*, *L*, and *G* are known quantities, and the solution of Eq. 3-9 yields *J*. This fixes the size of the required shaft (see Eqs. 3-2 and 3-4). Note that for stiffness requirements, *J*, rather than *J/c* of the strength requirement, is the significant parameter. This equation is used in torsional vibration analyses. The term *JG* is referred to as the *torsional stiffness* of the shaft.

Another application of Eq. 3-9 is found in the laboratory. There a shaft can be subjected to a known torque *T*, *J* can be computed from the dimensions of the specimen, and the relative angular rotation ϕ between two planes a distance *L* apart can be measured. Then, by using Eq. 3-9, the shearing modulus of elasticity in the elastic range can be computed, i.e., $G = TL/J\phi$.

In using Eq. 3-9, note particularly that angle ϕ must be expressed in radians. Also observe the similarity of Eq. 3-9 to Eq. 2-4, $\Delta = PL/AE$, formerly derived for axially loaded rods.

EXAMPLE 3-7

Consider the stepped shaft shown in Fig. 3-10 attached to a wall at *E*, and determine the rotation of the end *A* when the two torques at *B* and at *D* are applied. Assume the shearing modulus *G* to be 80×10^9 N/m², a typical value for steels.

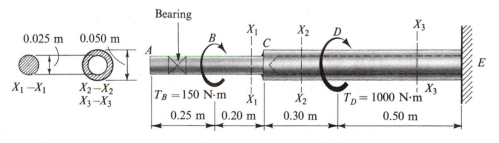

Fig. 3-10

SOLUTION

From Eq. 3-2

$$J_{AB} = J_{BC} = \frac{\pi d^4}{32} = \frac{\pi (2.5 \times 10^{-2})^4}{32} = 3.83 \times 10^{-8} \text{ m}^4$$

From Eq. 3-4

$$J_{CD} = J_{DE} = \frac{\pi}{32}(d_o^4 - d_i^4) = \frac{\pi}{32}(5^4 - 2.5^4)10^{-8} = 57.5 \times 10^{-8} \text{ m}^4$$

where subscripts indicate the range of applicability of a given value. Then by passing arbitrary sections X_1-X_1, X_2-X_2, and X_3-X_3, and each time considering a portion of the shaft to the left of such sections, the internal torques for the various intervals are found to be

$$T_{AB} = 0, \qquad T_{BD} = T_{BC} = T_{CD} = 150 \text{ N·m}, \qquad T_{DE} = 1\,150 \text{ N·m}$$

To find the rotation of the end A, Eq. 3-8 is applied with the limits of integration broken at points where T or J changes its value abruptly.

$$\phi = \int_E^A \frac{T_x\, dx}{J_x G} = \int_E^D \frac{T_{DE}\, dx}{J_{DE} G} + \int_D^C \frac{T_{CD}\, dx}{J_{CD} G} + \int_C^B \frac{T_{BC}\, dx}{J_{BC} G} + \int_B^A \frac{T_{AB}\, dx}{J_{AB} G}$$

In the last group of integrals, T's and J's are constant between the limits considered, so each integral reverts to a known solution, Eq. 3-9. Hence

$$\phi = \frac{T_{DE} L_{DE}}{J_{DE} G} + \frac{T_{CD} L_{CD}}{J_{CD} G} + \frac{T_{BC} L_{BC}}{J_{BC} G} + \frac{T_{AB} L_{AB}}{J_{AB} G}$$

$$= \frac{(1\,150)(0.5)}{(57.5 \times 10^{-8})80 \times 10^9} + \frac{(150)(0.3)}{(57.5 \times 10^{-8})80 \times 10^9}$$

$$+ \frac{(150)(0.2)}{(3.83 \times 10^{-8})80 \times 10^9} + 0$$

$$= 0.012\,5 + 0.001\,0 + 0.009\,8$$

$$= 0.023\,3 \text{ radian} \qquad \text{or} \qquad (360/2\pi)(0.023\,3) = 1.33°$$

The part AB of the shaft contributes nothing to the value of the angle ϕ as no internal torque acts through it. It rotates as much as the section at B. Little is contributed to ϕ by the shaft from C to D because a small internal torque and a large J are associated with this segment. No doubt there is a disturbance in the strains at the step, but this local effect plays a small role in the overall rotation.

The angle computed would hold equally true for a relative rotation of sections for an analogous problem of a rotating shaft.

*3-8. SHEARING STRESSES AND DEFORMATIONS IN CIRCULAR SHAFTS IN THE INELASTIC RANGE

The torsion formula for circular sections derived above is based on Hooke's law. Therefore, it applies only up to the point where the proportional limit of a material in shear is reached in the outer annulus of a shaft.

Now the solution will be extended to include inelastic behavior of a material. As before, the equilibrium requirements at a section must be met. The deformation assumption of linear strain variation from the axis remains applicable. Only the difference in material properties affects the solution.

A section through a shaft is shown in Fig. 3-11(a). The linear strain variation is shown schematically on the same figure. Some possible mechanical properties of materials in shear, obtained, for example, in experiments with thin tubes in torsion, are as shown in Figs. 3-11(b), (c), and (d). The corresponding shearing-stress distribution is shown to the right in each case. The stresses are determined from the strain. For example, if the strain is a at an interior annulus, Fig. 3-11(a), the corresponding stress is found from the stress-strain diagram. This procedure is applicable to solid shafts as well as to integral shafts made of concentric tubes of different materials,

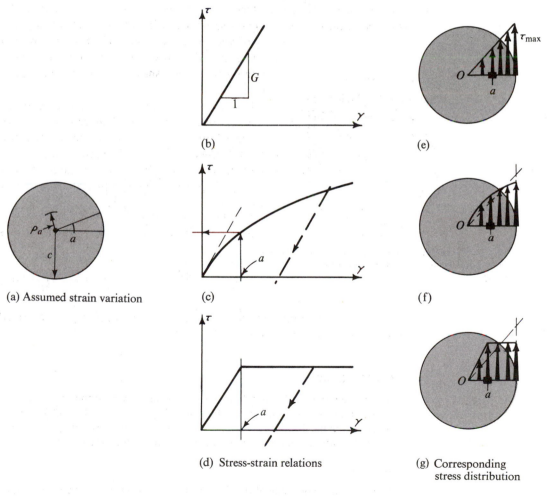

(b)

(e)

(a) Assumed strain variation

(c)

(f)

(d) Stress-strain relations

(g) Corresponding stress distribution

Fig. 3-11. Stress in circular shafts.

providing the corresponding stress-strain diagrams are used. The derivation for a linearly elastic material is simply a special case of this approach.

After the stress distribution is known, the torque T carried by these stresses is found as before, i.e.,

$$T = \int_A [\tau(dA)]\rho \qquad (3\text{-}10)$$

Either analytical or graphical procedures can be used for evaluating this integral.

Although the shearing-stress distribution after the elastic limit is exceeded is nonlinear and the elastic torsion formula Eq. 3-3 does not apply, it is sometimes used to calculate a fictitious stress for the ultimate torque. The computed stress is called the *modulus of rupture;* see the largest ordinates of the dashed lines on Figs. 3-11(f) and (g). It serves as a rough index of the ultimate strength of a material in torsion. For a thin-walled tube the stress distribution is very nearly the same regardless of the mechanical properties of the material, Fig. 3-12. For this reason experiments with thin-walled tubes are widely used in establishing the shearing stress-strain $(\tau\text{-}\gamma)$ diagrams.

If a shaft is strained into the plastic range and the applied torque is then removed, every "imaginary" annulus rebounds elastically. Because of the differences in the strain paths, which cause permanent set in the material, residual stresses develop. This process will be illustrated in one of the examples that follow.

For determining the rate of twist of a circular shaft or tube, Eq. 3-7 can be used in the following form:

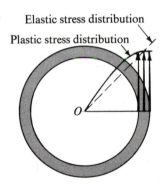

Elastic stress distribution

Plastic stress distribution

Fig. 3-12. For thin-walled tubes the difference between elastic and plastic stresses is small.

$$\frac{d\phi}{dx} = \frac{\gamma_{max}}{c} = \frac{\gamma_a}{\rho_a} \qquad (3\text{-}11)$$

Here either the maximum shearing strain at c or the strain at ρ_a determined from the stress-strain diagram must be used.

EXAMPLE 3-8

A solid steel shaft of 24 mm diameter is so severely twisted that only an 8 mm diameter elastic core remains on the inside, Fig. 3-13(a). If the material properties can be idealized as shown in Fig. 3-13(b), what residual stresses and residual rotation will remain upon release of the applied torque?

SOLUTION

To begin, the magnitude of the initially applied torque and the corresponding angle of twist must be determined. The stress distribution corresponding to

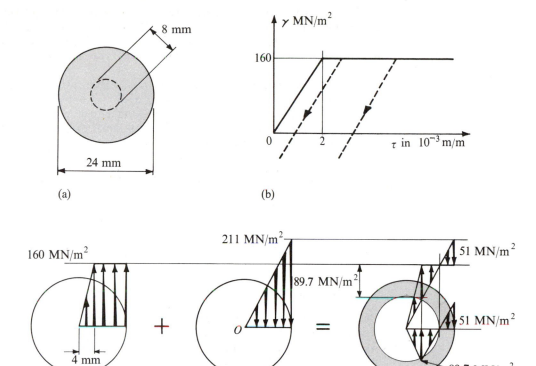

(a)

(b)

160 MN/m²

4 mm

(c) Elastic-plastic stress
distribution

+

211 MN/m²

89.7 MN/m²

O

(d) Elastic rebound
stresses

=

51 MN/m²

51 MN/m²

89.7 MN/m²

(e) Residual stresses

Fig. 3-13

the given condition is shown in Fig. 3-13(c). The stresses vary linearly from 0 to 160 MN/m² when $0 \leq \rho \leq 4$ mm; the stress is a constant 160 MN/m² for $\rho > 4$ mm. Equation 3-10 can be used to determine the applied torque T. The release of the torque T causes elastic stresses, and Eq. 3-3 applies, Fig. 3-13(d). The difference between the two stress distributions, corresponding to no external torque, gives the residual stresses.

$$T = \int_A \tau \rho \, dA = \int_0^c 2\pi\tau\rho^2 \, d\rho = \int_0^{0.004} \left[\frac{\rho}{0.004} 160 \right] 2\pi\rho^2 \, d\rho$$

$$+ \int_{0.004}^{0.012} (160) \, 2\pi\rho^2 \, d\rho = (16 + 557)10^{-6} \text{ MN·m}$$

$$= 573 \times 10^{-6} \text{ MN·m} = 573 \text{ N·m}$$

(Note the smallness of the contribution of the first integral.)

$$\tau_{\max} = \frac{Tc}{J} = \frac{573 \times 0.012}{(\pi/32)(0.024)^4} = 211 \times 10^6 \text{ N/m}^2 = 211 \text{ MN/m}^2$$

At $\rho = 12$ mm, $\tau_{\text{residual}} = 211 - 160 = 51$ MN/m².

Two diagrams of the residual stresses are shown in Fig. 3-13(e). For clarity the initial results are replotted from the horizontal line. In the entire shaded portion of the diagram, the residual torque is clockwise; an exactly equal residual torque acts in the opposite direction in the inner portion of the shaft.

The initial rotation is best determined by calculating twist of the elastic core. At $\rho = 4$ mm, $\gamma = 2 \times 10^{-3}$. The elastic rebound of the shaft is given by Eq. 3-9. The difference between the inelastic and the elastic twists gives the residual rotation per unit length of shaft. If the initial torque is re-applied in the same direction, the shaft responds elastically.

$$\text{Inelastic:} \quad \frac{d\phi}{dx} = \frac{\gamma_a}{\rho_a} = \frac{2 \times 10^{-3}}{0.004} = 0.5 \text{ rad/m}$$

$$\text{Elastic:} \quad \frac{d\phi}{dx} = \frac{T}{JG} = \frac{573}{(\pi/32)(0.024)^4(80) \times 10^9} = 0.22 \text{ rad/m}$$

$$\text{Residual:} \quad \frac{d\phi}{dx} = 0.5 - 0.22 = 0.28 \text{ rad/m}$$

EXAMPLE 3-9

Determine the torque carried by a solid circular shaft of mild steel when shearing stresses above the proportional limit are reached essentially everywhere. For mild steel, the shearing stress-strain diagram can be idealized to that shown in Fig. 3-14(a). The shearing yield-point stress τ_{yp} is to be taken as being the same as the proportional limit in shear τ_{pl}.

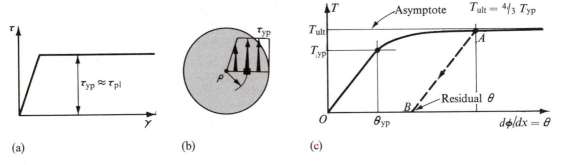

(a)　　　　　(b)　　　　　(c)

Fig. 3-14

SOLUTION

If a large torque is imposed on a member, large strains take place everywhere except near the center. Corresponding to the large strains for the idealized material considered, the yield-point shearing stress will be reached everywhere except near the center. However, the resistance to the applied torque offered by the material located near the center of the shaft is negligible as the corresponding ρ's are small, Fig. 3-14(b). (See the contribution to the torque T by the elastic action in Example 3-8.) Hence, it can be assumed with a sufficient degree of accuracy that a constant shearing stress τ_{yp} is acting everywhere on the section considered. The torque corresponding to this condition may be considered the ultimate or limit torque. (Figure 3-14(c) gives a firmer basis for this statement.) Thus

$$T_{ult} = \int_A (\tau_{yp}\, dA)\, \rho = \int_0^c 2\pi \rho^2 \tau_{yp}\, d\rho = \frac{2\pi c^3}{3}\tau_{yp}$$

$$= \frac{4}{3}\frac{\tau_{yp}}{c}\frac{\pi c^4}{2} = \frac{4}{3}\frac{\tau_{yp} J}{c} \tag{3-12}$$

Note that the maximum elastic torque capacity of a solid shaft is $T_{yp} = \tau_{yp} J/c$ according to Eq. 3-3. Therefore since T_{ult} is $\frac{4}{3}$ times this value, only $33\frac{1}{3}$ per cent of the torque capacity remains after τ_{yp} is reached at the extreme fibers of a shaft. A plot of torque T vs. θ, the angle of twist per unit distance, as full plasticity develops is in Fig. 3-14(c). Point A corresponds to the results found in the preceding example; line AB is the elastic rebound; and point B is the residual θ for the same problem.

It should be noted that in machine members, because of the fatigue properties of materials, the ultimate static capacity of the shafts as evaluated here is often of minor importance.

*3-9. STRESS CONCENTRATIONS

Equations 3-3, 3-3 a , and 3-5 apply only to solid and tubular shafts while the material behaves elastically. Moreover, the cross-sectional areas along the shaft should remain reasonably constant. If a *gradual* variation in the diameter takes place, the above equations give satisfactory solutions. On the other hand, for stepped shafts where the diameters of the adjoining portions change abruptly, large perturbations of shearing stresses take place. High *local* shearing stresses occur at points away from the center of the shaft. Methods of determining these local concentrations of stress are beyond the scope of this text. However, by forming a ratio of the true maximum shearing stress to the maximum stress given by Eq. 3-3, a torsional-stress-concentration factor can be obtained. An analogous method was used for obtaining the stress-concentration factors in axially loaded members (Art. 2-11). The stress-concentration factors depend only on the geometry of the member. Stress-concentration factors for various proportions of stepped round shafts are shown in Fig. 3-15.*

To obtain the actual stress at a geometrical discontinuity of a stepped shaft, a curve for a particular D/d is selected in Fig. 3-15. Then, corresponding to the given $r/(d/2)$ ratio, the stress-concentration factor K is read from the curve. Lastly, from the definition of K, the actual maximum shearing stress is obtained from the modified Eq. 3-3, i.e.,

$$\tau_{max} = K\frac{Tc}{J} \tag{3-3b}$$

where the shearing stress Tc/J is determined for the smaller shaft.

*This figure is adapted from a paper by L. S. Jacobsen, "Torsional-Stress Concentrations in Shafts of Circular Cross-section and Variable Diameter," *Trans. ASME.*, 1926, vol. 47, p. 632.

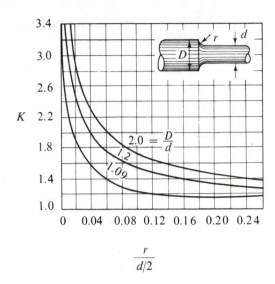

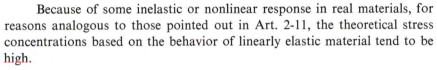

Fig. 3-15. Torsional stress-concentration factors in circular shafts of two diameters.

A study of stress-concentration factors shown in Fig. 3-15 emphasizes the need for a generous fillet radius r at all sections where a transition in the shaft diameter is made.

Considerable stress raisers also occur in shafts at oil holes and at keyways for attaching pulleys and gears to the shaft. A shaft prepared for a key, Fig. 3-16, is no longer a circular member. However, according to the procedures suggested by the ASME, for a usual design, computations for shafts with keyways are made using Eq. 3-3 or 3-5, but the allowable shearing stress is *reduced* by 25%. This supposedly compensates for the stress concentration and reduction in cross-sectional area.

Fig. 3-16. Circular shaft with a keyway.

Because of some inelastic or nonlinear response in real materials, for reasons analogous to those pointed out in Art. 2-11, the theoretical stress concentrations based on the behavior of linearly elastic material tend to be high.

*3-10. SOLID NONCIRCULAR MEMBERS

The analytical treatment of solid noncircular members in torsion is beyond the scope of this book. Mathematically the problem is complicated.* The first two assumptions stated in Art. 3-3 do not apply for noncircular members. Sections perpendicular to the axis of a member warp when a torque is applied. The nature of the distortions that take place in a rectangular section

---

*This problem remained unsolved until a famous French elastician, B. de St. Venant, developed a solution for such problems in 1853. The general torsion problem is sometimes referred to as the St. Venant problem.

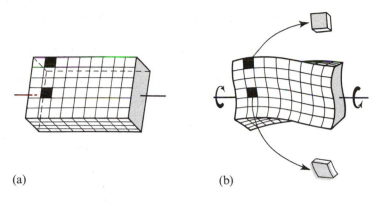

(a)   (b)

Fig. 3-17. Rectangular shaft (a) before (b) after a torque is applied.

can be seen from Fig. 3-17.* For a rectangular member, oddly enough, the corner elements do not distort at all. Shearing stresses at the corners are zero, and they are maximum at the midpoints of the long sides. Figure 3-18 shows the shearing-stress distribution along three radial lines emanating from the center. Note particularly the difference in this stress distribution compared with that of a circular section. For the latter, the stress is a maximum at the most remote point, but for the former, the stress is zero at the most remote point. This situation can be explained by considering a corner element as shown in Fig. 3-19. If a shearing stress τ existed at the corner, it could be resolved into two components parallel to the edges of the bar. However, as shears always occur in pairs acting on mutually perpendicular planes, these components would have to be met by shears lying in the planes of the outside surfaces. The latter situation is impossible as outside surfaces are free of all stresses. Hence τ must be zero. Similar considerations can be applied to other

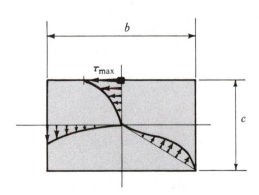

Fig. 3-18. Shearing-stress distribution in a rectangular shaft subjected to a torque.

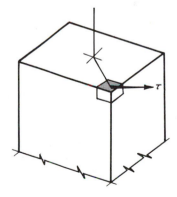

Fig. 3-19. The shearing stress shown cannot exist.

*An experiment with a rubber eraser on which a rectangular grating is ruled demonstrates this type of distortion.

points on the boundary. All shearing stresses in the plane of a cut near the boundaries act parallel to them.

Analytical solutions for torsion of rectangular, elastic members have been obtained.* The methods used are beyond the scope of this book. The final results of such analysis, however, are of interest. For the maximum shearing stress (see Fig. 3-18) and the angle of twist, these results can be put into the following form:

$$\tau_{max} = \frac{T}{\alpha bc^2} \quad \text{and} \quad \phi = \frac{TL}{\beta bc^3 G} \tag{3-13}$$

T as before is the applied torque; b is the long side and c is the short side of the rectangular section. The values of parameters α and β depend upon the ratio b/c. A few of these values are recorded in the table below. For thin sections, when b is much greater than c, the values of α and β approach $\frac{1}{3}$.

TABLE OF COEFFICIENTS FOR RECTANGULAR SHAFTS

b/c	1.00	1.50	2.00	3.00	6.00	10.0	∞
α	0.208	0.231	0.246	0.267	0.299	0.312	0.333
β	0.141	0.196	0.229	0.263	0.299	0.312	0.333

Formulas as above are available for many other types of cross-sectional areas in more advanced books. For cases that cannot be conveniently solved mathematically, a remarkable method has been devised.† It happens that the solution of the partial differential equation that must be solved in the elastic torsion problem is mathematically the same as the equation for a thin membrane, such as a soap film, lightly stretched over a hole. This hole must be geometrically similar to the cross section of the shaft being studied. Light air pressure must be kept on one side of the membrane. Then the following facts can be shown.

1. The shearing stress at any point is proportional to the slope of the stretched membrane at the same point, Fig. 3-20.
2. The direction of a particular shearing stress at a point is at right angles to the slope of the membrane at the same point, Fig. 3-20.
3. Twice the volume enclosed by the membrane is proportional to the torque carried by the section.

The foregoing analogy is called the *membrane analogy*. In addition to its value in experimental applications, it is a very useful mental tool for visual-

*S. Timoshenko and J. N. Goodier, *Theory of Elasticity* (3rd ed.), New York: McGraw-Hill, 1970, p. 312.

†This analogy was introduced by a German engineering scientist, L. Prandtl, in 1903.

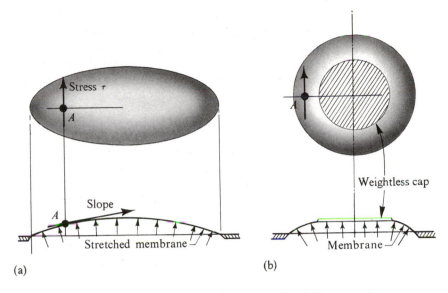

Fig. 3-20. Membrane analogy: (a) simply connected region, (b) multiply connected (tubular) region.

izing stresses and torque capacities of members. For example, all the sections shown in Fig. 3-21 can carry approximately the same torque at the same maximum shearing stress (same maximum slope of the membrane) since the volume enclosed by the membranes would be nearly the same in all cases. (For all these shapes, $b = L$ and $c = t$ in Eq. 3-13.) However, use of a little imagination will convince the reader that the contour lines of a soap film will "pile up" at a for the angular section. Hence, high local stresses will occur at that point.

Another analogy, the *sand-heap analogy*, has been developed for plastic torsion.* Dry sand is poured onto a raised flat surface having the shape of the cross section of the member. The surface of the sand heap so formed assumes a constant slope. For example, a cone is formed on a circular disc, or a

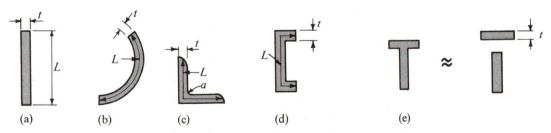

Fig. 3-21. Members of equal cross-sectional areas of the same thickness carrying the same torque.

*A. Nadai, *Theory of Flow and Fracture of Solids*, vol. 1 (2nd ed.), New York: McGraw-Hill, 1950.

pyramid on a square base. The constant maximum slope of the sand corresponds to the limiting surface of the membrane in the previous analogy. The volume of the sand heap, hence its weight, is proportional to the fully plastic torque carried by a section. The other items in connection with the sand surface have the same interpretation as those in the membrane analogy.

*3-11. THIN-WALLED HOLLOW MEMBERS

Unlike solid noncircular members, thin-walled tubes of any shape can be rather simply analyzed for the magnitude of the shearing stresses and the angle of twist caused by a torque applied to the tube. Thus, consider a tube of an arbitrary shape with varying wall thickness, such as shown in Fig. 3-22(a), subjected to a torque T. Isolate an element from this tube, as shown to an enlarged scale in Fig. 3-22(b). This element must be in equilibrium under the action of the forces F_1, F_2, F_3, and F_4. These forces are equal to the shearing stresses acting on the cut planes multiplied by the respective areas.

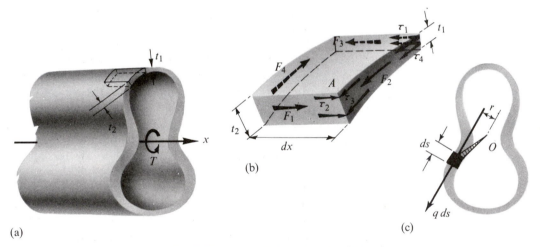

(a)

(b)

(c)

Fig. 3-22. Thin-walled member of variable thickness.

From $\sum F_x = 0$, $F_1 = F_3$; but $F_1 = \tau_2 t_2 \, dx$, and $F_3 = \tau_1 t_1 \, dx$, where τ_2 and τ_1 are shearing stresses acting on the respective areas $t_2 \, dx$ and $t_1 \, dx$. Hence, $\tau_2 t_2 \, dx = \tau_1 t_1 \, dx$, or $\tau_1 t_1 = \tau_2 t_2$. However, since the longitudinal cutting planes were taken an arbitrary distance apart, it follows from the above relations that the product of the shearing stress and the wall thickness is the same, i.e., constant, on any such planes. This constant will be denoted by q, and if the shearing stress is measured in newtons per square meter and the thickness of the tube in meters, q is measured in newtons per meter (N/m).

In Art. 2-9, Eq. 2-8, it was established that shearing stresses on mutually perpendicular planes are equal at a corner of an element. Hence at a corner such as A in Fig. 3-22(b), $\tau_2 = \tau_3$; similarly, $\tau_1 = \tau_4$. Therefore $\tau_4 t_1 = \tau_3 t_2$, or in general q is constant in the plane of a cut perpendicular to the axis of a member. On this basis an analogy can be formulated. The inner and outer boundaries of the wall can be thought of as being the boundaries of a channel. Then one can imagine a constant quantity of water steadily circulating in this channel. In this arrangement the quantity of water flowing through a plane across the channel is constant. Because of this analogy the quantity q has been termed the *shear flow*.

Next consider the cross section of the tube as shown in Fig. 3-22(c). The force per meter of the perimeter of this tube, by virtue of the previous argument, is constant and is the shear flow q. This shear flow multiplied by the length ds of the perimeter gives a force $q\,ds$ per differential length. The product of this infinitesimal force $q\,ds$ and r around some convenient point such as O, Fig. 3-22(c), gives the contribution of an element to the resistance of the applied torque T. Adding or integrating this,

$$T = \oint rq\,ds$$

where the integration process is carried around the tube along the center line of the perimeter. Since for a tube q is a constant, this equation may be written as

$$T = q\oint r\,ds$$

Instead of carrying out the actual integration, a simple interpretation of the above integral is available. It can be seen from Fig. 3-22(c) that $r\,ds$ is twice the value of the shaded area of an infinitesimal triangle of altitude r and base ds. Hence the complete integral is twice the whole area bounded by the center line of the perimeter of the tube. Defining this area by a special symbol $\textcircled{A}$, one obtains

$$T = 2\textcircled{A}q \qquad \text{or} \qquad q = T/(2\textcircled{A}) \tag{3-14}$$

This equation* applies only to thin-walled tubes. The area $\textcircled{A}$ is approximately an average of the two areas enclosed by the inside and the outside surfaces of a tube, or, as noted above, it is an area enclosed by the center line of the wall's contour. Equation 3-14 is not applicable at all if the tube is slit.

Since for any tube the shear flow q given by Eq. 3-14 is constant, from the definition of shear flow, the shearing stress at any point of a tube where the wall thickness is t is

$$\tau = q/t \tag{3-15}$$

In the elastic range, Eqs. 3-14 and 3-15 are applicable to any shape of tube. For inelastic behavior, Eq. 3-15 applies only if the thickness t is constant. The analysis of tubes of more than one cell is beyond the scope of this book.

*Equation 3-14 is sometimes called Bredt's formula in honor of the German engineer who developed it.

For linearly elastic material the angle of twist of a hollow tube can be found by applying the principle of conservation of energy. The angle of twist per unit distance, θ, is then given by*

$$\theta = \frac{d\phi}{dx} = \frac{T}{4\textcircled{A}^2 G} \oint \frac{ds}{t} \qquad (3\text{-}16)$$

EXAMPLE 3-10

Rework Example 3-3 using Eqs. 3-14 and 3-15. The tube has outside and inside radii of 10 mm and 8 mm, respectively, and the applied torque is 40 N·m.

SOLUTION

The mean radius of the tube is 9 mm and the wall thickness is 2 mm. Hence

$$\tau = \frac{q}{t} = \frac{T}{2\textcircled{A}t} = \frac{40}{2\pi(0.009)^2(0.002)} = 39.3 \times 10^6 \ \text{N/m}^2.$$

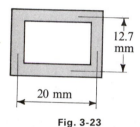

12.7 mm

20 mm

Fig. 3-23

Note that by using Eqs. 3-14 and 3-15, only one shearing stress is obtained and that it is just about the average of the two stresses computed in Example 3-3. The thinner the walls, the more accurate the answer, or vice versa.

It is interesting to note that a rectangular tube, shown in Fig. 3-23, with a wall thickness of 2 mm, for the same torque will have nearly the same shearing stress as the above circular tube. This is so because its enclosed area is about the same as the $\textcircled{A}$ of the circular tube. However, some local stress concentrations will be present at the corners of a square tube.

*3-12. SHAFT COUPLINGS

Frequently situations arise where the available lengths of shafting are not long enough. Likewise, for maintenance or assembly reasons, it is often desirable to make up a long shaft from several pieces. To join the pieces of the shaft together, the so-called flanged shaft couplings of the type shown in Fig. 3-24 are used. When bolted together, such couplings are termed *rigid*, to differentiate them from another type called *flexible* that provides for misalignment of adjoining shafts. The latter type is almost universally used to join the shaft of a motor to the equipment driven. Here only rigid-type couplings are considered. The reader is referred to machine design texts and manufacturer's catalogues for the other type.

For rigid couplings it is customary to assume that shearing strains in the bolts vary directly (linearly) as their distance from the axis of the shaft. Friction between the flanges is neglected. Therefore, analogous to the torsion problem of circular shafts, if the bolts are of the same material, elastic

*See, for example, E. P. Popov, *Introduction to Mechanics of Solids*, Englewood Cliffs, N.J.: Prentice-Hall, 1968.

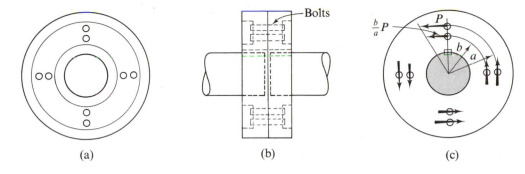

Fig. 3-24. Flanged shaft coupling.

shearing *stresses* in the bolts also vary linearly as their respective distances from the center of a coupling. The shearing stress in any one bolt is assumed to be *uniform* and is governed by the distance from its center to the center of the coupling. Then, if the shearing stress in a bolt is multiplied by its cross-sectional area, the force in a bolt is found. On this basis, for example, for bolts of *equal size* in two "bolt circles," the forces on the bolts located by the respective radii a and b are as shown in Fig. 3-24(c). The moment of the forces developed by the bolts around the axis of a shaft gives the torque capacity of a coupling.

The above reasoning is the same as that used in deriving the torsion formula for circular shafts, except that, instead of a continuous cross section, a discrete number of points is considered. This analysis is crude, since stress concentrations are undoubtedly present at the points of contact of the bolts with the flanges of a coupling. A conversion of the torsion formula for this use and for analyzing more difficult cases than couplings is discussed in Chapter 14.

The above method of analysis is valid only for the case of a coupling in which the bolts act primarily in shear. However, in some couplings the bolts are tightened so much that the coupling acts in a different fashion. The initial tension in the bolts is great enough to cause the entire coupling to act in friction. Under these circumstances the above analysis is not valid, or is valid only as a measure of the ultimate strength of the coupling should the stresses in the bolts be reduced. However, if high tensile strength bolts are used, there is little danger of this happening, and the strength of the coupling may be greater than it would be if the bolts had to act in shear.*

EXAMPLE 3-11

Estimate the torque-carrying capacity of a steel coupling forged integrally with the shaft, shown in Fig. 3-25, as controlled by an allowable shearing stress of 40 000 kN/m² in the eight bolts. The bolt circle is 0.24 m in diameter.

*See "Symposium on High-Strength Bolts," Part I, by L. T. Wyly, and Part II by E. J. Ruble, *Proceedings AISC*, 1950. Also see Art. 14-2.

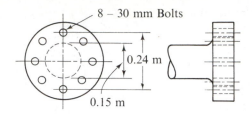

8 – 30 mm Bolts

0.24 m

0.15 m

Fig. 3-25

SOLUTION

Area of one bolt:

$$A = (1/4)\pi(30)^2 = 706 \text{ mm}^2 = 7.06 \times 10^{-4} \text{ m}^2$$

Allowable force for one bolt:

$$P_{allow} = A\tau_{allow} = 7.06 \times 10^{-4}(40 \times 10^3) = 28.2 \text{ kN}$$

Since eight bolts are available at a distance of 0.12 m from the central axis,

$$T_{allow} = (28.2)(0.12)(8) = 27.1 \text{ kN} \cdot \text{m}$$

PROBLEMS FOR SOLUTION

3-1. Find the shearing stress developed in the extreme fibers of a 75 mm diameter steel shaft due to an applied torque of 5 500 N·m. Assuming that the torque is applied in the direction shown in Fig. 3-5(a), indicate on a suitable sketch the directions of the computed stress.

3-2. A hollow shaft is of 4 in. outside diameter and 3 in. inside diameter. If the allowable shearing stress is 8,000 psi, what torque can it transmit? What is the stress at the inner surface of the shaft when the allowable torque is applied? *Ans:* 68,700 in.-lb.

3-3. A shaft of Douglas Fir is to be used in a certain process industry. If the allowable shearing stress parallel to the grain of the wood is 840 kN/m², calculate the maximum torque that can be transmitted by a 200 mm round shaft with the grain of the wood parallel to the axis.

3-4. A 6 in. diameter core, i.e., an axial hole of 3 in. radius, is bored out from a 9 in. diameter solid circular shaft. What percentage of the torsional strength is lost by this operation? *Ans:* 19.6%.

3-5. The solid cylindrical shaft of variable size shown in the figure is acted upon by the torques indicated. What is the maximum torsional stress in the shaft, and between what two pulleys does it occur? *Ans:* 17.9 MPa.

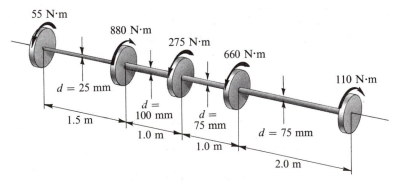

55 N·m

880 N·m

275 N·m

660 N·m

110 N·m

$d = 25$ mm

1.5 m

$d = 100$ mm

1.0 m

$d = 75$ mm

1.0 m

$d = 75$ mm

2.0 m

PROB. 3 – 5

3-6. A 150 mm diameter solid steel shaft is transmitting 600 Hp at 1.5 Hz. Compute the maximum shearing stress. Find the change that would occur in the shearing stress if the speed were increased to 6.0 Hz. *Part. Ans: 72 MPa.*

3-7. Two shafts, one a hollow steel shaft with an outside diameter of 90 mm and an inside diameter of 30 mm, the other a solid shaft with a diameter of 90 mm, are to transmit 75 hp each. Compare the shearing stresses in the two shafts if both operate at 3 Hz. *Part. Ans: 21 MPa.*

3-8. The solid 50 mm diameter steel line shaft shown in the figure is driven by a 30 hp motor at 3 Hz. Find the maximum torsional stresses in the sections *AB, BC, CD,* and *DE* of the shaft.

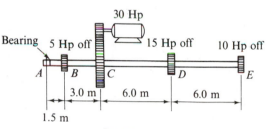

PROB. 3 – 8

3-9. A motor, through a set of gears, drives a line shaft as shown in the figure, at 630 rpm. Thirty hp are delivered to a machine on the right; 90 hp on the left. Select a solid round shaft of the same size throughout. The allowable shearing stress is 5,750 psi. *Ans: 2-in. diameter.*

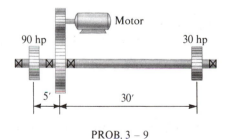

PROB. 3 – 9

3-10. Design a hollow steel shaft to transmit 300 hp at 75 rpm without exceeding a shearing stress of 6,000 psi. Use 1.2:1 as the ratio of the outside diameter to the inside diameter. *Ans: 6.22 in.*

3-11. Find the total angle of twist between *A* and *E* for the shaft in Prob. 3-8. $G = 84\,000$ MN/m². *Ans: 8.6°.*

3-12. What must the length of a 5 mm diameter aluminum wire be so that it could be twisted through one complete revolution without exceeding a shearing stress of 42 000 kN/m²? $G = 27\,000$ MN/m². *Ans: 10.1 m.*

3-13. A hollow steel rod 6 in. long is used as a torsional spring. The ratio of inside to outside diameters is $\frac{1}{2}$. The required stiffness for this spring is $\frac{1}{12}$ of a degree per one inch-pound of torque. Determine the outside diameter of this rod. $G = 12 \times 10^6$ psi. *Ans: 0.25 in.*

3-14. A solid aluminum shaft 1.0 m long and of 50 mm outside diameter is to be replaced by a tubular steel shaft of the same length and the same outside diameter so that either shaft could carry the same torque and have the same angle of twist over the total length. What must the inner radius of the tubular steel shaft be? $G_{St} = 84\,000$ MN/m² and $G_{Al} = 28\,000$ MN/m². *Ans: 22.6 mm.*

3-15. A 2 in. diameter shaft of 3 ft in length is clamped at one end and is free at the other end. A $1\frac{1}{2}$ in. diameter concentric bore is to be made in this shaft from the free end extending inward. Determine the required length of the bore such that the shaft would twist a total of 0.120° due to the application of an end torque of 900 in.-lb. Let $G = 12 \times 10^6$ psi. *Ans: 17 in.*

3-16. A 100-hp motor is driving a line shaft through gear *A* at 26.3 rpm. Bevel gears at *B* and *C* drive rubber-cement mixers. If the power requirement of the mixer driven by gear *B* is 25 hp and that of *C* is 75 hp, what are the required shaft diameters? The allowable shearing stress in

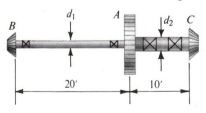

PROB. 3 – 16

the shaft is 6,000 psi. A sufficient number of bearings is provided to avoid bending. If G is 12×10^6 psi, what is the angle of twist under load in the left section of the shaft? State answer in degrees. *Ans:* $d_1 = 3.71$ in., $d_2 = 5.34$ in., and $\phi = 3.72°$.

3-17. Two gears are attached to two 50 mm diameter steel shafts as shown in the figure. The gear at B has a 200 mm pitch diameter; the gear at C a 400 mm pitch diameter. Through what angle will the end A turn if at A a torque of 560 N·m is applied and the end D of the second shaft is prevented from rotating? $G = 84\,000$ MN/m².

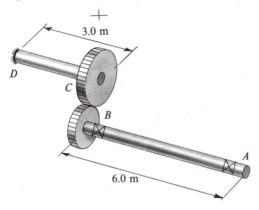

PROB. 3 – 17

3-18. In Example 3-7, find the magnitude of a torque that applied at A alone would cause the same angular rotation at A as do the two torques applied at B and D. *Ans:* 282 N·m.

3-19. (a) Determine the maximum shearing stress in the shaft subjected to the torques shown

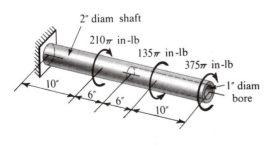

PROB. 3 – 19

in the figure. (b) Find the angle of twist in degrees between the two ends. Let $G = 12 \times 10^6$ psi. *Ans:* (a) 900 psi, (b) 0.11°.

3-20. A dynamometer is employed to calibrate the required power input to operate an exhaust fan at 20 Hz. The dynamometer consists of a 12 mm diameter solid shaft and two disks attached to the shaft 300 mm apart as shown in the figure. One disk is fastened through a tube at the input end; the other is near the output end. The relative displacement of these two disks as viewed in stroboscopic light was found to be 6° 0′. Compute the power input in hp required to operate the fan at the given speed. Let $G = 84\,000$ MN/m². *Ans:* 10 hp.

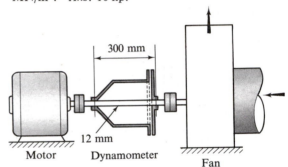

PROB. 3 – 20

3-21. A solid, tapered steel shaft is rigidly fastened to a fixed support at one end and is subjected to a torque T at the other end (see figure). Find the angular rotation of the free end if $d_1 = 6$ in.; $d_2 = 2$ in.; $L = 20$ in.; and $T = 27,000$ in.-lb. Assume that the usual assumptions of strain in prismatic circular shafts subjected to torque apply, and let $G = 12 \times 10^6$ psi. *Ans.* 0.264°.

PROB. 3 – 21

* 3-22. A thin-walled elastic frustum of a cone has the dimensions shown in the figure. Determine the torsional stiffness of this member, i.e., the magnitude of torque per unit angle of twist. The shearing modulus for the material is G.

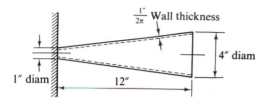

PROB. 3 – 22

* 3-23. A 150 mm diameter shaft of a linearly elastic material has in it a conical bore that is 600 mm long as shown in the figure. The shaft is rigidly attached to a fixed support at one end and is subjected to a torque T at the free end. Determine the maximum angular rotation of the shaft.

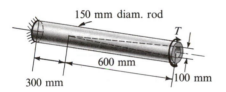

PROB. 3 – 23

* 3-24. Assume that during a drilling operation a shaft of constant torsional rigidity JG is loaded by a concentrated torque $T_1 = -1,000$ in.-lb and a distributed torque $t_x = 100$ in.-lb per inch as shown in the figure. Find the angular rotation of the free end. Plot the torque $T(x)$ and the angle-of-twist $\phi(x)$ diagrams. *Ans:* $\phi_{max} = 10,000/GJ$.

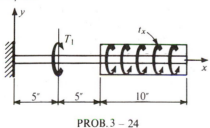

PROB. 3 – 24

3-25. A tube of 50 mm outside diameter and 2 mm thickness is attached at the ends by means of rigid flanges to a solid shaft of 25 mm diameter as shown in the figure. (All dimensions in mm.) If both the tube and the shaft are made of the same linearly elastic material, what part of the applied torque T is carried by the tube? *Ans:* 83.7%.

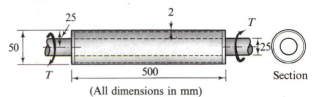

(All dimensions in mm)

PROB. 3 – 25

3-26. If the outside tube in the preceding problem is made of aluminum and the shaft is made of steel, what torque can be applied to the assembly such that the shearing stress in the aluminum tube would not exceed 100 MPa? Let $G_{St} = 84$ GPa, and $G_{Al} = 28$ GPa. What would the angle of twist be in the 500 mm length of the aluminum tube for the above torque?

3-27. A specimen of an SAE 1060 steel bar of 20 mm diameter and 450 mm in length failed at a torque of 900 N·m. What is the modulus of rupture of this steel in torsion? *Ans:* 573 MPa.

* 3-28. A 2 in. diameter steel bar is 100 in. long. One end of bar is fixed; the other is rotated through an angle $\phi = 17.19°$. What torque T was applied at the free end to produce this rotation? Idealized mechanical properties for the material of the shaft are shown in the figure. *Ans:* 30.4 k-in.

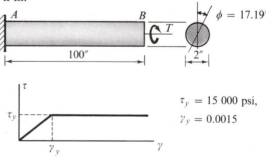

$\tau_y = 15\ 000$ psi,
$\gamma_y = 0.0015$

PROB. 3– 28

3-29. A solid circular shaft of 150 mm diameter is machined down to a 75 mm diameter along a part of the shaft. If at the transition point of the two diameters the fillet radius is 12 mm, what maximum shearing stress is developed when a torque of 2 700 N·m is applied to the shaft? What will the maximum shearing stress be if the fillet radius is reduced to 3 mm?

3-30. Find the required fillet radius for the juncture of a 6 in. diameter shaft with a 4 in. diameter segment if the shaft transmits 110 hp at 100 rpm and the maximum shearing stress is limited to 8,000 psi. *Ans: 0.31 in.*

3-31. Compare the maximum shearing stress and angle of twist for members of equal length having a square section, a rectangular section, and a circular section of equal area. All members are subjected to the same torque. The circular section is 100 mm in diameter and the rectangular section is 25 mm wide. For the square section, $\alpha = 0.208$ and $\beta = 0.141$; for the rectangular section, $\alpha \approx \beta \approx \frac{1}{3}$.

3-32. Compare the torsional strength and stiffness of thin-walled tubes of circular cross section of linearly elastic material with and without a longitudinal slot (see figure). *Ans: $3R/t$, $t^2/(3R^2)$.*

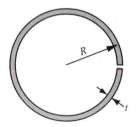

PROB. 3 –32

3-33 through 3-35. Find the maximum shearing stresses developed in members having the cross sections shown in the figures due to an applied torque of 500 in.-lb in each case. Neglect stress concentrations. *Ans: Prob. 3-33; 5,560 psi.*

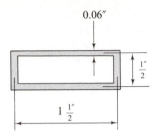

0.06″

$\frac{1''}{2}$

$1\frac{1''}{2}$

PROB. 3 – 33

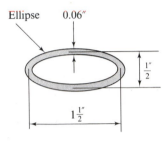

Ellipse   0.06″

$\frac{1''}{2}$

$1\frac{1''}{2}$

PROB. 3 – 34

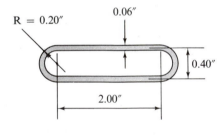

0.06″

$R = 0.20''$

0.40″

2.00″

PROB. 3 – 35

3-36. For a member having the cross section shown in the next figure, find the maximum shearing stresses and angles of twist per unit length due to an applied torque of 1,000 in-lb. Neglect stress concentrations. Comment on the advantage gained by the increase in the wall thickness over part of the cross section. *Ans: 11.1 psi, 0.691/G rad/in.*

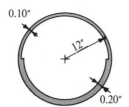

0.10″
12″
0.20″

PROB. 3 - 36

3-37. A rigid coupling with six 25 mm diameter bolts in a 200 mm diameter bolt circle is subjected to a torque of 20 000 N·m. Compute the shearing stress in the bolts. *Ans:* 67.9 MPa.

3-38. A coupling is made with eight $\frac{3}{4}$ in. diameter high strength bolts located on a 10 in. diameter bolt circle. (a) Calculate the torque that can be transmitted by this coupling if the allowable stress in the bolts is 10,500 psi. (b) Find the hp that can be transmitted when the shaft and couplings are rotating at 250 rpm. *Ans:* (b) 737 hp.

3-39. A flange coupling has 6 bolts having a cross-sectional area of 0.2 in.² each in an 8 in. diameter bolt circle, and 6 bolts having a cross-sectional area of 0.5 in.² each in a 5 in. diameter bolt circle. If the allowable shearing stress in the bolt is 16 ksi, what is the torque capacity of this coupling? *Ans:* 152 k-in.

*__3-40.__ Six 20 mm diameter bolts in the outer bolt circle of 175 mm radius are aluminum, and six 20 mm diameter bolts in the inner bolt circle of 125 mm radius are steel. What is the torque capacity of the coupling? Assume the allowable shearing stress for both materials at 40 000 kN/m² and use $G_{Al} = 28\,000$ MN/m² and $G_{St} = 84\,000$ MN/m². *Ans:* 15.6 kN·m.

4 Axial Force — Shear — and Bending Moment

4-1. INTRODUCTION

The effect of axial forces on straight members was treated in Chapters 1 and 2. Torsion of straight members was discussed in Chapter 3. It should be intuitively clear to the reader that these are not the only types of forces to which a member may be subjected. In fact, in many engineering structures members resist forces applied laterally or transversely to their axes. This type of member is termed a *beam*. Numerous applications of beams can be found in structural and machine parts. The main members supporting floors of buildings are beams, just as an axle of a car is a beam. Many shafts of machinery act simultaneously as torsion members and as beams. With modern materials, the beam is a dominant member of construction. The determination of the system of internal forces necessary for equilibrium of any beam segment will be the main objective of this chapter.*

Beams may be straight or curved, but this chapter will concentrate on a study of straight beams. Straight beams occur more frequently in practice; moreover, the system of forces at a section of a straight beam is the same as in a curved one. Hence, if the behavior of a straight beam is understood, little needs to be added regarding curved beams. To simplify the work of this chapter,† the forces applied to the beams will be assumed to lie in the same plane, i.e., a "planar" beam problem will be discussed exclusively. Further, although in actual installations a straight beam may be vertical, inclined, or horizontal, for *convenience*, the beams discussed here will be shown in a horizontal position. All beams considered will be statically determinate, i.e., reactions can always be determined by applying the equations of static equilibrium.

For the axially loaded or torsion members previously considered, only one internal force was required at a section to satisfy the conditions of

---

*The contents of this chapter may be familiar to some students. Nevertheless, it is well to review the material presented here. A thorough knowledge of this material must be had prior to the study of the chapters that follow.

†See Chapter 7 for treatment of the more general problem.

equilibrium. However, in general, a system of *three* internal force components can be recognized at a section of a beam. These quantities will be determined in this chapter by isolating segments of a beam and applying the equilibrium conditions to them. The analysis relating these forces to the stresses that they cause in the beam will be discussed in the next two chapters.

4-2. DIAGRAMMATIC CONVENTIONS FOR SUPPORTS

In studying beams it is imperative to adopt diagrammatic conventions for their supports and loadings inasmuch as several kinds of supports and a great variety of loads are possible. A thorough mastery of and *adherence* to such conventions avoids much confusion and minimizes the chances of making mistakes. These conventions form the pictorial language of engineers. As stated in the introduction, for *convenience*, the beams will usually be shown in a horizontal position.

Three types of supports are recognized for beams loaded with forces acting in the same plane. These are identified by the kind of resistance they offer to the forces. One type of support is physically realized by a *roller* or a *link*. It is capable of resisting a force in only *one specific line of action*. The link shown in Fig. 4-1(a) can resist a force only in the direction of line AB. The roller in Fig. 4-1(b) can resist only a vertical force, while the rollers in Fig. 4-1(c) can resist only a force that acts perpendicular to the plane CD. This type of support will be usually represented in this text by rollers as shown in Figs. 4-1(b) and (c), and it will be understood that *a roller support is capable of resisting a force in either direction** along the line of action of the reaction. To avoid this ambiguity, a schematic link will be occasionally employed to indicate that the reactive force may act in either direction (see Fig. 4-4). A reaction of this type corresponds to a single unknown when

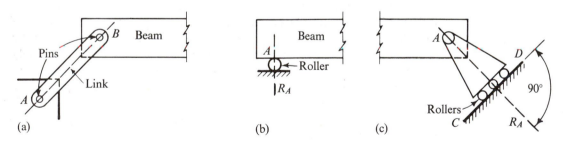

Fig. 4-1. Link and roller type of support. (The only possible line of action of the reaction is shown by the dashed lines.)

*This implies that in the actual design a link must be provided if the reaction acts away from the beam, in other words, the beam is not allowed to lift off from the support at A in Fig. 4-1(b). In this figure it may be helpful to show the roller on top of the beam in the case of a downward reaction in order to make it clear that the beam is constrained against moving up vertically at the support. This practice will be followed usually in the text.

equations of statics are applied. For inclined reactions the *ratio* between the two components is fixed (see Example 1-3).

Another type of support that may be used for a beam is a *pin*. In construction such a support is realized by using a detail as shown in Fig. 4-2(a). In this text such supports will be represented diagrammatically as shown in Fig. 4-2(b). A pinned support is capable of resisting a force acting in *any* direction of the plane. Hence, in general, the reaction at such a support may have two components, one in the horizontal and one in the vertical direction. Unlike the ratio applying to the roller or link support, that between the reaction components for the pinned support is *not fixed*. To determine these two components, two equations of statics must be used.

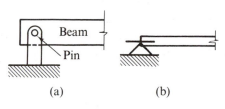

Fig. 4-2. Pinned support: (a) actual, (b) diagrammatic.

The third type of support used for beams is capable of resisting a force in any direction *and is also capable of resisting a couple or a moment*. Physically, such a support is obtained by building a beam into a brick wall, casting it into concrete, or welding the end of a beam to the main structure. A system of *three* forces can exist at such a support, two components of force and a moment. Such a support is called a *fixed support*, i.e., the built-in end is fixed or prevented from rotating. The standard convention for indicating it is shown in Fig. 4-3.

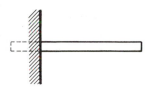

Fig. 4-3. Fixed support.

To differentiate fixed supports from the roller and pin supports, which are not capable of resisting moment, the latter two are termed *simple supports*. Figure 4-4 summarizes the foregoing distinctions between the three types of supports and the kind of resistance offered by each type. Practicing engineers normally assume the supports to be of one of the three types by "judgment," although in actual construction, supports for beams do not always clearly fall into these classifications. A more refined investigation of this aspect of the problem is beyond the scope of this text.

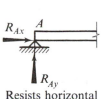

Resists horizontal
and vertical forces

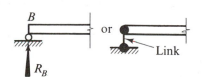

Resists vertical forces only

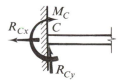

Resists horizontal
and vertical forces
and moment

Simple Supports

Fixed support

Fig. 4-4. The three common types of support.

4-3. DIAGRAMMATIC CONVENTIONS FOR LOADING

Beams are called upon to support a variety of loads. Frequently a force is delivered to the beam through a post, a hanger, or a bolted detail as shown in Fig. 4-5(a). Such arrangements apply the force over a very limited portion of the beam and are idealized for the purposes of beam analysis as *concentrated* forces. These are shown diagrammatically in Fig. 4-5(b). On the other hand, in many instances the forces are applied over a considerable portion of the beam. In a warehouse, for example, goods may be piled up along the length of a beam. Such forces are termed *distributed* loads.

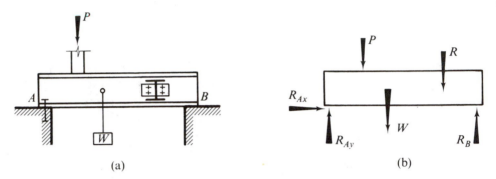

(a)   (b)

Fig. 4-5. Concentrated loading on a beam (a) actual, (b) idealized.

Many types of distributed loads occur. Among these, two kinds are particularly important: the *uniformly distributed* loads and the *uniformly varying* loads. The first could easily be an idealization of the warehouse load just mentioned, where the same kind of goods are piled up to the same height along the beam. Likewise the beam itself, if of constant cross-sectional area, is an excellent illustration of the same kind of loading. A realistic situation and a diagrammatic idealization are shown in Fig. 4-6. This load is usually expressed as force per unit length of the beam, unless specifically noted otherwise. In SI units it may be given as newtons per meter (N/m); in the English, as pounds per inch (lb/in.), as pounds per foot (lb/ft), or as kilopounds per foot (k/ft).

Uniformly varying loads act on the vertical and inclined walls of a

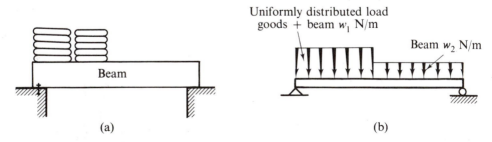

(a)   (b)

Fig. 4-6. Distributed loading on a beam (a) actual, (b) idealized.

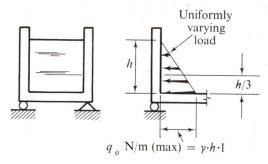

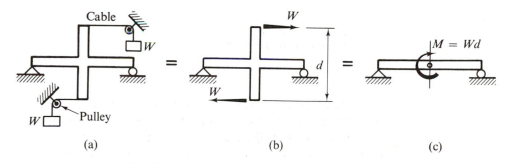

vessel containing liquid. This is illustrated in Fig. 4-7 where it is assumed that the vertical beam is *one* meter *wide* and γ (N/m³) is the unit weight of the liquid. For this type of loading, it should be carefully noted that the maximum intensity of the load of q_0 N/m is applicable only to an *infinitesimal length* of the beam. It is twice as large as the average intensity of pressure. Hence the total force exerted by such a loading on a beam is $(q_0 h/2)$ N, and its resultant acts at a distance $h/3$ above the vessel's bottom. Horizontal bottoms of vessels containing liquid are loaded uniformly.

Finally, it is conceivable to load a beam with a concentrated moment applied to the beam essentially at a point. One of the possible arrangements for applying a concentrated moment is shown in Fig. 4-8(a), and its diagrammatic representation to be used in this text is shown in Fig. 4-8(c).

Fig. 4-7. Hydrostatic loading on a vertical wall.

Fig. 4-8. A method of applying a concentrated moment to a beam.

The necessity for a complete understanding of the foregoing symbolic representation for supports and forces cannot be overemphasized. Note particularly the kind of resistance offered by the different types of supports and the manner of representation of the forces at such supports. These notations will be used to construct free-body diagrams for beams.

4-4. CLASSIFICATION OF BEAMS

Beams are classified into several groups, depending primarily on the kind of support used. Thus, if the supports are at the ends and are either pins or rollers, the beams are *simply supported* or *simple* beams, Fig. 4-9(a) and (b). The beam becomes a *fixed* beam or *fixed-ended* beam, Fig. 4-9(c), if the ends have fixed supports. Likewise, following the same scheme of nomenclature, the beam shown in Fig. 4-9(d) is a beam fixed at one end and simply supported at the other. Such beams are also called *restrained* beams as an end is

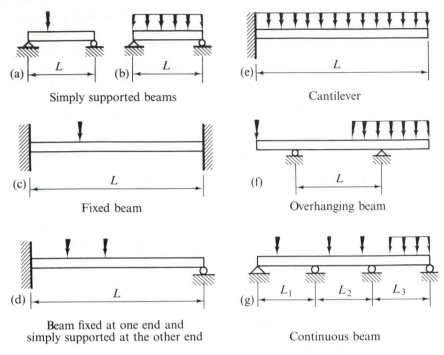

(a) L (b) L (e) L

Simply supported beams   Cantilever

(c) L (f) L

Fixed beam   Overhanging beam

(d) L (g) L_1 L_2 L_3

Beam fixed at one end and
simply supported at the other end   Continuous beam

Fig. 4-9. Types of beams.

"restrained" from rotation. A beam fixed at one end and completely free at the other has a special name, a *cantilever* beam, Fig. 4-9(e).

If the beam projects beyond a support, the beam is said to have an *overhang*. Thus the beam shown in Fig. 4-9(f) is an overhanging beam. If intermediate supports are provided for a physically continuous member acting as a beam, Fig. 4-9(g), the beam is termed a *continuous* beam.

For all beams the distance L between supports is called a *span*. In a continuous beam there are several spans that may be of varying lengths.

In addition to classifying beams on the basis of supports, descriptive phrases pertaining to the loading are often used. Thus the beam shown in Fig. 4-9(a) is a simple beam with a concentrated load, while the one in Fig. 4-9(b) is a simple beam with a uniformly distributed load. Other types of beams are similarly described.

For most of the work in mechanics of materials it is also meaningful to further classify beams into statically determinate and statically indeterminate beams. If the beam, loaded in a plane, is statically determinate, the number of unknown reaction components does not exceed three. These unknowns can always be determined from the equations of static equilibrium. The next article will briefly review the methods of statics for computing reactions for statically determinate beams. An investigation of statically indeterminate beams will be postponed until Chapter 11.

4-5. CALCULATION OF BEAM REACTIONS

All subsequent work with beams will begin with determination of the reactions. When all of the forces are applied in one plane, three equations of static equilibrium are available for this purpose. These are $\sum F_x = 0$, $\sum F_y = 0$, and $\sum M_z = 0$, and have already been discussed in Chapter 1. For straight beams in the horizontal position, the x-axis will be taken in a horizontal direction, the y-axis in the vertical direction, and the z-axis normal to the plane of the paper. The application of these equations to several beam problems is illustrated below and is intended to serve as a review of this important procedure. The deformation of beams, being small, can be neglected when the above equations are applied. For stable beams the small amount of deformation that does take place changes the points of application of the forces imperceptibly.

EXAMPLE 4-1

Find the reactions at the supports for a simple beam loaded as shown in Fig. 4-10(a). Neglect the weight of the beam.

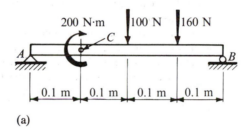

(a)

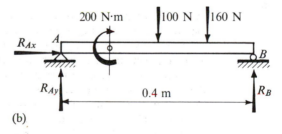

(b)

Fig. 4-10

SOLUTION

The loading of the beam is already given in diagrammatic form. The nature of the supports is examined next, and the unknown components of these reactions are boldly indicated on the diagram. The beam, with the unknown reaction components and all the applied forces, is redrawn in Fig. 4-10(b) to deliberately emphasize this important step in constructing a free-body diagram.

At A, *two* unknown reaction components may exist, since the end is pinned. The reaction at B can act only in a vertical direction since the end is on a roller. The points of application of all forces are carefully noted. After a free-body diagram of the beam is made, the equations of statics are applied to obtain the solution.

$$\Sigma F_x = 0 \qquad\qquad\qquad\qquad R_{Ax} = 0$$

$$\Sigma M_A = 0 \circlearrowleft +, \qquad 200 + (100)(0.2) + (160)(0.3) - R_B(0.4) = 0$$

$$R_B = +670 \text{ N} \uparrow$$

$$\Sigma M_B = 0 \circlearrowleft +, \qquad R_{Ay}(0.4) + 200 - (100)(0.2) - (160)(0.1) = 0$$

$$R_{Ay} = -410 \text{ N} \downarrow$$

Check:

$$\Sigma F_y = 0 \uparrow +, \qquad -410 - 100 - 160 + 670 = 0$$

Note that $\Sigma F_x = 0$ uses up one of the three independent equations of statics, thus only two additional reaction components can be determined from statics. If more unknown reaction components or moments exist at the support, the problem becomes statically indeterminate. In Fig. 4-9 the beams shown in parts (c), (d), and (g) are statically indeterminate beams as may be proved by examining the number of unknown reaction components. (Verify this statement.)

Note that the concentrated moment applied at C enters only into the expressions for the summation of moments. The positive sign of R_B indicates that the direction of R_B has been correctly assumed in Fig. 4-10(b). The inverse is the case of R_{Ay}, and the vertical reaction at A is downward. Note that a check on the arithmetical work is available if the calculations are made as shown.

ALTERNATE SOLUTION

In computing reactions some engineers prefer to make calculations in the manner indicated in Fig. 4-11. Fundamentally this involves the use of the same principles. Only the details are different. The reactions for every force are determined one at a time. The total reaction is obtained by summing these reactions. This procedure permits a running check of the computations as they are performed. For every force the sum of its reactions is equal to the force itself. For example, for the 160 N force, it is easy to see that the upward forces of 40 N and 120 N total 160 N. On the other hand, the concentrated moment at C, being a couple, is resisted by a couple. It causes an *upward* force of 500 N at the right reaction and a *downward* force of 500 N at the left reaction.

EXAMPLE 4-2

Find the reactions for the partially loaded beam with a uniformly varying load shown in Fig. 4-12(a). Neglect the weight of the beam.

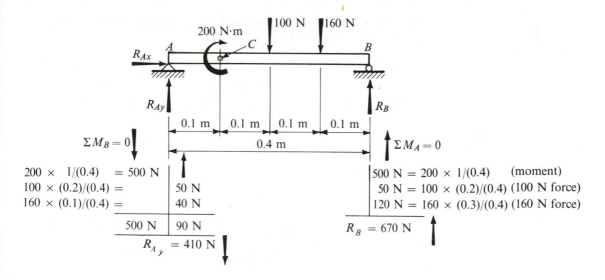

Fig. 4-11

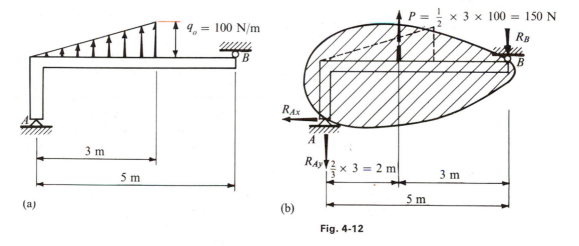

(a)

(b)

Fig. 4-12

SOLUTION

An examination of the supporting conditions indicates that there are three unknown reaction components, hence the beam is statically determinate. These and the applied load are shown in Fig. 4-12(b). Note particularly that the configuration of the member is not important for computing the reactions. A crudely shaped outline bearing no resemblance to the actual beam is indicated to emphasize this point. However, this new body is supported at points A and B in the same manner as the original beam.

For calculating the reactions the distributed load is replaced by an equivalent concentrated force P. It acts through the centroid of the distributed forces. These pertinent quantities are marked on the working sketch, Fig.

4-12(b). After a free-body diagram is prepared, the solution follows by applying the equations of static equilibrium.

$$\Sigma F_x = 0 \qquad\qquad R_{Ax} = 0$$
$$\Sigma M_A = 0 \circlearrowleft +, \qquad +(150)(2) - R_B(5) = 0, \qquad R_B = 60\text{N} \downarrow$$
$$\Sigma M_B = 0 \circlearrowleft +, \qquad -R_{Ay}(5) + (150)(3) = 0, \qquad R_{Ay} = 90\text{N} \downarrow$$

Check:

$$\Sigma F_y = 0 \uparrow +, \qquad -90 + 150 - 60 = 0$$

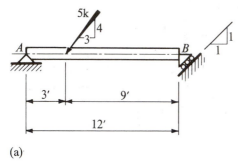

(a)

EXAMPLE 4-3

Determine the reactions at A and B for the "weightless" beam shown in Fig. 4-13(a). The applied loads are given in kilo-pound or 1,000-lb units called kips, which are designated here by k.

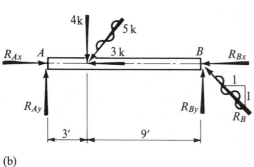

(b)

Fig. 4-13

SOLUTION

A free-body diagram is shown in Fig. 4-13(b). At A there are two unknown reaction components, R_{Ax} and R_{Ay}. At B the reaction R_B acts normal to the supporting plane and constitutes a single unknown. It is expedient to replace this force by the two components R_{By} and R_{Bx}, which in this particular problem are numerically equal. Similarly, it is best to replace the inclined force with the two components shown. These steps reduce the problem to one where all forces are either horizontal or vertical. This is of great convenience in applying the equations of static equilibrium.

$$\Sigma M_A = 0 \circlearrowleft +, \qquad +4(3) - R_{By}(12) = 0, \qquad R_{By} = 1\text{ k} \uparrow = |R_{Bx}|$$
$$\Sigma M_B = 0 \circlearrowleft +, \qquad +R_{Ay}(12) - 4(9) = 0, \qquad R_{Ay} = 3\text{ k} \uparrow$$
$$\Sigma F_x = 0 \rightarrow +, \qquad + R_{Ax} - 3 - 1 = 0, \qquad R_{Ax} = 4\text{ k} \rightarrow$$

$$R_A = \sqrt{4^2 + 3^2} = 5\text{ k}$$

$$R_B = \sqrt{1^2 + 1^2} = \sqrt{2}\text{ k}$$

Check:

$$\Sigma F_y = 0 \uparrow +, \qquad +3 - 4 + 1 = 0$$

4-6. APPLICATION OF METHOD OF SECTIONS

The main object of this chapter is to establish means for determining the forces that exist at a section of a beam. To obtain these forces, the method of sections, the basic approach of mechanics of materials, will now be applied.

The analysis of any beam is begun by preparing a free-body diagram. The reactions can always be computed using the equations of equilibrium, provided the beam is statically determinate. The complete system of forces that maintains the beam in equilibrium is thus established, and *no distinction need be made between the applied and reactive forces* in the subsequent steps of analysis. The method of sections can then be applied at any section of the beam by employing the previously used concept that if a *whole body* is in equilibrium, *any part* of it is likewise in equilibrium.

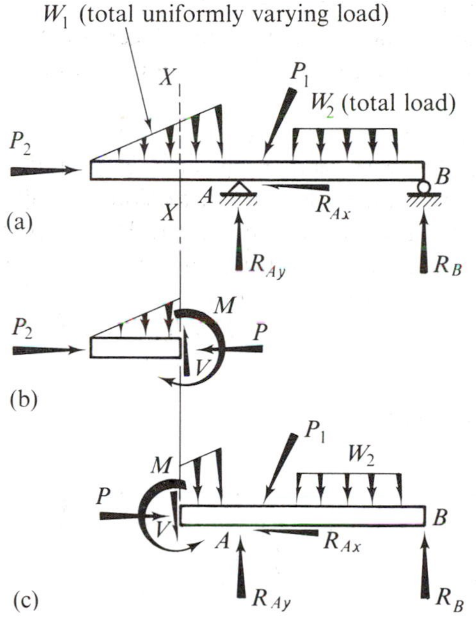

To be more specific, consider a beam, such as the one shown in Fig. 4-14(a), with certain concentrated and distributed forces acting on it. The reactions are also presumed to be known since they may be computed as in the examples considered earlier in Art. 4-5. The externally applied forces and the reactions at the support keep the *whole body* in equilibrium. Now consider an imaginary cut X-X *normal* to the axis of the beam, which separates the beam into two segments as shown in Figs. 4-14(b) and (c). Note particularly that the imaginary section goes through the distributed load and separates it too. Each of these beam segments is a free body that must be in equilibrium. However, the conditions of equilibrium require the existence of a system of forces at the cut section of the beam. In general, at a section of a beam a vertical force, a horizontal force, and a moment are necessary to maintain the part of the beam in equilibrium. These quantities take on a special significance in beams and therefore will be discussed separately.

Fig. 4-14. An application of the method of sections to a statically determinate beam.

4-7. SHEAR IN BEAMS

To maintain a segment of a beam such as shown in Fig. 4-14(b) in equilibrium there must be an internal vertical force V at the cut to satisfy the equation $\sum F_y = 0$. This internal force V, acting *at right angles* to the axis of the beam, is called the *shear* or the *shearing force*. The shear is numerically equal to the algebraic sum of all the vertical components of the external forces acting on the isolated segment, but it is opposite in direction. Given the qualitative data shown in Fig. 4-14(b), V is opposite in direction to the downward load to the left of the section. The shear at the cut may also be computed by considering the right-hand segment shown in Fig. 4-14(c). It is then equal numerically and is opposite in direction to the sum of all the vertical forces, including the reaction components, to the right of the section. Whether the right-hand segment or the left is used to determine the shear at a section is immaterial—arithmetical simplicity governs. Shears at *any other section* may be computed similarly.

At this time a significant observation must be made. The *same* shear shown in Fig. 4-14(b) and (c) at the section X-X is opposite in direction in the two diagrams. For that *part* of the downward load W_1 to the left of section X-X, the beam at the section provides an upward support to maintain vertical forces in equilibrium. Conversely, the loaded portion of the beam exerts a downward force *on* the beam as in Fig. 4-14(c). At a section "two directions" of shear must be differentiated, depending upon *which segment* of the beam is considered. This follows from the familiar action-reaction concept of statics and has occurred earlier in the case of an axially loaded rod, and again in the torsion problem.

The direction of the shear at section X-X would be reversed in *both* diagrams if the distributed load W_1 were acting upward. Frequently a similar reversal in the direction of shear takes place at one section or another along a beam for reasons which will become apparent later. The adoption of a sign convention is necessary to differentiate between the two possible directions of shear. The definition of positive shear is illustrated in Fig. 4-15(a).* A downward internal force acting on the left side of a cut or an

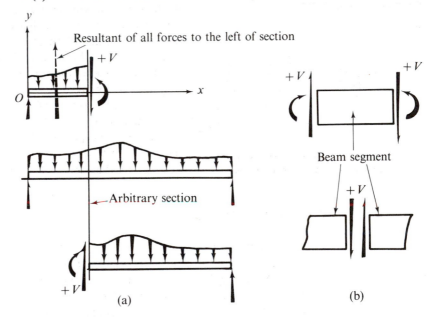

Fig. 4-15. Definition of positive shear.

*This definition of positive shear is contrary to a rigorous mathematical treatment associated with a right-hand rectangular Cartesian system. It will be used throughout this book, however, since it is the sign convention which is predominant in the technical literature. For a fully consistent treatment, see for example, E. P. Popov, *Introduction to Mechanics of Solids*, Englewood Cliffs, New Jersey: Prentice-Hall, 1968.

upward force acting on the right side of the same cut corresponds to a positive shear. Positive shears are shown in Fig. 4-15(b) for an element isolated from a beam by two sections. The shear at section X-X of Fig. 4-14(a) is a negative shear.

4-8. AXIAL FORCE IN BEAMS

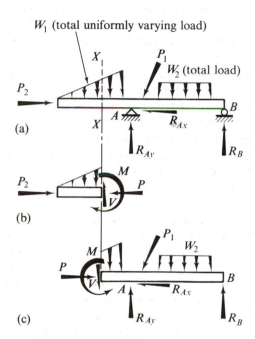

Fig. 4-16. An application of the method of sections to a statically determinate beam (Repeated)

In addition to the shear V, a horizontal force such as P, shown in Fig. 4-16(b) or (c), may be necessary at a section of a beam to satisfy the conditions of equilibrium. The magnitude and sense of this force follows from a particular solution of the equation $\sum F_x = 0$. If the horizontal force P acts toward the cut, it is called a *thrust*; if away from the cut, it is termed *axial tension*. In referring to either of these forces the term *axial force* is used. The effect of an axial force on a section of a member has already been discussed in Chapter 1. It was shown there that it is imperative to apply this force through the *centroid* of the cross-sectional area of a member to avoid bending. Similarly, here *the line of action of the axial force will always be directed through the centroid of the beam's cross-sectional area.*

Any section along a beam may be examined for the magnitude of the axial force in the above manner. The tensile force at a section is customarily taken positive. The axial force (thrust) at section X-X in Fig. 4-16(b) and (c) is equal to the horizontal force P_2.

4-9. BENDING MOMENT IN BEAMS

The determination of the shear and axial force at a section of a beam completes two of the requirements of statics which a segment must fulfill. These forces satisfy the equations $\sum F_x = 0$ and $\sum F_y = 0$. The remaining condition of static equilibrium for a planar problem is $\sum M_z = 0$. This, in general, can be satisfied only by developing a couple or an *internal resisting moment* within the cross-sectional area of the cut to counteract the moment caused by the external forces. The internal resisting moment must act in a direction opposite to the external moment to satisfy the governing equation $\sum M_z = 0$. Likewise it follows from the same equation that *the magnitude of the internal resisting moment equals the external moment.* These moments tend to bend a beam in the plane of the loads and are usually referred to as *bending moments.*

The internal bending moment M is indicated in Fig. 4-16(b). It can be developed only within the cross-sectional area of the beam and is equivalent to a couple. To determine this moment necessary to maintain the equilibrium of a segment, the sum of the moments caused by the forces may be made around any point in the plane; of course, *all* forces times their arms must be included in the sum. The internal forces V and P form *no* exception. To exclude the moments caused by these forces from the sum, it is usually most convenient in numerical problems to *select the point of intersection of these two internal forces as the point around which the moments are summed*. Both V and P have arms of zero length at this point, which is located *on the centroid of the cross-sectional area* of the beam.

Instead of considering the segment to the left of section X-X, the right-hand segment of the beam, Fig. 4-16(c), may also be used to determine the internal bending moment. As explained above, this internal moment is equal to the external moment of the applied forces (including reactions). The summation of moments is made most conveniently around the *centroid* of the section *at* the cut. In Fig. 4-16(b) the resisting moment may be physically interpreted as a pull on the top fibers of the beam and a push on the lower ones. The same interpretation applies to the same moment shown in Fig. 4-16(c).

If the load W_1 in Fig. 4-16(a) were acting in the opposite direction, the resisting moments in Figs. 4-16(b) *and* (c) would reverse. This and similar situations necessitate the adoption of a sign convention for the bending moments. This convention is associated with a definite physical action of the beam. For example, in Figs. 4-16(b) and (c), the internal moments shown pull on the top portion of the beam and compress the lower. This tends to increase the length of the top surface of the beam and to contract the lower surface. A continuous occurrence of such moments along the beam makes the beam

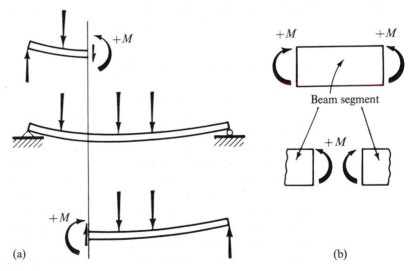

Fig. 4-17. The definition of a positive bending moment.

deform convex upwards, i.e., "shed water." Such bending moments are assigned a *negative sign*. Conversely, a positive moment is defined as one that produces compression in the top part and tension in the lower part of a beam's cross section. Under such circumstances the beam assumes a shape that "retains water." For example, a simple beam supporting a group of downward forces deflects down as shown in *exaggerated* form in Fig. 4-17(a), a fact immediately suggested by physical intuition. In such a beam, a detailed investigation of bending moments along the beam shows that all of them are positive. The sense of a positive bending moment at a section of a beam is defined in Fig. 4-17(b).

4-10. SHEAR, AXIAL-FORCE, AND BENDING-MOMENT DIAGRAMS

By the methods discussed above, the magnitude and sense of shears, axial forces, and bending moments may be obtained at many sections of a beam. Moreover, with the sign conventions adopted for these quantities, a plot of their values may be made on *separate* diagrams. On such diagrams, from a base line representing the length of a beam, ordinates may be laid off equal to the computed quantities. When these ordinate points are plotted and interconnected by lines, graphical representations of the functions are obtained. These diagrams, corresponding to the kind of quantities they depict, are called respectively *the shear diagram, the axial-force diagram, or the bending-moment diagram*. With the aid of such diagrams, the magnitudes and locations of the various quantities become immediately apparent. It is convenient to make these plots directly below the free-body diagram of the beam, using the same horizontal scale for the length of the beam. Draftsmanlike precision in making such diagrams is usually unnecessary, although the significant ordinates are generally marked with their numerical value.

The axial-force diagrams are not as commonly used as the shear and the bending-moment diagrams. This is so because the majority of beams investigated in practice are loaded by forces that act perpendicular to the axis of the beam. For such loadings of a beam, there are no axial forces at any section.

Shear and moment diagrams are exceedingly important. From them a designer sees at a glance the kind of performance that is desired from a beam at every section. In Chapter 10 on the design of members, methods of constructing these diagrams in a rapid manner will be discussed. However, the procedure discussed above of sectioning a beam and finding the system of forces at the section is the most fundamental approach. It will be used in the following illustrative examples.

EXAMPLE 4-4

Construct shear, axial-force, and bending-moment diagrams for the weightless beam shown in Fig. 4-18(a) subjected to the inclined force $P = 5$ kN.

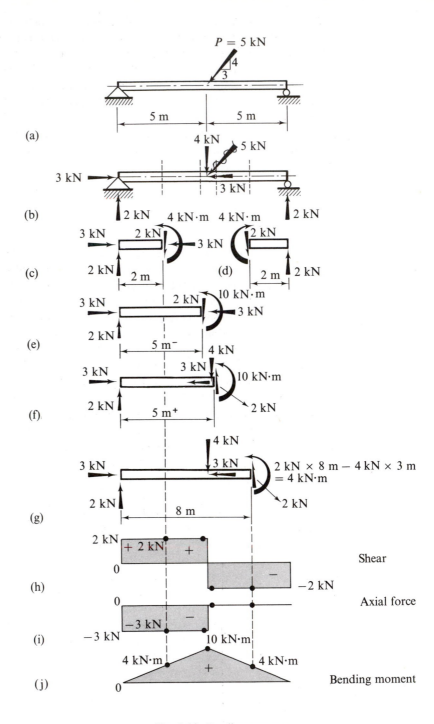

(a)

(b)

(c)

(d)

(e)

(f)

(g)

(h) Shear

(i) Axial force

(j) Bending moment

Fig. 4-18. Bending moment.

SOLUTION

A free-body diagram of the beam is shown in Fig. 4-18(b). Reactions follow from inspection after the applied force is resolved into the two components. Then several sections through the beam are investigated, as shown in Figs. 4-18(c), (d), (e), (f), and (g). In every case the same question is posed: *What are the necessary internal forces to keep the segment of the beam in equilibrium?* The corresponding quantities are recorded on the respective free-body diagrams of the beam segment. The ordinates for these quantities are indicated by heavy dots in Figs. 4-18(h), (i), and (j), with due attention paid to their signs.

Note that the free bodies shown in Figs. 4-18(d) and (g) are alternates, as they furnish the same information, and normally both would not be made. Note that a section *just to the left* of the applied force has one sign of shear, Fig. 4-18(e), while *just to the right*, Fig. 4-18(f), it has another. This indicates the importance of determining shears on either side of a concentrated force. For the condition shown, the beam *does not resist* a shear which is equal to the whole force. The bending moment in both cases is the same.

In this particular case, after a few individual points have been established on the three diagrams in Figs. 4-18(h), (i), and (j), the behavior of the respective quantities across the whole length of the beam may be reasoned out. Thus, although the segment of the beam shown in Fig. 4-18(c) is 2 m long, it may vary in length anywhere from zero to *just to the left* of the applied force, and *no change in the shear and the axial force occurs*. Hence the ordinates in Figs. 4-18(h) and (i) *remain* constant for this segment of the beam. On the other hand, the bending moment depends directly on the distance from the supports, hence it varies linearly as shown in Fig. 4-18(j). Similar reasoning applies to the segment shown in Fig. 4-18(d), enabling one to complete the three diagrams on the right-hand side. The use of the free body of Fig. 4-18(g) for completing the diagram to the right of center yields the same result.

EXAMPLE 4-5

Construct shear and bending-moment diagrams for the beam loaded with the forces shown in Fig. 4-19(a).

SOLUTION

An arbitrary section at a distance x from the left support isolates the beam segment shown in Fig. 4-19(b). This section is applicable for any value of x just to the left of the applied force P. The shear, regardless of the distance from the support, remains constant and is $+P$. The bending moment varies linearly from the support, reaching a maximum of $+Pa$.

An arbitrary section applicable anywhere *between* the two applied forces is shown in Fig. 4-19(c). No shearing force is necessary to maintain equilibrium of a segment in this part of the beam. Only a constant bending moment of $+Pa$ must be resisted by the beam in this zone. Such a state of bending or flexure is called *pure* bending.

Shear and bending-moment diagrams for this loading condition are

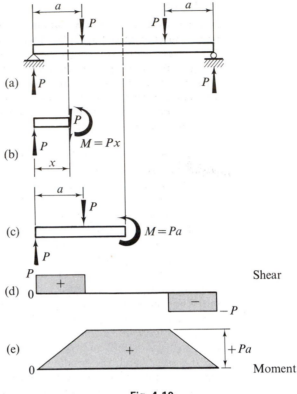

Fig. 4-19

shown in Figs. 4-19(d) and (e). No axial-force diagram is necessary, as there is no axial force at any section of the beam.

EXAMPLE 4-6

Plot a shear and a bending-moment diagram for a simple beam with a uniformly distributed load, Fig. 4-20(a).

SOLUTION

The best way of solving this problem is to write down algebraic expressions for the quantities sought. For this purpose an arbitrary section taken at a distance x from the left support is used to isolate the segment shown in Fig. 4-20(b). Since the applied load is continuously distributed along the beam, this section is typical and applies to *any section* along the length of the beam. In more difficult cases several zones of a beam may have to be investigated depending on the distribution of the applied loads. In some instances it is even advisable to resort to several origins of x to simplify the form of the algebraic functions.

    The shear V is equal to the left upward reaction *less* the load to the left of the section. The internal bending moment M resists the moment caused by the reaction on the left *less* the moment caused by the forces to the left of

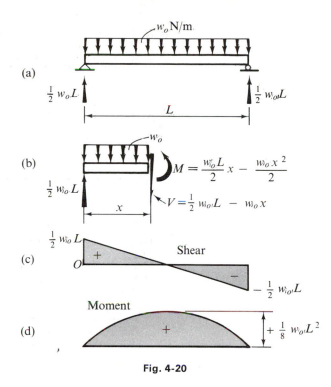

(a)

w_o N/m

$\frac{1}{2} w_o L$ $\frac{1}{2} w_o L$

L

(b)

$-w_o$

$M = \dfrac{w_o L}{2} x - \dfrac{w_o x^2}{2}$

$\frac{1}{2} w_o L$

$V = \frac{1}{2} w_o L - w_o x$

x

(c)

$\frac{1}{2} w_o L$

O +   Shear

$- \frac{1}{2} w_o L$

Moment

(d)

+ $+ \frac{1}{8} w_o L^2$

Fig. 4-20

the same section. The summation of moments is performed around an axis *at the section.* Although it is customary to isolate the left-hand segment, similar expressions may be obtained by considering the right-hand segment of the beam, with due attention paid to sign conventions. The plot of the V and M functions is shown in Figs. 4-20(c) and (d).

EXAMPLE 4-7

Determine shear, axial-force, and bending-moment diagrams for the cantilever loaded with an inclined force at the end, Fig. 4-21(a).

SOLUTION

First the inclined force is replaced by the two components shown in Fig. 4-21(b) and the reactions are determined. The *three* unknowns at the support follow from the familiar equations of statics. This completes the free-body diagram shown in Fig. 4-21(b). *Completeness in indicating all of these forces is of the utmost importance.*

A segment of the beam is shown in Fig. 4-21(c); from this segment it may be seen that the shearing force and the axial force remain the same regardless of the distance x. On the other hand, the bending moment is a variable quantity. A summation of moments around C gives $(PL - Px)$ acting in the direction shown. This represents a *negative* moment. The moment at the support is likewise a *negative* bending moment as it tends to pull on the *upper* fibers of the beam. The three diagrams are plotted in Figs. 4-21(d), (e), and (f).

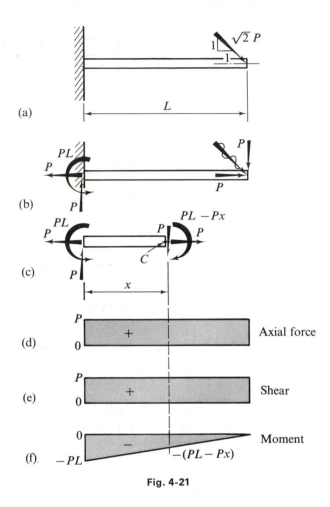

Fig. 4-21

EXAMPLE 4-8

Given a curved beam whose centroidal axis is bent into a semicircle of 0.2 m radius as shown in Fig. 4-22(a). If this member is being pulled by the 1000 N forces shown, find the axial force, the shear, and the bending moment at the section A-A, $\alpha = 45°$. The centroidal axis and the applied forces all lie in the same plane.

SOLUTION

There is no essential difference in the method of attack in this problem compared with that in a straight-beam problem. The body as a whole is examined for conditions of equilibrium. From the conditions of the problem here, such is already the case. Next, a segment of the beam is isolated, Fig. 4-22(b). *Section A-A is taken perpendicular to the axis of the beam.* Before determining the quantities wanted at the cut, the applied force P is resolved into components parallel and perpendicular to the cut. These directions are taken respectively as the y- and x-axes. This resolution replaces P by the

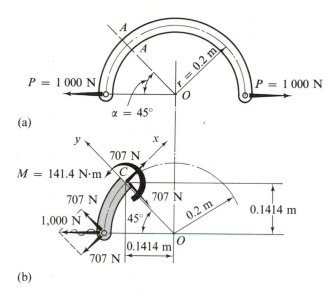

(a)

(b)

Fig. 4-22

components shown in Fig. 4-22(b). From $\sum F_x = 0$, the axial force at the cut is $+707$ N. From $\sum F_y = 0$, the shear is 707 N in the direction shown. The bending moment at the cut can be determined in several different ways. For example, if $\sum M_O = 0$ is used, note that the lines of action of the applied force P and the shear at the section pass through O. Therefore only the axial force at the centroid of the cut times the radius needs to be considered, and the *resisting* bending moment is $707(0.2) = 141.4$ N·m, acting in the direction shown. An alternative solution may be obtained by applying $\sum M_C = 0$. At C, a point lying on the centroid, the axial force and the shear intersect. The bending moment is then the product of the applied force P and the 0.1414 m arm. In both of these methods of determining bending moment, use of the components of the force P is avoided as this is more involved arithmetically.

It is suggested that the reader complete this problem in terms of a general angle α. Several interesting observations may be made from such a general solution. The moments at the ends will vanish for $\alpha = 0°$ and $\alpha = 180°$. For $\alpha = 90°$ the shear vanishes and the axial force becomes equal to the applied force P. Likewise the maximum bending moment is associated with $\alpha = 90°$.

4-11. STEP-BY-STEP PROCEDURE

In beam analysis it is exceedingly important to be able to determine the shear, the axial force, and the bending moment at any section. The technique of obtaining these quantities is unusually clear-cut and systematic. To lend further emphasis, the steps used in all such problems are summarized. This summary is intended to aid the student in an orderly analysis of problems. Sheer memorization of this procedure is discouraged.

1. Make a good sketch of the beam on which *all* of the applied forces are clearly noted and located by dimension lines from the supports.

2. Boldly indicate the unknown reactions (colored pencil may be used to advantage). Remember that a roller support has *one* unknown, a pinned support has *two* unknowns, and a fixed support has *three* unknowns.

3. Replace all of the inclined forces (known and unknown) by components acting parallel and perpendicular to the beam.*

4. Apply the equations of statics to obtain the reactions.† A check on the reactions computed in the manner indicated in Examples 4-1, 4-2, and 4-3 is highly desirable.

5. Pass a section at the desired location through the beam perpendicular to its axis. This imaginary section cuts *only the beam* and isolates the forces that act on the segment.

6. Select a segment to either side of the proposed section and redraw this segment, indicating all external forces acting on it. This must include all the reaction components.

7. Indicate the three possible unknown quantities at the cut section, i.e., show P, V, and M, assuming their directions.

8. Apply the equations of equilibrium to the segment and solve for P, V, and M. If the solution indicates any of these quantities to be a negative value, then the originally assumed direction at the cut must be reversed.

This procedure enables one to determine the shear, the axial force, and the bending moment at any section of a beam. Signs for these quantities follow from the definitions given earlier. If diagrams for this system of internal forces are wanted, several sections may have to be investigated. Do not fail to determine the abrupt change in shear at concentrated forces and the abrupt change in bending-moment value at points where concentrated moments are introduced. Algebraic expressions for the same quantities sometimes are also necessary.

In the above discussion the construction of shear and moment diagrams was illustrated principally for horizontal members. For inclined members, except for directing the coordinate axes along and perpendicular to the axis of a bar, the procedure is the same. In curved and in spatial structural systems the directions of the axes are along the axes of the member or members. In such cases one of the coordinate axes is taken tangent to the axis of the member—as shown for example in Fig. 4-22. To conform with the diagrammatic scheme used in this text for horizontal beams, the ordinates for bending moment in curved and spatial systems should be plotted on the compression side‡ of a section.

At this time it is suggested that Art. 1-9 on the basic approach of mechanics of materials be reviewed, as a better appreciation may now be had of the contents of that article.

*More ingenuity may be required for curved beams.

†This step can be avoided in cantilevers by proceeding from the free end.

‡In some texts on structural analysis the opposite scheme is used.

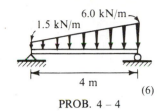

PROB. 4 – 4   (6)

(*Note:* In addition to beams, simple frames are included in the following problems. For the quantities asked, the analysis of these frames is analogous to that of beams.)

4-1. Show that the effect on a structure of the tensile forces acting in a flexible cable going over a frictionless pulley is the same as that of the same two forces applied at the center of the axle.

4-2. Compute the reactions at the hinged supports A and B. *Ans:* $R_{Ax} = 18.75$ kN, $R_{Ay} = 75$ kN.

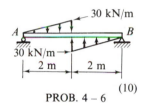

PROB. 4 – 5   (−20)

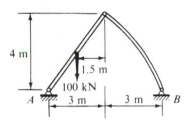

PROB. 4 – 2

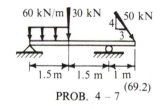

PROB. 4 – 6   (10)

4-3. For the beam loaded as shown in the figure determine the magnitude and direction of the reactions. *Ans:* $R_A = 4$ k.

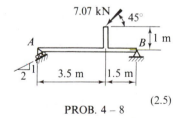

PROB. 4 – 7   (69.2)

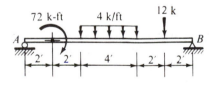

PROB. 4 – 3

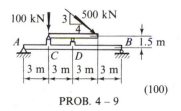

PROB. 4 – 8   (2.5)

4-4 through 4-13. For the planar structures loaded as shown in the figures determine the reactions or all reaction components. All structures are to be assumed weightless. *A correctly drawn free-body diagram is an essential part of each problem.* *Ans:* Upward reaction component for the left reaction is given in parentheses by the figures in the units of the applied loads.

PROB. 4 – 9   (100)

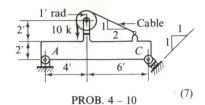

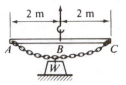

PROB. 4 – 10 (7)

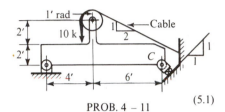

PROB. 4 – 11 (5.1)

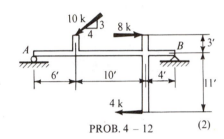

PROB. 4 – 12 (2)

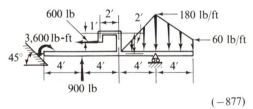

PROB. 4 – 13 (−877)

4-14. For the beam loaded as shown in the figure, find the shear and the bending moment at the center of the span caused by the applied load. *Ans: V = −1 k, M = −13.5 k-ft.*

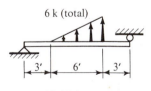

PROB. 4 – 14

4-15. A chain block used for raising 100 kN weights by means of a spreader beam is shown in the figure. The chain *AB* is 2.4 m long; chain *BC* is 3.2 m long. Neglecting the weight of the as-

sembly, find the components of all forces acting parallel and perpendicular to the beam when in use. *Ans: $R_{Ax} = 36$ kN.*

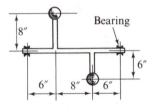

PROB. 4 –15

4-16. Two 3 lb weights are attached to a shaft by means of rigid arms as shown in the figure. Neglecting the weight of the shaft and the arms, find the reactions at the bearings if the shaft rotates at 600 rpm. *Ans: $R_{\text{left}} = 116$ lb.*

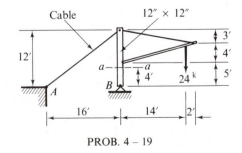

PROB. 4 – 16

4-17. Determine the bending moment at the support *B* in Prob. 4-5. *Ans:* 30 kN·m.

4-18. Determine the shear and the bending moment at a section midway between *C* and *D* in the beam *AB* of Prob. 4-9. *Ans:* +300 kN, +1 350 kN-m.

4-19. Compute the reaction components at *A* and *B*, and calculate the axial force, shear and bending moment at section *a-a* of the 12 by 12 inch timber mast. *Ans: $P_a = −45$ k.*

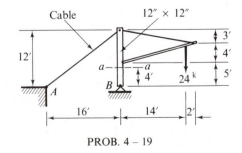

PROB. 4 – 19

4-20. For the planar structure shown in the figure determine the axial force, the shear and the bending moment at section *a-a*. *Ans:* −6 k-ft.

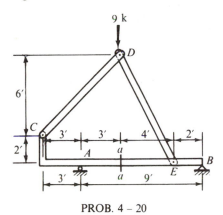

PROB. 4 – 20

4-21. A hydraulic jack exerts a downward force of 5 400 N on the linkage shown. What are the axial force, the shear, and the bending moment at section *a-a* caused by the application of this force? All dimensions are in meters.

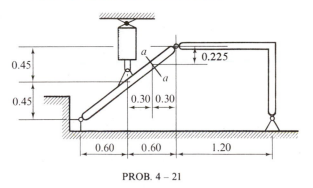

PROB. 4 – 21

4-22 through 4-33. For the planar structures shown in the figures, determine the axial force, the shear, and the bending moment at sections *a-a*, *b-b*, *c-c*, and *d-d*, wherever they apply. Neglect the weight of members. *In every case, draw a free body of the isolated part of the structure and clearly show on it the sense of the computed quantities.* Some sections are shown close together. In these cases, determine the quantities asked for just to the left and just to the right of the point in question, assuming that the widths of members are negligibly small. *Ans:* The answers for some problems are given in the following order: axial force, shear, and moment. The signs of shear and moment apply only for horizontal members.

    Prob. 4-22. −30 kN, +20 kN, +30 kN·m.
    Prob. 4-25. +10 kN, +5 kN, +2.5 kN·m.
    Prob. 4-26. At *a-a:* −28 k, −152 k-ft. At *b-b:* −8 k, −152 k-ft. At *c-c:* −8 k, −176 k-ft. At *d-d:* −8 k, +24 k-ft.
    Prob. 4-27. At *a-a:* 0, +44.4 k, +77.2 k-ft. At *b-b:* +50 k, −5.6 k, −222.8 k-ft. At *c-c:* +50 k, +17.4 k, +111. k-ft.
    Prob. 4-28. At *a-a:* −3.43 k, −1.71 k, +61.7 k-in.
    Prob. 4-32. At *a-a:* −7.2 k, 9.6 k, 24 k-ft. At *b-b:* − .15 k, 12.6 k, 30 k-ft.
    Prob. 4-33. At *a-a:* +1.21 k, 0, 17.1 k-ft.

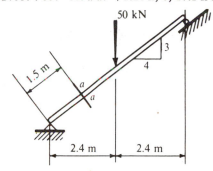

PROB. 4 – 22

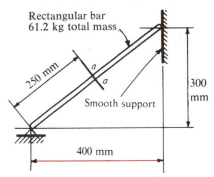

PROB. 4 – 23

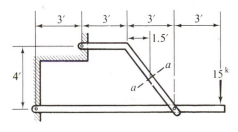

PROB. 4 – 24

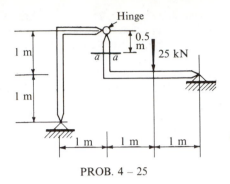

PROB. 4 – 25

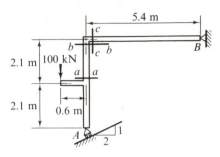

PROB. 4 – 29

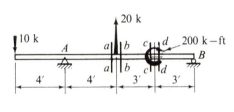

PROB. 4 – 26

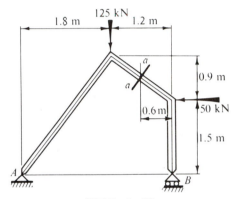

PROB. 4 – 30

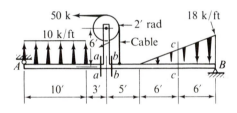

PROB. 4 – 27

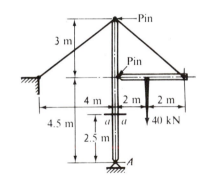

PROB. 4 – 31

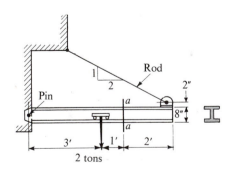

PROB. 4 – 28

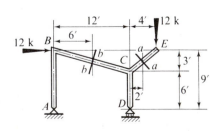

PROB. 4 – 3 2

CHAP. 4 AXIAL FORCE—SHEAR—AND BENDING MOMENT   **116**

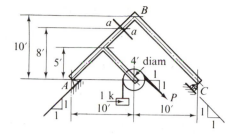

PROB. 4 – 33

4-34. Determine the axial force, the shear, and the bending moment at section *a-a*. There is no connection between members *AD* and *BC* at point *E*.

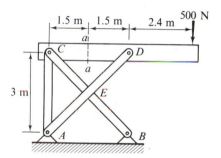

PROB. 4 – 34

4-35 through 4-40. Plot the shear and moment diagrams for the beams loaded as shown in the figures. *Ans:* Max. moment in parentheses by the figure.

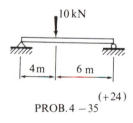

(+24)

PROB. 4 – 35

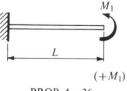

(+M_1)

PROB. 4 – 36

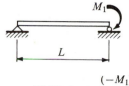

(−M_1)

PROB. 4 – 37

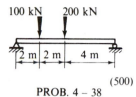

(500)

PROB. 4 – 38

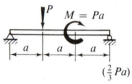

PROB. 4 – 39   (80)

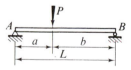

($\frac{2}{3}Pa$)

PROB. 4 – 40

4-41 through 4-43. For beams loaded as shown in the figures, express the shear and bending moments by algebraic expressions for the interval *AB*. *Ans:* Prob. 4-42: $M = \frac{1}{6}$ k $(L^2x - x^3)$.

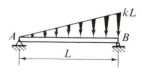

PROB. 4 – 41

PROB. 4 – 42

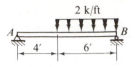

2 k/ft

PROB. 4 – 43

4-44. Establish general algebraic equations for the axial force, shear, and bending moment for the curved beam of Example 4-8. *Ans: M = Pr sin α.*

4-45. A rectangular bar bent into a semicircle is built in at one end and is subjected to an internal radial pressure of p lb per unit length (see figure). Write the general expressions for $P(\theta)$, $V(\theta)$, and $M(\theta)$, and plot the results on a polar diagram. Show positive directions assumed for P, V, and M on a free-body diagram. *Ans: $M = pr^2 (1 - \cos \theta)$.*

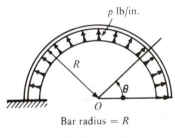

p lb/in.

Bar radius = R

PROB. 4 – 45

4-46. A bar is made in the shape of a right angle as shown in the figure and is built in at one of its ends. (a) Write the general expressions for V, M, and T (torque) caused by the application of a force F normal to the plane of the bent bar. Plot the results. (b) If in addition to the applied force F the weight of the bar w lb per unit length is also to be considered, what system of internal force components develops at the built-in end? *Ans:* (a) $M = -F(L - x)$, (b) $M = -(F + aw + \frac{1}{2}wL)L$.

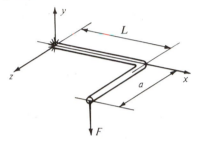

PROB. 4 – 46

(*Note:* For additional problems see Chapter 10.)

5 Pure Bending of Beams

5-1. INTRODUCTION

The system of forces that may exist at a section of a beam was discussed in the previous chapter. This was found to consist of an axial force, a shearing force, and a bending moment. The effect of one of these forces, the axial force, on a member was discussed in Chapters 1 and 2. In this chapter another element of the force system that may be present at a section of a member, the internal bending moment, will be considered. Moreover, since in some cases a segment of a beam may be in equilibrium under the action of a moment alone, a condition called *pure bending or flexure*, this in itself represents a complete problem. It is the purpose of this chapter to relate the internal bending moment to the stresses it causes in a beam. If, in addition to the internal bending moment, an axial force and a shear also act simultaneously, complex stresses arise. These will be treated in Chapters 7, 8, and 9. The deflection of beams due to bending will be discussed in Chapter 11.

A major part of this chapter will be devoted to methods for determining the stresses in straight homogeneous beams caused by bending moments. Topics on beams made from two or more different materials, curved beams, and stress concentrations are also included.

5-2. SOME IMPORTANT LIMITATIONS OF THE THEORY

Just as in the case of axially loaded rods and in the torsion problem, all forces applied to a beam will be assumed to be steady and delivered to the beam without shock or impact. Shock or impact problems will be considered in Chapter 15. Moreover, all of the beams will be assumed to be stable under the applied forces. A similar point was brought out in Chapter 1, where it was indicated that a rod acting in compression cannot be too slender, or its behavior will not be governed by the usual compressive strength criterion. In such cases the *stability* of the member becomes important. As an example, consider the possibility of using a sheet of paper on edge as a beam. Such a beam has a substantial depth, but even if it is used to carry a force over a small span, it will buckle sideways and collapse. The same phenomenon may take place in more substantial members which may likewise collapse under an applied force. Such unstable beams do not come within the scope of this chapter. All the beams considered here will be assumed to be sufficiently stable laterally by virtue of their proportions, or to be thoroughly braced in the transverse direction. A better understanding of this important phenomenon will result after the study of the chapter on columns. The majority of beams used in structural framing and machine parts are such that the flexural theory to be developed here is applicable. This is indeed fortunate as the theory governing the stability of members is more complex.

5-3. BASIC ASSUMPTIONS

For the present it is assumed that only *straight* beams having constant cross-sectional areas *with an axis of symmetry* are to be included in the discussion. Moreover, it is assumed that the applied bending moments lie in a plane containing this axis of symmetry and the beam axis. Let it be further agreed that for the sake of simplicity in making sketches, the axis of symmetry will be taken vertically. Several cross-sectional areas of beams satisfying these conditions are shown in Fig. 5-1. A generalization of this problem will be made in Art. 5-7.

A segment of a beam fulfilling the above requirements is shown in Fig. 5-2(a), and its cross-section is shown in Fig. 5-2(b). For such a beam a line through the centroid of all cross-sections will be referred to as the axis of the beam. Next, imagine that two planes are passed through the beam perpendicular to its axis. The intersections of these planes with a longitudinal plane passing through the beam axis and the axis of symmetry is shown by lines *AB* and *CD*. Then it is not difficult to imagine that when this segment is subjected to the bending moments *M* at its ends as shown in Fig. 5-2(c), the beam bends, and the planes perpendicular to the beam axis tilt slightly. Moreover, the lines *AB* and *CD remain* straight.* This can be satisfactorily verified experimentally.† Generalizing this observation for the whole beam,

---

*This can be demonstrated by using a rubber model with a ruled grating drawn on it. Alternatively, thin vertical rods passing through the rubber block can be used. In the immediate vicinity of the applied moments the deformation is more complex. However, in accord with the St. Venant's principle (Art. 2-11), this is only a local phenomenon which rapidly dissipates.

†Rigorous solutions from the Mathematical Theory of Elasticity show that slight warpage of these lines may take place. Such warpage occurs if a beam carries a shear in addition to a bending moment. However, the warpage of the *adjoining sections* is exceedingly similar in shape. Thus the distance between any two points such as *A* and *C* on the adjoining sections remains practically the same whether warped or straight lines *AB* and *CD* are considered. And since the distance between the adjoining sections is the basis for establishing the elementary flexure theory, the foregoing assumption forms an excellent working hypothesis for all cases. Moreover, a conclusion of far-reaching importance is that the existence of a shear at a section does not invalidate the expressions to be derived in this chapter. This will be implied in the subsequent work.

Fig. 5-1. Beam cross sections with a vertical axis of symmetry

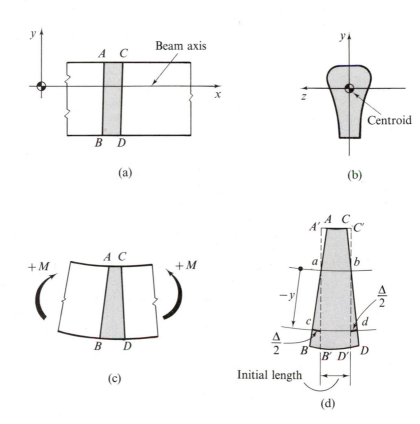

(a)

(b)

(c)

(d)

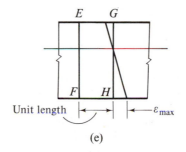

(e)

Fig. 5-2. Behavior of a beam in bending

one obtains the *most fundamental hypothesis** of the flexure theory, based on the geometry of deformations. It may be stated thus:

1. Plane sections through a beam, taken normal to its axis, *remain plane* after the beam is subjected to bending.

This means that in a bent beam two planes normal to the beam axis and initially parallel cease to be parallel. In a side view, the behavior of two such planes corresponds to the behavior of lines AB and CD of Figs. 5-2(a) and (c). An element of the beam contained between these planes is shown in Fig. 5-2(d). Under the action of the moments of the sense shown, the distance AC becomes smaller than BD. Further, because of the internal moment, a push must exist on the upper part of the beam and a pull on the lower. Hence, the undistorted beam element must be related to the distorted one, as $A'C'D'B'$ is to $ACDB$, shown in more detail in Fig. 5-2(d). From this diagram it is seen that the fibers or "filaments" of the beam along the surface† ab do not change in length. Hence, *the fibers in the surface ab are not stressed at all*, and, as the element selected was an arbitrary one, fibers free of stress exist continuously over the whole length and width of the beam. These fibers lie in a surface which is called the *neutral surface* of the beam. Its intersection with a right section through the beam is termed the *neutral axis* of the beam. Either term implies a location of *zero stress* in the member subjected to bending.

The precise location of the neutral surface in a beam will be determined in the next article. First, a study of the nature of the strains in fibers parallel to the neutral surface will be made. Thus, consider a typical fiber such as cd parallel to the neutral surface and located at a distance‡ $-y$ from it. During bending it elongates an amount Δ. If this elongation is divided by the initial length L of the fiber, the *strain* ε in that fiber is obtained. Next, note that from the geometrical assumption made earlier, elongations of different fibers vary *linearly* from the neutral axis since these elongations are fixed by the triangles aBB', bDD', aAA', and bCC'. On the other hand, the initial length of all fibers is the same. Hence the original fundamental assumption may be restated§ thus:

1a. In a beam subjected to bending, strains in its fibers vary linearly or directly as their respective distances from the neutral surface.

This situation is analogous to the one found earlier in the torsion problem where the *shearing* strains vary linearly from the axis of a circular shaft. In a beam, strains vary linearly from the *neutral surface*. This variation is represented diagrammatically in Fig. 5-2(e). These *axial* strains are

---

*This hypothesis with an inaccuracy was first introduced by Jacob Bernoulli (1645–1705), a Swiss mathematician. In the correct form it dates back to the writings of the French engineering educator M. Navier (1785–1836).

†A rigorous solution shows that this surface is slightly cylindrical in two directions. In the present treatment this surface is assumed to be curved only in the direction shown.

‡Positive direction of y is taken upward from the neutral axis.

§Experimentally, this assumption may be more easily verified than assumption (1).

associated with stresses which act *normal* to the section of a beam. The above corollary to the original assumption is applicable in the elastic as well as in the inelastic range of the material's behavior.* For the present this generality will be limited by introducing the second fundamental assumption of the flexure theory:

2. Hooke's law is applicable to the individual fibers, i.e., stress is proportional to strain. The same elastic modulus E is assumed to apply to material in tension as well as in compression. The Poisson effect and the interference of the adjoining differently stressed fibers are ignored.

Combining the foregoing assumptions, the basis for establishing the flexural theory for the *elastic* case is obtained:

On a section of a beam, normal stresses resulting from bending vary linearly as their respective distances from the neutral axis.

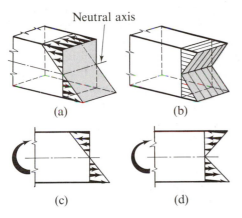

Neutral axis

(a)   (b)

(c)   (d)

Fig. 5-3. Stress distribution at a section of a beam resisting a bending moment

It should be firmly fixed in the reader's mind that these stresses act normal to the section of a beam. They are the result of *axial* elongation or contraction of the various beam fibers. Their linear variation from the neutral axis, to repeat, is due to the linear variation of the strains and to the proportionality of stress to strain. The distance to the various fibers of the beam is measured *vertically* from the neutral axis. Figures 5-3(a) and (b) illustrate the nature of the stress distribution in a beam resisting a bending moment. Two alternative schemes of representing this three-dimensional problem in a plane are shown in Figs. 5-3(c) and (d). In subsequent work these will be the usual forms for showing the flexural stress distribution at a section of a beam.

5-4. THE FLEXURE FORMULA

After the nature of the stress distribution in the elastic range at a section of a beam is understood, quantitative expressions relating bending moment to stress may be established. For this purpose the neutral surface is first located from considerations of static equilibrium.

Consider a beam segment subjected to a positive bending moment M as shown in Fig. 5-4(a). At section X-X this applied moment is resisted by stresses which vary linearly from the neutral axis. The highest stresses occur at the points most remote from the neutral axis. For the beam shown this occurs along the line *ed*, Fig. 5-4(b). This stress, being a normal stress, is designated by σ_{max}, Fig. 5-4(a). Any other normal stress acting on the cross-sectional area is related to this stress by a ratio of distances from the neutral

*Beams stressed beyond the elastic limit are treated in Art. 5-8.

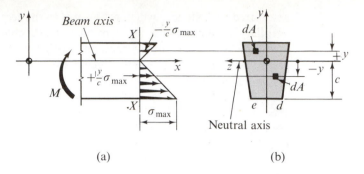

Fig. 5-4. A beam in pure flexure

(a) (b)

axis. Thus, on an infinitesimal area dA, Fig. 5-4(b), at a distance y from the neutral axis, the stress is $-(y/c)\,\sigma_{max}$, where distance c is measured from the neutral axis to the most remote fiber of the beam. Stresses below the neutral surface are given by a similar relation; the sign automatically reverses as y's are measured down from the neutral axis. This reversal of sign corresponds to the reversal in stress from compression to tension. Note that for a positive bending moment, the normal stresses at a section are positive (tension) for negative values of y, and negative (compression) for positive values of y. Hence the expression $-(y/c)\,\sigma_{max}$ is a general expression for the normal stress on *any* infinitesimal area of the beam's section at a distance y from the neutral axis.

Since the segment of the beam shown in Fig. 5-4(a) must be in equilibrium, the sum of all forces in the x-direction, which is taken horizontally, must vanish, i.e., $\sum F_x = 0$. Therefore as the beam's segment resists only a couple, the sum (or integral) of all forces acting *on the section* of the beam must vanish. Thus

$$\underbrace{\int_A \underbrace{\left(-\frac{y}{c}\sigma_{max}\right)}_{\text{(stress)}} \underbrace{dA}_{\text{(area)}}}_{\text{(force)}} = 0$$

where the subscript A of the integral indicates that the summation must be carried out over the entire cross-sectional area of the beam. At a section, however, σ_{max} and c are constants, so the integral may be rewritten as

$$-\frac{\sigma_{max}}{c} \int_A y\,dA = 0$$

Since in a stressed beam neither c nor σ_{max} can be zero, it follows that $\int_A y\,dA$ $= 0$. But by definition $\int_A y\,dA = \bar{y}A$, where $\bar{y}$ is the distance from a base line (neutral axis in the case considered) to the centroid of the area A, so $\bar{y}A = 0$. Then since A is not zero, $\bar{y}$ must be. Therefore the distance from the neutral axis to the centroid of the area must be zero, and *the neutral axis passes through the centroid of the cross-sectional area of the beam.* Hence the neutral axis may be quickly and easily determined for any beam by simply finding the centroid of the cross-sectional area.

Next, the remaining significant equation of static equilibrium will be applied to the beam segment shown in Fig. 5-4(a) to evaluate the magnitudes of the normal stresses. This equation is $\sum M_z = 0$, which for the present purpose is more conveniently stated as: *The external moment M is resisted by or equal to the internal bending moment developed by the flexural stresses at a section.* The latter quantity is determined by summing forces acting on infinitesimal areas dA, Fig. 5-4(b), multiplied by their respective arms from the neutral axis. By formulating these statements mathematically, the following equality is obtained:

$$M = \int_A \underbrace{\underbrace{\left(-\frac{y}{c}\sigma_{max}\right)}_{\text{(stress)}} \underbrace{dA}_{\text{(area)}}}_{\underbrace{\text{(force)}}} \qquad \underbrace{y}_{\text{(arm)}} = -\frac{\sigma_{max}}{c}\int_A y^2\, dA$$

(moment)

where (as before) σ_{max}/c is a constant, hence it appears outside the integral sign. The integration must be performed over the entire cross-sectional area A of the beam.

The integral $\int_A y^2\, dA$ depends only on the geometric properties of the cross-sectional area. In mechanics of materials this integral is called the *moment of inertia* of the cross-sectional area about the centroidal axis, when y is measured from such an axis. It is a definite constant for any particular area, and in this text it will be designated by I. With this notation the foregoing expression may be written more compactly as

$$M = -\frac{\sigma_{max}}{c}I \qquad \text{or} \qquad \sigma_{max} = -\frac{Mc}{I}$$

It is customary to dispense with the sign for the normal stress as its sense can be found by inspection. At any section the normal stresses must act in such a manner as to build up a couple statically equivalent to the resisting bending moment, the sense of which is known. Hence the above equation can be written simply as

$$\sigma_{max} = \frac{Mc}{I} \qquad\qquad (5\text{-}1)$$

Equation 5-1 is the *flexure formula** for beams. It gives the *maximum*

---

*It took nearly two centuries to develop this seemingly simple expression. The first attempts to solve the flexure problem were made by Galileo in the seventeenth century. In the form in which it is used today the problem was solved in the early part of the nineteenth century. Generally, Navier of France is credited for this accomplishment. However, some maintain that credit should go to Coulomb, who also derived the torsion formula.

normal stress in a beam subjected to a bending moment M. Moreover, since stress σ on any point of a cross-section is $-(y/c)\,\sigma_{max}$, the general expression for normal stresses at a section is given as

$$\sigma = -\frac{My}{I} \qquad (5\text{-}1a)$$

These formulas are of unusually great importance in mechanics of materials. In these formulas, M is the internal or resisting bending moment, which is equal to the external moment at the section where the stresses are sought. The bending moment is expressed in *newton·meter* (N·m) or in *inch-pound* units for use in these formulas. The distance y from the neutral axis of the beam to the point on a section where the normal stress σ is wanted is measured perpendicular to the neutral axis and should be expressed either in meters or inches. When it reaches its maximum value (measured either up or down) it corresponds to c, and as y approaches this maximum value, the normal stress σ approaches σ_{max}. In this equation I is the moment of inertia of the *whole* cross-sectional area of the beam *about its neutral axis*. To avoid confusion with the *polar* moment of inertia, I is sometimes referred to as the *rectangular* moment of inertia. It has the dimensions of m⁴ or in.⁴. Its evaluation for various areas will be discussed in the next article. The use of consistent units as indicated above makes the units of stress σ, [N·m] [m]/[m⁴] = N/m² = Pa, or [in.-lb][in.]/[in.⁴] = [lb per in.²] = psi.

The student is urged to reflect on the meanings of the terms used in the derived equations. The stresses given by these equations indicate that they act perpendicular to the section and vary linearly from the neutral axis. These facts are very significant. Likewise, the three-dimensional aspect of the problem must be kept in mind.

The foregoing discussion applies only to cases where the material behaves *elastically*. The important concepts used in deriving the flexure formula may be summaried as follows:

1. *Deformation* was assumed giving the linear variation of strain from the neutral axis.
2. *Properties of materials* were used to relate strain and stress.
3. *Equilibrium* conditions were used to locate the neutral axis and to determine the internal stresses.

These are the same concepts as were used to derive the torsion formula.

5-5. COMPUTATION OF THE MOMENT OF INERTIA

In applying the flexure formula, the moment of inertia I of the cross-sectional area about the neutral axis must be determined. Its value is defined by the integral of $y^2\,dA$ over the entire cross-sectional area of a member, and it must be emphasized that for the flexure formula the moment of inertia must be computed around the neutral axis of the cross-sectional area. This axis, according to Art. 5-4, passes through the centroid of the cross-sectional area. For symmetrical sections the neutral axis is perpendicular to the axis of symmetry. Such an axis is one of the *principal axes** of the cross-sectional area. Most readers should already be familiar with the method of determining the moment of inertia I. However, the necessary procedure is reviewed below.

The first step in evaluating I for an area is to find the centroid of the area. An integration of $y^2\,dA$ is then performed with respect to the horizontal axis passing through the area's centroid. Actual integration over areas is necessary for only a few elementary shapes, such as rectangles, triangles, etc. After this is done, most cross-sectional areas used in practice may be broken down into a combination of these simple shapes. Values of moments of inertia for some simple shapes may be found in any standard civil or mechanical engineering handbook (also see Table 2 of the Appendix). To find I for an area composed of several simple shapes, the *parallel-axis theorem* (sometimes called the *transfer formula*) is necessary; the development of it follows.

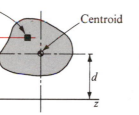

Fig. 5-5. Shaded area used in deriving the parallel-axis theorem

The area shown in Fig. 5-5 has a moment of inertia I_o around the horizontal axis passing through its own centroid, i.e., $I_o = \int y^2\,dA$, where y is measured from the centroidal axis. The moment of inertia I_{zz} of the same area around another horizontal axis z-z by definition is

$$I_{zz} = \int_A (d + y)^2\,dA$$

where as before y is measured from the axis through the centroid. Squaring the quantities in the parentheses and placing the constants outside the integral signs

$$I_{zz} = d^2 \int_A dA + 2d \int_A y\,dA + \int_A y^2\,dA = Ad^2 + 2d \int_A y\,dA + I_o$$

However, since the axis from which y is measured passes through the centroid of the area, $\int y\,dA$ or $\bar{y}A$ is zero. Hence

$$I_{zz} = I_o + Ad^2 \tag{5-2}$$

This is the parallel-axis theorem. It can be stated as follows: The moment of inertia of an area around any axis is equal to the moment of inertia of the same area around a parallel axis passing through the area's centroid, plus

---

*By definition the principal axes are those about which the rectangular moment of inertia is a maximum or a minimum. Such axes are always mutually perpendicular. The product of inertia, defined by $\int yz\,dA$ vanishes for the principal axes. An axis of symmetry of a cross-section is always a principal axis. For further details see the appendix to Chapter 8.

the product of the same area and the square of the distance between the two axes.*

The following examples illustrate the method of computing I directly by integration for two simple areas. Then an application of the parallel-axis theorem to a composite area is given. Values of I for commercially fabricated steel beams, angles, and pipes are given in Tables 3 to 8 of the Appendix.

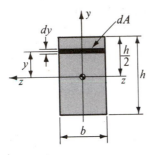

Fig. 5-6

EXAMPLE 5-1

Find the moment of inertia around the horizontal axis passing through the centroid for the rectangular area shown in Fig. 5-6.

SOLUTION

The centroid of this section lies at the intersection of the two axes of symmetry. Here it is convenient to take dA as $b\,dy$. Hence

$$I_{zz} = I_o = \int_A y^2\,dA = \int_{-h/2}^{+h/2} y^2 b\,dy = b\,\frac{y^3}{3}\bigg|_{-h/2}^{+h/2} = \frac{bh^3}{12}. \qquad (5\text{-}3)$$

Similarly

$$I_{yy} = \frac{b^3 h}{12}$$

These expressions are used frequently, as rectangular beams are commonly employed in practice.

EXAMPLE 5-2

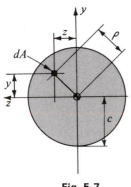

Fig. 5-7

Find the moment of inertia about a diameter for a circular area of radius c, Fig. 5-7.

SOLUTION

Since there is some chance of confusing I with J for a circular section, it is well to refer to I as the *rectangular* moment of inertia of the area in this case.

To find I for a circle, first note that $\rho^2 = z^2 + y^2$, as may be seen from the figure. Then using the definition of J, noting the symmetry around both axes, and using Eq. 3-2

$$J = \int_A \rho^2\,dA = \int_A (y^2 + z^2)\,dA = \int_A y^2\,dA + \int_A z^2\,dA$$

$$= I_{zz} + I_{yy} = 2I_{zz}$$

$$I_{zz} = I_{yy} = \frac{J}{2} = \frac{\pi c^4}{4} \qquad (5\text{-}4)$$

*A similar theorem can be proved for the product of intertia (for definition of the product of inertia see the appendix to Chapter 8): $I_{yz} = I_{y_c z_c} + Ad_1 d_2$ where $I_{y_c z_c}$ is the product of inertia of an area A about the centroidal axes y_c, z_c and I_{yz} is the product of inertia of the same area about a set of parallel axes y, z. The distance between y and y_c is d_1, and that between z and z_c is d_2. Subsequently this relation will be referred to as Eg. 5-2a.

In mechanical applications circular shafts often act as beams, hence Eq. 5-4 will be found useful. For a tubular shaft, the moment of inertia of the hollow interior must be subtracted from the above expression.

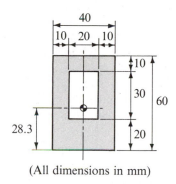

(All dimensions in mm)

Fig. 5-8. (All dimensions in mm)

EXAMPLE 5-3

Determine the moment of inertia I around the horizontal axis for the area shown in Fig. 5-8 for use in the flexure formula.

SOLUTION

As the moment of inertia wanted is for use in the flexure formula, it must be obtained around the axis through the centroid of the area. Hence the centroid of the area must be found first. This is most easily done by treating the entire outer section and deducting from it the hollow interior. For convenience, the work is carried out in tabular form. Then the parallel-axis theorem is used to obtain I.

Area	A [mm^2]	y [mm] (from bottom)	Ay
Entire area	$40(60) = 2400$	30	72 000
Hollow interior	$-20(30) = -600$	35	$-21\,000$
	$\sum A = 1800$ mm^2		$\sum Ay = 51\,000$ mm^3

$$\bar{y} = \frac{\sum Ay}{\sum A} = \frac{51\,000}{1\,800} = 28.3 \text{ mm from bottom}$$

For entire area:
$$I_o = \frac{bh^3}{12} = \frac{40(60)^3}{12} = 72 \times 10^4 \text{ mm}^4$$

$$Ad^2 = 2400(30 - 28.3)^2 = \underline{0.69 \times 10^4 \text{ mm}^4}$$
$$I_{zz} = \overline{72.69 \times 10^4 \text{ mm}^4}$$

For hollow interior:
$$I_o = \frac{bh^3}{12} = \frac{20(30)^3}{12} = 4.50 \times 10^4 \text{ mm}^4$$

$$Ad^2 = 600(35 - 28.3)^2 = \underline{2.69 \times 10^4 \text{ mm}^4}$$
$$I_{zz} = \overline{7.19 \times 10^4 \text{ mm}^4}$$

For composite section: $I_{zz} = (72.69 - 7.19)10^4 = 65.50 \times 10^4 \text{ mm}^4$

Note particularly that in applying the parallel-axis theorem, each element of the composite area contributes two terms to the total I. One term is the moment of inertia of an area around its own centroidal axis, the other term is due to the transfer of its axis to the centroid of the whole area. Methodical work is the prime requisite in solving such problems correctly.

5-6. REMARKS ON THE FLEXURE FORMULA

The bending stress at any point of a beam section is given by Eq. 5-1a, $\sigma = -My/I$. The largest stress at the same section follows from this relation by taking $|y|$ at a maximum, which leads to Eq. 5-1 $\sigma_{max} = Mc/I$. In most practical problems the maximum stress given by Eq. 5-1 is the quantity sought; thus it is desirable to make the process of determining σ_{max} as simple as possible. This can be accomplished by noting that both I and c are constants for a given section of a beam. Hence I/c is a constant. Moreover, since this ratio is only a function of the cross-sectional dimensions of a beam, it can be uniquely determined for any cross-sectional area. This ratio is called the *elastic section modulus* of a section and will be designated by S. With this notation Eq. 5-1 becomes

$$\sigma_{max} = \frac{Mc}{I} = \frac{M}{\dfrac{I}{c}} = \frac{M}{S} \tag{5-5}$$

or stated otherwise

$$\text{maximum bending stress} = \frac{\text{bending moment}}{\text{elastic section modulus}}$$

If the moment of inertia I is measured in in.^{4} (or m^4) and c in in. (or m), S is measured in in.^{3} (or m^3). Likewise, if M is measured in inch-pounds (or N·m), the units of stress, as before, become pounds per square inch (or N/m^2). It bears repeating that the distance c as used here is measured from the neutral axis to the most remote fiber of the beam. This makes $I/c = S$ a minimum, and consequently M/S gives the maximum stress. The efficient sections for resisting bending have as large an S as possible for a given amount of material. This is accomplished by locating as much of the material as possible far from the neutral axis.

The use of the term *elastic section modulus* in Eq. 5-5 corresponds somewhat to the use of the area term A in Eq. 1-1 ($\sigma = P/A$). However, only the maximum flexural stress on a section is obtained from Eq. 5-5, but the stress computed from Eq. 1-1 holds true across the whole section of a member.

Equation 5-5 is widely used in practice because of its simplicity. To facilitate its use, section moduli for many manufactured cross sections are tabulated in handbooks. Values for a few steel sections are given in Tables 3 to 8 in the Appendix. Equation 5-5 is particularly convenient for the design of beams. Once the maximum bending moment for a beam is determined and an allowable stress is decided upon, Eq. 5-5 may be solved for the required section modulus. This information is sufficient to select a beam.

However, a detailed consideration of beam design will be delayed until Chapter 10. This is necessary inasmuch as a shearing force, which in turn causes stresses, usually also acts at a beam section. The interaction of the various kinds of stresses must be considered first to gain complete insight into the problem.

The application of the flexure formulas to particular problems should cause little difficulty if the meaning of the various terms occurring in them has been thoroughly understood. The following two examples illustrate investigations of bending stresses at specific sections.

EXAMPLE 5-4

A 0.3 m by 0.4 m wooden cantilever beam weighing 0.75 kN/m carries an upward concentrated force of 20 kN at the end, as shown in Fig. 5-9(a). Determine the maximum bending stresses at a section 2 m from the free end.

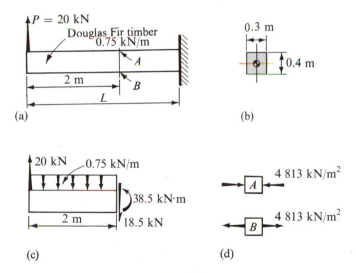

Fig. 5-9

SOLUTION

A free-body diagram for a 2 m segment of the beam is shown in Fig. 5-9(c). To keep this segment in equilibrium requires a shear of $20 - 0.75(2) = 18.5$ kN and a bending moment of $20(2) - 0.75(2)1 = 38.5$ kN·m at the cut section. Both of these quantities are shown with their proper sense in Fig. 5-9(c). By inspecting the cross-sectional area, the distance from the neutral axis to the extreme fibers is seen to be 0.2 m, hence $c = 0.2$ m. This is applicable to both the tension and the compression fibers.

From Eq. 5-3: $I_{zz} = \dfrac{bh^3}{12} = \dfrac{(0.3)(0.4)^3}{12} = 16 \times 10^{-4}$ m^4

From Eq. 5-1: $\sigma_{max} = \dfrac{Mc}{I} = \dfrac{(38.5)(0.2)}{16 \times 10^{-4}} = \pm 4\ 813$ kN/m^2

From the sense of the bending moment shown in Fig. 5-9(c) the top fibers of the beam are seen to be in compression, and the bottom ones in tension. In the answer given, the positive sign applies to the tensile stress, the negative sign applies to the compressive stress. Both of these stresses decrease at a linear rate toward the neutral axis where the bending stress is zero. The normal stresses acting on infinitesimal elements at A and B are shown in Fig. 5-9(d). It is important to learn to make such a representation of an element as it will be frequently used in Chapters 7, 8, and 9.

ALTERNATE SOLUTION

If only the maximum stress is desired, the equation involving the section modulus may be used. The section modulus for a rectangular section in algebraic form is

$$S = \frac{I}{c} = \frac{bh^3}{12}\frac{2}{h} = \frac{bh^2}{6} \tag{5-6}$$

In this problem, $S = (0.3)(0.4)^2/6 = 8 \times 10^{-3}$ m^3, and by Eq. 5-5

$$\sigma_{max} = \frac{M}{S} = \frac{38.5}{8 \times 10^{-3}} = 4\,813 \text{ kN/m}^2 \quad \text{or} \quad \text{kPa.}$$

Both solutions lead to identical results.

EXAMPLE 5-5

Find the maximum tensile and compressive stresses acting normal to the section A-A of the machine bracket shown in Fig. 5-10(a) caused by the applied force of 8 kips.

SOLUTION

The shear and bending moment of proper magnitude and sense to maintain the segment of the member in equilibrium are shown in Fig. 5-10(c). Next the neutral axis of the beam must be located. This is done by locating the centroid of the area shown in Fig. 5-10(b) (see also Fig. 5-10(d)). Then the moment of inertia about the neutral axis is computed. In both these calculations the legs of the cross section are assumed rectangular, neglecting fillets. Then, keeping in mind the sense of the resisting bending moment and applying Eq. 5-1, one obtains the desired values.

Area Number	A [in.^{2}]	y [in.] (from ab)	Ay
1	4.0	0.5	2.0
2	3.0	2.5	7.5
3	3.0	2.5	7.5
	$\sum A = 10.0$ in.^{2}		$\sum Ay = 17.0$ in.^{3}

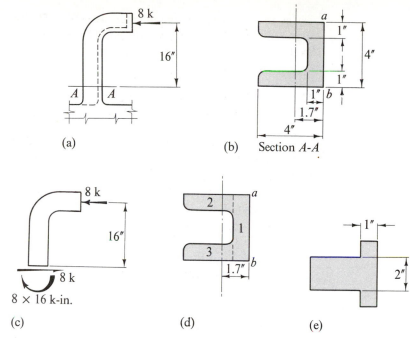

(a) (b) Section *A-A*

(c) (d) (e)

Fig. 5-10

$$\bar{y} = \frac{\sum Ay}{\sum A} = \frac{17.0}{10.0} = 1.70 \text{ in.} \qquad \text{from the line } ab$$

$$I = \sum (I_o + Ad^2) = \frac{4(1)^3}{12} + 4(1.2)^2 + \frac{(2)1(3)^3}{12} + 2(3)(0.8)^2$$

$$= 14.43 \text{ in.}^4$$

$$\sigma_{max} = \frac{Mc}{I} = \frac{(8)(16)(2.3)}{14.43} = 20.4 \text{ ksi} \qquad \text{(compression)}$$

$$\sigma_{max} = \frac{Mc}{I} = \frac{(8)(16)(1.7)}{14.43} = 15.1 \text{ ksi} \qquad \text{(tension)}$$

These stresses vary linearly toward the neutral axis and vanish there. If for the same bracket the direction of the force *P* were reversed, the sense of the above stresses would also reverse. The results obtained would be the same if the cross-sectional area of the bracket were made *T*-shaped as shown in Fig. 5-10(e). The properties of this section about the significant axis are the same as those of the channel. Both these sections have an axis of symmetry.

The above example shows that members resisting flexure may be proportioned so as to have a different maximum stress in tension than in compression. This is significant for materials having different strengths in tension and compression. For example, cast iron is strong in compression and weak in tension. Thus, the proportions of a cast-iron member may be so set as to have a low maximum tensile stress. The potential capacity of the material may thus be better utilized. This matter will be further considered in the chapter on the design of beams.

*5-7. PURE BENDING OF BEAMS WITH UNSYMMETRICAL SECTION

Pure bending of elastic beams having an axis of symmetry was discussed in the preceding article. The applied moments were assumed to act in the plane of symmetry. These limitations, although expedient in developing the flexural theory, are too severe and may be greatly relaxed. The same formulas can be used for any beam in pure bending, providing the bending moments are applied in a plane parallel to either principal axis of the cross-sectional area. The previous derivation can be repeated identically. Stresses vary linearly from the neutral axis passing through the centroid. As before, the stress on any elementary area dA, Fig. 5-11, is $\sigma = -(y/c)\,\sigma_{max}$. Hence

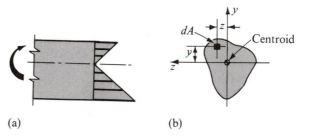

Fig. 5-11. A beam with an unsymmetrical cross-sectional area

$-(y/c)\,\sigma_{max}\,dA$ is an infinitesimal force acting on an element. The sum of the moments of these internal forces around the z axis develops the internal moment. However, as symmetry is lacking, these internal forces may build up a moment around the y axis. This must be reconciled.

The moment arms of forces acting on infinitesimal areas around the y axis are equal to z. Thus a possible moment M_{yy} around the y axis is

$$M_{yy} = \int_A -\frac{y}{c}\sigma_{max}\,dA\,z = -\frac{\sigma_{max}}{c}\int_A yz\,dA$$

The last integral represents the product of inertia of the cross-sectional area. It is equal to zero if the axes selected are the principal axes of the area.* Since these axes are the ones used here and are assumed to be the principal axes, $M_{yy} = 0$, and thus the formulas derived earlier apply to a beam with any shape of cross section.

*See the appendix to Chapter 8.

If a bending moment is applied without being parallel to either principal axis, the procedures which will be discussed in Chapter 7 must be followed.

*5-8. INELASTIC BENDING OF BEAMS

The elastic flexure formula derived earlier is valid only while stress is proportional to strain. A more general theory will now be discussed for material that does not obey Hooke's law.

The basic kinematic assumption of the flexure theory, as stated in Art. 5-3, asserts the plane sections through a beam taken normal to its axis remain plane after the beam is subjected to bending. This assumption remains applicable even if the material behaves inelastically. With no further assumptions, it means that strains in the fibers of a beam subjected to bending vary directly as their respective distances from the neutral axis. With this as a basis, together with equilibrium requirements and nonlinear stress-strain relations, the generalized theory of flexure is constructed.

Consider a segment of a prismatic beam subjected to bending moments. The cross-sectional area of this beam has a vertical axis of symmetry, Fig. 5-12(c). The linear variation of the strains from the neutral axis is diagrammatically represented for such a beam in Fig. 5-12(b). At the neutral axis, which as yet is undetermined, the strain is zero. Strains at points away

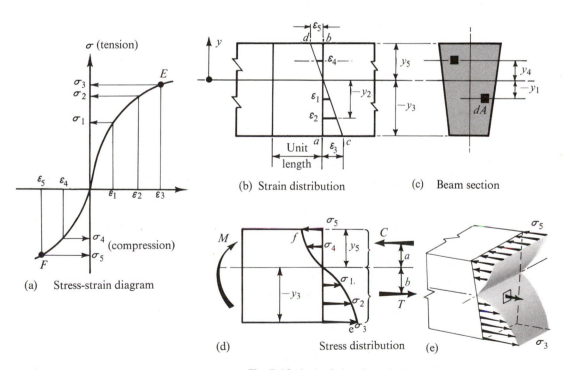

(a) Stress-strain diagram

(b) Strain distribution

(c) Beam section

(d)

Stress distribution

(e)

Fig. 5-12. Inelastic bending of a beam

from the neutral axis correspond to the horizontal distances from the line *ab* to the line *cd* in Fig. 5-12(b). For example, the strain of a fiber at a distance $-y_2$ from the neutral axis is ε_2. Such distances define the axial strain of every fiber in a beam.

To make the argument general, the material will be assumed to have a different stress-strain curve in tension and compression. A possible curve for such a material is shown in Fig. 5-12(a). Such curves may be obtained from axial-load experiments.

If the Poisson effect is neglected, the longitudinal fibers of a beam in bending behave independently. Each one of these may be thought of as an infinitesimal axially loaded rod, stressed to a level dependent upon its strain. Since the variation of the strain in a beam is set by the assumption, the stress pattern may be formulated from the stress-strain curve, Fig. 5-12(a). For example, corresponding to the tensile strain ε_1, at a distance $-y_1$ from the neutral axis, a tensile stress σ_1 acts in the beam. Similarly, ε_4 is associated with $-\sigma_4$, a compressive stress. The same thing applies to any other fiber of the beam. This determines the stress distribution shown in Figs. 5-12(d) and (e); it resembles the shape of the stress-strain curve (compare *EF* with *ef* by turning it clockwise through 90°).

Since the beam acts in pure bending, the same equations of statics will be used here as were used in establishing the elastic flexure formula. The two applicable relations as before are

$$\sum F_x = 0 \qquad \text{or} \qquad \int_A \sigma \, dA = 0 \qquad \qquad (5\text{-}7)$$

$$\sum M_z = 0 \qquad \text{or} \qquad -\int_A \sigma y \, dA = M \qquad \qquad (5\text{-}8)$$

where σ is a normal stress acting on an infinitesimal element dA of the cross-sectional area A of the beam, and y is the distance from the neutral axis to an element dA. It should be noted that for a positive M the integral in Eq. 5-8 becomes positive since for positive y's the stress σ is negative, whereas for negative y's the stress σ is positive.

The solution of the most general problem in inelastic bending, i.e., the satisfaction of the equilibrium Eqs. 5-7 and 5-8, requires a trial-and-error procedure. Initially the location of the neutral axis is unknown. A possible method consists of assuming a strain distribution, thus locating a trial neutral axis and giving the stress distribution shown in Fig. 5-12(d). Such trials must be continued until the sum of the forces C on the compression side of the beam is equal to the sum of the forces T on the tension side of the beam. When such a condition is fulfilled, the neutral axis of the beam is located. Note particularly that in inelastic flexure the neutral axis of a beam may not coincide with the centroidal axis of the cross-sectional area. It does so only if the cross-sectional area has two axes of symmetry and the stress-strain diagram is identical in tension and compression.

After the neutral axis is located and the magnitudes of C and T are known, their line of action may be determined. This is possible since the stress distribution on the cross-sectional area is known. Finally, the resisting moment is $T(a + b)$ or $C(a + b)$. The foregoing process is equivalent to the integration indicated by Eq. 5-8. However, the resisting moment so computed, based upon assumed strains, may not be equal to the applied moment. Hence, the process must be repeated by initially assuming greater or smaller strain at the extreme fibers until the resisting moment becomes equal to the applied moment.

The foregoing method of solving a general problem is tedious, and accelerated procedures for arriving at a solution have been developed.* However, the above discussion should be sufficient to give a picture of the behavior of a beam in flexure beyond the elastic limit. As a simple example, consider a beam of rectangular cross section subjected to bending. Let the stress-strain diagram of the beam material be alike in tension and compression as shown in Fig. 5-13(a). Then, as progressively increasing bending moments are applied to the beam, the strains will increase as exemplified by ε_1, ε_2, and ε_3 in Fig. 5-13(b). Corresponding to these strains and their linear variation from the neutral axis, the stress distribution will look as shown in Fig. 5-13(c). The neutral axis coincides with the centroidal axis

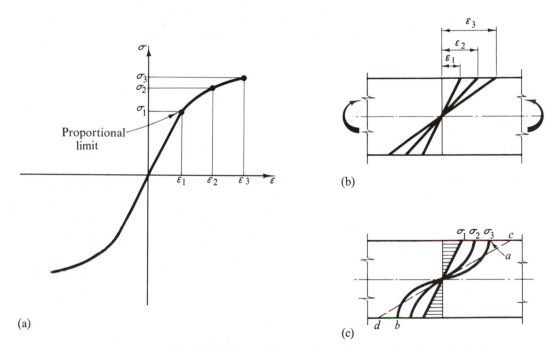

Fig. 5-13. Rectangular beam in bending exceeding the proportional limit of the material

*A. Nadai, *Theory of Flow and Fracture of Solids*, vol. I, New York: McGraw-Hill, 1950, p. 356.

in this case, as the section has two axes of symmetry and the stress-strain diagram is alike in tension and compression.

If σ_3 corresponds to the ultimate strength of the material in axial tension, the ultimate bending moment which the beam is capable of resisting may be predicted. It is associated with the stress distribution given by the curved line *ab* shown in Fig. 5-13(c). An equivalent resistance to the bending moment based on the assumption of linear stress distribution from the neutral axis is shown by the line *cd* in the same figure. Since both of these stress distributions supposedly resist the same moment and in the latter case lower stresses act near the neutral axis, higher stresses must act near the outer fibers. The stress in the extreme fibers, computed on the basis of the elastic flexure formula for the experimentally determined ultimate bending moment, is called the *rupture modulus* of the material in bending. It is higher than the true stress. For materials whose stress-strain diagrams approach a straight line all the way up to the ultimate strength, the discrepancy between the true maximum stress and the rupture modulus is small. On the other hand, the discrepancy is large for materials with a pronounced curvature in the stress-strain curve.

As another important example of inelastic bending, consider a rectangular beam of elastic-plastic material, Fig. 5-14. In such an idealization of material behavior a sharp separation of the member into distinct elastic and plastic zones is possible. For example, if the strain in the extreme fibers is double that at the beginning of yielding, only the middle half of the beam remains elastic, Fig. 5-14(a). In this case the outer quarters of the beam yield. The magnitude of the moment M_1 corresponding to this condition can be readily computed (see Example 5-7). At higher strains the elastic zone or core diminishes. Stress distribution corresponding to this situation is shown in Figs. 5-14(b) and (c).

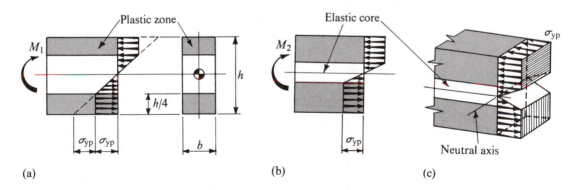

Fig.5-14 Elastic-plastic beam at a moderate level of straining

EXAMPLE 5-6

Determine the plastic or the ultimate capacity in flexure of a mild steel beam of rectangular cross section. Consider the material to be ideally elastic-plastic.

SOLUTION

The idealized stress-strain diagram is in Fig. 5-15(a). It is assumed that the material has the same properties in tension and compression. The strains that can take place during yielding are much greater than the maximum elastic strain (15 to 20 times the latter quantity). Therefore, since unacceptably large deformations of the beam occur along with very large strains, the plastic moment may be taken as the ultimate moment.

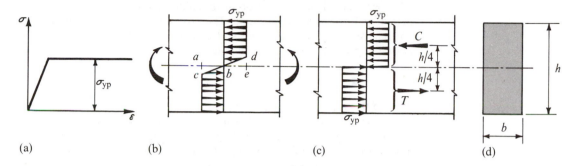

(a)   (b)   (c)   (d)

Fig. 5-15

The stress distribution shown in Fig. 5-15(b) applies after a large amount of deformation takes place. In computing the resisting moment the stresses corresponding to the triangular areas abc and bde may be neglected without unduly impairing the accuracy. They contribute little resistance to the applied bending moment because of their short moment arms. Hence the idealization of the stress distribution to that shown in Fig. 5-15(c) is permissible and has a simple physical meaning. The whole upper half of the beam is subjected to a uniform compressive stress σ_{yp}, while the lower half is all under a uniform tension σ_{yp}. That the beam is divided evenly into a tension and a compression zone follows from symmetry. Numerically

$$C = T = \sigma_{yp}(bh/2), \quad \text{i.e., (stress)} \times \text{(area)}$$

Each one of these forces acts at a distance $h/4$ from the neutral axis. Hence the plastic or ultimate resisting moment of the beam is

$$M_p \equiv M_{ult} = C\left(\frac{h}{4} + \frac{h}{4}\right) = \sigma_{yp}\frac{bh^2}{4}$$

where b is the breadth of the beam and h is its height.

The same solution may be obtained by directly applying Eqs. 5-7 and 5-8. Noting the sign of stresses, one can conclude that Eq. 5-7 is satisfied by taking the neutral axis through the middle of the beam. By taking $dA = b\,dy$ and noting the symmetry around the neutral axis, one changes Eq. 5-8 to

$$M_p \equiv M_{ult} = -2\int_0^{h/2} (-\sigma_{yp})yb\,dy = \sigma_{yp}bh^2/4 \tag{5-9}$$

The resisting bending moment of a beam of rectangular section when the outer fibers just reach σ_{yp}, as given by the elastic flexure formula, is

$$M_{yp} = \sigma_{yp}I/c = \sigma_{yp}(bh^2/6) \qquad \text{therefore} \qquad M_p/M_{yp} = 1.50$$

The ratio M_p/M_{yp} depends only on the cross-sectional properties of a member and is called the *shape factor*. The shape factor above for the rectangular beam shows that M_{yp} may be exceeded by 50 per cent before the ultimate capacity of a rectangular beam is reached.

For static loads such as occur in buildings, ultimate capacities can be determined using plastic moments. The procedures based on such concepts are referred to as *the plastic method of analysis* or *design*. For such work *plastic section modulus Z* is defined as follows:

$$M_p = \sigma_{yp}Z \tag{5-10}$$

For the rectangular beam analyzed above $Z = bh^2/4$.

The *Steel Construction Manual** provides a table of plastic section moduli for many common steel shapes. An abridged list of plastic section moduli for steel sections is given in Table 9 of the Appendix. For a given M_p and σ_{yp} the solution of Eq. 5-10 for Z is very simple.

The method of limit or plastic analysis is unacceptable in machine design in situations where fatigue properties of the material are important.

EXAMPLE 5-7

Find the residual stresses in a rectangular beam upon removal of the ultimate bending moment.

SOLUTION

The stress distribution associated with an ultimate moment is shown in Fig. 5-16(a). The magnitude of this moment has been determined in the preceding example and is $M_p = \sigma_{yp}bh^2/4$. Upon release of this plastic moment M_p every fiber in the beam can rebound elastically. The elastic range during the unloading is double that which could take place initially. There-

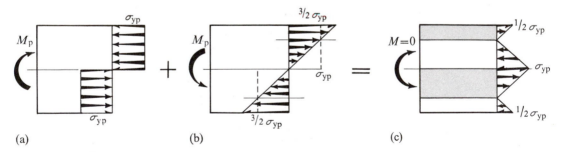

Fig. 5-16. Residual stress distribution in a rectangular bar.

*American Institute of Steel Construction, *AISC Steel Construction Manual*, (7th Ed.), New York: AISC, Inc., 1970, pp. 2–14 through 2–20.

fore, since $M_{yp} = \sigma_{yp}bh^2/6$ and the moment being released is $\sigma_{yp}(bh^2/4)$ or $1.5M_{yp}$, the maximum stress calculated on the basis of elastic action is $\frac{3}{2}\sigma_{yp}$ as shown in Fig. 5-16(b). Superimposing the initial stresses at M_p with the elastic rebound stresses due to the release of M_p, one finds the residual stresses, Fig. 5-16(c). Note that both tensile and compressive longitudinal microresidual stresses remain in the beam. The tensile zones are shaded in the figure. If such a beam were machined by gradually reducing its depth, the release of the residual stresses would cause undesirable deformations of the bar.

EXAMPLE 5-8

Determine the moment resisting capacity of an elastic-plastic rectangular beam.

SOLUTION

To make the problem more definite consider a cantilever loaded as in Fig. 5-17(a). If the beam is made of ideal elastic-plastic material and the applied

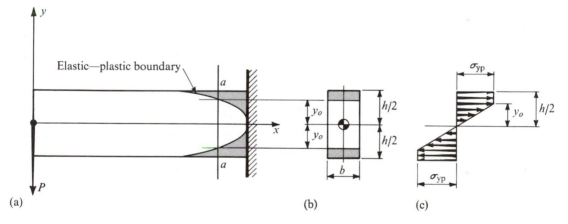

Fig. 5-17. Elastic-plastic cantilever

force P is large enough to cause yielding, plastic zones will be formed (shown shaded in the figure). At an arbitrary section a-a the corresponding stress distribution will be as shown in Fig. 5-17(c). The elastic zone extends over the depth of $2y_o$. Noting that within the elastic zone the stresses vary linearly and that everywhere in the plastic zone the axial stress is σ_{yp}, one finds that the resisting moment M is

$$M = -2\int_0^{y_o}\left(-\frac{y}{y_o}\sigma_{yp}\right)(b\,dy)y - 2\int_{y_o}^{h/2}(-\sigma_{yp})(b\,dy)y$$

$$= \sigma_{yp}\frac{bh^2}{4} - \sigma_{yp}\frac{by_o^2}{3} = M_p - \sigma_{yp}\frac{by_o^2}{3}$$

(5-11)

where the last simplification is done in accordance with Eq. 5-9. It is interesting to note that, in this general equation, if $y_o = 0$, the moment capacity becomes equal to the plastic or ultimate moment. On the other hand, if $y_o = h/2$, the moment reverts back to the limiting elastic case where $M = \sigma_{yp} b h^2/6$. When the applied bending moment along the span is known, the elastic-plastic boundary can be determined by solving Eq. 5-11 for y_o. As long as an elastic zone or core remains, the plastic deformations cannot progress without a limit. This is a case of contained plastic flow.

*5-9. STRESS CONCENTRATIONS

The flexure theory developed in the preceding articles applies only to beams of constant cross-sectional area. Such beams may be referred to as *prismatic* beams. If the cross-sectional area of the beam varies gradually, no significant deviation from the stress pattern discussed earlier takes place. However, if notches, grooves, rivet holes, or an abrupt change in the cross-sectional area of the beam occur, high *local* stresses arise. This situation is analogous to the ones discussed earlier for axial and torsion members. Again it is very difficult to obtain analytical expressions for the actual stress. In the past most of the information regarding the actual stress distribution came from accurate photoelastic experiments. Numerical methods employing finite elements are now used extensively in conjunction with computers.

Fortunately, as in the other cases discussed, only the geometric proportions of the member affect the local stress pattern. Moreover, since interest is in the maximum stress, the idea of the stress-concentration factor may be used to advantage. The ratio K of the actual maximum stress to the nominal maximum stress in the *minimum* section as given by Eq. 5-1 is defined as the stress-concentration factor in bending. This concept is illustrated in Fig. 5-18. Hence, in general,

$$(\sigma_{max})_{actual} = K \frac{Mc}{I} \qquad (5\text{-}12)$$

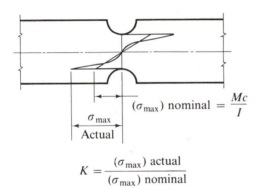

$$(\sigma_{max}) \text{ nominal} = \frac{Mc}{I}$$

$$K = \frac{(\sigma_{max}) \text{ actual}}{(\sigma_{max}) \text{ nominal}}$$

Fig. 5-18. Stress-concentration factor in bending

Figures 5-19 and 5-20 are plots of stress-concentration factors for two representative cases.* The factor K, depending on the proportions of the member, may be obtained from these diagrams. A study of these graphs indicates the desirability of generous fillets and the elimination of sharp notches to reduce local stress concentrations. These remedies are highly desirable in machine design. In structural work, particularly where ductile materials are used and the applied forces are not fluctuating, stress concentrations are ignored.

*These figures are reproduced from a paper by M. M. Frocht, "Factors of Stress Concentration Photoelastically Determined," *Trans. ASME*, vol. 57, p. A-67.

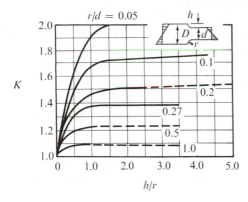

Fig. 5-19. Stress-concentration factors in pure bending for flat bars with various fillets

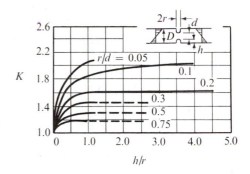

Fig. 5-20. Stress-concentration factors in bending for grooved flat bars

If the cross-sectional area of a beam is irregular itself, stress concentrations also occur. This becomes particularly significant if the cross-sectional area has re-entrant angles. For example, high localized stresses occur at the point where the flange* and the web of an I-beam meet. To minimize these, commercially rolled shapes have a generous fillet at all such points.

In addition to stress concentrations caused by changes in the cross-sectional area of a beam, another effect is significant. Forces often are applied over a limited area of a beam. Moreover, the reactions act only locally on a beam at the points of support. In the previous treatment, all such forces were idealized as concentrated forces. In practice the average bearing pressure between the member delivering such a force and the beam are computed at the point of contact of such forces with the beam. This bearing pressure or stress acts normal to the neutral surface of a beam and is at *right angles to the bending stresses discussed in this chapter.* A more detailed study of the effect of such forces shows that they cause a disturbance of all stresses on a local scale, and the bearing pressure as normally computed is a crude approximation. The stresses at right angles to the flexural stresses behave more nearly as shown in Fig. 2-16. An investigation of the disturbance caused in the bending-stress distribution by the bearing stresses is beyond the scope of this book.†

The reader must remember that the stress-concentration factors apply only while the material behaves elastically. Inelastic behavior of material tends to reduce these factors.

---

*The *web* is a thin vertical part of a beam. Thin horizontal parts of a beam are called *flanges*.

†By virtue of St. Venant's principle (Art. 2-11), at distances away from the concentrated forces comparable with the cross-sectional dimensions of a member, the formulas developed in this text are accurate, but the usual formulas are not applicable for short, stubby beams such as gear teeth.

*5-10. BEAMS OF TWO MATERIALS

So far, the beams analyzed were assumed to be of one homogeneous material. Important uses of beams made of several different materials occur in practice. Beams of two materials are especially common. Wooden beams are often reinforced by metal straps, and concrete beams are reinforced with steel rods. The fundamental theory underlying the elastic analysis of such beams will be discussed in this article. Extension of the analysis into the inelastic range follows the procedures discussed in Art. 5-8. A solution for a beam behaving inelastically is given in one of the examples that follow.

Consider a symmetrical beam of two materials with a cross section as shown in Fig. 5-21(a). The outer material (material 1) has an elastic modulus

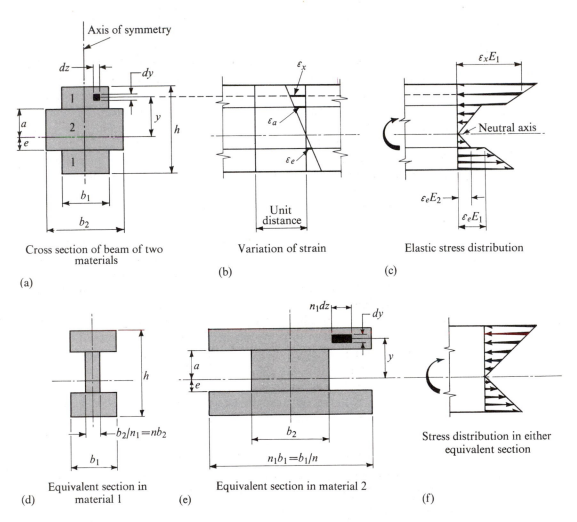

Cross section of beam of two materials

(a)

(b) Variation of strain

(c) Elastic stress distribution

(d) Equivalent section in material 1

(e) Equivalent section in material 2

(f) Stress distribution in either equivalent section

Fig. 5-21. Beam of two materials

E_1, and the modulus of the inner material (material 2) is E_2. If such a beam is subjected to bending, the basic deformation assumption used in the flexure theory remains valid. Plane sections at right angles to the axis of a beam remain plane. Therefore the strains must vary linearly from the neutral axis, as shown in Fig. 5-21(b). For the elastic case, stress is proportional to strain, and the stress distribution, assuming $E_1 > E_2$, is as shown in Fig. 5-21(c). Note that at the surfaces of contact of the two materials a break in the intensity of stress is indicated. Although the strain in both materials at such surfaces is equal, a greater stress develops in the stiffer material. The stiffness of a material is measured by the elastic modulus E. The foregoing information is sufficient to solve any beam problem of two (or more) materials by using a trial-and-error solution similar to the one discussed in Art. 5-8. However, a considerable simplification of the procedure is possible. In formally applying $\sum F_x = 0$ to locate the neutral axis and $\sum M_z = 0$ to obtain the resisting moment, only the correct magnitudes and locations of the resisting forces (not stresses) are significant. The new technique consists of constructing a section of one material on which the resisting forces are the same as on the original section. Such a section is termed an *equivalent* or *transformed cross-sectional area*. After a beam of several materials is reduced to an equivalent beam of one material, the usual elastic flexure formula applies.

The transformation of a section is accomplished by changing dimensions of a cross section parallel to the neutral axis in the ratio of elastic moduli of the materials. For example, if the equivalent section is wanted in material 1, the dimensions corresponding to material 1 do not change. The horizontal dimensions of material 2 are changed by a ratio n, where $n = E_2/E_1$, Fig. 5-21(d). On the other hand, if the transformed section is to be of material 2, the horizontal dimension of the other material is changed by a ratio $n_1 = E_1/E_2$, Fig. 5-21(e). The ratio n_1 is the reciprocal of n.

The legitimacy of transforming sections is seen by comparing the forces acting on the original and on the equivalent sections. The force from a known strain ε_x acting on an elementary area $dz\,dy$ in Fig. 5-21(a) is $\varepsilon_x E_1\,dz\,dy$. The same element of area in Fig. 5-21(e) is $n_1\,dz\,dy$. The force acting on it is $\varepsilon_x E_2 n_1\,dz\,dy$. However, from the definition of n_1, $E_1 = n_1 E_2$. So the forces acting on both elements are the same, and both, by virtue of their location, contribute equally to the resisting moment.

In a beam with a transformed area, strains and stresses vary linearly from its neutral axis. The stresses calculated in the usual manner are correct for the material of which the transformed section is made. For the other material the computed stress must be multiplied by the ratio n or n_1 of the transformed to the actual area. For example, the force acting on $n_1\,dz\,dy$ in Fig. 5-21(e) actually acts on $dz\,dy$ of the real material.

EXAMPLE 5-9

Consider a composite beam of the cross-sectional dimensions shown in Fig. 5-22(a). The upper 150 mm by 250 mm part is wood, $E_w = 10\,000$ MPa; the

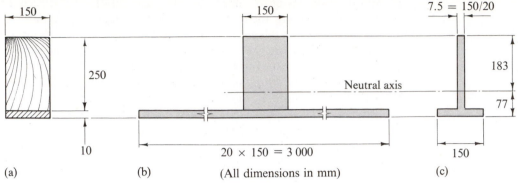

150

150

7.5 = 150/20

250

Neutral axis

183

77

10

20 × 150 = 3 000

150

(a)

(b)

(All dimensions in mm)

(c)

Fig. 5-22.

bottom 10 mm by 150 mm strap is steel, $E_s = 200\,000$ MPa. If this beam is subjected to a bending moment of 0.03 MN·m around a horizontal axis, what are the maximum stresses in the steel and wood?

SOLUTION

The ratio of the elastic moduli $E_s/E_w = 20$. Hence, using a transformed section of wood, the width of the bottom strip is $(0.15)(20) = 3$ m. The transformed area is shown in Fig. 5-22(b). Its centroid and moment of inertia around the centroidal axis are

$$\bar{y} = \frac{150(250)125 + 10(3\,000)255}{150(250) + 10(3\,000)} = 183 \text{ mm} = 0.183 \text{ m} \qquad \text{(from the top)}$$

$$I_{zz} = \frac{150(250)^3}{12} + (150)250(58)^2 + \frac{(3\,000)(10)^3}{12} + (10)(3\,000)(72)^2$$

$$= 478 \times 10^6 \text{ mm}^4 = 478 \times 10^{-6} \text{ m}^4$$

The maximum stress in the wood is

$$(\sigma_w)_{max} = \frac{Mc}{I} = \frac{(0.03)(0.183)}{478 \times 10^{-6}} = 11.5 \text{ MPa}$$

The maximum stress in the steel is

$$(\sigma_s)_{max} = n\sigma_w = 20\frac{(0.03)(0.077)}{478 \times 10^{-6}} = 96.5 \text{ MPa}$$

ALTERNATE SOLUTION

A transformed area in terms of steel may be used instead. Then the equivalent width of wood is $b/n = 150/20$ or 7.5 mm. This transformed area is shown in Fig. 5-22(c).

$$\bar{y} = \frac{(7.5)250(135) + 150(10)5}{(7.5)250 + 150(10)} = 77 \text{ mm} \qquad \text{(from the bottom)}$$

$$I_{zz} = \frac{(7.5)(250)^3}{12} + (7.5)250(58)^2 + \frac{150(10)^3}{12} + (150)10(72)^2$$

$$= 23.9 \times 10^6 \text{ mm}^4 = 23.9 \times 10^{-6} \text{ m}^4$$

$$(\sigma_s)_{max} = \frac{(0.03)(0.077)}{23.9 \times 10^{-6}} = 96.5 \text{ MPa}$$

$$(\sigma_w)_{max} = \frac{\sigma_s}{n} = \left(\frac{1}{20}\right)\frac{(0.03)(0.183)}{23.9 \times 10^{-6}} = 11.5 \text{ MPa}$$

Note that if the transformed section is an equivalent wooden section, the stresses in the actual wooden piece are obtained directly. Conversely, if the equivalent section is steel, stresses in steel are obtained directly. The stress in a material stiffer than the material of the transformed section is increased since, to cause the same unit strain, a higher stress is required.

EXAMPLE 5-10

Determine the maixmum stress in the concrete and the steel for a reinforced-concrete beam with the section shown in Fig. 5-23(a) if it is subjected to a positive bending moment of 50,000 ft-lb. The reinforcement consists of two #9 steel bars. (These bars are $1\frac{1}{8}$ in. in diameter and have a cross-sectional area of 1 in.^{2}). Assume the ratio of E for steel to that of concrete to be 15, i.e., $n = 15$.

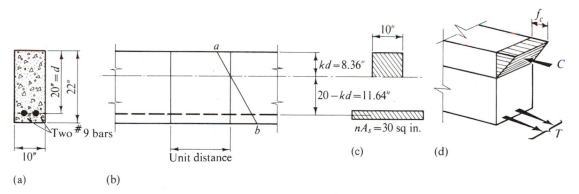

Fig. 5-23

SOLUTION

Plane sections are assumed to remain plane in a reinforced-concrete beam. Strains vary linearly from the neutral axis as shown in Fig. 5-23(b) by the line ab. A transformed section in terms of concrete is used to solve this problem. However, concrete is so weak in tension that there is no assurance that minute cracks will not occur in the tension zone of the beam. For this reason no credit is given to concrete for resisting tension. On the basis of this assumption, concrete in the tension zone of a beam only holds the reinforcing steel in place.* Hence in this analysis it virtually does not exist at all, and the transformed section assumes the form shown in Fig. 5-23(c). The cross section of concrete has its own shape above the neutral axis; below it no concrete is shown. Steel, of course, can resist tension, so it *is* shown as the transformed concrete area. For computation purposes, the steel is located by a single

---

*Actually it is used to resist shear and provide fireproofing for the steel.

dimension from the neutral axis to its centroid. There is a neglibile difference between this distance and the true distances to the various steel fibers.

So far the idea of the neutral axis has been used, but its location is unknown. However, it is known that this axis coincides with the axis through the centroid of the transformed section. It is further known that the first (or statical) moment of the area on one side of a centroidal axis is equal to the first moment of the area on the other side. Thus, let kd be the distance from the top of the beam to the centroidal axis as shown in Fig. 5-23(c), where k is an unknown ratio* and d is the distance from the top of the beam to the center of the steel. An algebraic restatement of the foregoing locates the neutral axis, about which I is computed and stresses are determined as in the preceding example.

$$\underbrace{10(kd)}_{\substack{\text{concrete}\\\text{area}}} \underbrace{(kd/2)}_{\text{arm}} = \underbrace{30}_{\substack{\text{transformed}\\\text{steel area}}} \underbrace{(20 - kd)}_{\text{arm}}$$

$$5(kd)^2 = 600 - 30(kd)$$

$$(kd)^2 + 6(kd) - 120 = 0$$

Hence $\qquad kd = 8.36 \text{ in.} \qquad$ and $\qquad 20 - kd = 11.64 \text{ in.}$

$$I = \frac{10(8.36)^3}{12} + 10(8.36)\left(\frac{8.36}{2}\right)^2 + 0 + 30(11.64)^2 = 6{,}020 \text{ in.}^4$$

$$(\sigma_c)_{\max} = \frac{Mc}{I} = \frac{(50{,}000)12(8.36)}{6{,}020} = 833 \text{ psi}$$

$$\sigma_s = n\frac{Mc}{I} = \frac{15(50{,}000)12(11.64)}{6{,}020} = 17{,}400 \text{ psi}$$

ALTERNATE SOLUTION

After kd is determined, instead of computing I, a procedure evident from Fig. 5-23(d) may be used. The resultant force developed by the stresses acting in a "hydrostatic" manner on the compression side of the beam must be located $kd/3$ below the top of the beam. Moreover, if b is the width of the beam, this resultant force $C = \frac{1}{2}(\sigma_c)_{\max}b(kd)$ (average stress times area). The resultant tensile force T acts at the center of the steel and is equal to $A_s\sigma_s$, where A_s is the cross-sectional area of the steel. Then if jd is the distance between T and C, and since $T = C$, the applied moment M is resisted by a couple equal to Tjd or Cjd.

$$jd = d - kd/3 = 20 - (8.36/3) = 17.21 \text{ in.}$$

$$M = Cjd = \frac{1}{2}b(kd)(\sigma_c)_{\max}(jd)$$

$$(\sigma_c)_{\max} = \frac{2M}{b(kd)(jd)} = \frac{2(50{,}000)12}{10(8.36)(17.21)} = 833 \text{ psi}$$

*This conforms with the usual notation used in books on reinforced concrete. In this text h is generally used to represent the height or depth of the beam.

$$M = T(jd) = A_s\sigma_s jd$$

$$\sigma_s = \frac{M}{A_s(jd)} = \frac{(50,000)12}{2(17.21)} = 17,400 \text{ psi}$$

Both methods naturally give the same answer. The second method is more convenient in practical applications. Since steel and concrete have different allowable stresses, the beam is said to have balanced reinforcement when it is designed so that the respective stresses are at their allowable level simultaneously. Note that the beam shown would become virtually worthless if the bending moments were applied in the opposite direction.

EXAMPLE 5-11

Determine the ultimate moment carrying capacity for the reinforced concrete beam of the preceding example. Assume that the steel reinforcement yields at 40,000 psi and that the ultimate strength of concrete $f'_c = 2,500$ psi.

SOLUTION

When the reinforcing steel begins to yield, large deformations commence. This is taken to be the ultimate capacity of steel: hence $T_{\text{ult}} = A_s\sigma_{\text{yp}}$.

At the ultimate moment, experimental evidence indicates that the compressive stresses can be approximated by the rectangular stress block shown in Fig. 5-24. It is customary to assume the average stress in this

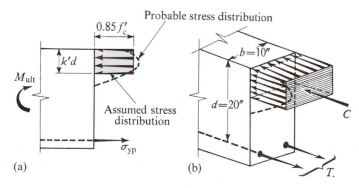

Fig. 5-24

compressive stress block to be $0.85f'_c$. On this basis, keeping in mind that $T_{\text{ult}} = C_{\text{ult}}$, one has

$$T_{\text{ult}} = \sigma_{\text{yp}}A_s = 40,000 \times 2 = 80,000 \text{ lb} = C_{\text{ult}}$$

$$k'd = \frac{C_{\text{ult}}}{0.85f'_c b} = \frac{80,000}{0.85 \times 2,500 \times 10} = 3.77 \text{ in.}$$

$$M_{\text{ult}} = T_{\text{ult}}\left(d - \frac{k'd}{2}\right) = 80,000\left(20 - \frac{3.77}{2}\right)\frac{1}{12} = 121,000 \text{ ft-lb}$$

*5-11. CURVED BEAMS

The flexure theory for curved bars is developed in this article. Attention is confined to beams having an axis of symmetry of the cross section, with this axis lying in one plane along the length of the beam. Only the elastic case is treated,* with the usual proviso that the elastic modulus is the same in tension and compression.

Consider a curved member such as shown in Figs. 5-25(a) and (b). The outer fibers are at a distance of r_o from the center of curvature O. The

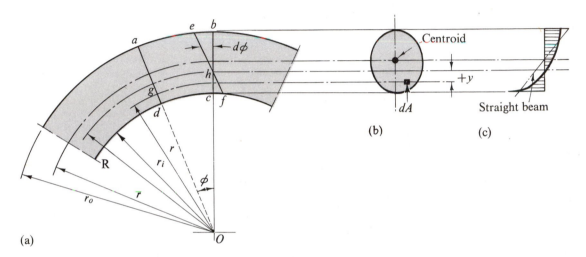

(b)   (c)

Fig. 5-25. Curved bar in pure bending

inner fibers are at a distance of r_i. The distance from O to the centroidal axis is $\bar{r}$. The solution† of this problem is again based on the familiar assumption: Sections perpendicular to the axis of the beam remain plane after a bending moment M is applied. This is diagrammatically represented by the line ef in relation to an element of the beam $abcd$. The element is defined by the central angle ϕ.

Although the basic deformation assumption is the same as for straight beams, and, from Hooke's law, the normal stress $\sigma = E\varepsilon$, a difficulty is encountered. The initial length of a beam fiber such as gh depends upon the distance r from the center of curvature. Thus, although the total deformation of beam fibers (described by the small angle $d\phi$) follows a linear law,

*For plastic analysis of curved bars see for example H. D. Conway, "Elastic-Plastic Bending of Curved Bars of Constant and Variabel Thickness," *Journal of Applied Mechanics*, **27**, no. 4 (Dec. 1960), 733–34.

†This approximate solution was developed by E. Winkler in 1858. The exact solution of the same problem by the methods of the mathematical theory of elasticity is due to M. Golovin, who solved it in 1881.

strains do not. The elongation of a generic fiber gh is $(R - r)d\phi$, where R is the distance from O to the neutral surface (not yet known), and its initial length is $r\phi$. The strain ε of any arbitrary fiber is $(R - r)(d\phi)/r\phi$, and the normal stress σ on an element dA of the cross-sectional area is

$$\sigma = E\varepsilon = E\frac{(R - r)\,d\phi}{r\phi} \tag{5-13}$$

For future use note also that

$$\frac{\sigma r}{R - r} = \frac{E\,d\phi}{\phi} \tag{5-14}$$

Equation 5-13 gives the normal stress acting on an element of area of the cross section of a curved beam. The location of the neutral axis follows from the condition that the summation of the forces acting perpendicular to the section must be equal to zero, i.e.

$$\sum F_n = 0, \qquad \int_A \sigma\,dA = \int_A \frac{E(R - r)\,d\phi}{r\phi}\,dA = 0$$

However, since E, R, ϕ, and $d\phi$ are constant at any one section of a stressed beam, they may be taken outside the integral sign and a solution for R obtained. Thus:

$$\frac{E\,d\phi}{\phi} \int_A \frac{R - r}{r}\,dA = \frac{E\,d\phi}{\phi} \left[R \int_A \frac{dA}{r} - \int_A dA \right] = 0$$

$$R = \frac{A}{\displaystyle\int_A dA/r} \tag{5-15}$$

where A is the cross-sectional area of the beam and R locates the neutral axis. Note that the neutral axis so found does not coincide with the centroidal axis. This differs from the situation found to be true for straight elastic beams.

Now that the location of the neutral axis is known, the equation for the stress distribution is obtained by equating the external moment to the internal resisting moment built up by the stresses given by Eq. 5-13. The summation of moments is made around the z axis, which is normal to the plane of the figure shown in Fig. 5-25(a).

$$\sum M_z = 0, \qquad M = \int_A \underbrace{\sigma\,dA}_{\text{(force)}}\ \underbrace{(R - r)}_{\text{(arm)}} = \int_A \frac{E(R - r)^2\,d\phi}{r\phi}\,dA$$

Again remembering that E, R, ϕ, and $d\phi$ are constant at a section, by using

Eq. 5-14, and performing the algebraic steps indicated, the following is obtained:

$$M = \frac{E\,d\phi}{\phi} \int_A \frac{(R-r)^2}{r}\,dA = \frac{\sigma r}{R-r} \int_A \frac{(R-r)^2}{r}\,dA$$

$$= \frac{\sigma r}{R-r} \int_A \frac{R^2 - Rr - Rr + r^2}{r}\,dA$$

$$= \frac{\sigma r}{R-r} \left[R^2 \int_A \frac{dA}{r} - R \int_A dA - R \int_A dA + \int_A r\,dA \right]$$

Here, since R is a constant, the first two integrals vanish as may be seen from the bracketed expression appearing just before Eq. 5-15. The third integral is A, and the last integral, by definition, is $\bar{r}A$. Hence

$$M = \frac{\sigma r}{R-r}(\bar{r}A - RA)$$

whence the normal stress acting on a curved beam at a distance r from the center of curvature is

$$\sigma = \frac{M(R-r)}{rA(\bar{r}-R)} \tag{5-16}$$

If positive y is measured toward the center of curvature from the neutral axis, and $\bar{r} - R = e$, Eq. 5-16 may be written in a form which more closely resembles the flexure formula for straight beams

$$\sigma = \frac{My}{Ae(R-y)} \tag{5-17}$$

These equations indicate that the stress distribution in a curved bar follows a hyperbolic pattern. The maximum stress is always on the inner (concave) side of the beam. A comparison of this result with the one that follows from the formula for straight bars is shown in Fig. 5-25(c). Note particularly that in the curved bar the neutral axis is pulled toward the center of the curvature of the beam. This results from the higher stresses developed below the neutral axis. The theory developed applies of course only to elastic stress distribution and only to beams in pure bending. For a consideration of situations where an axial force is also present at a section see Art. 7-2.

EXAMPLE 5-12

Compare stresses in a 50 mm by 50 mm rectangular bar subjected to end couples of 2 083 N·m in three special cases: (a) straight beam, (b) beam curved to a radius of 250 mm along the centroidal axis (i.e., $\bar{r} = 250$ mm Fig. 5-26(a)) and (c) beam curved to $\bar{r} = 75$ mm.

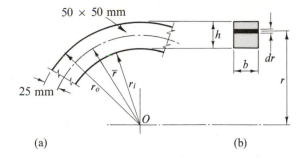

(a)   (b)

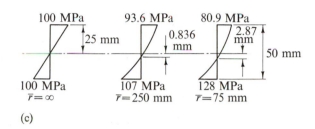

(c)

Fig. 5-26

SOLUTION

Case (a) follows directly by applying Eqs. 5-6 and 5-5 in that order.

$$S = \frac{bh^2}{6} = \frac{50(50)^2}{6} = 20.83 \times 10^3 \text{ mm}^3 = 20.83 \times 10^{-6} \text{ m}^3$$

$$\sigma_{max} = \frac{M}{S} = \frac{2\,083}{20.83 \times 10^{-6}} = \pm 100 \times 10^6 \text{ Pa} = \pm 100 \text{ MPa}$$

This result is shown in Fig. 5-26(c). $\bar{r} = \infty$ since a straight bar has an infinite radius of curvature.

To solve parts (b) and (c) the neutral axis must be located first. This is found in general terms by integrating Eq. 5-15. For the rectangular section, the elementary area is taken as $b\,dr$, Fig. 5-26(b). The integration is carried out between the limits r_i and r_o, the inner and outer radii, respectively.

$$R = \frac{A}{\int_A dA/r} = \frac{bh}{\int_{r_i}^{r_o} b\,dr/r} = \frac{h}{\int_{r_i}^{r_o} dr/r}$$

$$= \frac{h}{|\ln r\,|_{r_i}^{r_o}} = \frac{h}{\ln\left(\frac{r_o}{r_i}\right)} = \frac{h}{2.3026 \log\left(\frac{r_o}{r_i}\right)} \qquad (5\text{-}18)$$

where h is the depth of the section, ln is the natural logarithm, and log is a logarithm with a base of 10 (common logarithm).

For Case (b), $h = 50$ mm, $\bar{r} = 250$ mm, $r_i = 225$ mm, and $r_o = 275$ mm. The solution is obtained by evaluating Eqs. 5-18 and 5-16. Subscript i refers to the normal stress σ of the inside fibers; o of the outside fibers.

$$R = \frac{50}{\ln(275/225)} = 249.164 \text{ mm}$$

$$e = \bar{r} - R = 250 - 249.164 = 0.836 \text{ mm}$$

$$\sigma_i = \frac{M(R - r_i)}{r_i A(\bar{r} - R)} = \frac{(2\,083)(0.249\,164 - 0.225)}{(0.225)(0.002\,5)(0.000\,836)}$$

$$= 107 \times 10^6 \text{ Pa} = 107 \text{ MPa}$$

$$\sigma_o = \frac{M(R - r_o)}{r_o A(\bar{r} - R)} = \frac{(2\,083)(0.249\,164 - 0.275)}{(0.275)(0.002\,5)(0.000\,836)}$$

$$= -93.6 \times 10^6 \text{ Pa} = -93.6 \text{ MPa}$$

The negative sign of σ_o indicates a compressive stress. These quantities and the corresponding stress distribution are shown in Fig. 5-26(c); $\bar{r} = 250$ mm.

Case (c) is computed in the same way. Here $h = 50$ mm, $\bar{r} = 75$ mm, $r_i = 50$ mm, and $r_o = 100$ mm. Results of the computation are shown in Fig. 5-26(c).

$$R = \frac{50}{\ln(100/50)} = \frac{50}{\ln 2} = 72.13 \text{ mm}$$

$$e = \bar{r} - R = 75 - 73.13 = 2.87 \text{ mm}$$

$$\sigma_i = \frac{(2\,083)(0.022\,13)}{(0.05)(0.002\,5)(0.002\,87)} = 128 \times 10^6 \text{ Pa} = 128 \text{ MPa}$$

$$\sigma_o = \frac{(2\,083)(-0.027\,87)}{(0.1)(0.002\,5)(0.002\,87)} = -80.9 \times 10^6 \text{ Pa} = -80.9 \text{ MPa}$$

Several important conclusions, generally true, may be reached from the above example. First, *the usual flexure formula is reasonably good for beams of considerable curvature.* Only 7 per cent error in the maximum stress occurs in Case (b) for $\bar{r}/h = 5$, an error tolerable for most applications. For greater ratios of $\bar{r}/h$ this error diminishes. As the curvature of the beam increases, the stress on the concave side rapidly increases over the one given by the usual flexure formula. When $\bar{r}/h = 1.5$ a 28 per cent error occurs. Second, the evaluation of the integral for R over the cross-sectional area may become very complex. Finally, calculations of R must be very accurate since differences between R and numerically comparable quantities are used in the stress formula.

The last two difficulties prompted the development of other methods of solution. One such method consists of expanding certain terms of the solution into a series,* another of building up a solution on the basis of a special transformed section. Yet another approach consists of working "in reverse." Curved beams of various cross-sections, curvatures, and applied moments are analyzed for stress; then these quantities are divided

*S. Timoshenko, *Strength of Materials* (3rd ed.), Part I, (Princeton, N.J.: D. Van Nostrand, 1955, p. 369 and p. 373.

by a flexural stress that would exist for the same beam *if it were straight.* These ratios are then tabulated.* Hence, conversely, if stress in a curved beam is wanted, it is given as

$$\sigma = K\frac{Mc}{I} \tag{5-19}$$

where the coefficient K is obtained from a table or a graph and Mc/I is computed as in the usual flexure formula.

An expression for the distance from the center of curvature to the neutral axis of a curved beam of circular cross-sectional area is now given for future reference:

$$R = \frac{\bar{r} + \sqrt{\bar{r}^2 - c^2}}{2} \tag{5-20}$$

where $\bar{r}$ is the distance from the center of curvature to the centroid and c is the radius of the circular cross-sectional area.

PROBLEMS FOR SOLUTION

5-1 through 5-5. For the cross-sectional areas with the dimensions shown in the figures, determine the moment of inertia for each section with respect to the horizontal centroidal axis. For properties of steel wide-flange beams, channels, and angles shown for problems 5-3, 5-4, and 5-5, see Appendix Tables 4, 5, and 7, respectively. *Ans: Prob. 5-3:* 11,080 in.⁴; *Prob. 5-4:* 1,120 in.⁴; *Prob. 5-5:* 410 in.⁴.

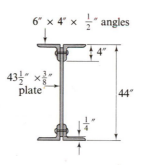

PROB. 5 – 3

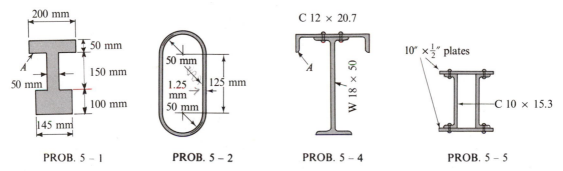

PROB. 5 – 1   PROB. 5 – 2   PROB. 5 – 4   PROB. 5 – 5

*Roark, R. J., *Formulas for Stress and Strain.* New York: McGraw-Hill, 1965, Fourth Edition, Table VII, p. 165.

5-6 through 5-10. The beams having the cross-sectional dimensions shown in the figures are each subjected to a positive bending moment of 54 000 N·m acting around the horizontal neutral axis. For each case, determine the bending stress acting on each of the three infinitesimal areas shown by the heavy dots. For properties of the wide-flange steel section shown in Prob. 5-8 see Table 4 of the Appendix. *Ans: Prob. 5-8:* 23.1 ksi (bottom); *Prob: 5-9:* −13.6 ksi (top).

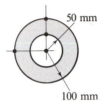

PROB. 5 – 6

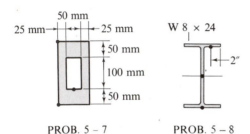

PROB. 5 – 7   PROB. 5 – 8

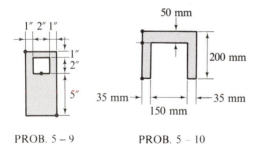

PROB. 5 – 9   PROB. 5 – 10

5-11. Verify the section moduli given in the Appendix for S 12 × 40.8, W 10 × 112, and C 12 × 20.7.

5-12. Determine the allowable bending moment for a rectangular wooden beam having a full-sized cross-section of 50 mm by 100 mm for an allowable bending stress of 8.4 MPa, (a) if bent around a neutral axis parallel to the 50 mm

side, (b) if bent around a neutral axis parallel to the 100 mm side. *Ans:* (a) 701 N·m.

5-13. If a pure bending moment of 23 kip-ft is to be resisted by a wide-flange section without exceeding a 22 ksi stress, (a) what size section should be used if the moment acts around the *X-X* axis, (b) around the *Y-Y* axis? *Ans:* W 8 × 17, W 10 × 45.

5-14. For a linearly elastic material, at the same maximum stress for a square member in the two different positions shown in the figure, determine the ratio of the bending moments. Bending takes place around the horizontal axis. *Ans:* $\sqrt{2}$.

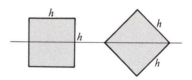

PROB. 5 – 14

5-15. The cast iron machine part, a section through which is shown in the figure, acts as a beam resisting a positive bending moment. If the allowable stress in tension is 3,000 psi and in compression 12,000 psi, what moment may be applied to this beam? *Ans:* 69.5 k-in.

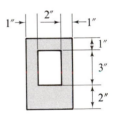

PROB. 5 – 15

5-16. A beam having a solid rectangular cross-section with the dimensions shown in the figure is subjected to a positive bending moment of 16 000 N·m acting around the horizontal axis. (a) Find the compressive force acting on the shaded area of the cross-section developed by the bending stresses. (b) Find the tensile force acting on the cross-hatched area of the cross-section.

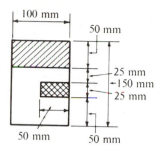

PROB. 5 – 16

5-17. Two 2 in. by 6 in., full-sized wooden planks are glued together to form a *T* section as shown in the figure. If a positive bending moment of 2,270 ft-lb is applied to such a beam acting around a horizontal axis, (a) find the stresses at the extreme fibers ($I = 136$ in.^{4}), (b) calculate the total compressive force developed by the normal stresses above the neutral axis because of the bending of the beam, (c) find the total force due to the tensile bending stresses at a section and compare it with the result found in (b). *Ans:* (b) 5,000 lb.

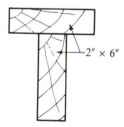

PROB. 5 – 17

5-18.* By integration, determine the force developed by the bending stresses and its position acting on the shaded area of the cross-section of the beam shown in the figure if the beam is subjected to a negative bending moment of 3 500 N·m acting around the horizontal axis.

PROB. 5 – 18

*5-19. A beam has the cross-section of an isoceles triangle as shown in the figure and is subjected to a negative bending moment of 4 000 N·m around a horizontal axis. Show by integration that $I_o = bh^3/36$. Determine the location and magnitude of the resultant tensile and compressive forces acting on a section.

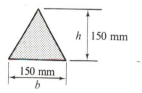

PROB. 5 – 19

5-20. A gray cast iron channel-shaped member, as shown in the figure, acts as a horizontal beam in a machine. When vertical forces are applied to this member the distance AB increases by 0.0010 in. and the distance CD decreases by 0.0090 in. What is the sense of the applied moment, and what normal stresses occur in the extreme fibers? $E = 15 \times 10^6$ psi. *Ans:* +5,620 psi at the top, −22,500 psi at the bottom.

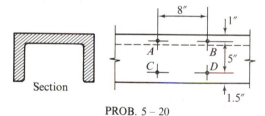

PROB. 5 – 20

5-21. A solid steel beam having the cross-sectional dimensions partially shown in the figure was loaded in the laboratory in pure bending. Bending took place around a horizontal neutral axis. Strain measurements showed that the top

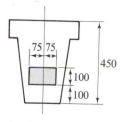

PROB. 5 – 21

fibers contracted 0.000 3 m/m longitudinally; the bottom fibers elongated 0.000 6 m/m longitudinally. Determine the total normal force in newtons which acted on the shaded area indicated in the figure at the time the strain measurements were made. $E = 200\ 000$ MN/m². All dimensions are in mm. *Ans:* 900 kN.

5-22. When two concentrated forces were applied to an W 18 × 50 steel beam as shown in the figure, an elongation of 0.0050 in. was observed between the gage points A and B. What was the magnitude of the applied forces? $E = 30 \times 10^6$ psi. *Ans:* 23.2 k.

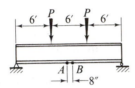

PROB. 5 – 22

5-23. As the screw of a large, steel C clamp, such as shown in the figure, is tightened down upon an object, the strain in the horizontal direction due to bending only is being measured by a strain gage at point B. If a strain of 900×10^{-6} in. per inch is noted, what is the load on the screw corresponding to the value of the observed strain? Let $E = 30 \times 10^6$ psi. *Ans:* 26.9 k.

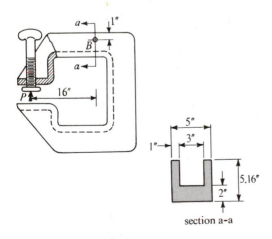

section a-a

PROB. 5 – 23

5-24. In Example 5-4, reverse the sense of the concentrated force and find the maximum bend-

ing stresses in the beam at the built in end if $L = 2.5$ m. *Ans:* 6.54 MN/m².

5-25. In a small dam, a typical vertical beam is subjected to the hydrostatic loading shown in the figure. Determine the stress at point D of section a-a due to the bending moment. *Ans:* 7.29 MPa.

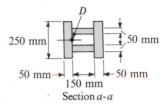

Section a-a

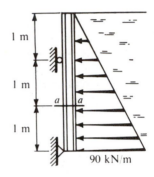

PROB. 5 – 25

5-26. Find the maximum flexural stress at a section 250 mm from the support for the cantilever beam loaded as shown in the figure. Show the result on an isolated element alongside the beam. The beam weighs approximately 350 N/m of length and $P = 450$ N. *Ans:* 2.18 MPa.

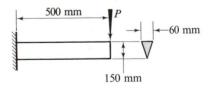

PROB. 5 – 26

5-27. At section a-a for the beam, loaded as shown in the figure, find: (a) the maximum normal stress; (b) the normal stress midway between the top and the bottom fibers. The beam weighs 230 lb/ft and $P = 2,100$ lb. *Ans:* (a) −572 psi.

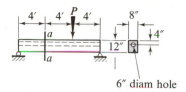

PROB. 5 – 27

5-28. A *T* beam shown in the figure is made of a material the behavior of which may be idealized as having a tensile proportional limit of 20 000 kN/m² and a compressive proportional limit of 40 000 kN/m². With a factor of safety of $1\frac{1}{2}$ on the initiation of yielding, find the magnitude of the largest force *F* which may be applied to this beam in a downward direction as well as in an upward direction. Base answers only on the consideration of the maximum bending stresses caused by *F*. *Ans:* 5.51 kN.

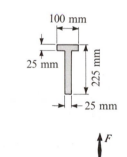

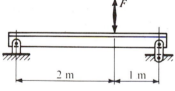

PROB. 5 – 28

5-29. Show that the maximum bending stress for a beam of rectangular cross-section is $\sigma_{max} = (Mc/I)[(2n + 1)/(3n)]$ if, instead of Hooke's law, the stress-strain relationship is $\sigma^n = E\varepsilon$, where *n* is a number dependent on the properties of the material.

* **5-30.** A 150 mm by 300 mm, rectangular section is subjected to a positive bending moment of 240 000 N·m around the "strong" axis. The material of the beam is nonisotropic and is such

that the modulus of elasticity in tension is $1\frac{1}{2}$ times as great as in compression, see figure. If the stresses do not exceed the proportional limit, find the maximum tensile and compressive stresses in the beam.

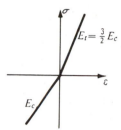

PROB. 5 – 30

5-31. Consider a linearly elastic beam subjected to a bending moment *M* around one of its principal axes for which the moment of inertia of the cross-sectional area around that axis is *I*. Show that for such a beam the normal force *F* acting on any part of the cross-sectional area A_1 is

$$F = MQ/I$$

where
$$Q = \int_{A_1} y \, dA = \bar{y} A_1$$

and $\bar{y}$ is the distance from the neutral axis of the cross section to the centroid of the area A_1 as shown in the figure.

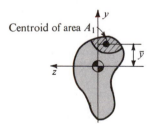

PROB. 5 – 31

5-32 through 5-36. Find the ratios M_{ult}/M_{yp} for mild steel beams resisting bending around the horizontal axes and having the cross-sectional dimensions shown in the figures. Assume the idealized stress-strain diagram used in Example 5-6. *Ans: Prob. 5-32:* 1.7. *Prob. 5-34:* 1.11. *Prob. 5-36:* 1.8.

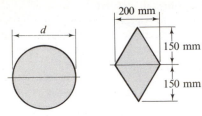

W 8 × 17

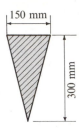

150 mm

300 mm

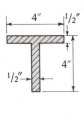

4″   ½″

4″

½″

PROB. 5 – 32   PROB. 5 – 33   PROB. 5 –'34   PROB. 5 – 35   PROB. 5 – 36

5-37 through 5-39. Composite beams having the cross-sectional dimensions shown in the figures are subjected to positive bending moments of 80 kN·m each. Materials are fastened together so that the beams act as a unit. Determine the maximum bending stress in each material. $E_{St} = 210\,000$ kN/m²; $E_{Al} = 70\,000$ kN/m²; $E_{Ci} = 105\,000$ kN/m² (*Hint for Prob. 5-39:* for an ellipse with semi-axes a and b, $I = \pi ab^3/4$ around the major centroidal axis.)

as a unit. $E_{St} = 30 \times 10^6$ psi; $E_w = 1.2 \times 10^6$ psi. The allowable bending stresses are $\sigma_{St} = 20$ ksi and $\sigma_w = 1.2$ ksi. *Ans: Prob. 5-40: 480 k-in., Prob. 5-41: 633 k-in.*

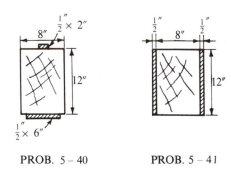

PROB. 5 – 40   PROB. 5 – 41

5-42. A reinforced concrete beam having a cross-section as shown in the figure is subjected to a positive bending moment of 8,000 ft-lb. Determine the maximum compressive stress in the concrete and the maximum stress in the steel. Assume $n = 15$. *Ans: +5,150 psi.*

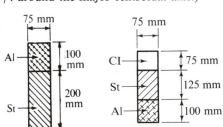

PROB. 5 - 37   PROB. 5 – 38

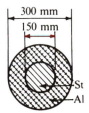

PROB. 5 – 39

5-40 and 5-41. Determine the allowable bending moment around horizontal neutral axes for the composite beams of wood and steel having the cross-sectional dimensions shown in the figures. Materials are fastened together so that they act

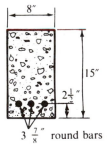

$3 \frac{7}{8}''$ round bars

PROB. 5 – 42

5-43. A 150 mm thick concrete slab is longitudinally reinforced with steel bars as shown in the figure. Determine the allowable bending moment per one meter width of this slab. Assume

$n = 12$ and the allowable stresses for steel and concrete as 150 MN/m² and 8 MN/m² respectively *Ans:* 16.3 kN·m/m.

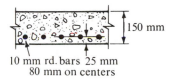

10 mm rd. bars 25 mm
80 mm on centers

PROB. 5 – 43

5-44.   A section of a hollow rectangular reinforced concrete beam is made as shown in the figure. The area of steel in tension is 9 in.², and $n = 10$. If the maximum compressive stress in the concrete caused by bending is known to be 1,000 psi, what is the stress in the steel and what bending movement is applied to the section? *Ans:* 9,830 k-in.

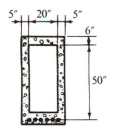

PROB. 5 – 44

5-45.   A beam has a cross-section as shown in the figure, and is subjected to a positive bending moment which causes a tensile stress in the steel of 20 ksi. If $n = 12$, what is the value of the bending moment? *Ans:* 122 k-ft.

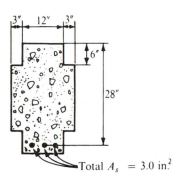

Total $A_s = 3.0$ in.²

PROB. 5 – 45

5-46.   Rework Example 5-12 by changing h to 100 mm.   *Ans:* (c) $+ 48.1$ MN/m².

5-47.   Derive Eq. 5-20.

5-48.   What is the largest bending moment which may be applied to a curved bar, such as shown in Fig. 5-25(a), with $\bar{r} = 3$ in., if it has a circular cross-sectional area of 2 in. diameter and the allowable stress is 12 ksi? *Ans:* 7.00 k-in.

6 Shearing Stresses in Beams

6-1. INTRODUCTION

It was shown in Chapter 4 that in a planar problem three elements of a force system may be necessary at a section of a beam to maintain the segment in equilibrium. These are an axial force, a shearing force, and a bending moment. The stress caused by an axial force was investigated in Chapter 1. In Chapter 5 the nature of the stresses caused by a bending moment in a beam was discussed. The stresses in a beam caused by the shearing force will be investigated in this chapter.

In all of previous derivations of the stress distribution in a member, the same sequence of reasoning was employed. First, a strain distribution was assumed across the section; next, properties of the material were brought in to relate these strains to stresses; and, finally, the equations of equilibrium were used to establish the desired relations. However, the development of the expression linking the shearing force and the cross-sectional area of a beam to the stress follows a different path. The procedure described above cannot be employed, as no simple assumption for the strain distribution due to the shearing force can be made. Instead, an indirect approach is used. *The stress distribution caused by flexure, as determined in the preceding chapter, is assumed, which, together with the equilibrium requirements, resolves the problem of the shearing stresses.*

First it will be necessary to establish that the shearing force is **inseparably** linked with a *change* in the bending moment at adjoining* sections through a beam. Thus, if a shear and a bending moment are present at one section through a beam, it will be shown that a *different* bending moment will exist at an adjoining section, although the shear may remain constant. This will lead to the establishment of the shearing stresses on the imaginary longitudinal planes through the members which are parallel to its axis. Then finally, since at a point equal shearing stresses exist on the mutually perpendicular

---

*Adjoining sections are parallel sections taken perpendicular to the axis of the beam a small distance apart.

planes, the shearing stresses whose direction is coincident with the shearing force at a section will be determined.

In this chapter the investigation of stresses will be limited to those in straight beams. The analysis of shearing stresses in curved beams is beyond the scope of this text. In the early part of this chapter only beams with a symmetrical cross section will be considered and the applied forces will be assumed to act in the plane containing an axis of symmetry and the axis of the beam. An illustration of shearing stress distribution in an elastic-plastic beam is also given. In addition to shearing stresses, the related problem of interconnection requirements for fastening together several longitudinal elements of built-up beams will also be considered.

6-2. RELATION BETWEEN SHEAR AND BENDING MOMENT

The interrelation of the bending moment and the shearing force must be established first. Later this will lead to the establishment of an expression for the shearing stresses in a beam.

Consider an element dx long, isolated from a beam by two adjoining sections taken perpendicular to the axis of the beam. Such an element is shown as a free body in Fig. 6-1. At the sections shown the shearing forces and bending moments act on the element as indicated. The elements of this force system are shown with a positive *sense* (for sign conventions see Figs. 4-15 and 4-17). Moreover, since the shear and the moment may each change from one section to the next, on the right-hand face of the element they are respectively noted as $V + dV$ and $M + dM$. As axial forces have no bearing on the problem considered here, they are not included.

The distributed load $q(x)$ acting on the beam and element is considered *positive* when acting in the *upward* direction. This load has the dimensions of

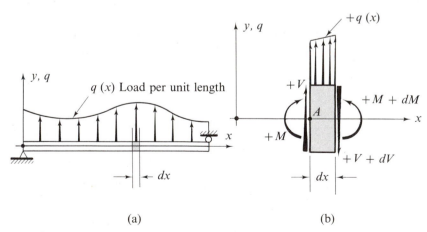

(a)   (b)

Fig. 6-1. Beam and an element cut out from it by the two adjoining sections dx apart

force per unit length (N/m or lb/in.). The selected positive direction for the applied load results in a positive differential relation, which will be found convenient in subsequent work (Chapters 10 and 11). However, for downward loads, such as a beam's own weight w_o, such loads must be taken with a negative sign, i.e., $q = -w_o$.

The element of a beam shown in Fig. 6-1 must be in equilibrium. Hence the summation of moments around the axis through point A perpendicular to the plane of the figure must be zero, i.e., $\sum M_A = 0$. Noting that from point A the arm of the distributed force is $dx/2$,

$$\sum M_A = 0 \circlearrowleft +, \qquad (M + dM) - M - (V + dV)\, dx + q\, dx \frac{dx}{2} = 0$$

By simplifying and ignoring* the infinitesimals of higher order, one can reduce this to $dM - V\, dx = 0$. Hence,

$$dM = V\, dx \qquad \text{or} \qquad \frac{dM}{dx} = V \qquad (6\text{-}1)$$

Equation 6-1 means that if a shear is acting at a section, there will be a *different* bending moment at an adjoining section. When shear is present, the difference between the bending moments on the adjoining sections is equal to $V\, dx$. If no shear acts at the adjoining sections of a beam, *no change in the bending moment occurs.* Conversely, the rate of change of the bending moment along the beam is equal to the shear. Thus although shear is treated in this chapter as an independent action on a beam, it is **inseparably** linked with a change in the bending moment along the beam's length.

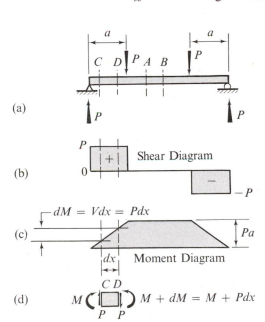

(a)

(b)   Shear Diagram

(c)   Moment Diagram

(d)

Fig. 6-2. Relation between shear and bending moment diagrams for the loading shown

As an illustration of the meaning of Eq. 6-1, the problem discussed in Example 4-5 for which the shear and bending moment diagrams were established is reproduced in Fig. 6-2. At any two sections such as A and B taken through the beam anywhere between the applied forces P, the bending moment is the same. *No shear* acts at these sections. On the other hand, between any two sections such as C and D near the support, a change in the bending moment does take place. Shearing forces act at these sections. These shears are shown acting on an element of the beam in Fig. 6-2(d). Note that in this zone of the beam the *change* in the bending

---

*This is *not* an approximation. Thus, consider an element of a beam Δx long, instead of dx. Then all quantities on adjoining faces vary by an amount Δ. Summing moments around the right-hand end, and simplifying the results, $\Delta M/\Delta x = V + q(x)\, \Delta x/2$ and by definition, $\lim_{\Delta x \to 0} \Delta M/\Delta x \equiv dM/dx = V$. The applied load $q(x)$ may vary in the interval considered.

moment in a distance dx is $P\,dx$ as the shear V is equal to P. The rate of change of the bending moment along the beam, dM/dx, is equal to the shear and is represented by the slope of the moment diagram. For subsequent discussion, the possibility of equal, as well as of different, bending moments on two adjoining sections through a beam must be appreciated.

Before a detailed analysis is given, a study of a sequence of photographs of a model (Fig. 6-3) may prove helpful. The model represents a segment of an I beam. In Fig. 6-3(a), in addition to the beam itself, blocks simulating stress distribution caused by bending moments may be seen. The moment on the right is assumed to be larger than the one on the left. This system of forces is in equilibrium providing vertical shears V (not seen in this view) also act on the beam segment. By separating the model along the neutral surface, one obtains two separate parts of the beam segment as in Fig. 6-3(b). Either one of these parts alone again must be in equilibrium.

If the upper and the lower segments of Fig. 6-3(b) are connected by a dowel or a bolt in an actual beam, the axial forces on either the upper or the lower part caused by the bending moment stresses must be maintained in equilibrium by a force in the dowel. The force which must be resisted can be evaluated by summing the forces in the axial direction caused by bending stresses. In performing such a calculation either the upper or the lower part of the beam segment can be used. The horizontal force transmitted by the dowel is the force needed to balance the net force caused by the bending stresses acting on the two adjoining sections. Alternatively, by subtracting the same bending stress on both ends of the segment, the same results can be obtained. This is shown schematically in Fig. 6-3(c), where assuming a zero bending moment on the left, only the normal stresses due to the increment in moment within the segment need be shown acting on the right.

If initially the I beam considered is one piece requiring no bolts or dowels, an imaginary plane can be used to separate the beam segment into two parts, Fig. 6-3(d). As before, the net force which must be developed across the cut area to maintain equilibrium can be determined. Dividing this force by the area of the imaginary horizontal cut gives average shearing stresses acting in this plane. In the analysis it is again expedient to work with the change in bending moment rather than with the total moments on the end sections.

After the shearing stresses on one of the planes are found (i.e., the horizontal one in Fig. 6-3(d)), shearing stresses on mutually perpendicular planes of an infinitesimal element also become known since they must be numerically equal. This approach establishes the shearing stresses in the plane of the beam section taken normal to its axis.

The process discussed above is quite general; two additional illustrations of separating the segment of the beam are in Figs. 6-3(e) and (f). In Fig. 6-3(e), the imaginary horizontal plane separates the beam just below the flange. Either the upper or the lower part of this beam can be used in calculating the shearing stresses in the cut. The imaginary vertical plane cuts off a part

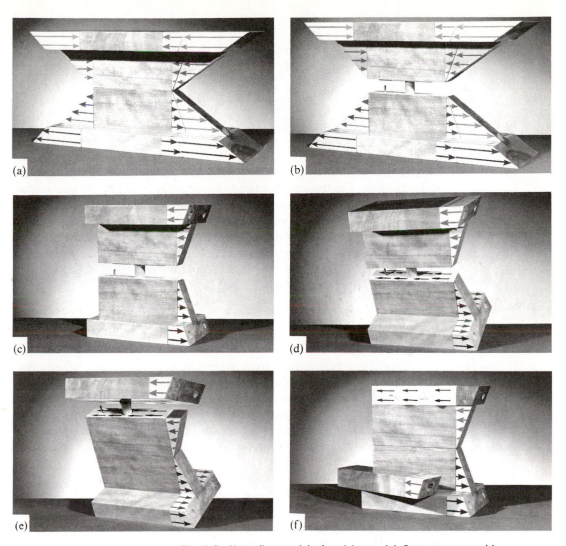

Fig. 6-3. Shear flow model of an I-beam. (a) Beam segment with bending stresses shown by blocks. (b) Shearing force transmitted through dowel. (c) For determining the force on a dowel only change in moment is needed. (d) The shearing force divided by the area of the cut yields shearing stress. (e) Horizontal cut below the flange for determining the shearing stress. (f) Vertical cut through the flange for determining the shearing stress.

of the flange in Fig. 6-3(f). This permits calculation of shearing stresses lying in a vertical plane in the figure.

Before finally proceeding with the development of equations for determining the shearing stresses in connecting bolts and in beams, an intuitively evident example is worthy of note. Consider a wooden plank placed on top of another as in Figs. 6-4(a) and (b). If these planks act as a

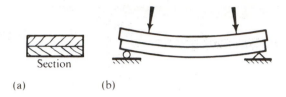

Fig. 6-4. Separate planks not fastened together slide on each other when loaded

beam and are not interconnected, sliding at the surfaces of their contact will take place. The tendency for this sliding may be visualized by considering the two loaded planks shown in Fig. 6-4(b). The interconnection of these planks with nails or glue is necessary to make them act as an integral beam. In the next article an equation will be derived for determining the required interconnection between the component parts of a beam to make them act as a unit. In the following article this equation will be modified to yield shearing stresses in initially solid beams.

6-3. SHEAR FLOW

Consider a beam made from several continuous planks whose cross section is shown in Fig. 6-5(a). For simplicity the beam has a rectangular cross section, but such a limitation is not necessary. To make this beam act as an integral member, it is assumed that the planks are fastened at intervals by vertical bolts. An element of this beam isolated by two parallel sections, both of which are perpendicular to the axis of the beam, is shown in Fig. 6-5(b).

If the element shown in Fig. 6-5(b) is subjected to a bending moment $+M_A$ at end A and to $+M_B$ at end B, bending stresses which act normal to the sections are developed. These bending stresses vary linearly from their respective neutral axes, and at any point at a distance y from the neutral axis are $-M_B y/I$ on the B end, and $-M_A y/I$ on the A end.

From the beam element, Fig. 6-5(b), isolate the top plank as in Fig. 6-5(c). The fibers of this plank nearest the neutral axis are located by the distance y_1. Then, since stress times area is equal to a force, the forces acting perpendicular to the ends A and B of this plank may be determined. At the end B the force acting on an infinitesimal area dA at a distance y from the neutral axis is $(-M_B y/I)dA$. The total force acting on the area $fghj$ (A_{fghj}) is the sum, or the integral, of these elementary forces over this area. Denoting the total force acting normal to the area $fghj$ by F_B and remembering that, at a section, M_B and I are constants, one obtains the following relation:

$$F_B = \int_{\substack{\text{area} \\ fghj}} -\frac{M_B y}{I}\, dA = -\frac{M_B}{I} \int_{\substack{\text{area} \\ fghj}} y\, dA = -\frac{M_B Q}{I} \qquad (6\text{-}2)$$

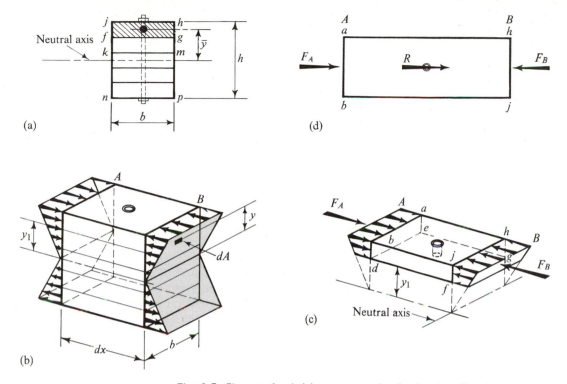

(a)

(d)

(b)

(c)

Fig. 6-5. Elements for deriving an expression for the shear flow in a beam

where

$$Q = \int_{\substack{\text{area} \\ fghj}} y\, dA = A_{fghj}\bar{y} \tag{6-3}$$

The integral defining Q is the first or the statical moment of area *fghj* around the neutral axis. By definition $\bar{y}$ is the distance from the neutral axis to the centroid of A_{fghj}.* Illustrations of the manner of determining Q are in Fig. 6-6. Equation 6-2 provides a convenient means of calculating the longitudinal force acting normal to any selected part of the cross-sectional area.

Next consider end A of the element in Fig. 6-5. One can then express the total force acting normal to the area *abde* as

$$F_A = -\frac{M_A}{I}\int_{\substack{\text{area} \\ abde}} y\, dA = -\frac{M_A Q}{I} \tag{6-4}$$

where the meaning of Q is the same as that in Eq. 6-2 since for prismatic beams an area such as *fghj* is equal to the area *abde*. Hence if the moments at A and B were equal, it would follow that $F_A = F_B$, and the bolt shown in

*Area *fgpn* and its $\bar{y}$ may be used for finding $|Q|$.

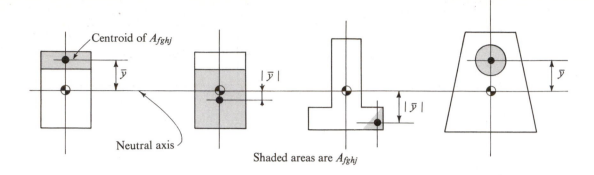

Centroid of A_{fghj}

$\bar{y}$

Neutral axis

$|\bar{y}|$

$|\bar{y}|$

$|\bar{y}|$

$\bar{y}$

Shaded areas are A_{fghj}

Fig. 6-6. Meaning of terms for finding $|Q|$

the figure would perform a nominal function of keeping the planks together and would not be needed to resist any known forces.

On the other hand, if M_A is not equal to M_B, which is always the case when shears are present at the adjoining sections, F_A is not equal to F_B. More push (or pull) develops on one end of a "plank" than on the other, as different normal stresses act on the section from the two sides. Thus if $M_A \neq M_B$, equilibrium of the horizontal forces in Fig. 6-5(c) may be attained only by developing a horizontal resisting force R in the bolt. If $M_B > M_A$, then $|F_B| > |F_A|$, and $|F_A| + R = |F_B|$, Fig. 6-5(d). The force $|F_B| - |F_A| = R$ tends to shear the bolt in the plane of the plank *edfg*.* If the shearing force acting across the bolt at the level *km* (Fig. 6-5(a)) were to be investigated, the two upper planks should be considered as one unit.

·If $M_A \neq M_B$ and the element of the beam is only dx long, the bending moments on the adjoining sections change by an infinitesimal amount. Thus if the bending moment at A is M_A, the bending moment at B is $M_B = M_A + dM$. Likewise, in the same distance dx the longitudinal forces F_A and F_B change by an infinitesimal force dF, i.e., $|F_B| - |F_A| = dF$. By substituting these relations into the expression for F_B and F_A found above, with areas *fghj* and *abde* taken equal, one obtains an expression for the differential longitudinal push (or pull) dF:

$$dF = |F_B| - |F_A| = \left(\frac{M_A + dM}{I}\right)Q - \left(\frac{M_A}{I}\right)Q = \frac{dM}{I}Q$$

In the final expression for dF the actual bending moments at the adjoining sections are eliminated. Only the difference in the bending moments dM at the adjoining sections remains in the equation.

Instead of working with a force dF which is developed in a distance

---

*The forces $(|F_B| - |F_A|)$ and R are not collinear, but the element shown in Fig. 6-5(c) is in equilibrium. To avoid ambiguity, shearing forces acting in the vertical cuts are omitted from the diagram.

dx, it is more significant to obtain a similar force per unit of beam length. This quantity is obtained by dividing *dF* by *dx*. Physically this quantity represents the difference between F_B and F_A for an element of the beam of unit length. The quantity *dF/dx* will be designated by *q* and will be referred to as the *shear flow*. Since force is measured in newtons or pounds, shear flow *q* has units of newtons per meter or pounds per inch. Then, recalling that *dM/dx = V*, one obtains the following expression for the shear flow in beams:

$$q = \frac{dF}{dx} = \frac{dM}{dx}\frac{1}{I}\int_{\substack{area \\ fghj}} y \, dA = \frac{V A_{fghj}\,\bar{y}}{I} = \frac{VQ}{I} \qquad (6\text{-}5)$$

In this equation *I* stands for the moment of intertia of the entire cross-sectional area around the neutral axis, just as it does in the flexure formula from which it came. The total shearing force at the section investigated is represented by *V*, and the integral of *y dA* for determining *Q* extends only over the cross-sectional area of the beam to one side of this area at which *q* is investigated.

In retrospect, note carefully that Eq. 6-5 was derived on the basis of the elastic flexure formula, but no term for a bending moment appears in the final expressions. This resulted from the fact that only the change in the bending moments at the adjoining sections had to be considered, and the latter quantity is linked with the shear *V*. The shear *V* was substituted for *dM/dx*, and this masks the origin of the established relations. Equation 6-5 is very useful in determining the necessary interconnection between the elements making up a beam. This will be illustrated by examples.

EXAMPLE 6-1

Two long wooden planks form a T section of a beam as shown in Fig. 6-7(a). If this beam transmits a constant vertical shear of 3 000 N, find the necessary spacing of the nails between the two planks to make the beam act as a unit. Assume that the allowable shearing force per nail is 700 N.

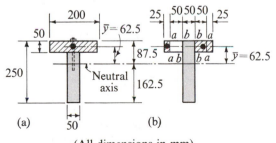

(All dimensions in mm)

Fig. 6-7.

SOLUTION

In attacking such problems the analyst must ask: What part of a beam has a tendency to slide longitudinally from the remainder? Here it is the plane of contact of the two planks; Eq. 6-5 must be applied to determine the shear flow in this plane. To do this, the neutral axis of the whole section and its moment of inertia around the neutral axis must be found. Then as V is known and Q is defined as the statical moment of the area of the upper plank around the neutral axis, q may be determined. The distance y_c from the top to the neutral axis is

$$y_c = \frac{50(200)25 + 50(200)150}{50(200) + 50(200)} = 87.5 \text{ mm}$$

$$I = \frac{200(50)^3}{12} + (50)200(62.5)^2 + \frac{50(200)^3}{12} + (50)200(62.5)^2$$

$$= 113.5 \times 10^6 \text{ mm}^4 = 113.5 \times 10^{-6} \text{ m}^4$$

$$Q = A_{fghj}\bar{y} = (50)200(87.5 - 25) = 625 \times 10^3 \text{ mm}^3 = 625 \times 10^{-6} \text{ m}^3$$

$$q = \frac{VQ}{I} = \frac{(3\,000)(625 \times 10^{-6})}{113.5 \times 10^{-6}} = 16\,500 \text{ N/m}$$

Thus a force of 16 500 N must be transferred from one plank to the other in every linear meter along the length of the beam. However, from the data given, each nail is capable of resisting a force of 700 N, hence one nail can take care of $700/(16\,500) = 0.043$ linear meter along the length of the beam. As shear remains constant at the consecutive sections of the beam, the nails should be spaced throughout at 43 mm intervals. In a practical problem a 40 mm spacing would probably be used.

SOLUTION FOR AN ALTERNATE ARRANGEMENT OF PLANKS

If, instead of using the two planks as above, a beam of the same cross section were made from five pieces, Fig. 6-7(b), a different nailing schedule would be required.

To begin, the shear flow between one of the 25 mm by 50 mm pieces and the remainder of the beam is found, and although the contact surface a-a is vertical, the procedure is the same. The push or pull on an element is built up in the same manner:

$$Q = A_{fghj}\bar{y} = (25)50(62.5) = 78 \times 10^3 \text{ mm}^3 = 78 \times 10^{-6} \text{ m}^3$$

$$q = \frac{VQ}{I} = \frac{(3\,000)(78 \times 10^{-6})}{113.5 \times 10^{-6}} = 2\,060 \text{ N/m}$$

If the same nails as before are used to join the 25 mm by 50 mm piece to the 50 mm by 50 mm piece they may be $700/(2\,060) = 0.340$ m apart. This nailing applies to both sections a-a.

To determine the shear flow between the 50 mm by 250 mm vertical piece and either one of the 50 mm by 50 mm pieces, the whole 75 mm by 50 mm area must be used to determine Q. It is the difference of pushes (or pulls) on this whole area that causes the unbalanced force that must be transferred at the surface b-b:

$$Q = A_{fghj}\bar{y} = (75)50(62.5) = 234 \times 10^3 \text{ mm}^3 = 234 \times 10^{-6} \text{ m}^3$$

$$q = \frac{VQ}{I} = \frac{(3\,000)(234 \times 10^{-6})}{113.5 \times 10^{-6}} = 6\,190 \text{ N/m}$$

Nails should be spaced at $700/(6\,190) = 0.113$ m, or, in practice, at 0.1 m intervals along the length of the beam in both sections b-b. These nails could be driven in first, then the 25 mm by 50 mm pieces put on.

EXAMPLE 6-2

A simple beam on a 6 m span carries a load of 3 kN/m including its own weight. The beam's cross section is to be made from several full-sized wooden pieces as is shown in Fig. 6-8(a). Specify the spacing of the 10 mm lag screws

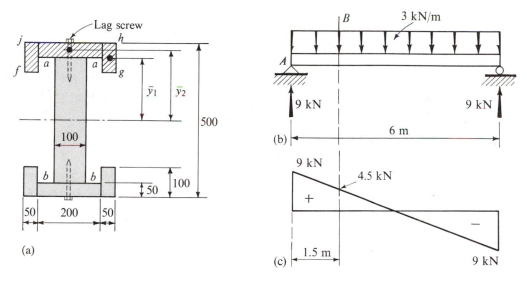

Fig. 6-8. (All dimensions in mm)

shown which is necessary to fasten this beam together. Assume that one 10 mm lag screw, as determined by laboratory tests, is good for 2 kN when transmitting lateral load parallel to the grain of the wood. For the *entire* section, I is equal to 2.36×10^{-3} m⁴.

SOLUTION

To find the spacing of the lag screws, the shear flow at section a-a must be determined. The loading on the given beam is shown in Fig. 6-8(b), and to show the variation of the shear along the beam, the shear diagram is constructed in Fig. 6-8(c). Next, to apply the shear flow formula, $\int_{\substack{area \\ fghj}} y\,dA = Q$ must be determined. This is done by considering the *shaded* area to one side of the cut a-a in Fig. 6-8(a). The statical moment of this area is most conveniently computed by multiplying the area of the *two* 50 mm by 100 mm pieces by the distance from their centroid to the neutral axis of the beam and adding to this product a similar quantity for the 50 mm by 200 mm piece. The largest shear flow occurs at the supports, as the largest vertical shears V of 9 kN act there:

$$Q = A_{fghj}\bar{y} = 2A_1\bar{y}_1 + A_2\bar{y}_2$$
$$= 2(50)100(200) + 50(200)225 = 4.25 \times 10^6 \text{ mm}^3 = 4.25 \times 10^{-3} \text{ m}^3$$
$$q = \frac{VQ}{I} = \frac{9(4.25 \times 10^{-3})}{2.36 \times 10^{-3}} = 16.2 \text{ kN/m}$$

At the supports the spacing of the lag screws must be $2/16.2 = 0.123$ m apart. This spacing of the lag screws applies only at a section where the shear V is equal to 9 kN. Similar calculations for a section where $V = 4.5$ kN gives $q = 8.1$ kN/m; and the spacing of the lag screws becomes $2/8.1 = 0.246$ m. Thus it is proper to specify the use of 10 mm lag screws at 120 mm on centers for a distance of 1.5 m nearest both of the supports and 240 mm spacing of the same lag acrews for the middle half of the beam. A greater refinement in making the transition from one spacing of fastenings to another may be desirable in some problems. The *same* spacing of lag screws should be used at the section *b-b* as at the section *a-a*.

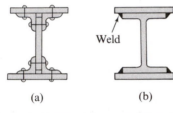

Weld

(a)   (b)

Fig. 6-9. Typical beam sections consisting of component parts: (a) plate girder, (b) I-beam reinforced with plates

In a manner analogous to the above, the spacing of rivets or bolts in composite beams made from continuous angles and plates, Fig. 6-9, may be determined. Welding requirements are established similarly. The *nominal* shearing stress in a rivet is determined by dividing the total shearing *force* transmitted by the rivet (shear flow times spacing of the rivets) by the cross-sectional area of the rivet. For a detailed analysis of rivets, bolts, and welds, see Chapter 14.

6-4. THE SHEARING STRESS FORMULA FOR BEAMS

The shearing stress formula for beams may be obtained by modifying the shear flow formula. Thus, analogous to the earlier procedure, an element of a beam may be isolated between two adjoining sections taken perpendicular to the axis of the beam. Then by passing *another imaginary section* through this element parallel to the axis of the beam, a new element is obtained, which corresponds to the element of one "plank" used in the earlier derivations. A side view of such an element is shown in Fig. 6-10(a) where the imaginary longitudinal cut is made at a distance y_1 from the neutral axis. The cross-sectional area of the beam is shown in Fig. 6-10(c).

If shearing forces exist at the sections through the beam, a different bending moment acts at section A than at B. Hence more push or pull is developed on one side of the *partial area fghj* than on the other, and, as before, this *longitudinal* force in a distance dx is

$$dF = \frac{dM}{I} \int_{\substack{\text{area} \\ fghj}} y \, dA = \frac{dM}{I} A_{fghj}\bar{y} = \frac{dM}{I} Q$$

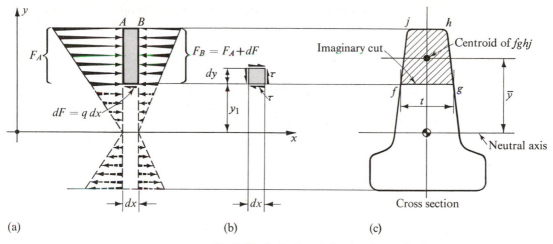

Fig. 6-10. Derivation of shearing stresses in a beam

In a *solid* beam the force resisting dF may be developed only *in the plane* of the longitudinal cut taken parallel to the axis of the beam. Therefore, assuming that the shearing stress τ is *uniformly distributed* across the cut of width t, the shearing stress in the *longitudinal plane* may be obtained by dividing dF by the area $t\,dx$. This yields the horizontal shearing stress τ. For an *infinitesimal* element, however, numerically equal shearing stresses act on the mutually perpendicular planes, Fig. 6-10(b). Hence the same relation gives *simultaneously* the longitudinal shearing stress and the shearing stress *in the plane of the vertical section at the longitudinal* cut.*

$$\tau = \frac{dF}{dx\,t} = \frac{dM}{dx}\frac{A_{fghj}\bar{y}}{It}$$

This equation may be simplified, since according to Eq. 6-1 $dM/dx = V$ and by Eq. 6-5, $q = VQ/I$. Hence

$$\tau = \frac{VA_{fghj}\bar{y}}{It} = \frac{VQ}{It} = \frac{q}{t} \tag{6-6}$$

Equation 6-6 is the important formula for the shearing stresses in a beam.† It gives the shearing stresses *at the longitudinal cut*. As before, V

---

*The appearance of $\bar{y}$ in this relation may be explained differently. If the shear is present at a section through a beam, the moments at the adjoining sections are M and $M + dM$. The magnitude of M is irrelevant for determination of the shearing stresses. Hence alternately, *no* moment need be considered at one section *if at the adjoining section a bending moment dM is assumed to act*. Then on a partial area of the section, such as the shaded area in Fig. 6-10(c), this bending moment dM will cause an *average normal stress* $(dM)\bar{y}/I$ as given by the flexure formula. In the latter relation $\bar{y}$ locates the fiber which is at an *average* distance from the neutral axis in the *partial* area of a section. Multiplying $(dM)\bar{y}/I$ by the partial area of the section leads to the same expression for dF as above.

†This formula was derived by D. I. Jouravsky in 1855. Its development was prompted by observing horizontal cracks in wood ties on several of the railroad bridges between Moscow and St. Petersburg.

is the *total* shearing force at a section, and I is the moment of inertia of the *whole* cross-sectional area about the neutral axis. Both V and I are constant at a section through a beam. Here Q is the statical moment around the neutral axis of the *partial* area of the cross section to one side of the imaginary longitudinal cut, and $\bar{y}$ is the distance from the neutral axis of the beam to the centroid of the partial area A_{fghj}. Finally, t is the width of the imaginary longitudinal cut, which is usually equal to the thickness or width of the member. The shearing stress at different longitudinal cuts through the beam assumes different values as the values of Q and t for such cuts differ.

Care must be exercised in making the longitudinal cuts preparatory for use in Eq. 6-6. The proper sectioning of some cross-sectional areas of beams is shown in Figs. 6-11(a), (b), (d), and (e). The use of inclined cutting planes

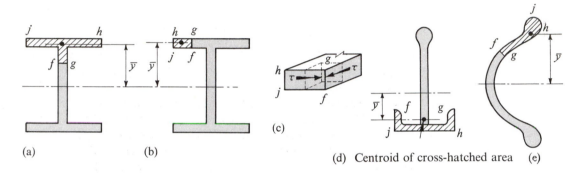

Fig. 6-11. Proper sectioning for partial areas of the cross section for computing the shearing stresses in beams

should be avoided *unless* the cut is made across a small thickness.* When the axis of symmetry of the cross-sectional area of the beam is vertical and in the plane of the applied forces, the longitudinal cuts are usually made horizontally. In such cases the solution of Eq. 6-6 gives simultaneous values of *horizontal and vertical* shearing stresses, as such planes are mutually perpendicular. The latter stresses act in the plane of the transverse section through the beam. Collectively, these shearing stresses resist the shearing force at the same section, thus satisfying the relation of statics $\sum F_y = 0$. The validity of this statement for a special case will be proved in Example 6-3.

For thin members only, Eq. 6-6 may be used to determine the shearing stresses with a cut such as *f-g* of Fig. 6-11(b). These shearing stresses act in a vertical plane and are directed perpendicularly to the plane of the paper. Matching shearing stresses act horizontally, Fig. 6-11(c). These shearing stresses act in *entirely different directions* than those obtained by making horizontal cuts, such as *f-g* in Figs. 6-11(a) and (d). As these shearing stresses

*Rigorous solutions of the problem indicate that *wide* inclined cuts through the section may lead to inconsistencies.

do not contribute directly to the resistance of the vertical shear V, their significance will be discussed in Art. 6-6.

The application of Eq. 6-6 to two *particularly important* types of cross-sectional areas of beams will now be illustrated.

EXAMPLE 6-3

Derive an expression for the shearing-stress distribution in a beam of solid rectangular cross section transmitting a vertical shear V.

SOLUTION

The cross-sectional area of the beam is shown in Fig. 6-12(a). A longitudinal cut through the beam at a distance y_1 from the neutral axis isolates the partial

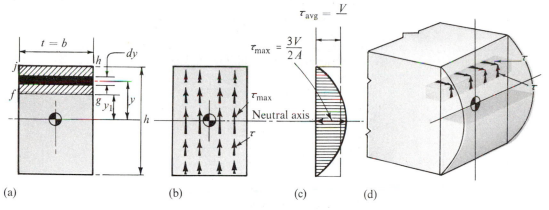

(a)   (b)   (c)   (d)

Fig. 6-12

area *fghi* of the cross section. Here $t = b$ and the infinitesimal area of the cross section may be conveniently expressed as $b\,dy$. By applying Eq. 6-6 the horizontal shearing stress is found *at* the level y_1 of the beam. *At* the same cut, numerically equal vertical shearing stresses act *in the plane of the cross section*.

$$\tau = \frac{VQ}{It} = \frac{V}{It} \int_{\substack{\text{area} \\ fghj}} y\,dA = \frac{V}{Ib} \int_{y_1}^{h/2} by\,dy$$

$$= \frac{V}{I} \left. \frac{y^2}{2} \right|_{y_1}^{h/2} = \frac{V}{2I} \left[\left(\frac{h}{2} \right)^2 - y_1^2 \right]$$

This equation shows that in a beam of rectangular cross section both the horizontal and the vertical shearing stresses vary parabolically. The maximum value of the shearing stress is obtained when y_1 is equal to zero. *In the plane of the cross section*, Fig. 6-12(b), this is diagrammatically represented by $\tau_{\max}$ at the neutral axis of the beam. At increasing distances from the neutral axis, the shearing stresses gradually diminish. At the upper and lower boundaries of the beam, the shearing stresses cease to exist as $y_1 = \pm h/2$. These values of the shearing stresses at the various levels of the beam may be represented by the parabola shown in Fig. 6-12(c). An isometric view of the beam with horizontal and vertical shearing stresses is shown in Fig. 6-12(d).

To satisfy the condition of statics $\sum F_y = 0$, at a section of the beam the sum of all the vertical shearing stresses τ times their respective areas dA must be equal to the vertical shear V. That this is the case may be shown by integrating $\tau \, dA$ over the *whole* cross-sectional area A of the beam, using the general expression for τ found above.

$$\int_A \tau \, dA = \frac{V}{2I} \int_{-h/2}^{+h/2} \left[\left(\frac{h}{2} \right)^2 - y_1^2 \right] b \, dy_1 = \frac{Vb}{2I} \left[\left(\frac{h}{2} \right)^2 y_1 - \left(\frac{y_1^3}{3} \right) \right]_{-h/2}^{+h/2}$$

$$= \frac{Vb}{\left(2\frac{bh^3}{12} \right)} \left[\left(\frac{h}{2} \right)^2 h - \frac{2}{3} \left(\frac{h}{2} \right)^3 \right] = V$$

As the derivation of Eq. 6-6 was indirect, this proof showing that the shearing stresses integrated over the section equal the vertical shear is reassuring. Moreover, since an agreement in signs is found, this result indicates that *the direction of the shearing stresses at the section through a beam is the same as that of the shearing force V*. This fact may be used to determine the sense of the shearing stresses.

As noted above, the maximum shearing stress in a rectangular beam occurs at the neutral axis, and for this case the general expression for $\tau_{\max}$ may be simplified by setting $y_1 = 0$.

$$\tau_{\max} = \frac{Vh^2}{8I} = \frac{Vh^2}{8\frac{bh^3}{12}} = \frac{3}{2} \frac{V}{bh} = \frac{3}{2} \frac{V}{A} \tag{6-7}$$

where V is the total shear and A is the *entire* cross-sectional area. The same result may be obtained more directly if it is noted that to make $VQ/(It)$ a maximum, Q must attain its largest value, as in this case V, I, and t are constants. From the property of the statical moments of areas around a centroidal axis, the maximum value of Q is obtained by considering one-half the cross-sectional area around the neutral axis of the beam. Hence, alternately,

$$\tau_{\max} = \frac{VQ}{It} = \frac{V \left(\frac{bh}{2} \right) \left(\frac{h}{4} \right)}{\left(\frac{bh^3}{12} \right) b} = \frac{3}{2} \frac{V}{A} \tag{6-7a}$$

Since beams of rectangular cross-sectional area are used frequently in practice, Eq. 6-7a is very useful. It is widely used in the design of wooden beams as the shearing strength of wood on planes parallel to the grain is small. Thus, although equal shearing stresses exist on mutually perpendicular planes, wooden beams have a tendency to split longitudinally along the neutral axis. Note that the maximum shearing stress is $1\frac{1}{2}$ times as great as the *average* shearing stress* V/A.

*Application of the expression $\tau = V/A$ is permissible only for rivets and bolts and must not be used in the design of beams.

EXAMPLE 6-4

Using the elementary theory, determine the shearing stress distribution due to shear V in the elastic-plastic zone of a rectangular beam.

SOLUTION

The situation in the problem occurs, for example, in a cantilever loaded as shown in Fig. 6-13(a). In the elastic-plastic zone, the external bending moment

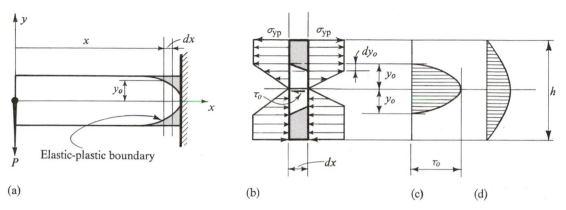

Fig. 6-13. Shearing stress distribution in a rectangular, elastic-plastic beam

$M = -Px$, whereas, according to Eq. 5-11, the internal resisting moment $M = M_p - \sigma_{yp}by_o^2/3$. Upon noting that y_o varies with x and differentiating the above equations, one notes the following equality:

$$\frac{dM}{dx} = -P = -\frac{2by_o\sigma_{yp}}{3}\frac{dy_o}{dx}$$

This relation will be needed later. First, however, proceeding as in the elastic case, consider the equilibrium of a beam element as shown in Fig. 6-13(b). Larger longitudinal forces act on the right side of this element than on the left. By separating it at the neutral axis and equating the force at the cut to the difference in the longitudinal force, one obtains

$$\tau_o\, dx\, b = \sigma_{yp}\, dy_o\, b/2$$

where b is the width of the beam. After substituting dy_o/dx from the relation found earlier and eliminating b, one finds the maximum horizontal shearing stress τ_o:

$$\tau_o = \frac{\sigma_{yp}}{2}\frac{dy_o}{dx} = \frac{3P}{4by_o} = \frac{3}{2}\frac{P}{A_o} \tag{6-8}$$

where A_o is the cross-sectional area of the elastic part of the cross section. The shearing stress distribution for the elastic-plastic case is shown in Fig. 6-13(c). This can be contrasted with that for the elastic case, shown in Fig. 6-13(d). Since equal and opposite normal stresses occur in the plastic zones, no unbalance in longitudinal forces occurs and no shearing stresses are developed.

This elementary solution has been refined by using a more carefully formulated criterion of yielding caused by the simultaneous action of normal and shearing stresses.*

EXAMPLE 6-5

An I-beam is loaded as in Fig. 6-14(a). If it has the cross section shown in Fig. 6-14(c), determine the shearing stresses at the levels indicated. Neglect the weight of the beam.

SOLUTION

A free-body diagram of a segment of the beam is in Fig. 6-14(b). It is seen from this diagram that the vertical shear at every section is 50 kips. Bending

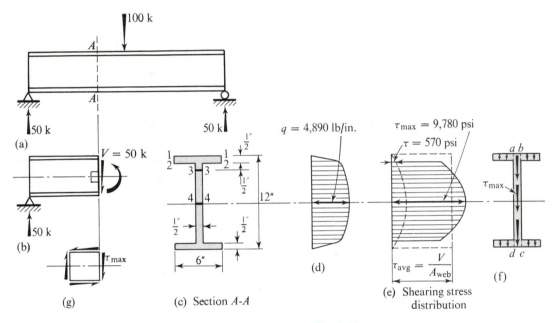

Fig. 6-14

moments do not enter directly into the present problem. The shear flow at the various levels of the beam is computed in the following table using Eq. 6-5. Since $\tau = q/t$ (Eq. 6-6), the shearing stresses are obtained by dividing the shear flows by the respective widths of the beam.

$$I = \frac{6(12)^3}{12} - \frac{(5.5)(11)^3}{12} = 254 \text{ in.}^4$$

For use in Eq. 6-5 the ratio $V/I = 50,000/254 = 197$ lb/in.^{4}

*D. C. Drucker, "The Effect of Shear on the Plastic Bending of Beams," *Journal of Applied Mechanics*, 1956, vol. 23, pp. 509–14.

Level	$A_{fghj}{}^a$	$\bar{y}{}^b$	$Q = A_{fghj}\bar{y}$	$q = VQ/I$	t	τ, psi
1-1	0	6	0	0	6.0	0
2-2	$(0.5)6 = 3.00$	5.75	17.25	3,400	6.0 0.5	570 6,800
3-3	$(0.5)6 = 3.00$ $(0.5)(0.5) = 0.25$	5.75 5.25	$\left.\begin{array}{l}17.25\\1.31\end{array}\right\}18.56$	3,650	0.5	7,300
4-4	$(0.5)6 = 3.00$ $(0.5)(5.5) = 2.75$	5.75 2.75	$\left.\begin{array}{l}17.25\\7.56\end{array}\right\}24.81$	4,890	0.5	9,780

$^a A_{fghj}$ is the partial area of the cross section above a given level in square inches.
$^b \bar{y}$ is the distance from the neutral axis to the centroid of the partial area in inches.

The positive signs of τ show that, for the section considered, the stresses act downward on the right face of the elements. The sense of the shearing stresses acting on the section coincides with the sense of the shearing force V. For this reason a strict adherence to the sign convention is often unnecessary. It is always true that $\int_A \tau \, dA$ is equal to V and has the same sense.

Note that at the level 2-2 two widths are used to determine the shearing stress—one just above the line 2-2, and one just below. A width of 6 in. corresponds to the first case, and 0.5 in. to the second. This transition point will be discussed in the next article. The results obtained, which by virtue of symmetry are also applicable to the lower half of the section, are plotted in Fig. 6-14(d) and (e). By a method similar to the one used in the preceding example, it may be shown that the curves in Fig. 6-14(e) are parts of a second-degree parabola.

The variation of the shearing stress indicated by Fig. 6-14(e) may be interpreted as is shown in Fig. 6-14(f). The maximum shearing stress occurs at the neutral axis; the vertical shearing stresses throughout the web of the beam are nearly of the same magnitude. The shearing stresses occurring in the flanges are very small. For this reason the maximum shearing stress in an I-beam is often approximated by dividing the total shear V by the cross-sectional area of the web (area $abcd$ in Fig. 6-14(f)). Hence

$$(\tau_{max})_{approx} = V/A_{web} \qquad (6\text{-}9)$$

In the example considered this gives

$$(\tau_{max})_{approx} = \frac{50,000}{(0.5)(12)} = 8,330 \text{ psi}$$

This stress differs by about 15 per cent from the one found by the accurate formula. For most cross sections a much closer approximation to the true maximum shearing stress may be obtained by dividing the shear by the web area between the flanges only. For the above example this procedure gives a stress of 9,091 psi, which is an error of only about 8 per cent. It should be clear from the above that division of V by the whole cross-sectional area of the beam to obtain the shearing stress is not permissible.

An element of the beam at the neutral axis is shown in Fig. 6-14(g). At levels 3-3 and 2-2, bending stresses, in addition to the shearing stresses, act on the vertical faces of the elements. No shearing stresses and only bending stresses act on the elements at level 1-1.

The maximum shearing stress was found to be at the neutral axis in both of the above examples. This is not always so. For example, if the sides of the cross-sectional area are not parallel, as for a triangular section, τ is a function of Q and t, and the maximum shearing stress occurs midway between the apex and the base, which does not coincide with the neutral axis.

The same procedure as above applies to the investigation of the longitudinal joints in beams in which such contact surfaces are glued. A knowledge of the shearing stresses in the glued joint aids in the selection of an adhesive of proper strength. If beams of two materials are investigated, transformed sections can be used to obtain a solution. In reinforced concrete beams, concrete between the neutral axis and the reinforcing steel, as well as that above the neutral axis, is assumed to resist the shearing stresses.

*6-5. LIMITATIONS OF THE SHEARING STRESS FORMULA

The shearing stress formula for beams is based on the flexure formula. Hence all of the limitations imposed on the flexure formula apply. The material is assumed to be elastic with the same elastic modulus in tension as in compression. The theory developed applies only to straight beams. Moreover, there are additional limitations which are not present in the flexure formula. Some of these will be discussed now.

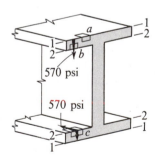

Fig. 6-15. Boundary conditions are not satisfied by the flange elements at the levels 2-2

Consider a section through the I-beam analyzed in Example 6-5. Some of the results of this analysis are reproduced in Fig. 6-15. The shearing stresses computed earlier for the level 1-1 apply to the infinitesimal element a. The vertical shearing stress is zero for this element. Likewise, *no* shearing stresses exist *on* the top plane of the beam. This is as it should be, since the top *surface* of the beam is a *free* surface. In mathematical phraseology this means that the conditions at the boundary are satisfied. For beams of rectangular cross section the situation at the boundaries is correct.

A different situation is discovered when the shearing stresses determined for the I-beam at the levels 2-2 are scrutinized. The shearing stresses were found to be 570 psi for the elements such as b or c shown in the figure. This requires matching horizontal shearing stresses on the inner planes of the flanges. However, the latter planes *must be free* of the shearing stresses as they are *free boundaries* of the beam. This leads to a contradiction which cannot be resolved by the methods of mechanics of materials. The more advanced techniques of the mathematical theory of elasticity must be used to obtain a correct solution.

Fortunately, the above defect of the shearing stress formula for beams is not too serious. The shearing stresses in the flanges which were considered are small.* The significant shearing stresses occur in the web and, for all practical purposes, are correctly given by Eq. 6-6. *No appreciable error is involved by using the relations derived in this chapter for thin-walled members*, and the majority of beams belong to this group. Moreover, as stated earlier, the solution for the shearing stresses for a beam with a rectangular cross section is satisfactory.

In mechanical applications circular shafts frequently act as beams. Hence beams having a solid circular cross section form an important class. These beams are not "thin-walled." An examination of the boundary conditions for circular members, Fig. 6-16(a), leads to the conclusion that when shearing stresses are present they must act parallel to the boundary. As no matching shearing stress can exist *on* the free surface of the beam, no shearing stress component can act normal to the boundary. However, according to Eq. 6-6, *vertical* shearing stresses of *equal* intensity act at every level such as *ac* in Fig. 6-16(b). This is incompatible with the boundary conditions for the elements as *a* and *c* at the boundary, and the solution indicated by Eq. 6-6 is inconsistent.† Fortunately, the *maximum* shearing stresses occurring at the neutral axis satisfy the boundary conditions and are very near their true value (within about 5%).‡

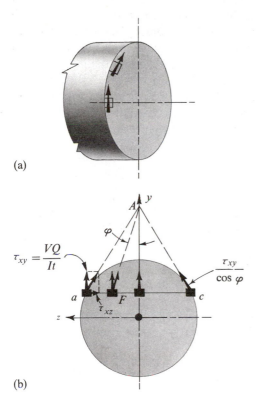

(a)

(b)

Fig. 6-16. Modification of a solution based on the shearing stress formula to satisfy the boundary conditions

---

*Other shearing stresses will be discussed in the next article.

†The exact solution of this problem is beyond the scope of this text. However, a better approximation of the true stresses may be obtained rather simply. First an assumption is made that the shearing stress as found by Eq. 6-6 gives a true *component* of the shearing stress acting in the *vertical direction*. Then, since at every level the shearing stresses at the boundary must act tangent to the boundary, the lines of action of these shearing stresses intersect at some point as *A* in Fig. 6-16(b). Thus a second assumption is made that all shearing stresses at a given level act in a direction toward a single point as *A* in Fig. 6-16(b). Whence the shearing stress at any point such as *F* becomes equal to $\tau_{xy}/\cos \phi$. The stress system found in the above manner is consistent.

‡A.E.H. Love, *Mathematical Theory of Elasticity* (4th ed.), New York: Dover, 1944, p. 348.

*6-6. FURTHER REMARKS ON THE DISTRIBUTION OF THE SHEARING STRESSES

In an I-beam the existence of shearing stresses lying in a longitudinal cut as *c-c* in Fig. 6-17(a) was indicated in Art. 6-4. These shearing stresses act perpendicular to the plane of the paper. Their magnitude may be found by applying Eq. 6-6, and their sense follows by considering the bending moments at the adjoining sections through the beam. For example, if for the beam shown in Fig. 6-17(b) *positive* bending moments increase toward the reader, larger *normal* forces act on the *nearer* cross section. For the elements shown, $\tau t\,dx$ or $q\,dx$ must aid the smaller force acting on the partial area of the cross section. This fixes the sense of the shearing stresses in the longitudinal cuts. However, numerically equal shearing stresses act on the mutually perpendicular planes of an *infinitesimal* element, and the shearing stresses on such planes either meet or part with their directional arrowheads at

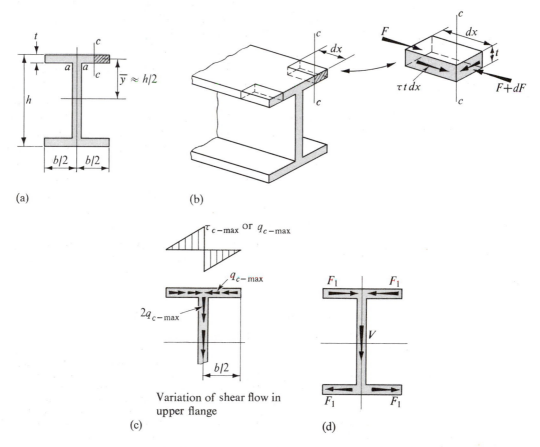

(a)　　　　(b)

(c) Variation of shear flow in upper flange

(d)

Fig. 6-17. Existence of shearing forces in the flange of an I-beam which act perpendicularly to the axis of symmetry

a corner. Hence the sense of the shearing stresses in the plane of the cross section becomes known also.

The magnitude of the shearing stresses varies for the different vertical cuts. For example, if the cut c-c in Fig. 6-17(a) is at the edge of the beam, the *shaded* area of the beam's cross section is zero. However, if the thickness of the flange is constant, and the cut c-c is made progressively closer to the web, the shaded area increases from zero at a linear rate. Moreover, as $\bar{y}$ remains constant for any such area, Q also increases linearly from zero toward the web. Therefore, since V and I are constant at any section through the beam, the shear flow $q_c = VQ/I$ follows the same variation. If the thickness of the flange remains the same, the shearing stress $\tau_c = VQ/It$ varies similarly. The same variation of q_c and τ_c applies on both sides of the axis of symmetry of the cross section. However, as may be seen from Fig. 6-17(b), these quantities in the plane of the cross section act in *opposite* directions on the two sides. The variation of these shearing stresses or shear flows is represented in Fig. 6-17(c), where for simplification it is assumed that the web is very thin.

In common with all stresses, the shearing stresses shown in Fig. 6-17(c), when integrated over the area on which they act, are equivalent to a force. The magnitude of the horizontal force F_1 for *one-half* of the flange, Fig. 6-17(d), is equal to the *average* shearing stress multiplied by *one-half of the whole area of the flange*, i.e.,

$$F_1 = \left(\frac{\tau_{c\text{-max}}}{2}\right)\left(\frac{bt}{2}\right) \quad \text{or} \quad F_1 = \left(\frac{q_{c\text{-max}}}{2}\right)\left(\frac{b}{2}\right)$$

If an I-beam transmits a vertical shear, these horizontal forces act in the upper and lower flanges. However, because of the *symmetry* of the cross section, these equal forces occur in pairs and *oppose* each other, and cause no apparent external effect.

To determine the shear flow at the juncture of the flange and the web (cut a-a in Fig. 6-17(a)), the *whole* area of the flange times $\bar{y}$ must be used in computing the value of Q. However, since in finding $q_{c\text{-max}}$ *one-half* the flange area times the same $\bar{y}$ has already been used, the *sum* of the *two horizontal shear flows* coming in from opposite sides gives the *vertical* shear flow* at the cut a-a. Hence, figuratively speaking, the horizontal shear flows "turn through 90° and merge to become the vertical shear flow." Thence the shear flows at the various horizontal cuts through the web may be determined in the manner explained in the preceding articles. Moreover, as the resistance to the vertical shear V in thin-walled I-beams is developed mainly in the web, it is so shown in Fig. 6-17(d). The sense of the shearing stresses and shear flows *in the web* coincides with the direction of the shear V. Note that the vertical shear flow "splits" upon reaching the lower flange.

*The same statement can*not* be made with regard to the shearing stresses as the thickness of the flange may differ from that of the web.

This is represented in Fig. 6-17(d) by the two forces F_1 which are the result of the horizontal shear flows in the flanges.

The shearing forces which act at a section of an I-beam are shown in Fig. 6-17(d), and, for equilibrium, *the applied vertical forces must act through the centroid of the cross-sectional area* to be coincident with V. If the forces are so applied, *no torsion* of the member will occur. This is true for all sections having cross-sectional areas with an axis of symmetry. To avoid torsion of such members, the applied forces must act in the plane of symmetry of the cross section and the axis of the beam. A beam with an unsymmetrical section will be discussed next.

*6-7. SHEAR CENTER

Consider a beam whose cross section is a channel, Fig. 6-18(a). The walls of this channel are assumed to be so thin that all computations may be based on the dimensions to the *center line* of the walls. Bending of this channel takes place around the horizontal axis and although this cross section does not have a vertical axis of symmetry, it will be *assumed* that the bending stresses are given by the usual flexure formula. Assuming further that this channel resists a vertical shear, the bending moments will vary from one section through the beam to another.

By taking an arbitrary cut as c-c in Fig. 6-18(a), q and τ may be found in the usual manner. Along the horizontal legs of the channel, these quantities vary linearly from the free edge, just as they do for one side of the flange in an I-beam. The variation of q and τ is parabolic along the web. The variation of these quantities is shown in Fig. 6-18(b), where they are plotted along the center line of the channel's section.

The *average* shearing stress $\tau_a/2$ multiplied by the area of the flange gives a force $F_1 = (\tau_a/2)bt$, and the sum of the vertical shearing stresses over the area of the web is the shear $V = \int_{-h/2}^{+h/2} \tau t \, dy$.* These shearing forces acting in the plane of the cross section are shown in Fig. 6-18(c) and indicate that a force V *and a couple* $F_1 h$ are developed at the section through the channel. Physically there is a tendency for the channel to twist around some longitudinal axis. To prevent twisting and thus maintain the applicability of the initially assumed bending stress distribution, the externally applied forces must be applied in such a manner as to *balance the internal couple* $F_1 h$. For example, consider the segment of a cantilever beam of negligible weight, shown in Fig. 6-18(d), to which a vertical force P is applied parallel to the web at a distance e from the web's *center line*. To maintain this applied force in equilibrium, an *equal and opposite* shearing force V must be developed in the web.

---

*When the thickness of a channel is variable, it is more convenient to find F_1 and V by using the respective shear flows, i.e., $F_1 = (q_a/2)b$ and $V = \int_{-h/2}^{+h/2} q \, dy$. Since the flanges are thin, the vertical shearing force carried by them is negligible.

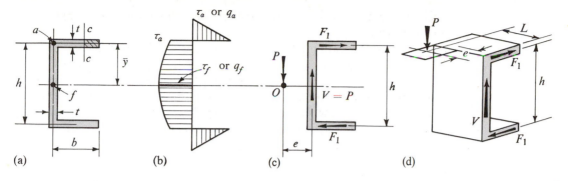

Fig. 6-18. Deriving the location of the shear center for a channel

Likewise, to cause *no twisting of the channel*, the couple *Pe* must *equal* the couple F_1h. At the same section through the channel, the bending moment *PL* is resisted by the *usual* flexural stresses (these are not shown in the figure).

An expression for the distance *e*, locating the plane in which the force *P* must be applied so as to cause *no twist* in the channel, may now be obtained. Thus, remembering that $F_1h = Pe$ and $P = V$,

$$e = \frac{F_1h}{P} = \frac{(1/2)\tau_a bth}{P} = \frac{bth}{2P}\frac{VQ}{It} = \frac{bth}{2P}\frac{Vbt(h/2)}{It} = \frac{b^2h^2t}{4I} \qquad (6\text{-}10)$$

Note that the distance *e* is independent of the magnitude of the applied force *P*, as well as of its location along the beam. The distance *e* is a property of a section and is measured outward from the *center* of the web to the applied force.

A similar investigation may be made to locate the plane in which the horizontal forces must be applied so as to cause no twist in the channel. However, for the channel considered, by virtue of symmetry, it may be seen that this plane coincides with the neutral plane of the former case. The intersection of these two mutually perpendicular planes with the plane of the cross section locates a point which is called the *shear center*. The shear center is designated by the letter *O* in Fig. 6-18(c). The shear center for any cross section lies on a longitudinal line parallel to the axis of the beam. *Any transverse force* applied through the shear center causes no torsion of the beam.* A detailed investigation of this problem shows that when a member of any cross-sectional area *is* twisted, the twist takes place around the shear center, which remains fixed. For this reason, the shear center is sometimes called the *center of twist*.

For cross-sectional areas having one axis of symmetry, the shear center is always located on the axis of symmetry. For those which have two axes of symmetry, the shear center coincides with the centroid of the cross-

---

*See Art. 7-6 for the method of analysis of the lateral forces which do not lie in a plane parallel to one of the principal planes of the cross-sectional area.

sectional area. This is the case for the I-beam that was considered in the previous article.

The exact location of the shear center for unsymmetrical cross sections of thick material is difficult to obtain and is known only in a few cases. If the material is *thin*, as has been assumed in the preceding discussion, relatively simple procedures may always be devised to locate the shear center of the cross section. The usual method consists of determining the shearing forces, as F_1 and V above, at a section, and then finding the location of the external force necessary to keep these forces in equilibrium.

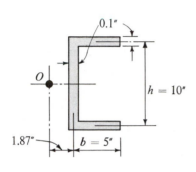

Fig. 6-19

EXAMPLE 6-6

Find the approximate location of the shear center for a beam with the cross section of the channel shown in Fig. 6-19.

SOLUTION

Instead of using Eq. 6-10 directly, some further simplifications may be made. The moment of inertia of a thin-walled channel around its neutral axis may be found with sufficient accuracy by neglecting the moment of inertia of the flanges *around their own axes* (only!). This expression for I may then be substituted into Eq. 6-10 and, after simplifications, a formula for e of channels is obtained.

$$I \approx I_{\text{web}} + (Ad^2)_{\text{flanges}} = \frac{th^3}{12} + 2bt\left(\frac{h}{2}\right)^2 = \frac{th^3}{12} + \frac{bth^2}{2}$$

$$e = \frac{b^2h^2t}{4I} = \frac{b^2h^2t}{4\left(\frac{bth^2}{2} + \frac{th^3}{12}\right)} = \frac{b}{2 + \frac{h}{3b}} \tag{6-10a}$$

Equation 6-10a shows that when the width of flanges b is very large, e approaches its maximum value of $b/2$. When h is very large, e approaches its minimum value of zero. Otherwise, e assumes an intermediate value between these two limits. For the numerical data given in Fig. 6-19,

$$e = \frac{5}{2 + \frac{10}{3(5)}} = 1.87 \text{ in.}$$

Hence the shear center O is $1.87 - 0.05 = 1.82$ in. from the outside vertical face of the channel. The answer would not be improved if Eq. 6-10 were used in the calculations.

EXAMPLE 6-7

Find the approximate location of the shear center for the cross section of the I-beam shown in Fig. 6-20(a). Note that the flanges are unequal.

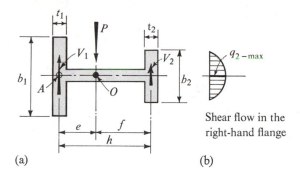

Shear flow in the
right-hand flange

(a)   (b)

Fig. 6-20

SOLUTION

This cross section has a horizontal axis of symmetry, and the shear center is located on it; where it is located remains to be answered. The applied force P causes significant bending and shearing stresses *only in the flanges*, and the contribution of the web to the resistance of the applied force P is negligible.

Let the shearing force resisted by the left-hand flange of the beam be V_1, and by the right-hand flange, V_2. For equilibrium, $V_1 + V_2 = P$. Likewise, to have no twist of the section, from $\sum M_A = 0$, $Pe = V_2h$ (or $Pf = V_1h$). Thus only V_2 remains to be determined to solve the problem. This may be done by noting that the right-hand flange is actually an ordinary rectangular beam. The shearing stress (or shear flow) in such a beam is distributed parabolically, Fig. 6-20(b), and since the area of a parabola is two-thirds of the base times the maximum altitude, $V_2 = \frac{2}{3}b_2(q_2)_{max}$. However, since the total shear $V = P$, by Eq. 6-5 $(q_2)_{max} = VQ/I = PQ/I$, where Q is the statical moment of the *upper half of the right-hand flange* and I is the moment of inertia of the *whole* section. Hence

$$Pe = V_2h = \frac{2}{3}b_2(q_2)_{max}h = \frac{\frac{2}{3}hb_2PQ}{I}$$

$$e = \frac{2hb_2}{3I}Q = \frac{2hb_2}{3I}\frac{b_2t_2}{2}\frac{b_2}{4} = \frac{h}{I}\frac{t_2b_2^3}{12} = \frac{hI_2}{I} \qquad (6\text{-}11)$$

where I_2 is the moment of inertia of the *right-hand flange* around the neutral axis. Similarly, it may be shown that $f = hI_1/I$, where I_1 applies to the *left-hand flange*. If the web of the beam is thin, as originally assumed, $I \approx I_1 + I_2$, and $e + f = h$, as is to be expected.

A similar analysis leads to the conclusion that the shear center for a symmetrical angle is located at the intersection of the center lines of its legs, as shown in Figs. 6-21(a) and (b). This follows since the shear flow at every section, as *c-c*, is directed along the center line of a leg. These shear flows yield two identical forces F_1 in the legs. The vertical components of these forces equal the vertical shear applied through O. An analogous situation is

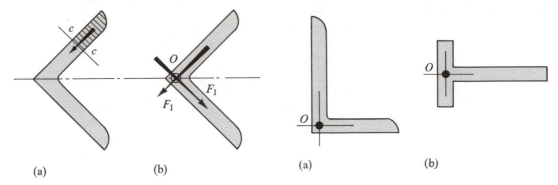

Fig. 6-21. Shear center for a symmetrical angle (equal legs) located by O.

Fig. 6-22. Shear center for the secitons shown located by O.

also found for any angle or T-section as shown in Figs. 6-22(a) and (b). The location of the shear center for various members is particularly important in aircraft applications.*

PROBLEMS FOR SOLUTION

6-1. Assuming that a member consists of five 2 in. by 6 in. full-sized wooden planks bolted together, as shown in the cross-sectional view in Fig. 6-5(a), show that $A_{fghj}\bar{y}_1 = A_{fgpn}\bar{y}_2$, where $\bar{y}_1$ is the distance from the centroid of the whole area to the centroid of area A_{fghj}, and $\bar{y}_2$ the corresponding distance for the area A_{fgpn}.

6-2. A cantilever 3 m long is fabricated from five full-sized 50 mm by 150 mm wooden planks. Its cross section is similar to the one shown in Fig. 6-5(a). The planks are fastened together by 20 mm diameter vertical bolts spaced 120 mm apart. This beam carries a uniformly distributed load including its own weight of 3.0 kN/m. Find the shearing stresses in a bolt which is located 1.5 m from the support. Make this investigation at all four planes of contact of the planks.

6-3. A hollow wooden beam is to be made from full-sized planks as shown in the figure. At the section considered, the total vertical shear in

the beam will be 930 lb and the bending moment 50 ft-lb. The nailing is to be done with 16 d (16 penny) box nails that are good for 50 lb each in shear. What must the spacing of nails be? *Ans:* 2 in.

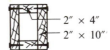

PROB. 6 – 3

6-4. A 15-ft wooden beam overhangs 10 ft and carries a concentrated force $P = 950$ lb at the end. (See figure). The beam is made up of full-sized, 2 in. thick boards nailed together with nails which have a shear resistance of 96 lb each. The moment of inertia of the whole cross section is approximately 1,900 in.[4] (a) What should the longitudinal spacing of the nails be connecting board A with boards B and C in the region of high shear? (b) For the same region, what should

---

*For further details, see E. F. Bruhn, *Analysis and Design of Flight Vehicle Structures*, Cincinnati, Ohio: Tri-State, 1965. See also, P. Kuhn, *Stresses in Aircraft and Shell Structures*, New York: McGraw-Hill, 1956.

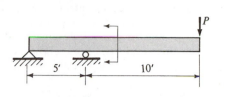

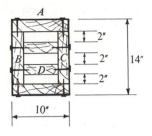

PROB. 6 – 4

be the longitudinal spacing of the nails connecting board *D* with boards *B* and *C*? In calculations neglect the weight of the beam. *Ans:* (a) 1.6 in., (b) 8 in.

6-5. A 10 in. square box beam is to be made from four wood pieces 2 in. thick. Two possible designs are considered as shown in the figure. Moreover, the design shown in (a) can be turned 90° in the application. (a) Select the design requiring the minimum amount of nailing for transmitting shear. (b) If the shear to be transmitted by this member is 620 lb, what must the nail spacing be for the best design? The nailing is to be done with 16 d (16 penny) box nails that are good for 50 lb each in shear. *Ans:* (b) 2.44 in.

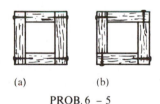

(a)   (b)

PROB. 6 – 5

6-6. A beam is fabricated from two channels and two cover plates as shown in the figure. If at the section investigated this member transmits a

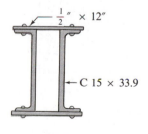

$\frac{1}{2}'' \times 12''$

C 15 × 33.9

PROB. 6 – 6

vertical shear of 150 kips and a moment of 450 kip-in., what must the spacing of $\frac{7}{8}$ in. rivets be in each row? Assume that one $\frac{7}{8}$ in. rivet is good for 9.02 kips in shear. $I = 1,351$ in.[4] *Ans:* 3.50 in.

6-7. Two W 8 × 31 beams are to be adequately fastened together with two rows of rivets, as shown in the figure, to make the two beams act as a unit. At the section considered, the total vertical shear is 40 kips and the bending moment is 2,700 ft-lb. Using $\frac{3}{4}$ in. rivets, which in single shear are good for 6.63 kips each, specify the proper rivet spacing. *Ans:* 4.66 in.

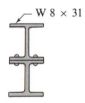

W 8 × 31

PROB. 6 – 7

6-8. A 40 mm by 50 mm rectangular bar is attached to a channel section by 10 mm machine screws at 150 mm on centers as shown in section in the figure. If the section transmits a vertical

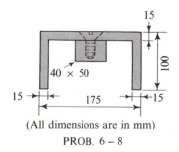

40 × 50

(All dimensions are in mm)

PROB. 6 – 8

shear of 20 kN, (a) determine the shearing stress in the machine screws; (b) determine the shearing stress at the horizontal juncture of the web with the flanges; (c) find the maximum shearing stress.

6-9. A plate girder is made up from two 14 in. by $\frac{1}{2}$ in. cover plates, four 6 in. by 4 in. by $\frac{1}{2}$ in. angles, and a $39\frac{1}{2}$ in. by $\frac{3}{8}$ in. web plate as shown in the figure. If at the section considered a total vertical shear of 150 kips is transmitted, what must the spacing of rivets A and B be? For the girder around the neutral axis, I is 14,560 in.[4]. Assume $\frac{3}{4}$ in. rivets and note that one rivet is good for 6.63 kips in single shear, 13.25 kips in double shear, and 11.3 kips in bearing on $\frac{3}{8}$ in. plate. *Ans:* Rivets A, 3.40 in. apart.

shown in the figure. If this beam is loaded with a downward concentrated force of 112 kips in the middle of the span, what is the shearing stress in the bolts? Neglect the weight of the beam. The moment of inertia I of the whole member around the neutral axis is 1,120 in.[4]. *Ans:* 12.6 ksi.

6-11. A T-flange girder is used to support a 900 kN load in the middle of a 7 m simple span. The dimensions of the girder are given in the figure in a cross-sectional view. If the 22 mm diameter rivets are spaced 125 mm apart longitudinally, what shearing stress will be developed in the rivets by the applied loading? The moment of inertia of the girder around the neutral axis is approximately $4\,300 \times 10^6$ mm[4]. *Ans:* 112 MPa.

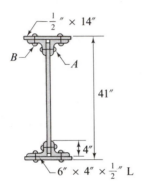

PROB. 6 – 9

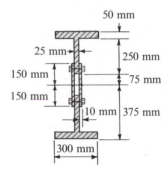

PROB. 6 – 11

6-10. A simply supported beam has a cross section consisting of a C 12 × 20.7 and a W 18 × 50 fastened together by $\frac{3}{4}$ in.-diameter bolts spaced longitudinally 6 in. apart in each row as

*6-12. A beam is made up from five separate timbers bolted together, as shown in the cross-sectional view. The bolts have a cross-sectional area of 320 mm[2] each and are spaced longitudinally 150 mm apart. If this beam spans 2.5 m and supports a load of 18 kN/m including its own weight, what is the maximum shearing stress

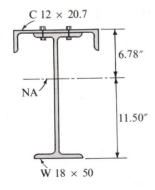

PROB. 6 – 10

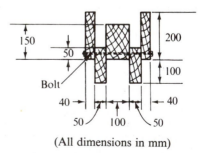

(All dimensions in mm)

PROB. 6 – 12

developed in the bolts? The moment of inertia of the beam's section is 243×10^6 mm⁴.

*6-13. A simply supported composite beam carrying a uniform load over a span of 10 m is made up by bolting a 100 mm by 300 mm timber to the upper flange of a steel beam as shown in the figure (all dimensions in the figure are in mm). Strain gages located at midspan on the inner surfaces of the steel flanges indicated strains as follows: Strain gage A: $\varepsilon = -420 \times 10^{-6}$; Strain gage B: $\varepsilon = +700 \times 10^{-6}$. (a) Compute the total force acting on the timber at this section. (b) If the bolts are placed two in a row at 600 mm apart uniformly, what is the average shear force carried by each bolt? Assume that each bolt contributes equally to the resistance of the total force. Let $E_{\text{steel}} = 200\,000$ MN/m², and $E_{\text{timber}} = 10\,000$ MN/m². *Ans:* 118 kN.

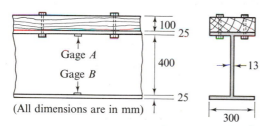

(All dimensions are in mm)

PROB. 6 – 13

6-14. If the allowable shearing stress for Douglas Fir is 700 kN/m², determine the capacity of a beam having a full-sized cross section of 50 mm by 100 mm to resist a vertical shear when placed with the 100 mm dimension vertical; when placed with the 50 mm dimension vertical.

6-15. Show that a formula, analogous to Eq. 6-7, for beams having a solid circular cross section of area A is $\tau_{\max} = \frac{4}{3}\,V/A$.

6-16. Show that a formula, analogous to Eq. 6-7, for thin-walled circular tubes acting as beams having a net cross-sectional area A is $\tau_{\max} = 2\,V/A$.

6-17. A cast iron beam has a T section as shown in the figure ($I = 53.1 \times 10^6$ mm⁴). If this beam transmits a vertical shear of 240 kN, find the shearing stresses at the levels indicated.

Report the results on a plot similar to the one shown in Fig. 6-14(e).

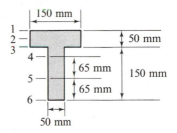

PROB. 6 – 17

6-18. A beam has a cross-sectional area in the form of an isosceles triangle for which the base b is equal to one-half its height h. (a) Using calculus and the conventional stress analysis formula, determine the location of the maximum shearing stress caused by a vertical shear V. Sketch the manner in which the shearing stress varies across the section. (b) If $b = 3$ in., $h = 6$ in., and $\tau_{\max}$ is limited to 100 psi, what is the maximum vertical shear V that this section may carry? *Ans:* (a) $h/2$; (b) 600 lb.

PROB. 6 – 18

6-19. A beam having the cross section of an isosceles triangle with a base $b = 150$ mm and a height $h = 450$ mm is subjected to a shear of 27 000 N at a section. Find the horizontal shearing stress at the neutral axis and at mid-height. After determining a few additional points, show the results on a plot similar to that of Fig. 6-12(c). (*Hint:* see Prob. 6-18.) *Ans:* $\tau_{\max} = 1.2$ MPa.

6-20. A beam is loaded so that the moment diagram for it varies as shown in the figure. (a) Find the maximum longitudinal shearing force in the $\frac{1}{2}$-in. diameter bolts spaced 12 in. apart. (b) Find the maximum shearing stress in the glued joint.

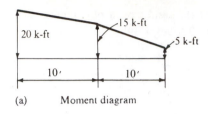

(a) Moment diagram

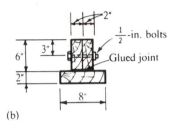

(b)

PROB. 6 – 20

6-21. A beam has a rhombic cross-section as shown in the figure. Assume that this beam transmits a vertical shear of 5 000 N, and investigate the shearing stresses at levels 50 mm apart beginning with the apex. Report the results on a plot similar to the one shown in Fig. 6-12(c).

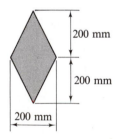

PROB. 6 – 21

6-22. A cast-iron beam has the cross-sectional dimensions shown in the figure. If the allowable stresses are 7 ksi in tension, 30 ksi in compression, and 8 ksi in shear, what is the maximum allowable shear and the maximum allowable bending moment for this beam? Consider only the vertical loading of the beam and confine calculations at the holes to section *a-a*.   *Ans:* 51.1 k, 197 k-in.

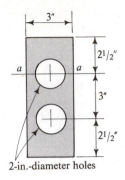

2-in.-diameter holes

PROB. 6 – 22

6-23. A box section having the dimensions shown in the figure is used for a beam on simple supports. In a certain region along the beam there is a constant, linear change in moment of 4 000 N·m/m along the axis of the beam. What is the maximum shearing stress at a section in this region?

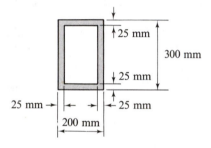

PROB. 6 – 23

6-24. A machine bracket having a rectangular cross section of 40 mm by 150 mm is subjected to a horizontal centrally applied force of 10 kN as shown in the figure. Determine the shearing stresses acting on infinitesimal elements located on a line perpendicular to the plane of the paper

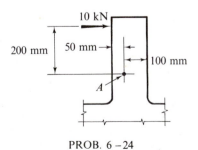

PROB. 6 – 24

at *A*. Show the results on an isolated element. Since this element is also subjected to a bending stress, without computing this stress, indicate the normal stresses of proper sense acting on the element. *Ans.* 2.22 MN/m².

6-25. Determine the maximum shearing stress at section *A-A* for the machine bracket in Example 5-5. Show the results on an infinitesimal element drawn near a free body of the bracket. Does bending stress also act on this element? *Ans:* 1.46 ksi.

6-26. A W 14 × 87 beam supports a uniformly distributed load of 4 kips/ft, including its own weight as shown in the figure. Using Eq. 6-6, determine the shearing stresses acting on the elements at *A* and *B*. Show the sense of the computed quantities on infinitesimal elements. If bending stresses also act on these elements, without additional computations, indicate the normal stresses of proper sense acting on the elements. *Ans:* $\tau_A = 3.68$ ksi.

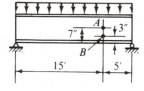

PROB. 6 – 26

*6-27. A solid beam having a 200 mm by 300 mm cross section is loaded as shown in the figure. From this beam isolate a segment 50 mm by 150 mm by 200 mm shown shaded in the figure. Then on a free-body diagram of this segment indicate the location, magnitude, and sense of all resultant forces acting on it caused by the

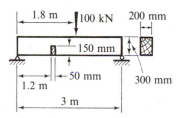

PROB. 6 – 27

bending and shearing stresses. Neglect the weight of the beam.

6-28. A girder fabricated from plywood has the cross-sectional dimensions shown in the figure. All longitudinal joints are glued. At a critical section this girder must resist a total vertical shear of 4.0 kips. Specify the quality of glue required by stating its capacity to resist shearing stress as controlled by the critical joint. For the whole section $I_{NA} = 3,648$ in.⁴ *Ans:* 30.7 psi.

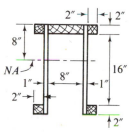

PROB. 6 – 28

6-29. A wooden I-beam is made with a narrow lower flange because of space limitations, as shown in the figure. The lower flange is fastened to the web with nails spaced longitudinally 1.5 in. apart, while the vertical boards in the lower flange are glued in place. Determine the stress in the glued joints and the force carried by each nail in the nailed joint if the beam is subjected to a vertical shear of 6 kips. The moment of inertia for the whole section around the neutral axis is 2,640 in.⁴ *Ans:* 60 psi, 565 lb.

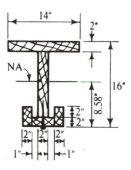

PROB 6 – 29

*6-30. A steel cantilever beam is made of two structural tees welded together as shown in the figure. Determine the allowable load *P* that the

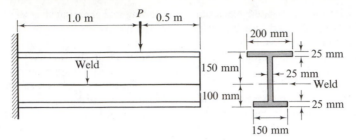

PROB. 6–30

beam can carry. Neglect the weight of the beam. The allowable stresses are: $\sigma = 150$ MN/m² in tension and compression, $\tau = 100$ MN/m² in shear on the rolled material, and $q = 2$ MN/m on the welded joint.   *Ans:* 144 kN.

6-31.   A beam is made up of four 50 mm by 100 mm full-sized Douglas Fir pieces which are glued to a 25 mm by 450 mm Douglas Fir plywood web as shown in the figure. Determine the maximum allowable shear and the maximum allowable bending moment which this section can carry if the allowable bending stress is 10 000 kN/m²; the allowable shearing stress in wood is 600 kN/m² and the allowable shearing stress in the glued joints is 300 kN/m². All dimensions in the figure are in mm.

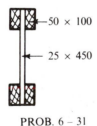

PROB. 6 – 31

6-32.   The AISI (American Iron and Steel Institute), in the *Design Manual for Light Gage Steel*, lists a beam section having the dimensions shown in the figure. Such beams are available in several gages of steel. (a) If the channels are made from No. 10 gage (0.135 in. thick) and the allowable bending stress is 18,000 psi, what is the maximum allowable bending moment for this section? (b) What is the maximum allowable

shear for the same beam if the allowable shearing stress is 12,000 psi? In computations disregard the small curvature of the plates at the corners.   *Ans:* (a) 67.2 k-in., (b) 15.8 k.

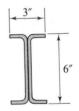

PROB. 6 – 32

6-33.   A steel welded box girder having the dimensions shown in the figure has to transmit a vertical shear force $V = 360$ k. Determine the shearing stresses at the sections a, b, and c. For this section $I = 45,000$ in.⁴.   *Ans:* $\tau_c = 8.22$ ksi.

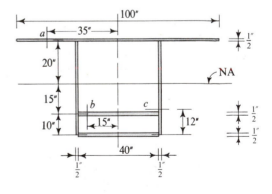

PROB. 6 – 33

6-34.   A beam is fabricated by slotting 4-in. standard steel pipes longitudinally and then securely welding them to a 23 in. by $\frac{3}{8}$ in. web plate as shown in the figure. I of the composite section around the neutral axis is 1,018 in.[4]. If at a certain section this beam transmits a vertical shear of 40 kips, determine the shearing stress in the pipe and in the web plate at a level 10 in. above the neutral axis. *Ans:* 1,500 psi, 634 psi.

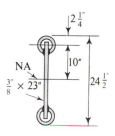

PROB. 6 – 34

6-35. Determine the shearing stress along the section A shown in the figure for a W 10 × 21 beam resisting a vertical shear of 20 kips and a bending moment of 30 kip-in. *Ans:* 1.80 ksi.

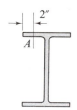

PROB. 6 – 35

***6-36.** A beam having a cross section with the dimensions shown in the figure is in a region where there is a constant, positive vertical shear

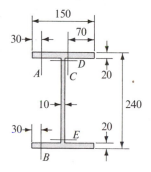

PROB. 6 – 36

of 100 kN. (a) Calculate the shear flow q acting at each of the five sections indicated in the figure. (b) Assuming a positive bending moment of 27 000 N·m at one section and a larger moment at the adjoining section 10 mm away, draw isometric sketches of each segment of the beam isolated by the sections 10 mm apart and the five sections (A, B, C, D, and E) shown in the figure, and on the sketches indicate all forces acting on the segments. Neglect vertical shearing stresses in the flanges. All dimensions are in mm.

6-37. A beam having the cross section with the dimensions shown in the figure transmits a vertical shear $V = 7$ kips applied through the shear center. Determine the shearing stresses at sections A, B, and C. I around the neutral axis is 35.7 in.[4]. The thickness of the material is $\frac{1}{2}$ in. throughout. *Ans:* 392 psi; 2,000 psi; 2,900 psi.

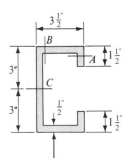

PROB. 6 – 37

6-38. A cantilever beam 1.2 m long is built up of two kinds of pine planks glued together as shown in the figure. An upward concentrated force of 4 000 N is to be applied to this beam in such a manner as not to cause any torsional stresses. Where must the force be applied? As-

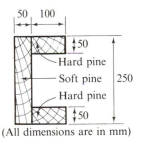

(All dimensions are in mm)

PROB. 6 – 38

sume that the planks may be considered thin. For hard pine $E = 10\ 000$ MN/m²; for soft pine $E = 7\ 000$ MN/m². (*Hint:* transform the section into an equivalent section of one material.)

*6-39 through 6-42. Determine the location of the shear center for the beams having the cross-sectional dimensions shown in the figures. In Prob. 6-41 assume that the cross-sectional area of the plate is negligible in comparison with the cross-sectional areas A of the flanges. *Ans: Prob. 6-41:* $e = (\alpha/\sin \alpha)a$ *from 0. Prob. 6-42:* $I = a^3 t\ (\alpha - \sin \alpha \cos \alpha)$ *and* $e = 2a[(\sin \alpha - \alpha \cos \alpha)/(\alpha - \sin \alpha \cos \alpha)]$.

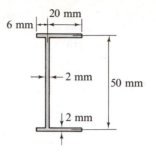

PROB. 6 – 40

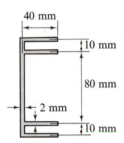

PROB. 6 – 39

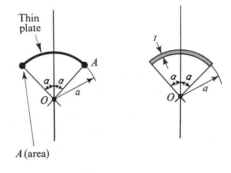

PROB. 6 – 41   PROB.6 – 42

7 Compound Stresses

7-1. INTRODUCTION

All fundamental formulas used in this text for the stress analysis of deformable bodies have been established in the previous chapters. These were derived under the assumption that only a single element of a force system is acting at a section of a member. For linearly elastic materials, these formulas are summarized below:

1. Normal stresses

   (a) due to an axial force $\qquad\qquad \sigma = \dfrac{P}{A}$ $\qquad\qquad$ (1-1)

   (b) due to bending·straight members $\qquad \sigma = -\dfrac{My}{I}$ $\qquad$ (5-1a)

   $\qquad\qquad$·symmetrical curved bars $\qquad \sigma = \dfrac{My}{Ae(R - y)}$ $\qquad$ (5-17)

2. Shearing stresses

   (a) due to torque·circular shaft $\qquad\qquad \tau = \dfrac{T\rho}{J}$ $\qquad$ (3-3a)

   $\qquad\qquad$·rectangular shaft $\qquad \tau_{\max} = \dfrac{T}{\alpha\, bc^2}$ $\qquad$ (3-13)

   $\qquad\qquad$·closed thin-walled tube $\qquad \tau = \dfrac{T}{2\,\textcircled{A}\, t}$ $\qquad$ (3-14, 3-15)

   (b) due to shearing force in a beam $\qquad \tau = \dfrac{VQ}{It}$ $\qquad$ (6-6)

In the case of inelastic material behavior, however, the large variety and complexity of the constitutive relations necessitates that each individual case be analyzed using the basic kinematic assumptions together with the appropriate stress-strain relations and equilibrium equations.

Attention will be directed to problems wherein several members of a force system may act simultaneously at a section of a member in order to satisfy the equilibrium conditions. The general problem of stress analysis arises when an axial force, a bending moment, a torque, and a shear occur simultaneously at a section through a member. However, a more limited objective is pursued in this chapter. To begin, the principle of superposition and its limitations are discussed, and so are the determination of normal stresses which arise from the simultaneous action of axial force and bending moment. This is followed by a discussion of eccentric loading, as well as unsymmetric bending problems. Shearing stresses due to the simultaneous action of torque and direct shear are then studied. Finally, at the end of the chapter, a special topic (a closely coiled helical spring) is considered.

7-2. SUPERPOSITION AND ITS LIMITATION

The basic stress analysis developed in this text thus far is completely predicated on small deformations of members. Situations such as those occuring in flexible rods, Fig. 7-1, are considered in Chapter 13. Superposition of several separately applied forces is not applicable if, for example, deflections significantly change the bending moments calculated on the basis of undeformed members. In Fig. 7-1(b), because of deflection v, an additional bending moment Pv is developed. In many problems, however, the deformation effect on stresses is small and can be neglected. This will be assumed in this chapter.

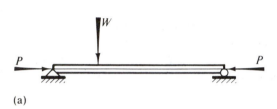

(a)

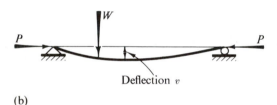

Deflection v

(b)

Fig. 7-1. Deflection in axially compressed beams induces an increase in bending moments

In members in which the overall deformations are small in the sense discussed above, superposition of the effects of separately applied forces is permissible. In considering this it is more basic to superpose strains than to superpose stresses as this enables one to treat both the elastic and the inelastic cases.

For a member simultaneously subjected to an axial force P and a bending moment M, strain superposition is shown schematically in Fig. 7-2. For clarity the strains are greatly exaggerated. Because of an axial force P a plane section perpendicular to the beam axis moves along it parallel to itself, Fig. 7-2(a). Because of a moment M applied around one of the principal axes a plane section rotates, Fig. 7-2(b). Superposition of strains due to P and M moves a plane section axially and rotates it as shown in Fig. 7-2(c). Note that if the axial force P causes a larger strain than that caused by M, the combined strains due to P and M will not change their sign within the member.

In addition to the moment which causes rotation of a plane section such as shown in Fig. 7-2, another moment acting around the other principal axis, the vertical axis in the diagram, can be applied. This second moment rotates the plane section around the vertical axis. The axial strain combined with strains caused by rotating the plane section around both principal axes is the most general case in an axially loaded bent member.

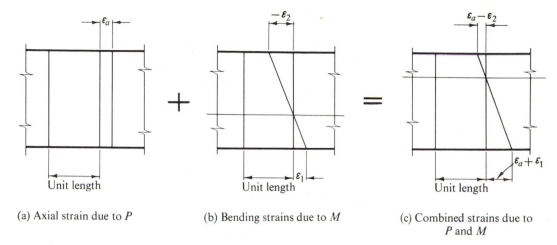

(a) Axial strain due to P (b) Bending strains due to M (c) Combined strains due to P and M

Fig. 7-2. Combined strains

By supplementing the above basic kinematic assumptions with the stress-strain relations and conditions of equilibrium, one can solve either elastic or inelastic problems. Except for the case of symmetrical sections, however, only linearly elastic problems will be considered here. The more general cases of inelastic behavior, although susceptible to the same type of analysis, are very cumbersome. Discussion of combined shearing stresses also will be limited to linearly elastic cases.

In linearly elastic problems, a linear relationship exists between stress and strain. Therefore, unlike the case in inelastic problems, not only strains but also stresses can be superposed. This means that if on the same element and for the same coordinate system two sets of stresses are known, algebraic addition of the components of the stress is possible, just as it is for vector components. It is important to note, however, that *superposition of stresses is applicable only in elastic problems where deformations are small.*

Four examples illustrating solutions for stress distribution in symmetrical members subjected to axial loads and bending moments follow. The solution of an elastic-plastic problem is given as the third example of this group.

EXAMPLE 7-1

A 50 mm by 75 mm, 1.5 m long bar of negligible weight is loaded as shown in Fig. 7-3(a). Determine the maximum tensile and compressive stresses acting normal to the section through the beam, i.e., find the maximum compound normal stresses. Assume elastic response of the material.

SOLUTION

To emphasize the method of superposition, this problem is solved by dividing it into two parts. In Fig. 7-3(b) the bar is shown subjected only to the axial

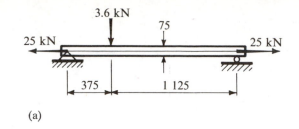

(a)

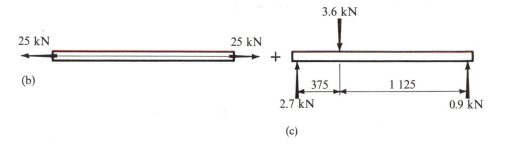

(b)

(c)

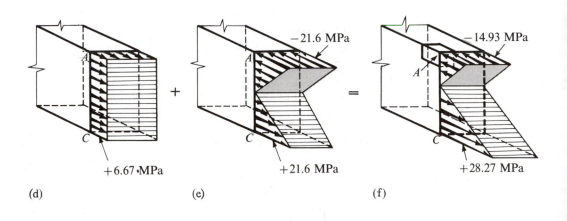

(d)   (e)   (f)

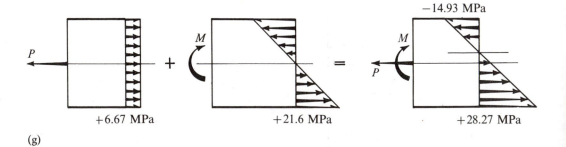

(g)

Fig. 7-3. (All dimensions in mm)

force, and in Fig. 7-3(c) the same bar is shown subjected only to the transverse force. For the axial force the normal stress throughout the length of the bar is

$$\sigma = \frac{P}{A} = \frac{25}{(0.05)(0.075)} = 6.67 \times 10^3 \text{ kN/m}^2 = 6.67 \text{ MPa} \qquad \text{(tension)}$$

This result is indicated in Fig. 7-3(d). The normal stresses due to the transverse force depend on the magnitude of the bending moment, and the maximum bending moment occurs at the applied force. As the left reaction is 2.7 kN, $M_{\max} = (2.7)(0.375) = 1.013 \text{ kN} \cdot \text{m}$. From the flexure formula, the maximum stresses at the extreme fibers caused by this moment are

$$\sigma = \frac{Mc}{I} = \frac{6M}{bh^2} = \pm 21.6 \times 10^3 \text{ kN/m}^2 = \pm 21.6 \text{ MPa}$$

These stresses act normal to the section of the beam and decrease linearly toward the neutral axis as in Fig. 7-3(e). Then, to obtain the compound stress for any particular element, bending stresses must be added algebraically to the direct tensile stress. Thus, as may be seen from Fig. 7-3(f), at point A the resultant normal stress is 14.93 MPa compression, and at C it is 28.27 MPa tension.

Side views of the stress vectors as commonly drawn are in Fig. 7-3(g). Although in this problem the given axial force is larger than the transverse force, bending causes higher stresses. However, the reader is cautioned not to regard slender members, particularly compression members, in the same light (see Fig. 7-1(b)).

Note that in the final result, the line of zero stress, which is located at the centroid of the section for flexure, moves upward. Also note that the local stresses, caused by the concentrated force, which act normal to the top surface of the beam, were not considered. Generally these stresses are treated independently as local bearing stresses.

The stress distribution shown in Figs. 7-3(f) and (g) would change if, for example, instead of the axial tensile forces of 25 kN applied at the ends, compressive forces of the same magnitude were acting on the member. The maximum tensile stress would be reduced to 14.93 MPa from 28.27 MPa, which would be desirable in a beam made of a material weak in tension and carrying a transverse load. This idea is utilized in prestressed construction. Tendons made of high-strength steel rods or cable passing through a beam with anchorages at the ends are used to precompress concrete beams. Such artifically applied forces inhibit the development of tensile stresses. Prestressing also has been used in racing-car frames.

EXAMPLE 7-2

A 50 mm by 50 mm elastic bar bent into a U shape as in Fig. 7-4(a) is acted upon by two opposing forces P of 8.33 kN each. Determine the maximum normal stress occurring at the section A-B.

SOLUTION

The section to be investigated is in the curved region of the bar, but this makes no essential difference in the procedure. First, a segment of the bar is taken as

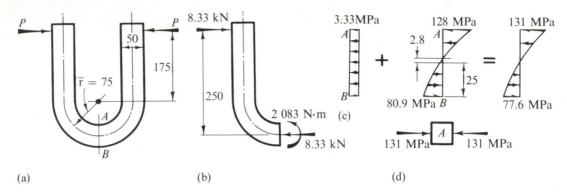

Fig. 7-4. (All dimensions in mm)

a free body, as shown in Fig. 7-4(b). At section A-B the axial force, applied at the centroid of the section, and the bending moment necessary to maintain equilibrium are determined. Then, each element of the force system is considered separately. The stress caused by the axial forces is

$$\sigma = \frac{P}{A} = \frac{8.33}{0.05(0.05)} = 3.33 \times 10^3 \text{ kPa} = 3.33 \text{ MPa} \qquad \text{(compression)}$$

and is shown diagrammatically in the first sketch of Fig. 7-4(c). The normal stresses caused by the bending moment may be obtained by using Eq. 5-16. However, for this bar, bent to a 75 mm radius, the solution is already known from Example 5-12. The stress distribution corresponding to this case is shown in the second sketch of Fig. 7-4(c). By superposing the results of these two solutions, the compound stress distribution is obtained. This is shown in the third sketch of Fig. 7-4(c). The maximum stress occurs at A and is a compressive stress of 131 MPa. An isolated element for the point A is shown in Fig. 7-4(d). Shearing stresses are absent at section A-B as no shearing force is necessary to maintain equilibrium of the segment shown in Fig. 7-4(b). The relative insignificance of the stress caused by the axial force is striking.

Problems similar to the above commonly occur in machine design. Hooks, C clamps, frames of punch presses, etc. illustrate the variety of situations to which the foregoing methods of analysis must be applied.

*EXAMPLE 7-3

Consider a rectangular elastic-plastic beam bent around the horizontal axis and simultaneously subjected to an axial tensile force. Determine the magnitudes of the axial forces and moments associated with the stress distributions shown in Figs. 7-5(a), (b), and (e).

SOLUTION

The stress distribution shown in Fig. 7-5(a) corresponds to the limiting elastic case, where the maximum stress is at the point of impending yielding. For

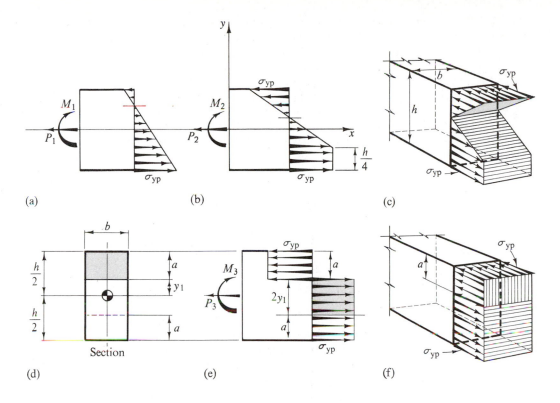

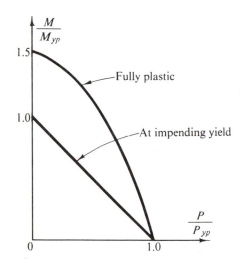

Fig. 7-5. Combined axial and bending stresses: (a) elastic stress distribution; (b) and (c) elastic plastic stress distribution; (e) and (f) fully plastic stress distribution.

Fig. 7-6. Interaction curves for P and M for a rectangular member

this case the stress-superposition approach can be used. Hence

$$\sigma_{\max} = \sigma_{yp} = \frac{P_1}{A} + \frac{M_1 c}{I} \qquad (7\text{--}1)$$

The force P at yield can be defined as $P_{yp} = A\sigma_{yp}$; from Eq. 5-5 the moment at yield is $M_{yp} = (I/c)\sigma_{yp}$. Dividing Eq. 7-1 by σ_{yp} and making use of the relations for P_{yp} and M_{yp}, one obtains

$$\frac{P_1}{P_{yp}} + \frac{M_1}{M_{yp}} = 1 \qquad (7\text{--}2)$$

This establishes a relationship between P_1 and M_1 so that the maximum stress just equals σ_{yp}. A plot of this equation corresponding to the case of impending yield is in Fig. 7-6. Plots of such relations are called *interaction curves*.

The stress distribution shown in Figs. 7-5(b) and (c) occurs after yielding has taken place in the lower quarter of the beam. With this stress distribution given, one can determine directly the magnitudes of P and M from the conditions of equilibrium. If on the other hand P and M were given, since superposition does not apply, a cumbersome process would be necessary to determine the stress distribution.

For the stresses given in Figs. 7-5(b) and (c), one simply applies Eqs. 5-7 and 5-8 developed for inelastic bending of beams, except that in Eq. 5-7 the sum of the normal stresses must equal the axial force P. Noting that in the elastic zone the stress can be expressed algebraically as $\sigma = (\sigma_{yp}/3) - [8\sigma_{yp}y/(3h)]$ and that in the plastic zone $\sigma = \sigma_{yp}$, one has

$$P_2 = \int_A \sigma \, dA = \int_{-h/4}^{+h/2} \frac{\sigma_{yp}}{3}\left(1 - \frac{8y}{h}\right)b \, dy + \int_{-h/2}^{-h/4} \sigma_{yp}b \, dy = \sigma_{yp}\frac{bh}{4}$$

$$M_2 = -\int_A \sigma y \, dA = -\int_{-h/4}^{+h/2} \frac{\sigma_{yp}}{3}\left(1 - \frac{8y}{h}\right)yb \, dy - \int_{-h/2}^{-h/4} \sigma_{yp}yb \, dy$$

$$= \frac{3}{16}\sigma_{yp}bh^2$$

Note that the axial force found above exactly equals the force acting on the plastic area of the section. The moment M_2 is greater than $M_{yp} = \sigma_{yp}bh^2/6$ and less than $M_{ult} = M_p = \sigma_{yp}bh^2/4$ (see Eq. 5-9).

The axial force and moment corresponding to the fully plastic case shown in Figs. 7-5(e) and (f) are simple to determine. As may be seen from Fig. 7-5(e) the axial force is developed by σ_{yp} acting on the area $2y_1b$. Because of symmetry, these stresses make no contribution to the moment. Forces acting on the top and the bottom areas $ab = [(h/2) - y_1]b$, Fig. 7-5(d), form a couple with a moment arm of $h - a = (h/2) + y_1$. Therefore

$$P_3 = 2y_1b\sigma_{yp} \qquad \text{or} \qquad y_1 = P_3/(2b\sigma_{yp})$$

and

$$M_3 = ab\sigma_{yp}(h - a) = \sigma_{yp}b\left(\frac{h^2}{4} - y_1^2\right) = M_p - \sigma_{yp}by_1^2$$

$$= \frac{3M_{yp}}{2} - \frac{P_3^2}{4b\sigma_{yp}}$$

Then dividing by $M_p = 3M_{yp}/2 = \sigma_{yp}bh^2/4$ and simplifying, one obtains

$$\frac{2M_3}{3M_{yp}} + \left(\frac{P_3}{P_{yp}}\right)^2 = 1 \tag{7-3}$$

This is a general equation for the interaction curve for P and M necessary to achieve the fully plastic condition in a rectangular member (see Fig. 7-6). Unlike the equation for the elastic case, the relation is nonlinear.

EXAMPLE 7-4

A small dam of triangular shape as shown in Fig. 7-7(a) is made from concrete, which weighs approximately 25 kN/m³. Find the normal stress distribution at section A-B when the water behind the dam is at the level indicated. For the purpose of calculation, consider one linear meter of the dam in

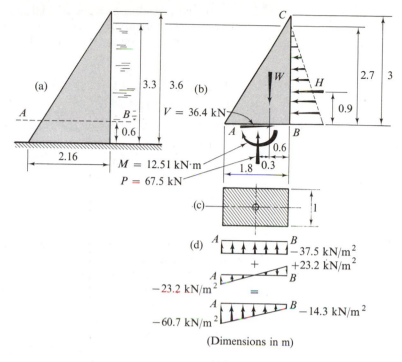

Fig. 7-7.

the direction perpendicular to the plane of the paper as an isolated beam. Use units of meters and kilonewtons.

SOLUTION

Pass a section at the required level* and isolate a part of the body as shown in Fig. 7-7(b). The hydrostatic pressure acting on the vertical upstream face of the dam causes the horizontal force H. This force is equal to the *average* pressure acting on the dam multiplied by the contact area, and as the pressure varies linearly from the free surface of the water, it acts 0.9 m above the A-B level. The unit weight of water is assumed to be 10 kN/m³.

The weight of the "beam" W, which is the weight of a meter section of the dam, is equal to the volume of concrete above the A-B level multiplied by the unit weight of concrete. The resultant of this gravity force acts through the *centroid* of the concrete volume, i.e., 0.6 m from the upstream face of the dam. The plan view of section A-B is shown in Fig. 7-7(c). The centroid and the neutral axis of *this cross-sectional area* are midway between the upstream and the downstream faces of the dam. Thus the usual free-body diagram for a segment of the beam is completed as shown in Fig. 7-7(b), where an "axial"

---

*This section is made horizontally. This departs from strict adherence to the method of sectioning discussed earlier. The axis of the dam bisects the angle ACB, and according to the procedures discussed earlier the section should be made perpendicular to this axis. However, it is customary to analyze masonry structures by taking *horizontal* sections. This follows from analogy with the section at the ground level which is horizontal.

force, a shear, and a bending moment are indicated. The problem of determining the compound normal stresses at section A-B is then solved by applying Eqs. 1-1 and 5-1 and superposing the results. This process is shown in Fig. 7-7(d).

$$H = V = [(\tfrac{1}{2})(2.7)(10)](2.7)1 = 36.4 \text{ kN}$$

$$W = P = [(\tfrac{1}{2})(1.8)(3)1]25 = 67.5 \text{ kN}$$

$$M = H(0.9) - W(0.3) = (36.4)(0.9) - (67.5)(0.3) = 12.5 \text{ kN} \cdot \text{m}$$

$$\sigma_B = -\frac{W}{A} + \frac{Mc}{I} = -\frac{67.5}{(1.8)(1)} + \frac{(12.5)(0.9)}{(1.8)^3/12} = -37.5 + 23.2$$

$$= -14.3 \text{ kN/m}^2 \quad \text{(compression)}$$

$$\sigma_A = -\frac{W}{A} - \frac{Mc}{I} = -37.5 - 23.2 = -60.7 \text{ kN/m}^2 \quad \text{(compression)}$$

The normal stresses caused by bending vary linearly from the neutral axis of the cross-section at A-B. When these stresses are superposed on the uniform stress caused by the force P, the final stress distribution shown in Fig. 7-7(d) results. Note that in this particular case all normal stresses at the section A-B are compressive, and no line of zero stress occurs within the investigated cross-section. Shearing stresses also exist at the level A-B, although their analysis is omitted here.

The results obtained are not exact since the horizontal cross-sectional area of the dam changes rapidly. Equations 1-1 and 5-1 apply only to prismatic members. Normal stresses at the downstream face of the dam do not satisfy the boundary conditions, since they act vertically, whereas the downstream face is inclined. Moreover, dams are rather *short* beams, and an analysis based on the ordinary formulas is approximate. An analysis for the shearing stresses on the basis of Eq. 6-6 leads to wrong results.*

*7-3. REMARKS ON PROBLEMS INVOLVING AXIAL FORCES AND BENDING MOMENTS: THE DAM PROBLEM

In the above example all normal stresses at the section A-B were found to be compressive. This resulted from the particular values of the axial force and the bending moment, and a different situation may easily arise. For example, if the base of the dam were narrower, the dam would weigh less, while the bending moment caused by the water pressure would remain the same. The compound stress in such a case may be *tensile* stress on the upstream face. However, some materials, such as concrete, are notably weak in resisting tensile stresses. This fact leads to a practically important problem: What must be the proportions of a concrete dam or pier so that no tension will exist in the material when forces are applied to the structure? To answer this question, consider a segment of a member above a sec-

*For further details see S. Timoshenko, and J. N. Goodier, *Theory of Elasticity* (3rd ed.), New York: McGraw-Hill, 1970, pp. 51 and 109.

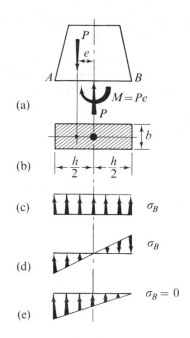

(a)

(b)

(c) σ_B

(d) σ_B

(e) $\sigma_B = 0$

Fig. 7-8. Establishing the location of the force P to cause zero stress at B.

tion *A-B* as shown in Fig. 7-8(a). The cross-sectional area at *A-B* is assumed to be rectangular as shown in Fig. 7-8(b). Let the *resultant* of all of the applied forces above the section *A-B* be a vertical force, and let it act at a distance *e* from the centroid of the cross-sectional area at *A-B*. Then at the section *A-B*, to maintain this segment in equilibrium, there must be an axial force *P* and a bending moment *Pe*, Fig. 7-8(a). The stress caused by the axial force is $\sigma = P/A = P/(bh)$, while the maximum stress caused by bending is $\sigma_{max} = Mc/I = M/S = 6Pe/(bh^2)$, where $bh^2/6$ is the section modulus of the *rectangular* cross section. The complete stress distributions across the section *A-B* corresponding to these two effects are shown in Figs. 7-8(c) and (d), respectively.

To satisfy the desired condition that the stress at *B* be zero, it follows that

$$\sigma_B = -\frac{P}{bh} + \frac{6Pe}{bh^2} = 0 \quad \text{or} \quad e = \frac{h}{6}$$

which means that if the force *P* is applied at a distance of *h*/6 from the centroidal axis of the cross section, the stress at *B* is just zero. The compound normal stress distribution across the whole section becomes "triangular" as is shown in Fig. 7-8(e). If the force *P* were applied closer to the centroid of the section, a smaller bending moment would be developed at the section *A-B*, and there would be some compression stress at *B*. The same argument may be repeated for the force acting to the right of the centroidal axis. Hence a practical rule, much used by designers of masonry structures, may be formulated thus: *If the resultant of all vertical forces acts within the middle third of the rectangular cross section, there is no tension in the material at that section.* It is understood that the resultant acts in a vertical plane containing one of the axes of symmetry of the rectangular cross-sectional area.

The foregoing discussion may be generalized in order to apply to any planar system of forces acting on a member. The resultant of these forces may be made to intersect the plane of the cross section as is shown in Fig. 7-9. At the point of intersection of this resultant with the section it may be resolved into horizontal and vertical components. If the vertical component of the resultant fulfills the conditions of the former problem, no tension will be developed at point *B*, as the horizontal component causes only shearing stresses. Hence, a more general "middle third" rule may be stated thus: there will be no tension at a section of a member of a *rectangular* cross section if the resultant of the forces above this section *intersects* one of the axes of symmetry of the section within the middle third.

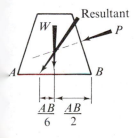

Fig. 7-9. The resultant shown acting at its extreme left-hand position to cause no tension at B.

*7-4. A SPECIAL LIMITATION: THE CHIMNEY PROBLEM

At times it is impossible to achieve a situation where the resultant would pass through the "middle third" of a rectangular cross section as discussed in the previous article. However, at some sections through a member, or between two different members, no tensile stresses may be transmitted. For example, no tensile stresses can be developed at the surface of contact of a concrete foundation for a tall chimney with the soil. On the other hand, the resultant of a horizontal force caused by the wind blowing on the chimney and the vertical force due to the weight of the chimney itself may pass outside the "middle third" of the *rectangular* foundation. Under these circumstances the method of analysis discussed in the preceding article must be modified.

Consider the weightless rectangular block shown in Fig. 7-10(a) to which is applied, outside of the "middle third," a vertical force P at a dis-

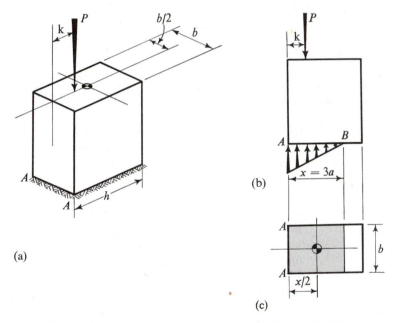

Fig. 7-10 Stress distribution between two surfaces that are unable to transmit tensile forces

tance k from one of the faces in the middle of the block's dimension b. Suppose next that at the contact surface of the block with the foundation no stresses exist to the right of the point B shown in Fig. 7-10(b). Thus, it is assumed that only the portion AB of the foundation, of length x and width b, is effective in resisting the applied force P. This corresponds to the shaded area in Fig. 7-10(c). The stress along the line B-B is zero by assumption.

Hence the following equation for the stress at B may be written.

$$\sigma_B = -\frac{P}{xb} + P\left(\frac{x}{2} - k\right)\frac{6}{bx^2} = 0$$

where $(x/2) - k$ is the eccentricity of the applied force with respect to the centroidal axis of the cross-hatched contact area, and $bx^2/6$ is its section modulus. Solving for x, it is found that $x = 3k$, and the pressure distribution will be "triangular" as shown in Fig. 7-10(b) (why?). As k decreases, the intensity of pressure on the line A-A increases; when k is zero, the block becomes unstable.

Similar reasoning can be applied to problems where a number of forces are acting on a member and the contact area is of any shape. Such problems are important in the design of foundations. However, since soil is never completely elastic and the solutions are based on a number of idealizations, they are approximate.

7-5. A FORCE APPLIED TO A PRISMATIC MEMBER ANYWHERE PARALLEL TO ITS AXIS

The compound stresses considered so far were caused by an axial force and a bending moment which acts around an axis perpendicular to the plane of symmetry of the cross section. The same procedure may be used in an analogous situation where the cross sectional area is *unsymmetrical*, provided the bending moment acts around one of the *principal axes* of the cross section (Art. 5-7).

Situations arise occasionally, however, where the bending moments are not applied in the above manner. For example, consider the case of a weightless rectangular block, Fig. 7-11(a), on which a force P, acting parallel to the axis of the member, is applied eccentrically with respect to the centroidal axis of the member.

The y and z axes shown in Fig. 7-11 are the axes of symmetry of the block's cross section,* and the location of the force P may be defined by the coordinates y_o and z_o. By applying two equal and opposite forces P at the centroid, as shown in Fig. 7-11(b), the problem is changed to that of an axial force P and a bending moment Pd in the plane of the applied force P and the axis of the member. This bending moment does not act in the plane of either of the principal axes, however, and no adequate formula has been derived in this text for such a loading condition. Hence, two additional equal and opposite forces of magnitude P are introduced at a point on the z axis at a distance z_o from the centroid of the section. The five forces now shown in Fig. 7-11(c) still represent the initial problem from the point of view of

*For an unsymmetrical cross section these axes should be the principal axes.

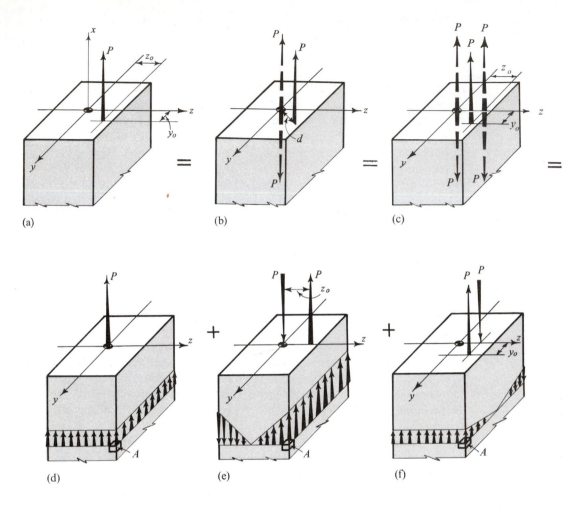

(a)   (b)   (c)

(d)   (e)   (f)

Fig. 7-11 Resolution of a problem into three problems, each one of which can be solved by the methods previously discussed

statics. However, this equivalent problem may now be conveniently resolved into three separate problems as indicated in Figs. 7-11(d), (e), and (f), which involve, respectively, an axial force P, a bending moment $M_{yy} = +Pz_o$ acting around the y-axis,* and a bending moment $M_{zz} = -Py_o$ acting around the z-axis. The usual axial and flexural stress formulas are applicable for these cases, and the compound normal stress at any point (y, z) of the cross section

---

*The moment is taken positive if it has the sense that would tend to advance a right-hand screw in the positive direction of the axis.

of an eccentrically loaded member can be found by superposition. Hence

$$\sigma_x = \pm \frac{P}{A} \pm \frac{M_{zz}y}{I_{zz}} \pm \frac{M_{yy}z}{I_{yy}} \qquad (7\text{-}4)$$

where A is the cross-sectional area of the member, and I_{zz} and I_{yy} are the moments of inertia of the cross-sectional area around the respective principal axes. In Eq. 7-4 positive signs correspond to tensile stresses, negative signs correspond to compressive stresses.

EXAMPLE 7-5

Find the stress distribution at the section $ABCD$ for the block shown in Fig. 7-12(a) if $P = 64$ kN. At the same section, locate the line of zero stress. Neglect the weight of the block.

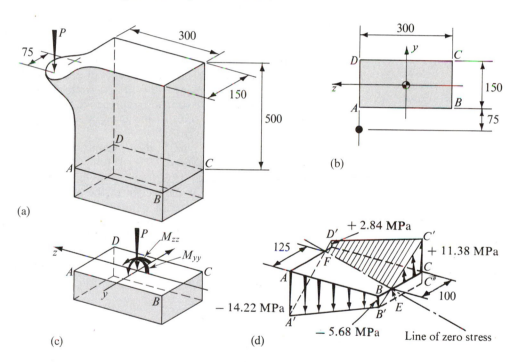

Fig. 7-12. (all dimensions in mm)

SOLUTION

The forces acting on the section $ABCD$, Fig. 7-12(c), are $P = -64$ kN, $M_{yy} = -64(0.15) = -9.6$ kN·m, and $M_{zz} = -64(0.075 + 0.075) = -9.6$ kN·m. The cross section of the block $A = (0.15)(0.3) = 0.045$ m^2, and the respective section moduli are $S_{zz} = (0.3)(0.15)^2/6 = 1.125 \times 10^{-3}$ m³ and $S_{yy} = (0.15)(0.3)^2/6 = 2.25 \times 10^{-3}$ m³. Hence, using a relation equivalent to Eq. 7-4 gives the compound normal stresses for the corner elements:

$$\sigma = \frac{P}{A} \mp \frac{M_{zz}}{S_{zz}} \pm \frac{M_{yy}}{S_{yy}} = -\frac{64}{45 \times 10^{-3}} \pm \frac{9.6}{1.125 \times 10^{-3}} \mp \frac{9.6}{2.25 \times 10^{-3}}$$

$$= (-1.42 \pm 8.53 \mp 4.27)10^3$$

Here the units of stress are kilonewtons per square meter or kilopascals. The sense of the forces shown in Fig. 7-12(c) determines the signs of stresses. Therefore if the subscript of the stress signifies its location, the corner normal stresses are:

$$\sigma_A = (-1.42 - 8.53 - 4.27)\,10^3 = -14\,220\,\text{kPA} = -14.22\,\text{MPa}$$

$$\sigma_B = (-1.42 - 8.53 + 4.27)\,10^3 = -5\,680\,\text{kPa} = -5.68\,\text{MPa}$$

$$\sigma_C = (-1.42 + 8.53 + 4.27)\,10^3 = +11\,380\,\text{kPa} = +11.38\,\text{MPa}$$

$$\sigma_D = (-1.42 + 8.53 - 4.27)\,10^3 = +2\,840\,\text{kPa} = +2.84\,\text{MPa}$$

These stresses are shown in Fig. 7-12(d). The ends of these four stress vectors at A', B', C', and D' lie in the plane $A'B'C'D'$. The vertical distance between the planes $ABCD$ and $A'B'C'D'$ defines the compound stress at any point on the cross section. The intersection of the plane $A'B'C'D'$ with the plane $ABCD$ locates the line of zero stress FE.

By drawing a line $B'C''$ parallel to BC, similar triangles $C'B'C''$ and $C'EC$ are obtained: thus the distance $CE = [11.38/(11.38 + 5.68)]150 = 100$ mm. Similarly, the distance AF is found to be 125 mm. Points E and F locate the line of zero stress. If the weight of the block is neglected, the stress distribution on any other section parallel to $ABCD$ is the same.

If the compound stresses at a section are all tensile or all compressive, the planes corresponding to the planes $ABCD$ and $A'B'C'D'$ may be extended beyond the actual cross-sectional area of the member until they intersect. This gives a *fictitious* line of zero stress. However, as in the former case, the stresses vary linearly from the line of zero stress as their respective perpendicular distances from this line.

EXAMPLE 7-6

Find the zone over which the vertical downward force P_o may be applied to the rectangular weightless block shown in Fig. 7-13(a) without causing any tensile stresses at the section A-B.

SOLUTION

The force $P = -P_o$ is placed at an arbitrary point in the first quadrant of the y-z coordinate system shown. Then the same reasoning used in the preceding example shows that with this position of the force the greatest tendency for a tensile stress exists at A. With $P = -P_o$, $M_{zz} = +P_o y$, and $M_{yy} = -P_o z$, setting the stress at A equal to zero fulfills the limiting condition of the

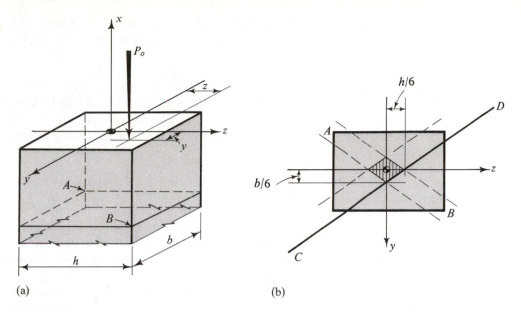

Fig. 7-13

problem. Using Eq. 7-4 allows the stress at A to be expressed as

$$\sigma_A = 0 = \frac{(-P_o)}{A} - \frac{(P_o y)(-b/2)}{I_{zz}} + \frac{(-P_o z)(-h/2)}{I_{yy}}$$

or

$$-\frac{P_o}{A} + \frac{P_o y}{b^2 h/6} + \frac{P_o z}{bh^2/6} = 0$$

Simplifying $[z/(h/6)] + [y/(b/6)] = 1$, which is an equation of a straight line. It shows that when $z = 0$, $y = b/6$; and when $y = 0$, $z = h/6$. Hence this line may be represented by the line CD in Fig. 7-13(b). A vertical force may be applied to the block anywhere on this line and the stress at A will be zero. Similar lines may be established for the other three corners of the section; these are shown in Fig. 7-13(b). If the force P is applied on any one of these lines or on any line parallel to such a line toward the centroid of the section, there will be no tensile stress at the corresponding corner. Hence the force P may be applied anywhere within the shaded area in Fig. 7-13(b) without causing tensile stress at any of the four corners or anywhere else. This zone of the cross-sectional area is called the *kern* of a section. By limiting the possible location of the force to the lines of symmetry of the rectangular cross section, the results found in this example verify the "middle third" rule discussed in Art. 7-3.

The foregoing method of analysis may also be used for tension members, but it applies only for *short* blocks in compression. Slender bars in compression require a special treatment (Chapter 13). Near the point of application of the force, the analysis used in this and the preceding example is incorrect. In the neighborhood of this point the stress distribution is somewhat similar to the one shown in Fig. 2-16. On the other hand, although the cross-sectional area of the block considered in this example was rectangular, the theory is not so limited. The same method of stress analysis may be used for a member with any cross-sectional area, provided the axes of the cross-sectional area around which bending moments are taken are the *principal axes*.* By definition the principal axes are those about which the rectangular moment of inertia is a maximum or a minimum. An axis of symmetry is always a principal axis, and the principal axes are mutually perpendicular.

7-6. UNSYMMETRICAL BENDING

In Chapter 5 on the flexure of members it was emphasized that the derived flexure formula is valid only if the applied bending moment acts around one or the other of the principal axes of the cross section. However, a member may be subjected to a bending moment which is inclined with respect to the principal axes.

An illustration of bending which does not fulfill the limitation imposed in Chapter 5 is shown in Fig. 7-14(a), where the plane in which the bending moment acts makes an angle α with the vertical axis. Since this type of bending does not occur in a plane of symmetry of the cross section, it is called *unsymmetrical bending*.† It represents the general case of bending or flexure for which a general flexure formula can be derived. Since such a formula is relatively complicated and can be avoided entirely, it will not be discussed in this text.‡ In principle, the general problem has already been solved in Art. 7-5 by the method of superposition. There, as may be seen from Fig. 7-11(a), in addition to the stresses caused by the axial force P, the bending stresses caused by the moment applied in an inclined plane were determined. This is an example of unsymmetrical bending and will be elaborated upon here.

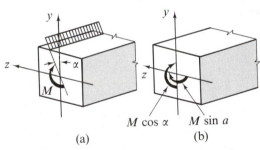

Fig. 7-14. (a) Bending moment in a plane that is not coincident with either of the principal axes, and (b) components of the bending moment in the planes of the principal axes.

From the principles of mechanics, any couple, which may be a bending moment at a section of a beam, may be resolved into components. Hence a bending moment such as shown in Fig. 7-14(a) may be resolved into the two components shown in Fig. 7-14(b). The component of the bending moment M acting around the z-axis is $M \cos \alpha$, while the one acting around the y-axis

*See the appendix to Chapter 8.

†Since the problem is more general than something lacking symmetry, it may be termed skew bending. This corresponds to the use of the words *schiefe* in German and *kosoi* in Russian, which mean inclined or skew.

‡See E. F. Bruhn, *Analysis and Design of Flight Vehicle Structures*, Cincinnati, Ohio: Tri-State, 1965.

is $|M \sin \alpha|$.* The sense of each component follows from the sense of the total moment M. This procedure may always be used to find the bending moments which act only around the principal axes of the cross section. Then these moments may be used *separately* in the usual flexure formula, and the compound normal stresses follow by superposition. If a member is subjected to *pure* unsymmetrical bending, this procedure gives all the stresses acting at a section through the member. On the other hand, if the bending moment is caused by a transverse force, shearing stresses will also be developed. These will be discussed later.

EXAMPLE 7-7

The 0.1 m by 0.15 m wooden beam shown in Fig. 7-15(a) is used to support a uniformly distributed load of 4 kN (total) on a simple span of 3 m. The applied load acts in a plane making an angle of 30° with the vertical, as shown in Fig. 7-15(b) and again in Fig. 7-15(c). Calculate the maximum bending stress at midspan, and, for the same section, locate the neutral axis. Neglect the weight of the beam.

SOLUTION

The maximum bending *in the plane of the applied load* occurs at the midspan, and according to Example 4-6 it is equal to $w_o L^2/8$ or $WL/8$, where W is the total load on the span L. Hence

$$M = \frac{WL}{8} = \frac{4(3)}{8} = 1.5 \text{ kN·m}$$

Next, this moment is resolved into components acting around the respective axes, i.e.,

$$M_{zz} = M \cos \alpha = 1.5(\sqrt{3}/2) = 1.3 \text{ kN·m}$$

$$|M_{yy}| = M \sin \alpha = 1.5(0.5) = 0.75 \text{ kN·m}$$

By considering the nature of the flexural stress distribution about both principal axes of the cross section, it may be concluded that the maximum tensile stress occurs at A. The maximum compressive stress is at C. The values of these stresses follow by superposing the stresses caused by each moment independently. Stresses at the other two corners of the cross section are similarly determined.

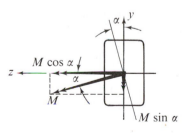

Fig. 7-A

*In vectorial representation of a couple, Fig. 7-A, the bending moment M is shown by a vector with a double-headed arrow, the sense of which coincides with the direction in which a right-hand screw would advance if turned in the direction of the sense of rotation of the moment. Thence resolution of this moment is carried out as for a force.

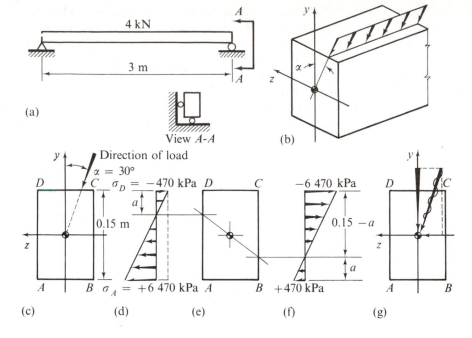

Fig. 7-15

$$\sigma_A = \frac{M_{zz}c_1}{I_{zz}} + \frac{M_{yy}c_2}{I_{yy}} = \frac{(1.3)(0.075)}{(0.1)(0.15)^3/12} + \frac{(0.75)(0.05)}{(0.15)(0.1)^3/12}$$

$$= +3\,470 + 3\,000 = +6\,470 \text{ kPa} \qquad \text{(tension)}$$

$$\sigma_B = +3\,470 - 3\,000 = +470 \text{ kPa} \qquad \text{(tension)}$$

$$\sigma_C = -3\,470 - 3\,000 = -6\,470 \text{ kPa} \qquad \text{(compression)}$$

$$\sigma_D = -3\,470 + 3\,000 = -470 \text{ kPa} \qquad \text{(compression)}$$

It is seen from this solution that the compound normal stresses at A and C are numerically equal.

The neutral axis, which is a line of zero stress in a beam, may be located as in Example 7-5. However, the simplified diagrams shown in Figs. 7-15(d) and (f) serve the same purpose. From similar triangles, $a/(0.15 - a) = 470/(6\,470)$ or $a = 0.0102$ m or 10.2 mm. This locates the neutral axis shown in Fig. 7-15(e). Note that since no axial force acts on the member, the line of zero stress passes through the centroid of the cross-sectional area.

When unsymmetrical bending of a beam is caused by applied transverse forces, another procedure equivalent to the above *is usually more convenient*. The applied forces are first resolved into components which act parallel to the principal axes of the cross-sectional area. Then the bending moments caused by these components around the respective axes are computed for use in the flexure formula. For the above example, such components of the applied load are shown in Fig. 7-15(g). Note that to avoid *torsional* stresses,

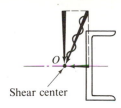

Fig. 7-16. Application of a lateral force through the shear center. No torsion is caused in the beam.

Shear center

the applied transverse forces must act through the shear center. For bilaterally symmetrical sections, e.g., a rectangle, a circle, an I-beam, etc., *the shear center coincides with the geometric center (centroid) of the cross section.* For other cross sections, such as a channel, the shear center lies elsewhere, as at O shown in Fig. 7-16, and it is at this point that the transverse force must be applied to prevent occurrence of torsional stresses. Single angles acting as beams must be treated similarly (see Fig. 6-22). For analysis of unsymmetrical bending, the applied forces must be resolved *at* the shear center parallel to the principal axes of the cross section.

7-7. SUPERPOSITION OF SHEARING STRESSES

In some situations, shearing stresses arise from torsion *and* a direct shearing force. The shearing stresses caused by each of these were discussed earlier in the text. In the chapter on torsion, only circular and thin-walled members were discussed in detail. This limits the type of problems which may be solved here. For problems where both torsional and direct shearing stresses may be determined, the compound *shearing* stress may be found by *superposition.* This procedure is analogous to the one used earlier for compound normal stresses. However, while normal stresses act only toward or away from an element of a member, shearing stresses in the plane of a cut may act in *any* direction. This results in a more difficult stress analysis problem, and the general solution is beyond the scope of this book. Attention will be directed to instances where the shearing stresses being superposed have the *same line of action.* This limitation excludes relatively few significant problems.

EXAMPLE 7-8

Find the maximum shearing stress due to the applied forces in the plane A-B of the 10 mm diameter high-strength shaft shown in Fig. 7-17(a).

SOLUTION

Since only the stresses due to the applied forces are wanted, the weight of the shaft need not be considered. The free body of a segment of the shaft is shown in Fig. 7-17(b). The system of forces at the cut necessary to keep this segment in equilibrium consists of a torque, $T = 20$ N·m, a shear, $V = 250$ N, and a bending moment, $M = 25$ N·m.

Due to the torque T, the shearing stresses in the cut A-B vary linearly from the axis of the shaft and reach the maximum value given by Eq. 3-3, $\tau_{max} = Tc/J$. These maximum shearing stresses, agreeing in sense with the *resisting* torque T, are shown at points A, B, D, and E in Fig. 7-17(c).

The "direct" shearing stresses caused by the shearing force V may be obtained by using Eq. 6-6, $\tau = VQ/(It)$. For the elements A and B, Fig. 7-17(d), $Q = 0$, hence $\tau = 0$. The shearing stress reaches its maximum value at the level ED. For this Q is equal to the cross-hatched area shown in Fig.

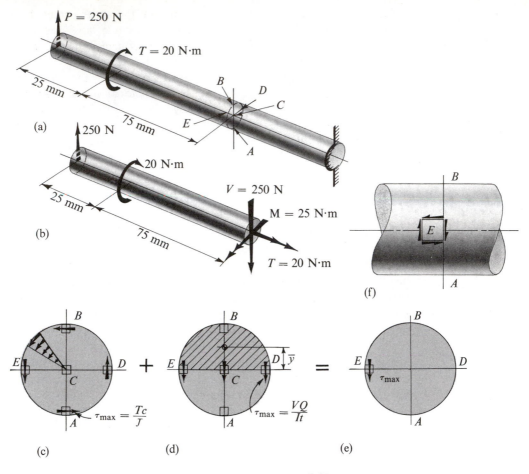

$P = 250$ N

$T = 20$ N·m

25 mm

75 mm

(a)

B D

C

E

A

250 N

20 N·m

25 mm

75 mm

(b)

$V = 250$ N

$M = 25$ N·m

$T = 20$ N·m

B

E

A

(f)

B

E C D

A

$\tau_{\max} = \dfrac{Tc}{J}$

(c)

$+$

B

E C D $\bar{y}$

A

$\tau_{\max} = \dfrac{VQ}{It}$

(d)

$=$

B

E D

$\tau_{\max}$

A

(e)

Fig. 7-17

7-17(d) multiplied by the distance from its centroid to the neutral axis. The latter quantity is $\bar{y} = 4c/(3\pi)$, where c is the radius of the cross-sectional area. Hence

$$Q = \left(\frac{\pi c^2}{2}\right)\left(\frac{4c}{3\pi}\right) = 2c^3/3$$

Moreover, since $t = 2c$, and $I = J/2 = \pi c^4/4$, the maximum direct shearing stress is

$$\tau_{\max} = \frac{VQ}{It} = \frac{V}{2c}\frac{2c^3}{3}\frac{4}{\pi c^4} = \frac{4V}{3\pi c^2} = \frac{4V}{3A}$$

where A is the *entire* cross-sectional area of the rod. (A similar expression was derived in Example 6-3 for a beam of rectangular section.) In Fig. 7-17(d) this shearing stress is shown acting down on the elementary areas at E, C, and D. This direction agrees with the direction of the shear V.

To find the maximum compound shearing stress in the plane A-B, the stresses shown in Figs. 7-17(c) and (d) are superposed. Inspection shows that the maximum shearing stress is at E, since in the two diagrams the shearing stresses at E act in the same direction. There are no direct shearing stresses at A and B, while at C there is no torsional shearing stress. The two shearing stresses act in *opposite* directions at D. The five points A, B, C, D, and E thus considered for the compound shearing stress are all that may be adequately treated by the methods developed in this text. However, this procedure selects the point where the maximum shearing stress acts.

$$J = \frac{\pi d^4}{32} = \frac{\pi (10)^4}{32} = 9.82 \times 10^2 \text{ mm}^4 = 9.82 \times 10^{-10} \text{ m}^4$$

$$I = \frac{J}{2} = 4.91 \times 10^{-10} \text{ m}^4$$

$$A = \frac{1}{4}\pi d^2 = 78.5 \text{ mm}^2 = 78.5 \times 10^{-6} \text{ m}^2$$

$$(\tau_{\text{max}})_{\text{torsion}} = \frac{Tc}{J} = \frac{20(0.005)}{9.82 \times 10^{-10}} = 102 \times 10^6 \text{ Pa}$$

$$(\tau_{\text{max}})_{\text{direct}} = \frac{VQ}{It} = \frac{4V}{3A} = \frac{4(250)}{3(78.5 \times 10^{-6})} = 4.25 \times 10^6 \text{ Pa}$$

$$\tau_E \approx (102 + 4)10^6 = 106 \times 10^6 \text{ Pa} = 106 \text{ MPa}$$

A planar representation of the shearing stress at E with the matching stresses on the longitudinal planes is shown in Fig. 7-17(f). No normal stress acts on this element as it is located on the neutral axis.

*7-8. STRESSES IN CLOSELY COILED HELICAL SPRINGS

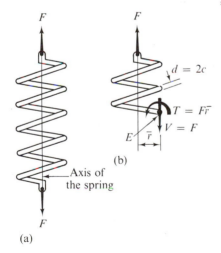

Fig. 7-18. Closely coiled helical spring

Helical springs such as the one shown in Fig. 7-18(a), are often used as elements of machines. With certain limitations, these springs may be analyzed for stresses by a method similar to the one used in the preceding example. The discussion will be limited to springs manufactured from rods or wires of circular cross section.* Moreover, *any one coil of such a spring will be assumed to lie in a plane which is nearly perpendicular to the axis of the spring*. This requires that the adjoining coils be close together. With this limitation, a section taken perpedicular to the axis of the spring's rod becomes *nearly vertical*.† Hence to maintain equilibrium of a segment of the spring, only a shearing force $V = F$

*For a complete discussion on springs see A. M., Wahl, *Mechanical Springs*, Cleveland, Ohio: Penton Publishing Co., 1944.

†This eliminates the necessity of considering an axial force and a bending moment at the section taken through the spring.

and a torque $T = F\bar{r}$ are required at *any* section through the rod, Fig. 7-18(b).*
Note that $\bar{r}$ is the distance from the axis of the spring to the *centroid of the rod's cross-sectional area.*

The maximum shearing stress at an arbitrary section through the rod could be obtained as in the preceding example, by superposing the torsional and the direct shearing stresses. This maximum shearing stress occurs at the inside of the coil at point E, Fig. 7-18(b). However, in the analysis of springs it has become *customary to assume that the shearing stress caused by the direct shearing force is uniformly distributed over the cross-sectional area of the rod.* Hence, the nominal direct shearing stress for any point on the cross section is $\tau = F/A$. Superposition of this *nominal* direct and the torsional shearing stress at E gives the maximum compound shearing stress. Thus since $T = F\bar{r}$, $d = 2c$, and $J = \pi d^4/32$,

$$\tau_{\max} = \frac{F}{A} + \frac{Tc}{J} = \frac{Tc}{J}\left(\frac{FJ}{ATc} + 1\right) = \frac{16F\bar{r}}{\pi d^3}\left(\frac{d}{4\bar{r}} + 1\right) \qquad (7\text{-}5)$$

It is seen from this equation that as the diameter of the rod d becomes small in relation to the coil radius $\bar{r}$, the effect of the direct shearing stress becomes small. On the other hand, if the reverse is true, the first term in the parenthesis becomes important. However, in the latter case the results indicated by Eq. 7-5 are considerably in error, and Eq. 7-5 should not be used, as it is based on the torsion formula for *straight rods.* As d becomes numerically comparable to $\bar{r}$, the length of the inside fibers of the coil differs greatly from the length of the outside fibers, and the assumptions of strain used in the torsion formula are not applicable.

The spring problem has been solved exactly† by the methods of the mathematical theory of elasticity, and while these results are complicated, for any one spring they may be made to depend on a single parameter $m = 2\bar{r}/d$, which is called the *spring index.* Thus Eq. 7-5 may be rewritten as

$$\tau_{\max} = K\frac{16F\bar{r}}{\pi d^3} \qquad (7\text{-}5a)$$

where K may be interpreted as a stress-concentration factor for closely coiled helical springs made from circular rods. A plot of K vs. the spring index is

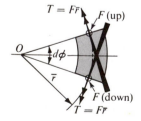

T = F\bar{r}, F (up), O, d\phi, \bar{r}, F (down), T = F\bar{r}

Fig. 7-B

*In previous work it has been reiterated that if a shear is present at a section, a change in the bending moment must take place along the member. Here a shear acts at every section of the rod, yet no bending moment nor a change in it appears to occur. This is so only because the rod is *curved.* An element of the rod viewed from the top is shown in the figure. At both ends of the element the torques are equal to $F\bar{r}$, and, using vectorial representation, act in the directions shown. The component of these vectors toward the axis of the spring O, resolved at the point of intersection of the vectors, $2F\bar{r}\,d\phi/2 = F\bar{r}\,d\phi$, opposes the couple developed by the vertical shears $V = F$, which are $\bar{r}\,d\phi$ apart.

†O. Goehner, "Die Berechnung Zylindrischer Schraubenfedern," *Zeitschrift des Vereins deutscher Ingenieure*, March 1932, vol. 76: 1, p. 269.

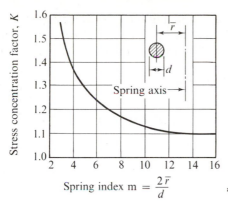

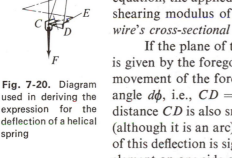

Fig. 7-19. Stress-concentraton factor for helical round-wire compression or tension springs

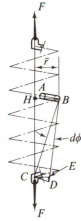

Fig. 7-20. Diagram used in deriving the expression for the deflection of a helical spring

shown* in Fig. 7-19. For heavy springs the spring index is small, hence the stress-concentration factor K becomes very important. For all cases the factor K accounts for the correct amount of direct shearing stress. Very high stresses are commonly allowed in springs because high-strength materials are used in their fabrication. For good quality spring steel, working shearing stresses range anywhere from 200 MPa to 700 MPa.

*7-9. DEFLECTION OF CLOSELY COILED HELICAL SPRINGS

As the subject of closely coiled helical springs was introduced above, for completeness, their deflection will be discussed in this article. Attention will be confined to closely coiled helical springs with a large spring index, i.e., the diameter of the wire will be assumed small in comparison with the radius of the coil. This permits the treatment of an element of a spring between two closely adjoining sections through the wire as a *straight circular bar in torsion*. The effect of direct shear on the deflection of the spring will be ignored. This is usually permissible as the latter effect is small.

Consider a helical spring such as shown in Fig. 7-20. A typical element AB of this spring is subjected throughout its length to a torque $T = F\bar{r}$. This torque causes a relative rotation between the two adjoining planes A and B, and with sufficient accuracy the amount of this rotation may be obtained by using Eq. 3-8, $d\phi = T\,dx/(JG)$, for straight circular bars. For this equation, the applied torque $T = F\bar{r}$, dx is the length of the element, G is the shearing modulus of elasticity, and J is the polar moment of inertia of the *wire's cross-sectional area*.

If the plane of the wire A is imagined fixed, the rotation of the plane B is given by the foregoing expression. The contribution of this element to the movement of the force F at C is equal to the distance BC multiplied by the angle $d\phi$, i.e., $CD = BC\,d\phi$. However, since the element AB is small, the distance CD is also small, and this distance may be considered perpendicular (although it is an arc) to the line BC. Moreover, only the vertical component of this deflection is significant, as in a spring consisting of many coils, for any element on one side of the spring there is a corresponding equivalent element on the other. The diametrically opposite elements of the spring balance out the horizontal component of the deflection and permit only the vertical

*An analytical expression which gives the value of K within 1 or 2% of the true value is frequently used. This expression in terms of the spring index m is $K_1 = (4m - 1)/(4m - 4) + 0.615/m$. It was derived by A. M. Wahl on the basis of some simplifying assumptions and is known as the *Wahl correction factor* for curvature in helical springs.

deflection of the force F. Therefore, by finding the *vertical* increment ED of the deflection of the force F due to an element of the spring AB and summing such increments for *all* elements of the spring, the deflection of the whole spring is obtained.

From similar triangles CDE and CBH,

$$\frac{ED}{CD} = \frac{HB}{BC} \quad \text{or} \quad ED = \frac{CD}{BC}HB$$

However, $CD = BC\,d\phi$, $HB = \bar{r}$, and ED may be denoted by $d\Delta$, as it represents an infinitesimal vertical deflection of the spring due to rotation of an element AB. Whence $d\Delta = \bar{r}\,d\phi$ and

$$\Delta = \int d\Delta = \int \bar{r}\,d\phi = \int_0^L \bar{r}\frac{T\,dx}{JG} = \frac{TL\bar{r}}{JG}$$

However, $T = F\bar{r}$, and for a closely coiled spring the *length L of the wire* may be taken with sufficient accuracy as $2\pi\bar{r}N$, where N is the number of *live* or *active* coils of the spring. Hence the deflection Δ of the spring is*

$$\Delta = \frac{2\pi F\bar{r}^3 N}{JG} \tag{7-6}$$

or if the value of J for the wire is substituted,

$$\Delta = \frac{64F\bar{r}^3 N}{Gd^4} \tag{7-6a}$$

Equations 7-6 and 7-6a give the deflection of a closely coiled helical spring along its axis when such a spring is subjected to either a tensile or compressive force F. In these formulas the effect of the direct shearing stress on the deflection is neglected, i.e., they give only the effect of torsional deformations.

The behavior of a spring may be conveniently defined by a force required to deflect the spring a unit distance. This quantity is known as the *spring constant*. It is designated in this text by k. From Eq. 7-6a the spring constant for a helical spring made from a wire with a circular cross section is

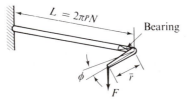

$$k = \frac{F}{\Delta} = \frac{Gd^4}{64\bar{r}^3 N} \quad \left[\frac{\text{N}}{\text{m}}\right] \quad \text{or} \quad \left[\frac{\text{lb}}{\text{in.}}\right]$$

*A convenient set-up for deriving the deflection of a helical spring consists of imagining the spring unwound as shown in the figure. The deflection $\bar{r}\varphi = \Delta$ of the force F due to the torsional effect on this rod is equivalent to the deflection of a helical spring with the same dimensions.

Fig. 7-C

7-1. A W 14 × 61 beam is loaded with a uniformly distributed load of 2 kips/ft, including its own weight and an axial tensile force of 120 kips. Determine the maximum normal stress if the beam spans 10 ft. (*Hint:* see Example 4-6.) *Ans.* 9.95 ksi.

7-2. A W 10 × 49 connecting beam 8 ft. long is subjected to a pull *P* of 100 kips as shown in the figure. At the ends where the pin connections are made the beam is reinforced with doubler plates, although parts of the flanges are removed. Determine the maximum flange stress in the middle of the member caused by the applied forces *P*. Qualitatively briefly discuss the load transfer at the ends. Where most likely would the highest stressed regions in this member be?

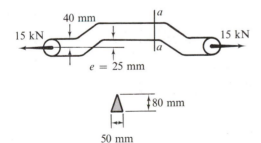

PROB. 7 – 2

7-3. A machine part for transmitting a pull of 15 kN is off-set as shown in the figure. Find the largest normal stress in the off-set portion of the member. *Ans:* −50.7 MN/m².

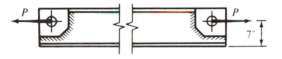

PROB. 7 – 3

7-4. An offset link is similar to the one shown in the preceding problem but is larger and has a

cross section in the form of a *T*, see figure. At the ends of the link the tensile forces *P* are applied 4 in. above the bottom of the flanges, and the offset $e = 2\frac{1}{2}$ in. from the line of action of these forces. Find the maximum stress if *P* = 40 kips and the material behaves elastically.

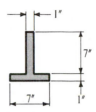

PROB. 7-4

7-5. A beam having the cross-sectional dimensions shown in the figure is subjected at a given section to the following forces: a bending moment of + 20 kN·m a total vertical shear of +20 kN, and an axial thrust of 30 kN. Determine the resultant normal force on the shaded part of the cross-section. *Ans:* −82.5 kN.

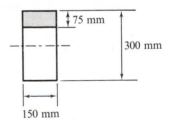

PROB. 7 – 5

7-6. Compute the maximum compressive stress acting normal to section *a–a* of the mast of Problem 4-19.

7-7. A large hook fabricated from a structural steel *T* (AISC designation: WT 4B) is loaded as shown in the figure. Determine the largest normal stress at the built-in end. For this section, *A* = 2.22 in.² and I_0 = 3.29 in.⁴ *Ans:* 20 ksi.

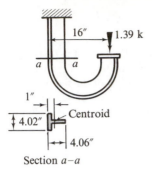

Section $a-a$

PROB. 7 – 6

7-8. A cast iron frame for a punch press has the proportions shown in the figure. What force P may be applied to this frame controlled by the stresses in the sections such as a–a, if the allowable stresses are 4,000 psi tension and 12,000 psi in compression? *Ans:* 9.12 k.

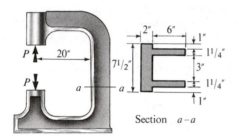

Section $a-a$

PROB. 7– 8

7-9. A short 100 mm square steel bar with a 50 mm diameter axial hole is built in at the base

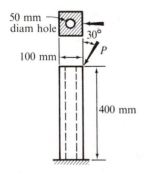

PROB. 7 – 9

and is loaded at the top as shown in the figure. Neglecting the weight of the bar, determine the value of the force P so that the maximum normal stress at the built-in end would not exceed 140 MPa. *Ans:* 129 kN.

7-10. An inclined beam having a cross-section of 0.20 m by 0.30 m supports a downward load as shown in the figure. Determine the maximum stress acting normal to the section A-A. Assume no eccentricity of the load or reactions and neglect the weight of the member.

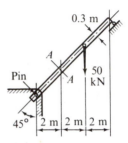

PROB. 7 – 10

7-11. A machine part having cross-sectional dimensions of 30 mm by 10 mm is loaded as shown in the figure. Determine the largest stress acting normal to the section A-A caused by the applied force. All dimensions given in the figure are in mm.

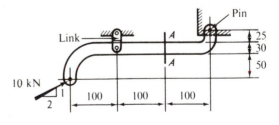

PROB. 7 – 11

7-12. A force of 169.8 kips is applied to a bar BC at C as shown in the figure. Find the maximum stress acting normal to the section A-A. The member BC is made from a piece of 6 in. by 6 in. steel bar. Neglect the weight of the bar. *Ans:* −18 ksi.

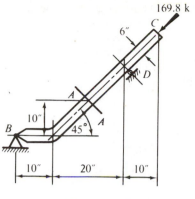

169.8 k

PROB. 7 – 12

7-13. A factory stairway having the centerline dimensions shown in the figure is made from two 9-in., 13.4-lb steel channels on edge separated by treads framing into them. The loading on each channel, including its own weight, is estimated to be 200 lb per foot of horizontal projection. Assuming that the lower end of the stairway is pinned and that the wall provides only horizontal support at the top, find the largest normal stress in the channels 5 ft above the floor level.   *Ans:* −2,600 psi.

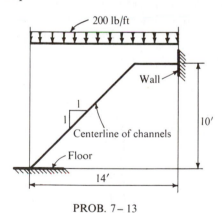

200 lb/ft

Wall

Centerline of channels

Floor

14′   10′

PROB. 7 – 13

7-14. Revise Prob. 7-13 by assuming that the upper ends of the stair channels are pinned, and that the lower ends can slide freely horizontally. *Ans:* −5,170 psi.

7-15. Calculate the maximum compressive stress acting on section *a–a* caused by the applied

load for the structure shown in the figure. The cross section at section *a–a* is that of a solid circular bar of 2-in. diameter. *Ans:* −1,110 psi.

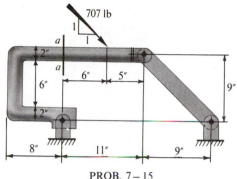

707 lb

PROB. 7 – 15

7-16. If in Prob. 4-31 the post is a circular timber 0.30 m. in diameter, what is the largest tensile stress acting normal to a section 2.5 m above the bottom of the post?

7-17. Compute the maximum compressive stress acting normal to the section *a–a* for the structure shown in the figure. The post *AB* has a 12-in.-by-12-in. cross section. Neglect the weight of the structure.   *Ans:* −174 psi.

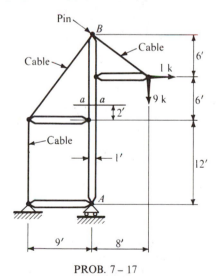

Pin

Cable

1 k

9 k

Cable

PROB. 7 – 17

7-18. Determine the largest stress acting normal to the *critical section* through the member *AB* caused by the applied force of 11.3 kips. The joint C is "rigid" and the member AB is made of an W 8 × 31 section. (*Hint:* check sections just above and just below C.) *Ans:* −13 ksi.

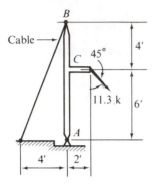

PROB. 7 – 18

7-19. A pin-joined bracket is loaded as shown in the figure. Determine the largest stress acting normal to the section A-A caused by the applied force of 18.4 kN. The cross-sectional area at A-A is 40 mm by 30 mm. All dimensions shown in the figure are in mm.

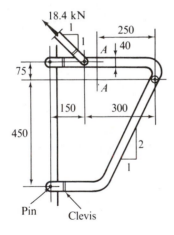

PROB. 7 – 19

7-20. A jib crane is made from an 8-in., 18.4-lb, steel I-shaped beam and a high-strength steel rod as shown in the figure. (a) Find the location of the movable load P that would cause the largest bending moment in the beam. Neglect the weight of the beam. (b) Using the load location

found in (a), how large may the load P be? Assume that the effect of shear in the beam is not significant, and let the allowable normal stress in the beam be 18,000 psi. Comment on the accuracy of the criterion established in (a). *Ans:* (a) 5 ft., (b) 7,940 lbs.

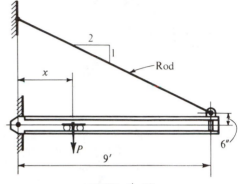

PROB. 7′– 20

7-21. A steel frame fabricated from W 8 × 17 steel sections supports a load P at a distance d from the center of the vertical column as shown in the figure. On the outside of the column at a distance 5 ft from the ground the following strains were measured: at A, $\varepsilon = 200 \times 10^{-6}$ in. per inch; and at B, $\varepsilon = -600 \times 10^{-6}$ in. per inch. What are the magnitudes of the load P and the distance d? let $E = 30 \times 10^6$ psi. *Ans:* 30 k, 5.63 in.

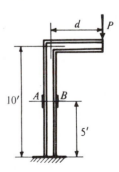

PROB. 7 – 21

7-22. A bar having a 0.10 m by 0.10 m cross-section is subjected to a single force F as shown in the figure. The longitudinal stresses on the extreme fibers at two sections 0.20 m apart are determined experimentally to be: $\sigma_A = 0$; $\sigma_B = -30$ MPa; $\sigma_C = -24$ MPa; $\sigma_D = -6$ MPa. Determine the magnitude of the vertical and horizontal components of the force F.

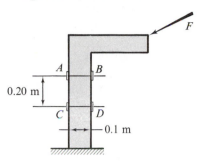

PROB. 7 – 22

7-23. In order to obtain the magnitude of an eccentric vertical load F on a tee-shaped steel column, strain gages are attached at A and B as shown in the figure. Determine the value of F if the longitudinal strain at A is -100×10^{-6} in./in. and at B is -800×10^{-6} in./in. Let $E = 30$

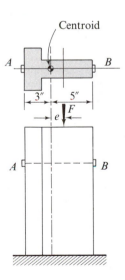

PROB. 7 – 23

$\times 10^6$ psi, $G = 12 \times 10^6$ 'psi. The cross-sectional area of the column is 24 in.^{2} *Ans:* 261 k.

7-24. A steel hook, having the proportions in the figure, is subjected to a downward load of 19 kips. The radius of the centroidal curved axis is 6 in. Determine the maximum stress in this hook.

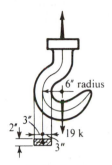

PROB. 7 – 24

7-25. A steel bar of 50 mm diameter is bent into a nearly complete circular ring of 300 mm outside diameter as shown in the figure. (a) Calculate the maximum stress in this ring caused by applying two 10 kN forces at the open end. (b) Find the ratio of the maximum stress found in (a) to the largest compressive stress acting normal to the same section.

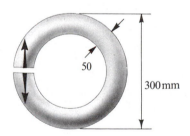

PROB. 7 – 25

7-26. The dimensions of a small concrete dam, retaining a water surface level with its crest, are shown in the figure. Assuming that concrete is capable of resisting some tension, determine the stresses acting normal to a horizontal section 2.0 m below the top. Assume that water weighs 10 kN/m^3 and concrete 25 kN/m^3.

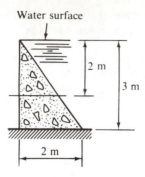

PROB. 7 – 26

7-27. What should the total height h of the dam shown in the cross-sectional view be so that the foundation pressure at A is just zero? Assume that water weighs 62.5 lb/ft³ and concrete 150 lb/ft³. *Ans: 27.6 ft.*

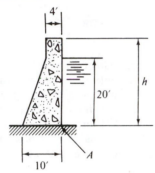

PROB. 7 – 27

7-28. What must the thickness t of a rectangular concrete dam 1.5 m high be in order to retain a water level even with its crest, as shown in the figure, without causing tension on the foundation at the upstream face? Assume the same unit weights as in Prob. 7-26. *Ans: 1.04 m.*

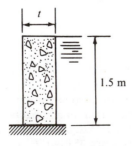

PROB. 7 – 28

7-29. A short block has cross-sectional dimensions in plan view as shown in the figure. Determine the range along the line A-A over which a downward vertical force could be applied to the top of the block without causing any tension at the base. Neglect the weight of the block. Dimensions given in the figure are in mm.

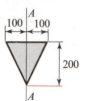

PROB. 7 – 29

7-30. The cross-sectional area in plan view of a short block is in the shape of an "arrow" as shown in the figure. Find the position of the vertical downward force on the line of symmetry of this section so that the stress at A is just zero. *Ans: 6.944 in. from A.*

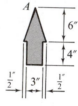

PROB. 7 – 30

7-31. Rework Example 7-5 by placing the vertical force P in line with the side AD and at a distance of 375 mm from the axis of symmetry.

7-32. If the block shown in Fig. 7-12(a) is made from steel weighing 75 kN/m³, find the magnitude of the force P necessary to cause zero stress at D. Neglect the weight of the small bracket supporting the load. For the same condition, locate the line of zero stress at the section $ABCD$.

7-33. An 8-ft-diameter steel stack, partially lined with brick on the inside, together with a 20-ft-by-20-ft concrete foundation pad weighs 160,000 lb. This stack projects 100 ft above the ground level, as shown in the figure, and is anchored to the foundation. If the horizontal wind pressure is assumed to be 20 lb per square foot of the projected area of the stack and the wind blows in the direction parallel to one of the sides of the square foundation, what is the maximum foundation pressure? *Ans:* 1.21 k per square foot.

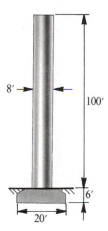

PROB. 7-33

7-34. A cast iron block is loaded as shown in the figure. Neglecting the weight of the block, determine the stresses acting normal to a section taken 0.5 m below the top and locate the line of zero stress. All dimensions given in the figure are in mm. *Ans:* ±24.6 MPa, ±9.6 MPa.

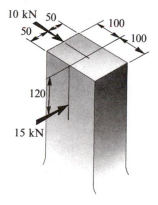

PROB. 7 – 34

7-35. An aluminum-alloy block is loaded as shown in the figure. The application of this load produces a tensile strain of 500×10^{-6} mm. per mm at A as measured by means of an electrical

strain gage. Compute the magnitude of the applied force P. Let $E = 10 \times 10^6$ kN/m². All dimensions given in the figure are in mm.

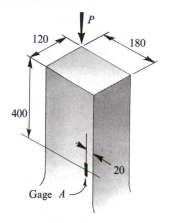

PROB. 7 – 35

7-36. If the application of an inclined force F at the centroid of the top surface of a member, as shown in the figure, causes a strain of -0.000100 at the gage A, what is the magnitude of the applied force P? Let $E = 210$ GPa.

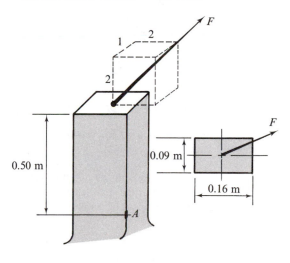

PROB. 7 – 36

7-37. A T-beam of linearly elastic-plastic material has the dimensions shown in the figure. (a) If the strain is $-\varepsilon_{yp}$ at the top of the flange and is zero at the juncture of the web with the flange, what axial force P and bending moment M act on the beam? Assume $\sigma_{yp} = 36$ ksi. (b) Determine the residual stress pattern which would develop upon removal of the forces in (a).

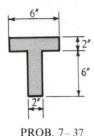

PROB. 7–37

7-38. A short compression member has the proportions shown in the figure; its $A = 72.9$ in.², $I_{zz} = 1,199$ in.⁴, and $I_{yy} = 633$ in.⁴ Determine the distance r along the diagonal where a longitudinal force P should be applied so that point A lies on the line of zero stress. Neglect the weight of the member. *Ans:* 53 in.

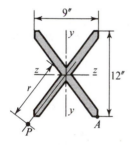

PROB. 7–38

7-39. Determine the kern for a member having a solid circular cross section.

7-40. A 6 m long, 150 mm by 200 mm beam is loaded in the middle of the span with an inclined concentrated force of 5 kN, as shown in the

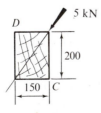

PROB. 7–40

cross-sectional view. Neglecting the weight of the beam, find the maximum bending stress and locate the neutral axis. *Ans: NA* passes through C and D.

7-41. A 6 m long, 150 mm by 200 mm beam is to be loaded in the middle of the span with an inclined concentrated force P, as shown in the cross-sectional view. If the maximum bending stress is 8 500 kN/m², neglecting the weight of the beam, what may the value of the force P be? All dimensions shown in the figure are in mm.

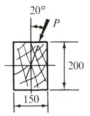

PROB. 7 – 41

7-42. A full-sized, 2 in. by 4 in. horizontal cantilever projects 4 ft from a concrete pier into which it is cast in a tilted position as shown in the figure. At the free end a vertical force of 100 lb is applied which acts through the centroid of the section. Determine the maximum flexural stress, caused by the applied force, in the beam at the built-in end and locate the neutral axis. Neglect the weight of the beam. *Ans:* ±1,423 psi, 3.34 in.

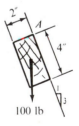

PROB. 7 – 42

7-43. An inclined force P acts on a cantilever beam in the plane shown in the cross-sectional view. At the section considered, the total resisting moment in the plane of the force is 8,000 ft-lb. Find the bending stress at point A. *Ans:* 325 psi.

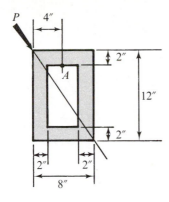

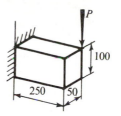

PROB. 7 – 45

PROB. 7 – 43

7-44. A tilted, simply supported beam with a depth to width ratio of 2 to 1 is to span 4 m and is to carry a uniformly distributed load of 15 kN per linear meter, including its own weight, applied as shown in the figure. (a) Determine the required dimensions of the beam so that the maximum stress due to bending does not exceed 10 MN/m². (b) Locate the neutral axis of the beam and show its position on the sketch.

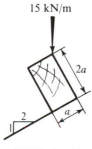

PROB. 7 – 44

7-45. A rectangular cantilever 250 mm long is loaded with $P = 50$ kN at the free end as shown in the figure. Determine the maximum shearing stress at the built-in end due to the direct shear and the torque. Show the result on a sketch analogous to Fig. 7-17(e). All dimensions shown in the figure are in mm. *Ans:* 35.3 MPa.

7-46. A helical compression spring is made from $\frac{1}{8}$ in. diameter phosphor-bronze wire and has an outside diameter of $1\frac{1}{4}$ in. If the allowable shearing stress is 30,000 psi, what force may be applied to this spring? Correct the answer for stress concentrations. *Ans:* 17.6 lb.

7-47. A helical valve spring is made of $\frac{1}{4}$ in. diameter steel wire and has an outside diameter of 2 in. In operation the compressive force applied to this spring varies from 20 lb minimum to 70 lb maximum. If there are 8 active coils, what is the valve lift (or travel), and what is the maximum shearing stress in the spring when in operation? $G = 11.6 \times 10^6$ psi. *Ans:* 0.38 in.

7-48. A helical spring is made of 12 mm diameter steel wire by winding it on a 120 mm diameter mandrel. If there are 10 active coils, what is the spring constant? $G = 82 \times 10^6$ kN/m². What force must be applied to the spring to elongate it 40 mm? *Ans:* 370 N.

7-49. If a helical tension spring consisting of 12 live coils of 6 mm steel wire and of 30 mm outside diameter is attached to the end of another helical tension spring of 18 live coils of 8 mm steel wire and of 40 mm outside diameter, what is the spring constant for this two spring system? What is the largest force that may be applied to these springs without exceeding a shearing stress of 480 MN/m²? $G = 82$ GN/m².

Note: Problems on bending of unsymmetric sections are at the end of Chapter 8.

8 Analysis of Plane Stress and Strain

8-1. INTRODUCTION

All fundamental formulas used in this text for determining the stresses at a section of a member have already been established. In the preceding chapter these formulas were classified into two groups. One group permits the determination of the normal stresses on the elements, the other, the shearing stresses. Superposition or compounding of like stresses was discussed in the same chapter. However, in some cases normal *and* shearing stresses may act *simultaneously* on an element of a member. For example, in a circular shaft that transmits torque with an axial force, all elements except those on the axis, simultaneously experience torsional shearing stresses and axial normal stresses. In fact, if an axial force, or a bending moment, acts with a shear, or a torque, some elements are subjected to both normal *and* shearing stresses. In these cases, both shearing and normal stresses are needed to describe completely all the stresses acting on an infinitesimal element, i.e., to define its *state of stress*. For example, the state of stress for an infinitesimal element A of a beam shown in Fig. 8-1(a) is given in Fig. 8-1(b). Using the procedures developed thus far, the planes that isolate this element are either parallel or perpendicular to the axis of the member. This method of isolating elements has been used throughout the text.

In this chapter it will be shown that by *changing the orientation* of an element, as defined by the angle θ for the element in Fig. 8-1(c), it is possible to describe the state of stress at a point in an *infinite number of ways*, which are all *equivalent*. In developing this procedure, a combination and resolution of normal and shearing stresses will be accomplished, whereas in the preceding chapter only the superposition of like stresses was treated.

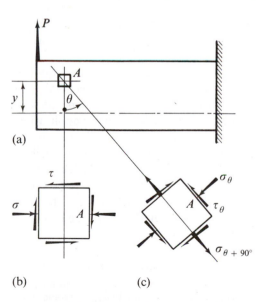

(a)

(b)

(c)

Fig. 8-1. The state of stress at a point described alternatively

For this, the *laws of transformation* of stress at a point will be developed, that is, equations will be derived that transform the information given by the conventional stress formulas into *equivalent stresses* acting on any plane through a given point. The planes where the normal or the shearing stresses reach their *maximum intensity* will also be determined, as the stresses associated with these planes have a particularly significant effect on materials.

The latter part of this chapter will deal with a topic parallel to the above for transformation of strains associated with one set of axes to a different set of axes. Some additional information on stress-strain relations for linearly elastic materials will also be presented.

8-2. THE BASIC PROBLEM

Although the components of stress at a point are vectorial in nature, *they are not ordinary vectors*. Mathematically they do not obey the laws of vector addition and subtraction. Stresses are vectors of a higher order,* because, in addition to having a magnitude and a sense, *they are also associated with the unit of area over which they act*. Hence, in combining the normal and the shearing stresses, the basic problem is solved by first converting the stresses into *forces*, which are vectors and consequently can be added or subtracted vectorially.

This procedure will be first illustrated on a numerical example. Then the developed approach will be generalized to obtain algebraic relations for a stress transformation, which enable one to obtain stresses on any inclined plane from a given state of stress. The methods used in these derivations do not involve properties of a material. Therefore, providing the initial stresses are given, the derived relations are applicable whether the material behaves elastically or plastically.

In deriving the laws of transformation of stress at a point, complete generality will be avoided in this text. Instead of treating a general three dimensional state of stress,† such as shown in Fig. 8-2(a), elements with stresses as shown in Fig. 8-2(b) will be considered. In practical applications this type of stress is particularly significant since it is usually possible to select at an outer boundary of a member one face of an element, such as *ABCD* in Fig. 8-2(b), which is free of significant surface stresses. On the other hand, the stresses acting on such elements right at the surface of a body are the highest ones stressed in a direction parallel to the surface. As before, for simplicity, the stresses acting on such elements will be shown as in Fig. 8-2(c).

---

*In the mathematical theory of elasticity they are termed second rank *tensors*.

†For a more general treatment of stress transformation the reader is referred to books on elasticity or plasticity. In the derivation developed here, in addition to σ_x, the normal stress σ_y is considered. The situation of having two normal stresses will be encountered in the next chapter in connection with thin shells.

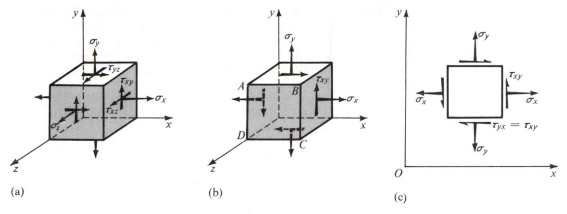

Fig. 8-2. Representation of stresses acting on an element

EXAMPLE 8-1

Let the state of stress for an element be as shown in Fig. 8-3(a). An alternative representation of the state of stress at the same point may be given on an infinitesimal wedge with an angle of $\alpha = 22\frac{1}{2}°$ as in Fig. 8-3(b). Find the stresses which must act on the plane AB of the wedge to keep the element in equilibrium.

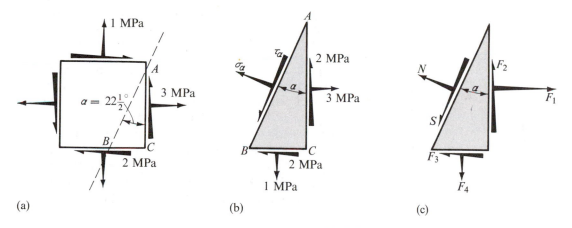

Fig. 8-3

SOLUTION

The wedge ABC is a part of the element in Fig. 8-3(a); therefore the stresses on the faces AC and BC are known. The unknown normal and shearing stresses acting on the face AB are designated in the figure by σ_α and τ_α, respectively. Their sense is assumed arbitrarily.

To determine σ_α and τ_α, for convenience only, let the area of the face defined by the line AB be 1 m². Then the area corresponding to the line AC is equal to (1) cos α = 0.924 m²; and that to BC is equal to (1) sin α = 0.383 m.² (More rigorously, the area corresponding to the line AB should be taken as *dA*, but this quantity cancels out in the subsequent algebraic expressions.) Forces F_1, F_2, F_3, and F_4, Fig. 8-3(c), can be obtained by multiplying the stresses by their respective areas. The unknown equilibrant forces N and S act respectively normal and tangential to the plane AB. Then, applying the equations of static equilibrium to the forces acting on the wedge gives the forces N and S.

$$F_1 = 3(0.924) = 2.78 \text{ MN} \qquad F_2 = 2(0.924) = 1.85 \text{ MN}$$

$$F_3 = 2(0.383) = 0.766 \text{ MN} \qquad F_4 = 1(0.383) = 0.383 \text{ MN}$$

$$\sum F_N = 0, \qquad N = F_1 \cos \alpha - F_2 \sin \alpha - F_3 \cos \alpha + F_4 \sin \alpha$$

$$= 2.78(0.924) - 1.85(0.383)$$

$$- 0.766(0.924) + 0.383(0.383)$$

$$= 1.29 \text{ MN}$$

$$\sum F_S = 0, \qquad S = F_1 \sin \alpha + F_2 \cos \alpha - F_3 \sin \alpha - F_4 \cos \alpha$$

$$= 2.78(0.383) + 1.85(0.924)$$

$$- 0.766(0.383) - 0.383(0.924)$$

$$= 2.12 \text{ MN}$$

The forces N and S act on the plane defined by AB, which was initially assumed to be 1 m². Their positive signs indicate that their assumed directions were chosen correctly. Dividing these forces by the area on which they act, the stresses acting on the plane AB are obtained. Thus $\sigma_\alpha = 1.29$ MPa and $\tau_\alpha = 2.12$ MPa and act in the direction shown in Fig. 8-3(b).

The foregoing procedure accomplished something remarkable. It transformed the description of the state of stress from one set of planes to another. Either system of stresses pertaining to an infinitesimal element describes the state of stress at the same point of a body.

The procedure of isolating a wedge and using the equations of the equilibrium of forces to determine stresses on inclined planes is fundamental. Ordinary sign conventions of statics suffice to solve any problem. The reader is urged to return to this approach whenever questions arise regarding the more advanced procedures developed in the remainder of this chapter.

8-3. EQUATIONS FOR THE TRANSFORMATION OF PLANE STRESS

Two algebraic expressions, one for the normal stress and one for the shearing stress, can be developed to give these stresses in terms of the initially known stresses and of an angle of inclination of the plane being investigated. The dependence of the stresses on the inclination of the plane thus becomes clearly apparent. The derivatives of these algebraic expressions with respect to the angle of inclination, when set equal to zero, locate the planes on which either the normal or the shearing stress reaches a maximum or minimum value. The stresses on these planes are of great importance in predicting the behavior of a given material.

The algebraic equations will be developed using an element, shown in Fig. 8-4(a), in a state of general plane stress. The normal tensile stresses are positive, and the compressive stresses are negative. Positive shearing stress is defined as *acting upward on the right face DE of the element*. The senses of the other shearing stresses follow from the equilibrium requirements. *This sign convention for shear stress is opposite to that established in Art. 4-7 and will be used only for this work.* Here the transformation of stresses is sought from the xy system of coordinate axes to the $x'y'$ system. The angle θ, which locates the x' axis, is positive when measured from the x axis toward the y axis in a counterclockwise direction.

By passing a plane BC normal to the x' axis through the element, the wedge in Fig. 8-4(b) is isolated. The plane BC makes an angle θ with the

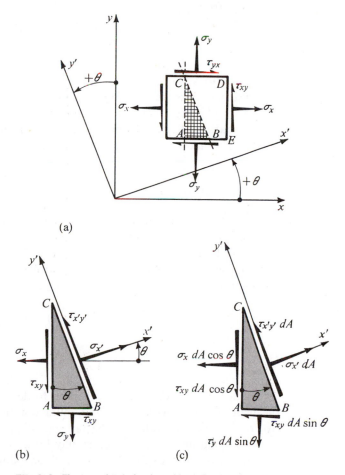

(a)

(b)   (c)

Fig. 8-4. Elements for derivation of formulas for stresses on an inclined plane

vertical axis, and, if this plane has an area dA, the areas of the faces AC and AB are $dA \cos \theta$ and $dA \sin \theta$, respectively. Multiplying the stresses by their respective areas, a diagram with the forces acting on the wedge can be constructed, Fig. 8-4(c). Then, by applying the equations of static equilibrium to the forces acting on the wedge, stresses $\sigma_{x'}$, and $\tau_{x'y'}$ are obtained:

$$\sum F_{x'} = 0, \qquad \sigma_{x'} dA = \sigma_x dA \cos \theta \cos \theta + \sigma_y dA \sin \theta \sin \theta$$
$$+ \tau_{xy} dA \cos \theta \sin \theta + \tau_{xy} dA \sin \theta \cos \theta$$
$$\sigma_{x'} = \sigma_x \cos^2 \theta + \sigma_y \sin^2 \theta + 2\tau_{xy} \sin \theta \cos \theta$$
$$= \sigma_x \frac{(1 + \cos 2\theta)}{2} + \sigma_y \frac{(1 - \cos 2\theta)}{2} + \tau_{xy} \sin 2\theta$$
$$\sigma_{x'} = \frac{\sigma_x + \sigma_y}{2} + \frac{\sigma_x - \sigma_y}{2} \cos 2\theta + \tau_{xy} \sin 2\theta \qquad (8\text{-}1)$$

Similarly, from $\sum F_{y'} = 0$,

$$\tau_{x'y'} = -\frac{\sigma_x - \sigma_y}{2} \sin 2\theta + \tau_{xy} \cos 2\theta \qquad (8\text{-}2)$$

Equations 8-1 and 8-2 are the general expressions for the normal and the shearing stress, respectively, on any plane located by the angle θ and caused by a known system of stresses. These relations are the equations for transformation of stress from one set of coordinate axes to another. Note particularly that σ_x, σ_y, and τ_{xy} are initially known stresses.

8-4. PRINCIPAL STRESSES

Interest often centers on the determination of the largest possible stress as given by Eqs. 8-1 and 8-2, and the planes on which such stresses occur will be found first. To find the plane for a maximum or a minimum normal stress, Eq. 8-1 is differentiated with respect to θ and the derivative set equal to zero, i.e.,

$$\frac{d\sigma_{x'}}{d\theta} = -\frac{\sigma_x - \sigma_y}{2} 2 \sin 2\theta + 2\tau_{xy} \cos 2\theta = 0$$

Hence $\qquad \tan 2\theta_1 = \dfrac{\tau_{xy}}{(\sigma_x - \sigma_y)/2} \qquad (8\text{-}3)$

where the subscript of the angle θ is used to designate the angle that defines the plane of the maximum or minimum normal stress. Equation 8-3 has two roots, since the value of the tangent of an angle in the diametrically opposite quadrants is the same, as may be seen from Fig. 8-5. These roots are 180° apart, and, as Eq. 8-3 is for a double angle, the roots of θ_1 are 90° apart. One of these roots locates a plane on which the maximum normal stress acts; the other locates the corresponding plane for the minimum nor-

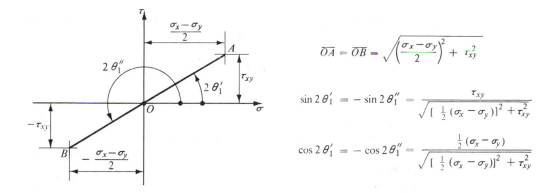

$$\overline{OA} = \overline{OB} = \sqrt{\left(\frac{\sigma_x - \sigma_y}{2}\right)^2 + \tau_{xy}^2}$$

$$\sin 2\theta_1' = -\sin 2\theta_1'' = \frac{\tau_{xy}}{\sqrt{[\frac{1}{2}(\sigma_x - \sigma_y)]^2 + \tau_{xy}^2}}$$

$$\cos 2\theta_1' = -\cos 2\theta_1'' = \frac{\frac{1}{2}(\sigma_x - \sigma_y)}{\sqrt{[\frac{1}{2}(\sigma_x - \sigma_y)]^2 + \tau_{xy}^2}}$$

Fig. 8-5. Angle functions for principal stresses

mal stress. To distinguish between these two roots, a prime and double prime notation is used.

Before evaluating the above stresses, carefully observe that if the location of planes on which no shearing stresses act is wanted, Eq. 8-2 must be set equal to zero. This yields the same relation as that in Eq. 8-3. Hence an important conclusion is reached: On planes on which maximum or minimum normal stresses occur, there are no shearing stresses. These planes are called the *principal planes* of stress, and the stresses acting on these planes— the maximum and minimum normal stresses—are called the *principal stresses*.

The magnitudes of the principal stresses can be obtained by substituting the values of the sine and cosine functions corresponding to the double angle given by Eq. 8-3 into Eq. 8-1. After this is done and the results are simplified, the expression for the maximum normal stress (denoted by σ_1) and the minimum normal stress (denoted by σ_2) becomes

$$(\sigma_{x'})_{\substack{\text{max} \\ \text{min}}} = \sigma_{1 \text{ or } 2} = \frac{\sigma_x + \sigma_y}{2} \pm \sqrt{\left(\frac{\sigma_x - \sigma_y}{2}\right)^2 + \tau_{xy}^2} \qquad (8\text{-}4)$$

where the positive sign in front of the radical must be used to obtain σ_1, and the negative sign to obtain σ_2. The planes on which these stresses act can be determined by using Eq. 8-3. A particular root of Eq. 8-3 substituted into Eq. 8-1 will check the result found from Eq. 8-4 and at the same time will locate the plane on which this principal stress acts.

8-5. MAXIMUM SHEARING STRESSES

If σ_x, σ_y, and τ_{xy} are known for an element, the shearing stress on any plane defined by an angle θ is given by Eq. 8-2, and a study similar to the one made above for the normal stresses may be made for the shearing stress. Thus, similarly, to locate the planes on which the maximum or the minimum shearing stresses act, Eq. 8-2 must be differentiated with respect to θ and the derivative set equal to zero. When this is carried out and the results are simplified, the operations yield

$$\tan 2\theta_2 = -\frac{(\sigma_x - \sigma_y)/2}{\tau_{xy}} \qquad (8\text{-}5)$$

where the subscript 2 is attached to θ to designate the plane on which the shearing stress is a maximum or a minimum. Like Eq. 8-3, Eq. 8-5 has two roots, which again may be distinguished by attaching to θ_2 a prime or a double prime notation. The two planes defined by this equation are mutually perpendicular. Moreover, the value of $\tan 2\theta_2$ given by Eq. 8-5 is a negative reciprocal of the value of $\tan 2\theta_1$ in Eq. 8-3. Hence the roots for the double angles of Eq. 8-5 are 90° away from the corresponding roots of Eq. 8-3. This means that the angles that locate the planes of maximum or minimum shearing stress form angles of 45° with the planes of the principal stresses. A substitution into Eq. 8-2 of the sine and cosine functions corresponding to the double angle given by Eq. 8-5 and determined in a manner analogous to that in Fig. 8-5 gives the maximum and the minimum values of the shearing stresses. These, after simplifications, are

$$\tau_{\substack{\max \\ \min}} = \pm\sqrt{\left(\frac{\sigma_x - \sigma_y}{2}\right)^2 + \tau_{xy}^2} \qquad (8\text{-}6)$$

Thus, the maximum shearing stress differs from the minimum shearing stress only in sign. Moreover, since the two roots given by Eq. 8-5 locate planes 90° apart, this result also means that the numerical values of the shearing stresses on the mutually perpendicular planes are the same. This concept was repeatedly used after being established in Art. 2-9. In this derivation the difference in sign of the two shearing stresses arises from the convention for locating the planes on which these stresses act. From the physical point of view these signs have no meaning, and for this reason the largest shearing stress regardless of sign will be called the *maximum shearing stress*.

The definite sense of the shearing stress can always be determined by direct substitution of the particular root of θ_2 into Eq. 8-2. A positive shearing stress indicates that it acts in the direction assumed in Fig. 8-4(b), and vice versa. The determination of the maximum shearing stress is of utmost importance for materials that are weak in shearing strength. This will be discussed further in the next chapter.

Unlike the principal stresses for which no shearing stresses occur on the principal planes, the maximum shearing stresses act on planes that are usually not free of normal stresses. Substitution of θ_2 from Eq. 8-5 into Eq. 8-1 shows that the normal stresses that act on the planes of the maximum shearing stresses are

$$\sigma' = \frac{\sigma_x + \sigma_y}{2} \qquad (8\text{-}7)$$

Therefore a normal stress acts simultaneously with the maximum shearing stress unless $\sigma_x + \sigma_y$ vanishes.

If σ_x and σ_y in Eq. 8-6 are the principal stress, τ_{xy} is zero and Eq. 8-6 simplifies to

$$\tau_{max} = \frac{\sigma_1 - \sigma_2}{2} \tag{8-8}$$

EXAMPLE 8-2

For the state of stress in Example 8-1, reproduced in Fig. 8-6(a), (a) rework the previous problem for $\theta = -22\frac{1}{2}°$, using the general equations for the transformation of stress; (b) find the principal stresses and show their sense on a properly oriented element; and (c) find the maximum shearing stresses with the associated normal stresses and show the results on a properly oriented element.

SOLUTION

Case (a). By directly applying Eqs. 8-1 and 8-2 for $\theta = -22\frac{1}{2}°$, with $\sigma_x = +3$ MPa, $\sigma_y = +1$ MPa, and $\tau_{xy} = +2$ MPa, one has

$$\sigma_{x'} = \frac{3+1}{2} + \frac{3-1}{2} \cos(-45°) + 2 \sin(-45°)$$

$$= 2 + 1(0.707) - 2(0.707) = +1.29 \text{ MPa}$$

$$\tau_{x'y'} = -\frac{3-1}{2} \sin(-45°) + 2 \cos(-45°)$$

$$= +1(0.707) + 2(0.707) = +2.12 \text{ MPa}$$

The positive sign of $\sigma_{x'}$ indicates tension; whereas the positive sign of $\tau_{x'y'}$ indicates that the shearing stress acts in the $+y'$ direction, as shown in Fig. 8-4(b). These results are shown in Fig. 8-6(b) as well as in Fig. 8-6(c).

Case (b). The principal stresses are obtained by means of Eq. 8-4. The planes on which the principal stresses act are found by using Eq. 8-3.

$$\sigma_{1 \text{ or } 2} = \frac{3+1}{2} \pm \sqrt{\left(\frac{3-1}{2}\right)^2 + 2^2} = 2 \pm 2.24$$

$$\sigma_1 = +4.24 \text{ MPa} \quad (\text{tension}), \qquad \sigma_2 = -0.24 \text{ MPa} \quad (\text{compression})$$

$$\tan 2\theta_1 = \frac{\tau_{xy}}{(\sigma_x - \sigma_y)/2} = \frac{2}{(3-1)/2} = 2$$

$$2\theta_1 = 63°26' \quad \text{or} \quad 63°26' + 180° = 243°26'$$

Hence $\qquad\qquad \theta_1' = 31°43' \quad$ and $\quad \theta_1'' = 121°43'$

This locates the two principal planes AB and CD, Figs. 8-6(d) and (e), on which σ_1 and σ_2 act. On which one of these planes the principal stresses act is unknown. So, Eq. 8-1 is solved by using, for example, $\theta_1' = 31°43'$. The stress found by this calculation is the stress that acts on the plane AB. Then,

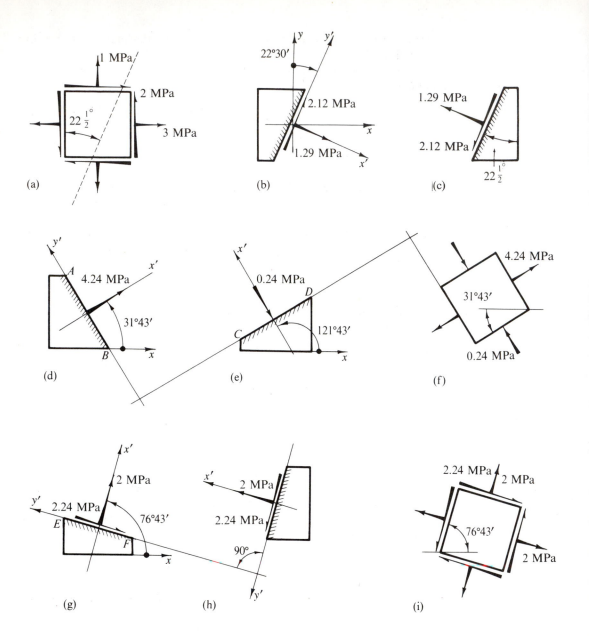

Fig. 8-6

since $2\theta'_1 = 63°26'$,

$$\sigma_{x'} = \frac{3+1}{2} + \frac{3-1}{2}\cos 63°26' + 2\sin 63°26' = +4.24\ \text{MPa} = \sigma_1$$

This result, besides giving a check on the previous calculations, shows that the maximum principal stress acts on the plane *AB*. The complete state of stress at the given point in terms of the principal stresses is shown in Fig. 8-6(f).

Case (c). The maximum shearing stress is found by using Eq. 8-6. The planes on which these stresses act are defined by Eq. 8-5. The sense of the shearing stresses is determined by substituting one of the roots of Eq. 8-5 into Eq. 8-2. Normal stresses associated with the maximum shearing stress are determined by using Eq. 8-7.

$$\tau_{max} = \sqrt{[(3-1)/2]^2 + 2^2} = \sqrt{5} = 2.24 \text{ MPa}$$

$$\tan 2\theta_2 = -\frac{(3-1)/2}{2} = -0.500$$

$$2\theta_2 = 153°26' \quad \text{or} \quad 153°26' + 180° = 333°26'$$

Hence $\quad\quad\quad \theta_2' = 76°43' \quad \text{and} \quad \theta_2'' = 166°43'$

These planes are shown in Figs. 8-6(g) and (h). Then, using $2\theta_2' = 153°26'$ in Eq. 8-2,

$$\tau_{x'y'} = -\frac{3-1}{2} \sin 153°26' + 2 \cos 153°26' = -2.24 \text{ MPa}$$

which means that the shear along the plane EF has an opposite sense to that in Fig. 8-4(b). From Eq. 8-7

$$\sigma' = \frac{3+1}{2} = 2 \text{ MPa}$$

The complete results are shown in Fig. 8-6(i).

The description of the state of stress can now be exhibited in three alternative forms: as the originally given data, and in terms of the stresses found in parts (b) and (c) of this problem. All these descriptions of the state of stress at the given point are equivalent.

8-6. AN IMPORTANT TRANSFORMATION OF STRESS

A significant transformation of one description of a state of stress at a point to another occurs when pure shearing stress is converted into principal stresses. For this purpose consider an element subjected only to shearing stresses τ_{xy} as in Fig. 8-7(a). Then from Eq. 8-4 the principal stresses $\sigma_{1 \text{ or } 2} = \pm\tau_{xy}$, i.e., numerically σ_1, σ_2, and τ_{xy} are all equal, although σ_1 is a tensile stress and σ_2 is a compressive stress. In this case, from Eq. 8-3 the principal planes are given by $\tan 2\theta_1 = \infty$, i.e., $2\theta_2 = 90°$ or $270°$. Hence $\theta_1' = 45°$ and $\theta_1'' = 135°$; the planes corresponding to these angles are shown in Fig. 8-7(b). To determine on which plane the tensile stress σ_1 acts, a substitution into Eq. 8-1 is made with $2\theta_1' = 90°$. This computation shows that $\sigma_1 = +\tau_{xy}$; hence the tensile stress acts perpendicular to the plane AB. Both principal stresses that are equivalent to the pure shearing stress are shown in Figs. 8-7(b) and (c). Therefore, whenever pure shearing stress is acting on an element it may be thought of as causing tension along one of

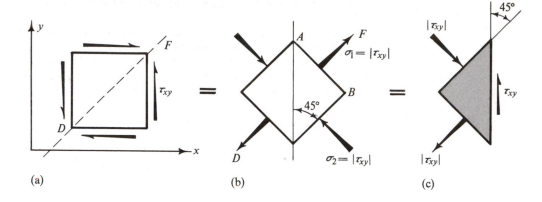

Fig. 8-7. Pure shearing stress is equivalent to tension-compression stresses acting on inclined planes at 45° to the shearing planes

the diagonals and compression along the other. The diagonal along which a tensile stress acts, such as *DF* in Fig. 8-7(a), is referred to as the *positive shear diagonal.*

From the physical point of view, the transformation of stress found agrees completely with intuition. The material "does not know" the manner in which its state of stress is described, and a little imagination should convince one that the tangential shearing stresses combine to cause pull along the positive shear diagonal and compression along the other diagonal.

8-7. MOHR'S CIRCLE OF STRESS

In this article the basic Eqs. 8-1 and 8-2 for the stress transformation at a point will be re-examined in order to interpret them graphically. In doing this, two objectives will be pursued. First, by graphically interpreting these equations, a greater insight into the general problem of stress transformation will be achieved. This is the main purpose of this article. Second, with the aid of graphical construction, a quicker solution of stress transformation problems can often be obtained. This will be discussed in the following article.

A careful study of Eqs. 8-1 and 8-2 shows that they represent a circle written in parametric form. That they do represent a circle is made clearer by first rewriting them as

$$\sigma_{x'} - \frac{\sigma_x + \sigma_y}{2} = \frac{\sigma_x - \sigma_y}{2} \cos 2\theta + \tau_{xy} \sin 2\theta \qquad (8\text{-}9)$$

$$\tau_{x'y'} = -\frac{\sigma_x - \sigma_y}{2} \sin 2\theta + \tau_{xy} \cos 2\theta \qquad (8\text{-}10)$$

Then by squaring both these equations, adding, and simplifying

$$\left(\sigma_{x'} - \frac{\sigma_x + \sigma_y}{2}\right)^2 + \tau_{x'y'}^2 = \left(\frac{\sigma_x - \sigma_y}{2}\right)^2 + \tau_{xy}^2 \qquad (8\text{-}11)$$

In every given problem σ_x, σ_y, and τ_{xy} are the three known constants, and $\sigma_{x'}$ and $\tau_{x'y'}$ are the variables. Hence Eq. 8-11 may be written in more compact form as

$$(\sigma_{x'} - a)^2 + \tau_{x'y'}^2 = b^2 \qquad (8\text{-}12)$$

where $a = (\sigma_x + \sigma_y)/2$ and $b^2 = [(\sigma_x - \sigma_y)/2]^2 + \tau_{xy}^2$ are constants.

This equation is the familiar expression of analytical geometry $(x - a)^2 + y^2 = b^2$ for a circle of radius b with its center at $(+a,0)$. Hence, if a circle satisfying this equation is plotted, the simultaneous values of a point (x, y) on this circle correspond to $\sigma_{x'}$ and $\tau_{x'y'}$ for a particular orientation of an inclined plane. The ordinate of a point on the circle is the shearing stress $\tau_{x'y'}$; the abscissa is the normal stress $\sigma_{x'}$. The circle so constructed is called a *circle of stress* or *Mohr's circle of stress*.*

A Mohr's circle based on the information for the given stresses in Fig. 8-8(a) is plotted in Fig. 8-8(b) with σ and τ as the coordinate axes. The center is located at $(a,0)$, and the radius equals b. Point A on the circle

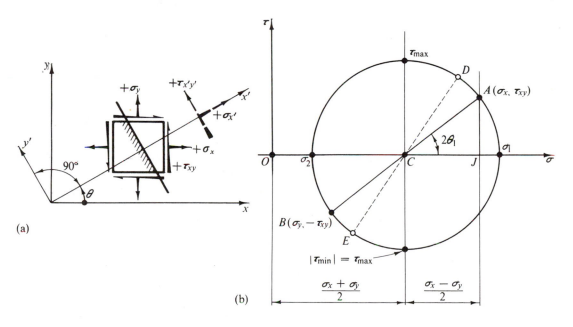

(a)

(b)

Fig. 8-8. Mohr's circle of stress

*It is so named in honor of Professor Otto Mohr of Germany, who in 1895 suggested its use in stress analysis problems.

corresponds to the stresses on the right face of the given element when $\theta = 0°$. For this point, $\sigma_{x'} = \sigma_x$, and $\tau_{x'y'} = \tau_{xy}$. As $AJ/CJ = \tau_{xy}/[(\sigma_x - \sigma_y)/2]$, according to Eq. 8-3, the angle ACJ is equal to $2\theta_1$.

With $\theta = 90°$ the x' axis is directed upward and the y' axis points to the left. From this orientation of the axes, the coordinates for point B on the circle are $\sigma_{x'} = \sigma_y$, and $\tau_{x'y'} = -\tau_{xy}$. The coordinates of points B and A satisfy Eq. 8-11. The same reasoning can be applied to any other pair of points, such as D or E, on the circle. The coordinates of such points give the stresses associated with a particular orientation of the $x'y'$ axes that define a plane passing through an element. All the possible ways of describing the stresses for an element for different θ's are represented by points on the Mohr's circle of stress. Therefore the following important conclusions regarding the state of stress at a point can be drawn:

1. The largest possible normal stress is σ_1; the smallest is σ_2. No shearing stresses exist together with either one of these principal stresses.
2. The largest shearing stress τ_{max} is numerically equal to the radius of the circle, also to $(\sigma_1 - \sigma_2)/2$. A normal stress equal to $(\sigma_1 + \sigma_2)/2$ acts on each of the planes of maximum shearing stress.
3. If $\sigma_1 = \sigma_2$, Mohr's circle degenerates into a point, and no shearing stresses at all develop in the xy plane.
4. If $\sigma_x + \sigma_y = 0$, the center of Mohr's circle coincides with the origin of the σ-τ coordinates, and the state of pure shear exists.
5. The sum of the normal stresses on any two mutually perpendicular planes is invariant, i.e.

$$\sigma_x + \sigma_y = \sigma_1 + \sigma_2 = \sigma_{x'} + \sigma_{y'} = \text{constant}$$

8-8. CONSTRUCTION OF MOHR'S CIRCLE OF STRESS

Mohr's circle of stress is widely used in practice for stress transformation. To be of value, the procedure must be rapid and simple. As an aid in application, the recommended procedure is outlined below. All the steps in constructing the circle can be justified on the basis of the previously developed relations. A typical Mohr's circle is in Fig. 8-9.

1. Make a sketch of the element for which the normal and the shearing stresses are known and indicate on this element the proper sense of these stresses. In an actual problem the faces of this element must have a precise relationship to the axes of a member being analyzed.
2. Set up a rectangular coordinate system of axes where the horizontal axis is the normal stress axis and the vertical axis is the shearing stress axis. Directions of positive axes are taken as usual, upward and to the right.
3. Locate the center of the circle, which is on the horizontal axis at a distance of $(\sigma_x + \sigma_y)/2$ from the origin. Tensile stresses are positive, compressive are negative.

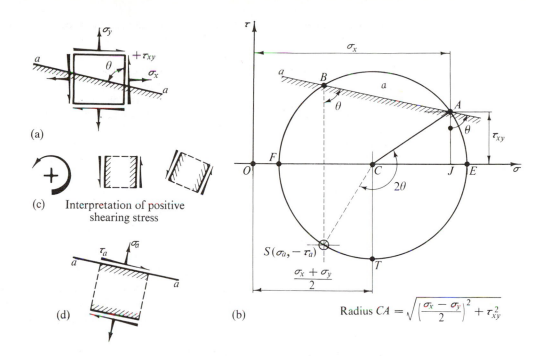

(a)

(c) Interpretation of positive shearing stress

(d)

(b)

$$\text{Radius } CA = \sqrt{\left(\frac{\sigma_x - \sigma_y}{2}\right)^2 + \tau_{xy}^2}$$

Fig. 8-9. Construction of Mohr's circle of stress

4. From the right face of the element prepared in Step (1), read the values for σ_x and τ_{xy} and plot the controlling point A on the circle. The coordinate distances to this point are measured from the origin. The sign of σ_x is positive if tensile, negative if compressive; that of τ_{xy} is positive if upward on the right face of the element, negative if downward.

5. Connect the center of the circle found in Step (3) with the point plotted in Step (4) and determine this distance, which is the radius of the circle.

6. Draw the circle using the radius found in Step (5). If only magnitudes and signs of stresses are of interest, this step completes the solution of the problem. The coordinates of points on the circle provide the required information.

7. To determine the direction and sense of the stresses acting on any inclined plane, draw through point A a line parallel to the inclined plane and locate point B on the circle. The coordinates of point S lying vertically on the opposite side of the circle from B give the stresses acting on the inclined plane. In Fig. 8-9(b) such stresses are identified as σ_a and $-\tau_a$.* A positive value of σ indicates a tensile stress, and vice versa. The sense of the shearing stress can be determined using the interpretation in Fig. 8-9(c). A tendency of the shearing stresses on two opposite faces of an element to cause counterclockwise rotation of the element is associated with a positive shearing

*Since the inclined plane a-a in Fig. 8-9(a) makes an angle θ with the right hand face of the element, the angle ABS in Fig. 8-9(b) is equal to θ. From the construction shown on this diagram it follows that the central angle ACS is equal to 2θ, and that σ_a and $-\tau_a$ satisfy Eqs. 8-1 and 8-2, respectively.

stress. On this basis the result $(+\sigma_a, -\tau_a)$ has the meaning shown in Fig. 8-9(d).

8. By proceeding in the reverse order, the plane on which the stresses associated with any point on the circle act can be found. Thus, drawing a line from A toward E or F, i.e., having the point corresponding to B coincide with one of these intercepts, determines the inclination of the plane on which the respective principal stresses act. For this special case the distance BS degenerates into a point. The principal stress given by the particular intercept (either E or F) acts normal to the line connecting this intercept point with point A. As before, positive stresses indicate tension, and vice versa.

By commencing with the highest or the lowest point on the circle, the planes on which the maximum shearing stresses and the associated normal stresses act can be found. For example, by imagining that point S is moved to T, the plane on which the stresses at T act is given by the new position of the line BA with point B moved to the highest point on the circle.

To solve the problems of stress transformation using Mohr's circle, the foregoing procedures can be applied graphically. However, it is recommended that trigonometric computations of the critical values be used in conjunction with the graphical construction. Then the work may be carried out on a crude sketch without scaling off the distances or angles, and the results will be accurate. Using Mohr's circle in this manner is equivalent to applying the basic equations of stress transformation.

There are other problems involving the rotation of axes that give rise to transformation equations that are *mathematically identical* to the stress transformation equations. This suggests the possibility of using the Mohr's circle construction by properly identifying the variables that correspond to σ_x, σ_y, τ_{xy}, $\sigma_{x'}$ and $\tau_{x'y'}$. The Mohr's circle of strain is one such example and is dealt with in Art. 8-12. Another useful application of the Mohr's circle is for finding the centroidal principal axes and the corresponding principal moments of inertia for an unsymmetrical cross section. The necessary equations are derived in the appendix to this chapter, where a numerical example is presented to illustrate the use of the Mohr's circle in this context.

EXAMPLE 8-3

Given the state of stress shown in Fig. 8-10(a), transform it (*a*) into the principal stresses, and (*b*) into the maximum shearing stresses and the associated normal stresses. Show the results for both cases on properly oriented elements.

SOLUTION

To construct Mohr's circle of stress, the following quantities are required.

1. Center of circle on the σ axis: $(-2 + 4)/2 = +1$ **MPa**

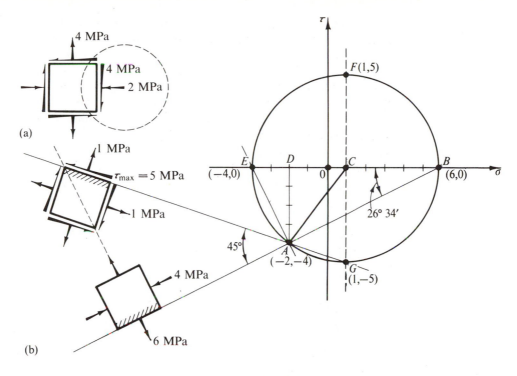

Fig. 8-10

2. Point A on circle from data on the right face of element: $(-2, -4)$ MPa

3. Radius of circle: $CA = \sqrt{CD^2 + DA^2} = 5$ MPa

After drawing the circle, one obtains $\sigma_1 = +6$ MPa, $\sigma_2 = -4$ MPa, and $\tau_{max} = 5$ MPa.

 Drawing a line from σ_1 at B toward A locates the plane on which the stress σ_1 acts. Similarly, beginning at point E and drawing a line toward A gives the plane on which the stress σ_2 acts. The maximum shearing stress τ_{max} and the associated normal stress σ' are given by the coordinates of point F. Directly downward at point G, the inclined line AG locates the plane on which $\tau_{max} = +5$ MPa and $\sigma' = +1$ MPa act.

 The complete results are shown on sketches in Fig. 8-10(b) on properly oriented elements. The angles shown are determined from suitable trigonometric relations. Thus since the tan $DBA = AD/DB = \frac{4}{8} = 0.5$, the angle $DBA = 26°34'$. The plane of maximum shear is located at $45°$ from the planes of principal stress. Of course the solution could have been made entirely by graphics.

 It is significant to note that the approximate direction of the algebraically larger principal stress found in the above example might have been anticipated. Instead of thinking in terms of the normal and the shearing stresses as given in the original data, Fig. 8-11(a), an equivalent problem in Fig. 8-11(b) may be considered. Here the shearing stresses have been replaced by the equivalent tension-compression stresses acting along the proper shear diagonals. Then, for qualitative reasoning, the outline of the original element may be obliterated, and the tensile stresses may be singled out as in Fig. 8-11(c). From this new diagram it is apparent that regardless of the magnitudes of the particular stresses involved, the resultant maximum tensile stress must act somewhere between the given tensile stress and the positive shear diagonal. In other words, *the line of action of the algebraically larger principal stress is "straddled" by the algebraically larger given normal stress and the positive shear diagonal.* The use of the negative shear diagonal, located at 90° to the positive shear diagonal, is helpful in visualizing this effect for cases where both given normal stresses are compressive, Figs. 8-11(d) and (e). This procedure provides a qualitative check on the orientation of an element for the principal stresses.

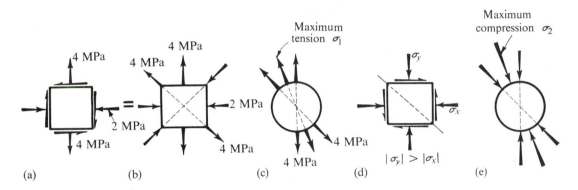

(a)   (b)   (c)   (d)   (e)

Fig. 8-11. A method for estimating the direction of the absolute maximum principal stresses

EXAMPLE 8-4

Using Mohr's circle, transform the stresses shown in Fig. 8-12(a) into stresses acting on the plane at an angle of $22\frac{1}{2}°$ with the vertical axis.

SOLUTION

Here the center of Mohr's circle is at $(3 + 1)/2 = +2$ MPa on the σ axis. The stresses on the right face of the element give (3,3) for the coordinates of point A on the circle. Therefore the radius of the circle is 3.16.

A line AB drawn parallel to the required inclined plane locates point B; directly above lies point D. The stresses acting on the required plane are given by the coordinates of point D. This solution is very easily accomplished by graphical construction; analytically the procedure is less direct. This type of graphical construction can be used effectively to provide a rapid qualitative check on analytical or experimental work.

For numerical work special trigonometric schemes may be devised in

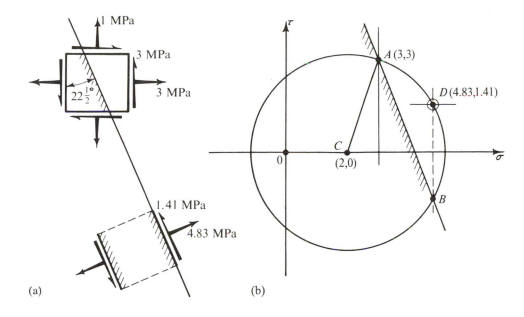

Fig. 8-12

each particular case. However, this approach often proves cumbersome, and direct application of Eqs. 8-1 and 8-2 is easier. Alternatively, one can always construct a wedge bounded by two axes and the inclined plane and solve the problem as illustrated in Example 8-1. In some instances the latter approach is least ambiguous.

*8-9. MOHR'S CIRCLE OF STRESS FOR THE GENERAL STATE OR STRESS

Stress transformation and the associated Mohr's circle of stress for it have been presented for a plane-stress problem. The treatment of the general three-dimensional stress transformation problem is beyond the scope of this book. However, some results of such an analysis are necessary for a more complete understanding of this subject. Therefore, several comments on transformation of the three-dimensional state of stress will now be made.

It is shown in books on elasticity and plasticity that any three dimensional state of stress (see Fig. 1-3 or 8-2(a)) can be transformed into three principal stresses acting in three orthogonal directions. This is a direct generalization of the case discussed earlier where two principal stresses were shown to act in two orthogonal directions in the plane-stress problem. An element after the appropriate stress transformation with three principal stresses acting on it is in Fig. 8-13(a). This element can be viewed from three different directions as in Fig. 8-13(b).

Corresponding to each projection of the element in Fig. 8-13(b) a Mohr's circle can be drawn using the procedures developed earlier. For

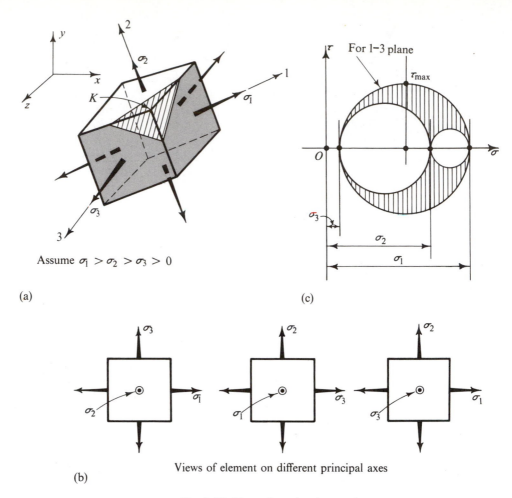

Assume $\sigma_1 > \sigma_2 > \sigma_3 > 0$

(a)

For 1–3 plane

τ_{max}

(c)

Views of element on different principal axes

(b)

Fig. 8-13. Three-dimensional state of stress

example, for an element situated in the 1–3 plane, the corresponding Mohr's circle to it passes through σ_1 and σ_3 as in Fig. 8-13(c). Analogous circles can be drawn for the 1–2 and the 2–3 planes. The three circles cluster together as in Fig. 8-13(c).

Next, suppose that instead of considering the planes on which principal stresses act, one considers an arbitrary plane such as the shaded plane K in Fig. 8-13(a). Then it can be shown* that the normal and shearing stresses acting on all such possible planes, when plotted as in Fig. 8-13(c), fall within the shaded part of the diagram. This means that the three circles already drawn give the limiting values of all possible stresses. This is an important fact and will be used in discussing material properties in a multiaxial state of stress.

*See O. Hoffman and G. Sachs, *Introduction to the Theory of Plasticity for Engineers*, New York: McGraw-Hill, 1953, p. 13.

In comparison with the general problem just presented, in the plane-stress problem $\sigma_3 = 0$. However, even in this less general problem the element is three-dimensional. Therefore, it is possible to study stresses on arbitrarily oriented planes corresponding to the plane K of Fig. 8-13(a). This has not been done earlier. With $\sigma_3 = 0$, three Mohr's circles are necessary to exhibit on a plot all the stresses on all the possible orientations of planes. For example, consider an element with $\sigma_1 = \sigma_2$, for which (from a two-dimensional point of view) Mohr's circle degenerates into a point. The same element, observed along different axes such as 1 and 3, with, for example, $\sigma_1 \neq 0$ and $\sigma_3 = 0$, generates a circle with a radius of $\sigma_1/2$. Thus, the direction from which an element is viewed is of the utmost importance.

*8-10. ANALYSIS OF PLANE STRAIN: GENERAL REMARKS

In the following articles the transformation of known strains associated with one set of axes or with known directions will be related to strains in any direction. It will be shown that the transformation of extensional and shearing strains completely resembles the transformation of normal and shearing stresses presented earlier. Thus, after establishing the strain transformation equations, Mohr's circle of strain will be introduced. Attention will be confined to the two-dimensional case, or more precisely to the plane-strain case, which means that $\varepsilon_z = \gamma_{zx} = \gamma_{zy} = 0$. The extension of the strain transformation to the general case involving Mohr's circle of strain for the three-dimensional problem will not be considered. Since the maximum strains usually occur on the free outer surfaces of a member, the two dimensional problem is by far the most important one.

In studying the strains at a point, only the relative displacement of the adjoining points is of importance. Translation and rotation of an element as a whole are of no consequence since these displacements are rigid-body displacements. For example, if the extensional strain of a diagonal ds of the element in Fig. 8-14(a) is being studied, the element in its deformed condition

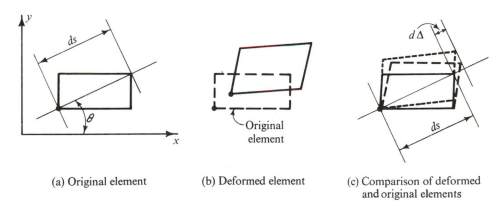

(a) Original element   (b) Deformed element   (c) Comparison of deformed and original elements

Fig. 8-14. Strains are determined on the basis of relative deformations

can be brought back for comparison purposes as in Fig. 8-14(c). It is immaterial whether the horizontal (dashed) or the vertical (dotted) sides of the deformed and the undeformed elements are matched to determine $d\Delta$. For the small strains considered throughout this text, the relevant quantity, elongation $d\Delta$ in the direction of the diagonal, is essentially the same regardless of the method of comparison employed.

In treating strains in the above manner, only kinematic questions have relevance. The mechanical properties of material do not enter the problem. However, after the main features of strain transformation have been presented, some additional relations between stress and strain for linearly elastic material will be given at the end of the chapter.

*8-11. EQUATIONS FOR THE TRANSFORMATION OF PLANE STRAIN

In establishing the equations for the transformation of strain, strict adherence to a sign convention is necessary. The sign convention used here is related to the one chosen for the stresses in Art. 8-3. The extensional strains ε_x and ε_y corresponding to elongations in the x and y directions, respectively, are taken positive. The shearing strain is considered positive if it elongates a diagonal having a positive slope in the xy coordinate system. For convenience in deriving the strain transformation equations, the element distorted by positive shearing strain will be taken as that shown in Fig. 8-15(a). As noted in the preceding article, this leads to perfectly general results providing the strains are small.

Next, suppose that the strains ε_x, ε_y, and γ_{xy} associated with the xy axes are known and extensional strain along some new x' axis is required. The new $x'y'$ system of axes is related to the xy axes as in Fig. 8-15(b). In these new coordinates, a length OA, which is dx' long, may be thought of as being a diagonal of a rectangular differential element dx by dy in the initial coordinates.

By considering the point O fixed in space, one can compute the displacements of point A caused by the imposed strains on a different basis in the two coordinate systems. The displacement in the x direction is $AA' = \varepsilon_x dx$; in the y direction, $A'A'' = \varepsilon_y dy$. For the shearing strain, assuming it causes the horizontal displacement shown in Fig. 8-15(a), $A''A''' = \gamma_{xy} dy$. The order in which these displacements occur is arbitrary. In Fig. 8-15(b), the displacement AA' is shown first, then $A'A''$, and finally $A''A'''$. By projecting these displacements onto the x' axis, one finds the displacement of point A along the x' axis. Then, recognizing that by definition $\varepsilon_x dx'$ in the $x'y'$ coordinate system is also the elongation of OA, one has the following equality:

$$\varepsilon_{x'} dx' = AA' \cos \theta + A'A'' \sin \theta + A''A''' \cos \theta$$

On substituting the appropriate expressions for the displacements and divid-

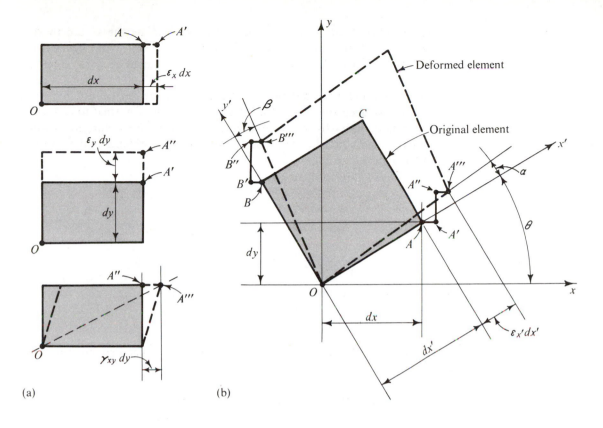

Fig. 8-15. Exaggerated deformations of elements for deriving strains along new axes

ing through by dx', one has

$$\varepsilon_{x'} = \varepsilon_x \frac{dx}{dx'} \cos\theta + \varepsilon_y \frac{dy}{dx'} \sin\theta + \gamma_{xy} \frac{dy}{dx'} \cos\theta$$

Since, however, $dx/dx' = \cos\theta$ and $dy/dx' = \sin\theta$

$$\varepsilon_{x'} = \varepsilon_x \cos^2\theta + \varepsilon_y \sin^2\theta + \gamma_{xy} \sin\theta \cos\theta \qquad (8\text{-}13)$$

Equation 8-13 is the basic expression for the strain in an arbitrary direction defined by the x' axis. To apply this equation, ε_x, ε_y, and γ_{xy} must be known. By use of the trigonometric identities already encountered in deriving Eq. 8-1, Eq. 8-13 may be rewritten as

$$\varepsilon_{x'} = \frac{\varepsilon_x + \varepsilon_y}{2} + \frac{\varepsilon_x - \varepsilon_y}{2} \cos 2\theta + \frac{\gamma_{xy}}{2} \sin 2\theta \qquad (8\text{-}14)$$

To complete the study of strain transformation at a point, shearing strain transformation must be also established. For this purpose, consider an element $OACB$ with sides OA and OB directed along the x' and the y' axes as in Fig. 8-15(b). By definition, the shearing strain for this element is the change in angle AOB. From the figure the change of this angle is $\alpha + \beta$.

For small deformations the small angle α can be determined by projecting the displacements AA', $A'A''$, and $A''A'''$ onto a normal to OA and dividing this quantity by dx'. In applying this approach the tangent of the angle is assumed equal to the angle itself. This is acceptable if the strains are small. Thus

$$\alpha \approx \tan \alpha = \frac{-AA' \sin \theta + A'A'' \cos \theta - A''A''' \sin \theta}{dx'}$$

$$= -\varepsilon_x \frac{dx}{dx'} \sin \theta + \varepsilon_y \frac{dy}{dx'} \cos \theta - \gamma_{xy} \frac{dy}{dx'} \sin \theta$$

$$= -(\varepsilon_x - \varepsilon_y) \sin \theta \cos \theta - \gamma_{xy} \sin^2 \theta$$

By analogous reasoning

$$\beta \approx -(\varepsilon_x - \varepsilon_y) \sin \theta \cos \theta + \gamma_{xy} \cos^2 \theta$$

Therefore, since the shearing strain $\gamma_{x'y'}$ of an angle included between the $x'y'$ axes is $\beta + \alpha$, one has

$$\gamma_{x'y'} = -2(\varepsilon_x - \varepsilon_y) \sin \theta \cos \theta + \gamma_{xy}(\cos^2 \theta - \sin^2 \theta)$$

or
$$\gamma_{x'y'} = -(\varepsilon_x - \varepsilon_y) \sin 2\theta + \gamma_{xy} \cos 2\theta \tag{8-15}$$

This is the second fundamental expression for the transformation of strain. Note that when $\theta = 0°$, the shearing strain associated with the xy axes is recovered.

Equations 8-14 and 8-15 for strain transformation are analogous to Eqs. 8-1 and 8-2 for stress transformation. This feature will be emphasized further in discussing Mohr's circle of strain.

*8-12. MOHR'S CIRCLE OF STRAIN

The two basic equations for the transformation of strains derived in the preceding article mathematically resemble the equations for the transformation of stresses derived in Art. 8-3. To achieve greater similarity between the appearances of the new equations and those of the earlier ones, Eq. 8-15 after division throughout by two is rewritten below as Eq. 8-16

$$\varepsilon_{x'} = \frac{\varepsilon_x + \varepsilon_y}{2} + \frac{\varepsilon_x - \varepsilon_y}{2} \cos 2\theta + \frac{\gamma_{xy}}{2} \sin 2\theta$$

$$\frac{\gamma_{x'y'}}{2} = -\frac{\varepsilon_x - \varepsilon_y}{2} \sin 2\theta + \frac{\gamma_{xy}}{2} \cos 2\theta \tag{8-16}$$

Since these strain transformation equations with the shearing strains divided by two are mathematically identical to the stress transformation Eqs. 8-1 and 8-2, Mohr's circle of strain can be constructed. In this construction every point on the circle gives two values: one for the extensional strain, the other for the shearing strain divided by two. Strains corresponding to elongation are positive; for contraction they are negative. Positive shearing strains distort the element as shown in Fig. 8-15(a). In plotting the circle the positive axes are taken in the usual manner, upward and to the right. The vertical axis is measured in terms of $\gamma/2$.

As an illustration of Mohr's circle of strain, consider that ε_x, ε_y, and $+\gamma_{xy}$ are given. Then on the $\varepsilon - \frac{1}{2}\gamma$ axes in Fig. 8-16 the center of the circle C is at $[(\varepsilon_x + \varepsilon_y)/2, 0]$ and, from the given data, point A on the circle is at $(\varepsilon_x, \gamma_{xy}/2)$. An examination of this circle leads to conclusions analagous to those reached before for the circle of stress.

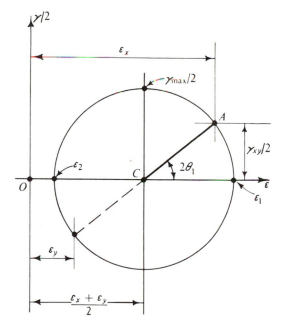

Fig. 8-16. Mohr's circle of strain

1. The maximum extensional strain is ε_1; the minimum is ε_2. These are the principal strains, and no shearing strains are associated with them. The directions of the extensional strains coincide with the directions of the principal stresses. As can be deduced from the circle, the analytical expression for the principal strains is

$$(\varepsilon_{x'})_{\substack{\max \\ \min}} = \varepsilon_{1\ \text{or}\ 2} = \frac{\varepsilon_x + \varepsilon_y}{2} \pm \sqrt{\left(\frac{\varepsilon_x - \varepsilon_y}{2}\right)^2 + \left(\frac{\gamma_{xy}}{2}\right)^2} \tag{8-17}$$

where the positive sign in front of the radical is to be used for ε_1, the maximum principal strain in the algebraic sense. The negative sign is to be used for ε_2, the minimum principal strain. The planes on which the principal strains act can be defined analytically from Eq. 8-16 by setting it equal to zero. Thus

$$\tan 2\theta_1 = \frac{\gamma_{xy}}{\varepsilon_x - \varepsilon_y} \tag{8-18}$$

since this equation has two roots, it is completely analogous to Eq. 8-3 and can be treated in the same manner.

2. The largest shearing strain γ_{max} is equal to two times the radius of the circle. Extensional strains of $(\varepsilon_1 + \varepsilon_2)/2$ in two mutually perpendicular directions are associated with the maximum shearing strain.

3. The sum of extensional strains in any two mutually perpendicular directions is invariant, i.e., $\varepsilon_1 + \varepsilon_2 = \varepsilon_x + \varepsilon_y = $ constant. Other properties of strains at a point can be established by studying the circle further.

EXAMPLE 8-5

It is observed that an element of a body contracts 0.00050 in. per inch along the x-axis, elongates 0.00030 in. per inch in the y direction, and distorts through an angle* of 0.00060 radian as in Fig. 8-17(a). Find the principal strains and determine the directions in which these strains act. Use Mohr's circle of strain to obtain the solution.

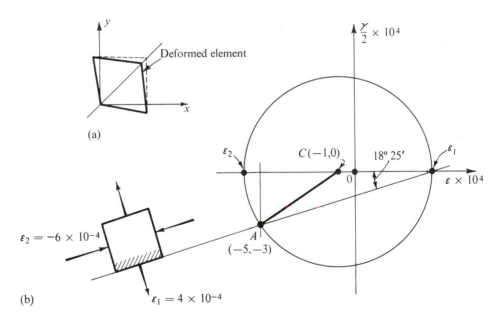

Fig. 8-17

*This measurement may be made by scribing a small square on a body, straining the body, and then measuring the change in angle which takes place. Photographic enlargements of grids have been used for this purpose.

The data given indicate that $\varepsilon_x = -5 \times 10^{-4}$, $\varepsilon_y = +3 \times 10^{-4}$, and $\gamma_{xy} = -6 \times 10^{-4}$. Hence, on a $\varepsilon - \frac{1}{2}\gamma$ system of axes, the center C of the circle is located at $(\varepsilon_x + \varepsilon_y)/2 = -1 \times 10^{-4}$ on the ε axis, Fig. 8-17. Point A is at $(-5 \times 10^{-4}, -3 \times 10^{-4})$. The radius of the circle AC is equal to 5×10^{-4}. Hence $\varepsilon_1 = +4 \times 10^{-4}$ in. per inch takes place in the direction perpendicular to the line $A - \varepsilon_1$; and $\varepsilon_2 = -6 \times 10^{-4}$ in. per inch occurs in the direction perpendicular to the line $A - \varepsilon_2$. From the geometry of the figure, $|\theta| = \tan^{-1}(0.0003/0.0009) = 18°25'$.

*8-13. STRAIN MEASUREMENTS; ROSETTES

Measurements of extensional strain are particularly simple to make, and highly reliable techniques have been developed for this purpose. In such work, these strains are measured along several closely clustered gage lines, diagrammatically indicated in Fig. 8-18(a) by lines *a-a*, *b-b*, and *c-c*. These

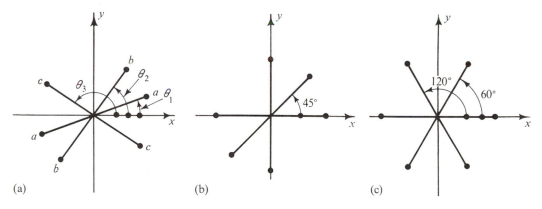

Fig. 8-18. (a) General strain rosette; (b) rectangular or 45° strain rosette; (c) equiangular or delta rosette

gage lines may be located on the member investigated with reference to some coordinate axes (as x and y) by the respective angles θ_1, θ_2, and θ_3. By comparing the initial distance between any two corresponding gage points with the distance in the stressed member, the elongation in the gage length is obtained. Dividing the elongation by the gage length gives the strain in the θ_1 direction, which will be designated ε_{θ_1}. By performing the same operation with the other gage lines, ε_{θ_2} and ε_{θ_3} are obtained. If the distances between the gage points are small, measurements approximating the strains at a point are obtained.

As an alternate to the foregoing experimental procedure, electric strain gages are unusually convenient to employ. These consist of very thin wires or foil which is glued to the member being investigated. As the forces are applied to a member, elongation or contraction of the wires or foil takes place concurrently with similar changes in the material. These changes in length alter the electrical resistance of the gage, which can be measured and calibrated to indicate the strain taking place.

Arrangements of gage lines at a point in a cluster, as shown in Fig. 8-18, are known as *strain rosettes*. If three strain measurements are taken at a rosette, the information is sufficient to determine the complete state of plane strain at a point.

If the angles θ_1, θ_2, and θ_3, together with the corresponding strains ε_{θ_1}, ε_{θ_2}, and ε_{θ_3} are known from measurements, three simultaneous equations patterned after Eq. 8-13 can be written. In writing these equations, it is convenient to employ the following notation: $\varepsilon_{x'} \equiv \varepsilon_{\theta_1}$, $\varepsilon_{x''} \equiv \varepsilon_{\theta_2}$, and $\varepsilon_{x'''} \equiv \varepsilon_{\theta_3}$.

$$\varepsilon_{\theta_1} = \varepsilon_x \cos^2 \theta_1 + \varepsilon_y \sin^2 \theta_1 + \gamma_{xy} \sin \theta_1 \cos \theta_1$$
$$\varepsilon_{\theta_2} = \varepsilon_x \cos^2 \theta_2 + \varepsilon_y \sin^2 \theta_2 + \gamma_{xy} \sin \theta_2 \cos \theta_2 \qquad (8\text{-}19)$$
$$\varepsilon_{\theta_3} = \varepsilon_x \cos^2 \theta_3 + \varepsilon_y \sin^2 \theta_3 + \gamma_{xy} \sin \theta_3 \cos \theta_3$$

This set of equations can be solved for ε_x, ε_y, and γ_{xy}, and the problem reverts back to the cases already considered.

To minimize computational work, the gages in a rosette are usually arranged in an orderly manner. For example, in Fig. 8-18(b), $\theta_1 = 0°$, $\theta_2 = 45°$, and $\theta_3 = 90°$. This arrangement of gage lines is known as the *rectangular* or the 45° *strain rosette*. By direct substitution into Eq. 8-19, it is found that for this rosette

$$\varepsilon_x = \varepsilon_{0°}, \qquad \varepsilon_y = \varepsilon_{90°}, \qquad \varepsilon_{45°} = \frac{\varepsilon_x}{2} + \frac{\varepsilon_y}{2} + \frac{\gamma_{xy}}{2}$$

or

$$\gamma_{xy} = 2\varepsilon_{45°} - (\varepsilon_{0°} + \varepsilon_{90°})$$

Thus ε_x, ε_y, and γ_{xy} become known.

Another arrangement of gage lines is shown in Fig. 8-18(c). This is known as the *equiangular*, or the *delta*, or the 60° *rosette*. Again, by substituting into Eq. 8-19 and simplifying, $\varepsilon_x = \varepsilon_{0°}$, $\varepsilon_y = (2\varepsilon_{60°} + 2\varepsilon_{120°} - \varepsilon_{0°})/3$, and $\gamma_{xy} = (2/\sqrt{3})(\varepsilon_{60°} - \varepsilon_{120°})$.

Other types of rosettes are occasionally used in experiments. The data from all rosettes can be analyzed by applying Eq. 8-19, solving for ε_x, ε_y, and γ_{xy}, and then applying Mohr's circle of strain.

Sometimes rosettes with more than three lines are used. An additional gage line measurement provides a check on the experimental work. For these rosettes, the invariance of the strains in the mutually perpendicular directions can be used to check the data.

The application of the experimental rosette technique in complicated problems of stress analysis is almost indispensable.

*8-14. ADDITIONAL LINEAR RELATIONS BETWEEN STRESS AND STRAIN AND AMONG E, G, AND v

Additional relations between stress and strain for linearly elastic, isotropic materials are discussed in this article. These relations are useful for obtaining stresses from planar strains and for finding volumetric changes in elastic materials subjected to uniform external pressure. The fundamental relation among the elastic constants E, D, and v is also established.

Relation between Principal Stresses and Strains

In many practical investigations, strains on the surface of a member are determined by means of rosettes. By using Mohr's circle of strain or strain transformation equations, the principal strains can be found. From these it is possible to determine the principal stresses directly. To establish the appropriate equations, it should be noted that in a plane-stress problem $\sigma_z = 0$, and Eq. 2-6 written in terms of the principal stresses simplifies to

$$\varepsilon_1 = \frac{\sigma_1}{E} - v\frac{\sigma_2}{E} \quad \text{and} \quad \varepsilon_2 = \frac{\sigma_2}{E} - v\frac{\sigma_1}{E}$$

Solving these equations simultaneously for the principal stresses, one obtains the required relations:

$$\sigma_1 = \frac{E}{1 - v^2}(\varepsilon_1 + v\varepsilon_2) \qquad \sigma_2 = \frac{E}{1 - v^2}(\varepsilon_2 + v\varepsilon_1) \qquad (8\text{-}20)$$

The elastic constants E and v must be determined from some appropriate experiments. With the aid of such experimental work, very complicated problems can be solved successfully.*

EXAMPLE 8-6

At a certain point on a steel machine part, measurements with an electric rectangular rosette indicate that $\varepsilon_{0°} = -0.00050$, $\varepsilon_{45°} = +0.0002$, and $\varepsilon_{90°} = +0.00030$. Assuming that $E = 200 \times 10^3$ MPa and $v = 0.3$ are accurate enough, find the principal stresses at the point investigated.

SOLUTION

From the data given, $\varepsilon_x = -0.00050$, $\varepsilon_y = +0.00030$, and

$$\gamma_{xy} = 2\varepsilon_{45°} - (\varepsilon_{0°} + \varepsilon_{90°})$$
$$= 2(+0.0002) - (-0.00050 + 0.00030) = +0.00060$$

*See M. Hetenyi, editor-in-chief, *Handbook of Experimental Stress Analysis*, Society for Experimental Stress Analysis, New York: Wiley, 1950.

The principal strains for these data were found in Example 8-5 and are $\varepsilon_1 = +0.00040$ and $\varepsilon_2 = -0.00060$. Hence, by Eq. 8-20, the principal stresses are

$$\sigma_1 = \frac{(200)(10)^3}{1-(0.3)^2}[+0.00040 + 0.3(-0.00060)] = +48.3 \text{ MPa}$$

$$\sigma_2 = \frac{(200)(10)^3}{1-(0.3)^2}[-0.00060 + 0.3(+0.00040)] = -105 \text{ MPa}$$

The tensile stress σ_1 acts in the direction of ε_1; see Fig. 8-17. The compressive stress σ_2 acts in the direction of ε_2.

Relation Among E, G, and v

The methods of transforming one description of the state of stress or strain into another have been established. In Art. 8-6, particular emphasis was placed on the fact that pure shearing stresses can be transformed into purely normal stresses. Therefore, one must conclude that the deformations caused by pure shearing stresses must be related to the deformations caused by the normal stresses. Based on this assertion, a fundamental relation among E, G, and v for linearly elastic, isotropic materials can be established.

According to Eq. 8-13, with only $\gamma_{xy} \neq 0$, for an x' axis at $\theta = 45°$, the linear strain $\varepsilon_{x'} = \gamma_{xy}/2$. This extensional strain $\varepsilon_{x'}$ can be related to the shearing stress τ_{xy}, since, according to Eq. 2-9, $\tau_{xy} = G\gamma_{xy}$. On this basis,

$$\varepsilon_{x'} = \tau_{xy}/(2G)$$

On the other hand, according to Art. 8-6, pure shearing stress τ_{xy} can be expressed alternatively in terms of the principal stresses $\sigma_1 = \tau_{xy}$, and $\sigma_2 = -\tau_{xy}$, acting at $45°$ to the directions of shearing stresses (see Fig. 8-7). So, by using Eq. 2-6, one finds that the linear strain along the x' axis at $\theta = 45°$ in terms of the principal stresses becomes

$$\varepsilon_{x'} = \varepsilon_1 = \frac{\sigma_1}{E} - v\frac{\sigma_2}{E} = \frac{\tau_{xy}}{E}(1 + v)$$

Equating the two alternative relations for the strain along the positive shear diagonal and simplifying,

$$G = \frac{E}{2(1 + v)} \tag{8-21}$$

This is the basic relation between E, G, and v; it shows that these quantities are not independent of one another. If any two of these are determined experimentally, the third can be computed. Note that the shearing modulus G is always less than the elastic modulus E, since the Poisson ratio v is a positive quantity. For most materials, v is in the neighborhood of $\frac{1}{4}$.

Dilatation; Bulk Modulus

By extending some of the established concepts, one can derive an equation for volumetric changes in elastic materials subjected to stress. In the process of doing this two new terms are introduced and defined.

The sides dx, dy, and dz of an infinitestimal element after straining become $(1 + \varepsilon_x)dx$, $(1 + \varepsilon_y)dy$, and $(1 + \varepsilon_z)dz$, respectively. After subtracting the initial volume from the volume of the strained element, the change in volume is determined. This is

$$(1 + \varepsilon_x)\,dx(1 + \varepsilon_y)\,dy(1 + \varepsilon_z)\,dz - dx\,dy\,dz \approx (\varepsilon_x + \varepsilon_y + \varepsilon_z)\,dx\,dy\,dz$$

where the products of strain $\varepsilon_x\varepsilon_y + \varepsilon_y\varepsilon_z + \varepsilon_z\varepsilon_x + \varepsilon_x\varepsilon_y\varepsilon_z$, being small, are neglected. Therefore, in the infinitesimal (small) strain theory, e, the change in volume per unit volume, often referred to as *dilatation*, is defined as

$$e = \varepsilon_x + \varepsilon_y + \varepsilon_z = \varepsilon_1 + \varepsilon_2 + \varepsilon_3 \qquad (8\text{-}22)$$

where the last equality follows from the fact that e is an invariant. A more restricted case of strain invariance was encountered in Art. 8-12 for the two-dimensional case, where it was shown that $\varepsilon_1 + \varepsilon_2 = \varepsilon_x + \varepsilon_y$. The shearing strains cause no change in volume.

Based on the generalized Hooke's law, the dilatation can be found in terms of stresses and material constants. For this purpose the first three parts of Eq. 2-6 must be added together. This yields

$$e = \varepsilon_x + \varepsilon_y + \varepsilon_z = \frac{1 - 2v}{E}(\sigma_x + \sigma_y + \sigma_z) \qquad (8\text{-}23)$$

which means that dilatation is proportional to the algebraic sum of all normal stresses. As a direct counterpart to the strain invariant, the sum $(\sigma_x + \sigma_y + \sigma_z)$ is the stress invariant.

If an elastic body is subjected to hydrostatic pressure of uniform intensity p, so that $\sigma_x = \sigma_y = \sigma_z = -p$, then from Eq. 8-23

$$e = -\frac{3(1 - 2v)}{E}p \qquad \text{or} \qquad \frac{-p}{e} = k = \frac{E}{3(1 - 2v)} \qquad (8\text{-}24)$$

The quantity k represents the ratio of the hydrostatic compressive stress to the decrease in volume and is called *modulus of compression* or *bulk modulus*.

* Appendix to Chapter 8

Transformation of Moments of Inertia of Areas to Different Axes

8A-1. TRANSFORMATION EQUATIONS FOR ROTATION OF AXES

By definition, the moments and product of inertia of an area are given as

$$I_x = \int y^2 \, dA, \qquad I_y = \int x^2 \, dA \qquad \text{and} \qquad I_{xy} = \int xy \, dA \quad \text{(8A-1)}$$

Let x, y be a pair of orthogonal axes about which the moments of inertia I_x and I_y, and the product of inertia I_{xy} are known for the area shown in Fig. 8A-1. Now consider a positive (counterclockwise) rotation of the

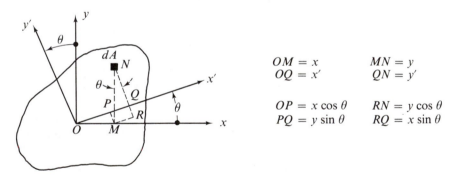

$$OM = x \qquad\qquad MN = y$$
$$OQ = x' \qquad\qquad QN = y'$$

$$OP = x \cos \theta \qquad RN = y \cos \theta$$
$$PQ = y \sin \theta \qquad RQ = x \sin \theta$$

Fig. 8A-1. Rotation of axes

axes through an angle θ and denote the new axes as x' and y'. The coordinates (x,y) of an infinitesimal area dA of the planar section can then be transformed to (x',y') and written as

$$x' = x \cos \theta + y \sin \theta$$
$$y' = y \cos \theta - x \sin \theta$$

$$\text{(8A-2)}$$

Then, beginning with an expression analogous to Eq. 8A-1,

$$I_{x'} = \int y'^2 \, dA = \int (y \cos \theta - x \sin \theta)^2 \, dA$$

$$= \cos^2 \theta \int y^2 \, dA + \sin^2 \theta \int x^2 \, dA - 2 \sin \theta \cos \theta \int xy \, dA$$

$$= I_x \cos^2 \theta + I_y \sin^2 \theta - I_{xy} \cdot 2 \sin \theta \cos \theta$$

$$= I_x \frac{(1 + \cos 2\theta)}{2} + I_y \frac{(1 - \cos 2\theta)}{2} - I_{xy} \sin 2\theta$$

or

$$I_{x'} = \frac{I_x + I_y}{2} + \frac{I_x - I_y}{2} \cos 2\theta + (-I_{xy}) \sin 2\theta \qquad \text{(8A-3)}$$

Similarly,

$$I_{y'} = \frac{I_x + I_y}{2} - \frac{I_x - I_y}{2} \cos 2\theta - (-I_{xy}) \sin 2\theta \qquad \text{(8A-4)}$$

and

$$-I_{x'y'} = -\frac{I_x - I_y}{2} \sin 2\theta + (-I_{xy}) \cos 2\theta \qquad \text{(8A-5)}$$

These equations relate the moments and product of inertia of areas (second moments) in the new $x'y'$-coordinates to the initial ones in the xy-coordinates for rotation of the axes through an angle θ.

8A-2. PRINCIPAL AXES AND PRINCIPAL MOMENTS OF INERTIA

The dependence of $I_{x'}$, $I_{y'}$, and $I_{x'y'}$ on the angle of rotation θ is exhibited in Eqs. 8A-3, 8A-4, and 8A-5. Since the principal axes* are defined as those around which the product of inertia vanishes, they may be located by setting $I_{x'y'}$ equal to 0 and solving Eq. 8A-5 for θ. However, a comparison shows that these transformation equations are identical to the stress transformation equations of Chapter 8 if σ_x, σ_y, τ_{xy}, σ'_x, and $\tau_{x'y'}$ are replaced by I_x, I_y, $-I_{xy}$, $I_{x'}$, and $-I_{x'y'}$. Hence, all the techniques using Mohr's circle as outlined in Art. 8-8 are equally applicable to the problem of finding the orientation of the principal axes, as well as the calculation of the corresponding principal moments of inertia.

EXAMPLE 8A-1

Using Mohr's circle, find the principal axes and principal moments of inertia for an angle section with unequal legs shown in Fig. 8A-2.

*If the origin is at the centroid, these are the *centroidal* principal axes.

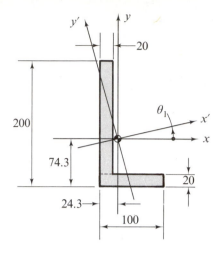

Fig. 8A-2. Angle with unequal legs (all dimensions in mm).

SOLUTION

The steel angle shown has dimensions of 200 mm × 100 mm × 20 mm thickness. The centroid of the section can be verified to lie at 74.3 mm from the bottom face and 24.3 mm from the left face. A pair of orthogonal axes x and y are set up as shown in Fig. 8A-2. The moments and product of inertia about these axes can be calculated by breaking up the angle section into two rectangles as shown in the figure and using the parallel axis theorems, Eqs. 5-2 and 5-2a.

$$I_x = (20/12)(180)^3 + (20)(180)(35.7)^2 + (100/12)(20)^3$$
$$+ (100)(20)(64.3)^2 = 22.64 \times 10^6 \text{ mm}^4$$
$$I_y = (180/12)(20)^3 + (20)(180)(14.3)^2 + (20/12)(100)^3$$
$$+ (100)(20)(25.7)^2 = 3.84 \times 10^6 \text{ mm}^4$$
$$I_{xy} = (20)(180)(-14.3)(35.7) + (100)(20)(25.7)(-64.3)$$
$$= -5.14 \times 10^6 \text{ mm}^4$$

The Mohr's circle can then be constructed as shown in Fig. 8A-3 with the horizontal axis representing the moments of inertia and the vertical axes the negative of the products of inertia.

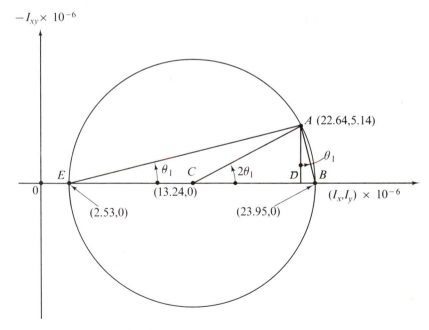

Fig. 8A-3. Mohr's circle for principal moments of inertia

1. Center of the circle: $\dfrac{I_x + I_y}{2} = 13.24 \times 10^6$ mm^4

2. Point A on the circle: $I_x = 22.64 \times 10^6$ mm^4, $-I_{xy} = 5.14 \times 10^6$ mm^4

3. Radius of circle: $\left[\left(\dfrac{I_x - I_y}{2}\right)^2 + (-I_{xy})^2\right]^{1/2} = 10.71 \times 10^6$ mm^4

After drawing the circle, one obtains the principal moments of inertia as 23.95×10^6 mm^4 and 2.53×10^6 mm^4. The principal axes are found by a positive rotation θ of the assumed pair of axes x and y, where θ can be obtained as

$$\tan \theta_1 = \frac{AD}{DE} = \frac{5.14}{22.64 - 2.53} = 0.256 \qquad \text{or} \qquad \theta_1 = 14.33°$$

The centroidal principal axes are shown in Fig. 8A-2, and the principal moments of inertia are $I_{x'} = 23.95 \times 10^6$ mm^4 and $I_{y'} = 2.53 \times 10^6$ mm^4.

Note that $I_x + I_y = I_{x'} + I_{y'}$, i.e., the sum of the moments of inertia around two mutually perpendicular centroidal axes is *invariant*.

These results can be obtained analytically by using the equations which are analogous to the ones for finding principal stresses, Eq. 8-4, and the corresponding directions, Eq. 8-3. In the notation for the moments of inertia, the required equations are

$$(I_{x'})_{\substack{\text{max} \\ \text{min}}} = \frac{I_x + I_y}{2} \pm \sqrt{\left(\frac{I_x - I_y}{2}\right)^2 + I_{xy}^2} \qquad \text{(8A-6)}$$

and

$$\tan 2\theta_1 = -\frac{I_{xy}}{(I_x - I_y)/2} \qquad \text{(8A-7)}$$

For details of applying these equations, see Art. 8-4.

PROBLEMS FOR SOLUTION

8-1. Infinitesimal elements A, B, C, D, and E are shown on the figures for two different members. Draw each one of these elements separately, and indicate on the isolated elements the stress acting on it. For each stress clearly show its direction and sense by arrows, and state the formula one would use in its calculation. Neglect the weight of the members.

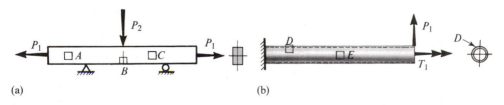

(a) (b)

PROB. 8 – 1

8-2 through 8-5.   For the infinitesimal elements shown in the figures, find the normal and shearing stresses acting on the indicated inclined planes. Use the "wedge" method of analysis discussed in Example 8-1.

Ans: Prob. 8-2: $\sigma = +40$ MPa, $\tau = 10$ MPa; $\sigma = 0, \tau = -10$ MPa.
Prob. 8-3: $\sigma = -1.2$ MPa, $\tau = 7.1$ MPa; $\sigma = +41.2$ MPa, $\tau = 7.1$ MPa.
Prob. 8-4: $\sigma = -400$ psi, $\tau = 4,970$ psi.
Prob. 8-5: $\sigma = -13.08$ ksi, $\tau = 0$; $\sigma = 7$ ksi, $\tau = 6.08$ ksi.

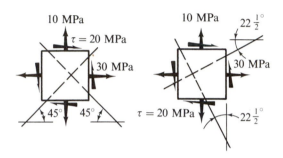

PROB. 8 – 2   PROB. 8 – 3

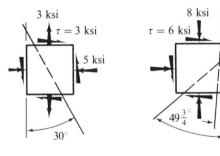

PROB. 8 – 4   PROB. 8 – 5

8-6.   Derive Eq. 8-2.

8-7.   Using Eqs. 8-1 and 8-2, rework Prob. 8-2.

8-8.   Using Eqs. 8-1 and 8-2, rework Prob. 8-3.

8-9.   Using the equations derived in Art. 8-4, find the principal stresses and show their sense on a properly oriented element for the following data: $\sigma_x = +8,000$ psi; $\sigma_y = +2,000$ psi; and $\tau = +4,000$ psi. The sign convention of the

stresses given follows that illustrated in Fig. 8-4(a). *Ans: 10 ksi, 0, $\theta = 26°34'$.*

8-10.   Rework Prob. 8-9 changing the data to $\sigma_x = -3$ ksi; $\sigma_y = +1$ ksi; $\tau = -2$ ksi.   *Ans. 1.83 ksi, -3.83 ksi, $22\frac{1}{2}°$.*

8-11.   Using the equations derived in Art. 8-5, find the maximum (principal) shearing stresses and the associated normal stresses for the data of Prob. 8-10. Show the results on a properly oriented element.   *Ans. 2.83 ksi, -1 ksi, $67\frac{1}{2}°$.*

The data given below for problems **8-12 through 8-15** follow the convention of signs for stresses established in Fig. 8-4(a). In each case show the data on an infinitesimal element. Then, using the formulas developed in Arts. 8-4 and 8-5, (a) find the principal stresses and show their sense on a properly oriented element; (b) find the maximum (principal) shearing stresses with the associated normal stresses and show the results on a properly oriented element.

8-12. $\sigma_x = +20$ ksi, $\sigma_y = 0$, $\tau = -10$ ksi. *Ans. 24.1 ksi, -4.1 ksi; 14.1 ksi.*

8-13. $\sigma_x = 0$, $\sigma_y = -4$ ksi, $\tau = -6$ ksi. *Ans. 4.33 ksi, -8.33 ksi; 6.33 ksi.*

8-14. $\sigma_x = -10$ MN/m², $\sigma_y = -40$ MN/m², $\tau = +20$ MN/m². *Ans: 0, -50 MN/m²; 25 MN/m².*

8-15. $\sigma_x = -10\,000$ kN/m², $\sigma_y = +20\,000$ kN/m², $\tau = -20\,000$ kN/m². *Ans: 30 kN/m², -20 kN/m²; 25 kN/m².*

8-16.   Using Mohr's circle of stress, solve for the (a) part of Prob. 8-14.

8-17.   Using Mohr's circle of stress, solve for the (a) part of Prob. 8-15.

8-18 through 8-21.   Draw Mohr's circle of stress for the states of stress given in the figures. (a) Clearly show the planes on which the principal stresses act, and for each stress indicate with arrows its direction and sense. (b) Same as (a) for the maximum shearing stresses and the associated normal stresses, *Ans: Prob.* 8-21. (a) 6 ksi, -4 ksi; (b) 5 ksi, 1 ksi.

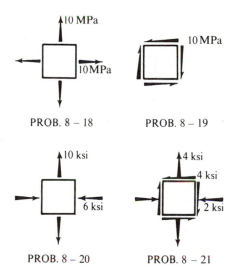

PROB. 8 – 18   PROB. 8 – 19

PROB. 8 – 20   PROB. 8 – 21

The data given below for problems **8-22 through 8-28** follow the convention of signs for stresses established in Figs. 8-4(a) and 8-9(a). In each case show the data on an infinitesimal element. Then, using Mohr's circle construction *and trigonometry*, (a) find the principal stresses and show their sense on a properly oriented element; (b) find the maximum (principal) shearing stresses with the associated normal stresses and show the results on a properly oriented element.

8-22. $\sigma_x = +60\,000$ kN/m², $\sigma_y = +30\,000$ kN/m², $\tau = +25\,000$ kN/m². *Ans:* $+74.2$ MPa, 15.8 MPa, $29\frac{1}{2}°$; 29.2 MPa, 45 MPa.

8-23. Same data as Prob. 8-9. *Ans.* $\tau_{max} = 5$ ksi, $\sigma' = 5$ ksi.

8-24. Same data as Prob. 8-10 *Ans.* $\tau_{max} = 2.83$ ksi, $\sigma' = -1$ ksi.

8-25. $\sigma_x = -30$ MN/m², $\sigma_y = -40$ MN/m², $\tau = +30$ MN/m². *Ans:* -4.5 MN/m², -65.5 MN/m², $40\frac{1}{4}°$; 30.5 MN/m², -35 MN/m².

8-26. $\sigma_x = -15$ MN/m², $\sigma_y = +35$ MN/m², $\tau = +60$ MN/m². *Ans:* 75 MPa, -55 MPa; 56.3°; 65 MPa, 10 MPa.

8-27. $\sigma_x = +20$ ksi, $\sigma_y = 0$, $\tau = -10$ ksi. *Ans:* see Prob. 8-12.

8-28. $\sigma_x = 0$, $\sigma_y = -4$ ksi, $\tau = -6$ ksi. *Ans:* see Prob. 8-13.

8-29. If $\sigma_x = \sigma_1 = 0$ and $\sigma_y = \sigma_2 = -4,000$ psi, using Mohr's circle of stress, find the stresses acting on a plane defined by $\theta = +30°$. *Ans:* -1 ksi, 1.73 ksi.

8-30. Using Mohr's circle of stress, for the data of Prob. 8-21, find the stresses acting on $\theta = 30°$.

8-31. Rework the above problem with $\theta = 20°$.

8-32. Using Mohr's circle of stress, rework Prob. 8-3.

8-33. At a particular point in a wooden member, the state of stress is as shown in the figure. The direction of the grain in the wood makes an angle of $+30°$ with the x-axis. The allowable shearing stress parallel to the grain is 150 psi for this wood. Is this state of stress permissible? Verify your answer by calculations.

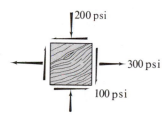

PROB. 8 – 33

8-34. A clevice transmits a force F to a bracket as shown in the figure. Stress analysis of this bracket gives the following stress components acting on the element A: 1,000 psi due to bending, ,1500 psi due to axial force, and 600 psi due to shear. (Note that these are stress magnitudes only, their directions and senses must be determined by inspection.) (a) Indicate the resultant stresses on a sketch of the isolated element A. (b) Using Mohr's circle for the state of stress found in (a), determine the principal stresses and the maximum shearing stresses with the associated normal stresses. Show the results on properly oriented elements. *Ans.* (b) 900 psi, −400 psi; 650 psi, 250 psi.

PROB. 8 – 34

8-35. At point A on an unloaded edge of an elastic body, oriented as shown in the figure with respect to the xy axes, the maximum shearing stress is 3 500 kN/m² (a) Find the principal stresses, and (b) determine the state of stress on an element oriented with its edges parallel to the xy axes. Show the results on a sketch of the element at A. (*Hint:* An effective solution may be obtained by constructing Mohr's circle of stress.) *Ans.* (b) $\sigma_x = 4\,480$ kN/m².

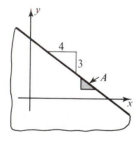

PROB. 8 – 35

8-36. The magnitudes and directions of the stresses on two planes intersecting at a point are as shown in the figure. Determine the directions

and magnitudes of the principal stresses at this point. Sketch the results on an element.

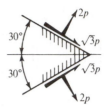

PROB. 8 – 36

8-37. Rederive Eq. 8-13 by assuming that the shearing deformation occurs first, then the deformation in the y-direction, and finally the deformation in the x-direction.

8-38. With the aid of Fig. 8-15, show that $\beta = -(\varepsilon_x - \varepsilon_y) \sin \theta \cos \theta + \gamma_{xy} \cos^2 \theta$.

8-39. If the unit strains are $\varepsilon_x = -0.00012$, $\varepsilon_y = +0.00112$, and $\gamma = -0.00020$, what are the principal strains and in which directions do they occur? Use Eqs. 8-17 and 8-18 or Mohr's circle of strain, as directed. *Ans.* $+0.00113$, -0.00013, $4°35'$.

8-40. If the unit strains are $\varepsilon_x = -0.00080$, $\varepsilon_y = -0.00020$, and $\gamma = +0.00080$, what are the principal strains and in which directions do they occur? Use Eqs. 8-17 and 8-18 or Mohr's circle, as directed. *Ans.* 0, -0.00100.

8-41. If the strain measurements given in the above problem were made on a steel member ($E = 200\,000$ MN/m² and $\nu = 0.3$), what are the principal stresses and in which direction do they act?

8-42. At a point in a stressed elastic plate the following information is known: maximum shearing strain $\gamma_{max} = 5 \times 10^{-4}$, and the sum of the normal stresses on two perpendicular planes passing through the point is 27 500 kN/m². The elastic properties of the plate are $E = 200\,000$ MN/m², $G = 80\,000$ MN/m², $\nu = 0.25$. Calculate the magnitude of the principal stresses at the point.

8-43. The data for a rectangular rosette, attached to a stressed steel member, are $\varepsilon_{0°} = -0.00022$, $\varepsilon_{45°} = +0.00012$, and $\varepsilon_{90°} = +0.00022$. What are the principal stresses and in which directions do they act? $E = 30 \times 10^6$ psi and $v = 0.3$. *Ans:* ± 5.76 ksi, $14°18'$.

8-44. The data for an equiangular rosette, attached to a stressed aluminum alloy member, are $\varepsilon_{0°} = +0.00040$, $\varepsilon_{60°} = +0.00040$, and $\varepsilon_{120°} = -0.00060$. What are the principal stresses and in which directions do they act? $E = 70000$ MN/m² and $v = 0.25$.

8-45. The data for a strain rosette with four gage lines attached to a stressed aluminum alloy member are $\varepsilon_{0°} = -0.00012$, $\varepsilon_{45°} = +0.00040$, $\varepsilon_{90°} = +0.00112$, and $\varepsilon_{135°} = +0.00060$. Check the consistency of the data. Then determine the principal stresses and the directions in which they act. Use the values of E and v given in Prob 8-44.

8-46. (a) Find the product of inertia for the triangular area shown in the figure with respect to the given axes. (b) For the same triangular area, using Eq. 5-2a and the result found in (a), determine the product of inertia with respect to vertical and horizontal axes through the centroid. *Ans:* (a) $b^2h^2/24$. (b) $-b^2h^2/72$.

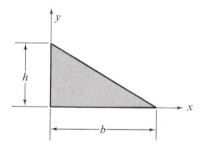

PROB. 8–46

8-47. (a) If in the above problem $b = 6$ in. and $h = 3$ in., find the principal axes and principal moments of inertia for the area. Make use of the results found in part (b) of the above problem, and use I_0 for a triangle around an axis as given in Table 2 of the Appendix. (b) If a beam having the above cross-section is subjected to a bending moment around the major principal axis, and the allowable bending stress $\sigma = 10$ ksi, what moment can be applied? (See Art. 7-6).

8-48. The angle having the dimensions shown in Fig. 8A-2 is simply supported at the ends on horizontal roller supports and spans 2 meters. What concentrated vertical force P may be applied at the mid-span if the allowable stress for the material is 200 MPa. Neglect the weight of the angle and use the pertinent results found in Example 8A-1. (Note that the vertical force must be applied through the shear center, see Fig. 6-22. Also see Art. 7-6.)

8-49. (a) Find the principal axes and principal moments of inertia for the cross-sectional area of an angle shown in the figure. Calculations are to be made using axes passing through the centroid. (b) The given dimensions of the cross-section, except for small radii at the ends and a fillet, correspond to the cross-sectional dimensions of an 8 in. by 6 in. by 1 in. angle listed in Table 7 of the Appendix. Using the information given in that table, calculate the principal moments of inertia and compare with the results found in (a). (*Hint:* Note that per Art. 13-7 and Example 13-1, $I_{min} = Ar^2_{min}$. The r listed in Table 7 for the z-axis is r_{min}. Further, from the invariance condition, $I_{min} + I_{max} = I_{x'} + I_{y'} = I_x + I_y \equiv I_{xx} + I_{yy}$, hence one can readily solve for I_{max}.)

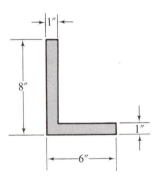

PROB. 8–49

8-50. Find the largest bending moment which an 8 in. by 6 in. by 1 in. angle can carry without exceeding a stress of 20 ksi. (See hint for part (b) of the above problem and Art. 7-6.) *Ans:* 358 k-in.

8-51. A 6 in. by 6 in. by $\frac{1}{2}$ in. steel angle with one of its legs placed in a horizontal position and its other leg directed downward is used as a cantilever 70.7 in. long. If an upward force of 1,000 lb is applied at the end of this cantilever through the shear center, what are the maximum tensile and compressive stresses at the built-in end?

Neglect the weight of the angle. See Art. 7-6 and hint for part (b) of Problem 8-49. (Properties of the angle can be found in Table 6 of the Appendix.) *Ans:* 14.7 ksi, −18.4 ksi.

9 Combined Stresses — Pressure Vessels — Failure Theories

9-1. INTRODUCTION

As should have become apparent from the preceding chapters, a description of the state of stress at a point of a stressed member can be found by using the conventional formulas and may involve normal *and* shearing stresses. Formal methods for an alternate description of the state of stress at the same point in terms of principal stresses, or the principal shearing stresses and the associated normal stresses, were treated in Chapter 8. In this chapter, the method of redescribing the state of stress in terms of the principal stresses will be applied to some particular cases of stressed members. As examples, the principal stresses occurring in a circular shaft and a beam when they are subjected to some familiar loading conditions will be examined. For these, the existence at many points of biaxial principal stresses, i.e., both principal stresses different from zero, will be shown. Moreover, since the earlier treatment of allowable stresses was based on the uniaxial principal stress determined from simple tension or compression tests, the resistance of materials to a biaxial stress will be discussed under the caption of strength or failure theories of materials. For the development of this topic, the stress formulas for thin-walled pressure vessels will be established, as experiments on pressure vessels provide information on the behavior of materials under biaxial stress. Early in this chapter, the stresses on inclined planes of a straight rod subjected to an axial force will also be determined.

9-2. INVESTIGATION OF STRESSES AT A POINT

To find the principal stresses or the stresses on any inclined plane at a point of a loaded member, the same basic procedure that was repeatedly used earlier must be employed (Art. 1-9). In statically determinate problems, the reactions are found first. Then a segment of the body is isolated by passing a section perpendicular to its axis *through the point* to be investigated, and the system of forces necessary to maintain the equilibrium of the segment is determined. The magnitudes of the stresses are determined next

275

by the conventional formulas. Then on an element isolated from the member, the computed stresses are indicated. The sense of the computed stresses is noted on this element by arrows agreeing with the sense of the internal forces at the cut. Two sides of this element are parallel and two sides are perpendicular to the axis of the member being investigated. *The definite relation of the sides of this element to the actual member must be clearly understood by the analyst.* After the sketch of an element is prepared and stresses of the same kind are compounded, the stresses may be determined on planes with any orientation through the same point. For this purpose, either analytical formulas or Mohr's circle of stress, discussed in the preceding chapter, can be used. The principal stresses, or the maximum shearing stress, are usually the quantities sought.

In the following three examples, an axially loaded rod, a circular shaft in torsion, and a rectangular beam with transversely applied force will be examined for principal stresses and stresses acting on inclined planes.

EXAMPLE 9-1

Find the stresses acting on an arbitrarily inclined plane in an axially loaded rod of constant cross-sectional area.

SOLUTION

Consider the prismatic bar subjected to axial tension in Fig. 9-1(a). By passing a section *X-X* perpendicular to the axis of the rod through a general point *G* and applying Eq. 1-1, one finds the stress $\sigma = P/A$, where A is the cross-sectional area of the rod. Moreover, since this normal stress is the only stress acting on the element, Fig. 9-1(b), it is the principal stress. Designating this stress as σ_y, and noting that $\sigma_x = 0$ and $\tau_{xy} = 0$, the normal and the shearing stresses acting on any inclined plane defined by the x' axis normal to this plane can be found using Eqs. 8-1 and 8-2:

$$
\left.
\begin{aligned}
\sigma_{x'} &= \sigma \sin^2 \theta = \frac{\sigma}{2} - \frac{\sigma}{2} \cos 2\theta \\[2mm]
\tau_{x'y'} &= +\frac{\sigma}{2} \sin 2\theta
\end{aligned}
\right\}
\tag{9-1}
$$

where the signs follow the conventions adopted in Fig. 8-4(b). From Eq. 9-1 it is seen that the maximum or the principal stress occurs when $\theta = 90°$. These equations emphasize the fact that the state of stress at a point can be described in an infinite number of ways.

ALTERNATE SOLUTION

Instead of solving this problem by the formulas already developed, it is instructive to rework it from basic principles. Thus, consider the same bar, Fig. 9-1(a), and pass through it two parallel planes *HJ* and *KL* inclined at an angle θ with the vertical. *Every* vertical fiber in the block *HJLK*, shown isolated in Fig. 9-1(c), elongates the same amount. All of these fibers are subjected to the same intensity of force. Hence, although this has not been

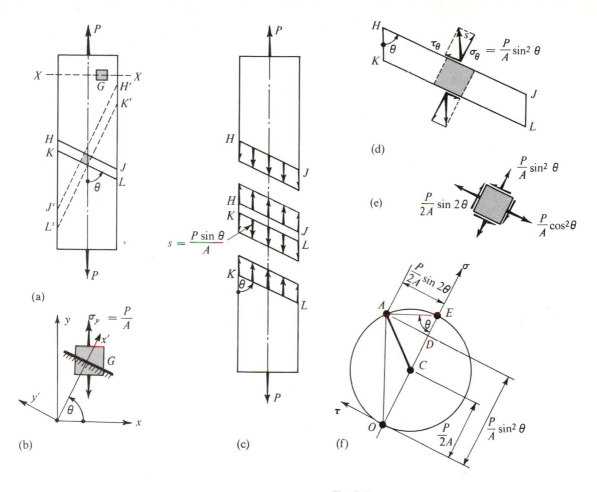

Fig. 9-1

done before in this text, the stress s acting in a *vertical* direction *on an inclined plane* may be said to be $s = P/(A/\sin \theta)$, since $A/\sin \theta$ is the *inclined cross-sectional area of the bar*. This manner of expressing the stress is unusual, and for this reason it is resolved into the normal and shearing stresses (Art. 1-3). This is done by direct resolution of s into the components, Fig. 9-1(d), since s, σ_θ, and τ_θ all act *on the same unit of area*. Thus

$$\sigma_\theta = s \sin \theta = \frac{P}{A} \sin^2 \theta$$

and
$$\tau_\theta = s \cos \theta = \frac{P}{A} \sin \theta \cos \theta = \frac{P}{2A} \sin 2\theta$$

These results agree with Eqs. 9-1.

It is instructive to carry the above solution a step further. By isolating the block $H'J'L'K'$ (shown dashed in Fig. 9-1(a)), whose sides are perpendicular to those of the block $HJLK$, the element shown in Fig. 9-1(e) is obtained. All normal and shearing stresses acting on this element may be determined by combining with the above solution an additional solution analogous to it for the block $H'J'L'K'$. The principal stresses for this element are obtained using Mohr's circle of stress as in Fig. 9-1(f). With the aid of trigonometry, it can be shown that the maximum principal stress $\sigma_1 = P/A$ and acts normal to the line AE in the vertical direction, which is to be expected. The minimum principal stress on the vertical side of the element is zero.

The above discussion also applies to stable compression members. In either tension or compression, in a state of pure uniaxial stress, *there are shearing stresses acting on some planes*. These shearing stresses reach their maximum value of $\sigma/2$ when $\theta = \pm 45°$, as can be seen from Eq. 9-1. For materials that are strong in tension or compression but *weak in shearing strength*, failures may be expected along the 45° planes. This is found to be so for several materials. Concrete and Duralumin are notable examples.*

EXAMPLE 9-2

Determine the principal stresses occurring in a solid circular shaft subjected to a torque T, Fig. 9-2(a).

SOLUTION

In a circular shaft, the maximum shearing stress occurs in the outermost thin lamina (at but not on the outer surface) and, by Eq. 3-3, is $\tau_{\max} = Tc/J$, where c is the radius of the shaft and J is its polar moment of inertia. This state of pure shearing stress is shown acting on an element in Fig. 9-2(a). However, according to Art. 8-6, a state of pure shearing stress transforms into tensile and compressive principal stresses, which are equal in magnitude to the shearing stress and act along the respective shear diagonals. Therefore the principal stresses are $\sigma_1 = +Tc/J$ and $\sigma_2 = -Tc/J$, acting in the direction shown in the figure.

The above stress transformation enables one to predict the type of failure that will take place in materials weak in tension. Such materials fail by tearing in a line perpendicular to the direction of σ_1. An example of such a failure for a sandstone sample is shown in Fig. 9-2(b); cast iron shafts fail in the same manner.† The failure takes place along a helix, shown in Fig. 9-2(a) by the dashed lines. Shafts made from materials weak in shearing strength, such as mild steel, break squarely across.

---

*Failures do not take place precisely on the 45° planes, since the normal stress existing simultaneously with the shearing stress also influences the breakdown of the material. This fact is rationalized in Mohr's theory of failure.

†Ordinary chalk behaves similarly. This may be demonstrated in a classroom by twisting a piece of chalk to failure.

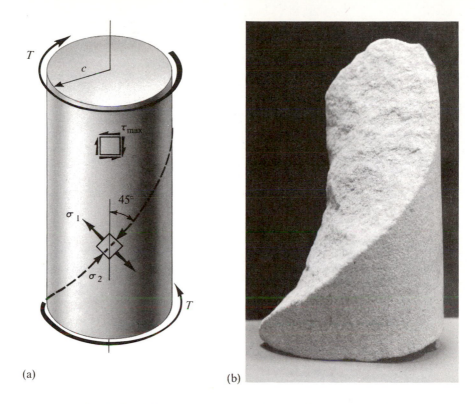

Fig. 9-2. (a) Alternative description of stresses for a shaft in torsion; (b) a core sample of sandstone after a torsion text. (Experiment by Professor D. Pirtz.)

EXAMPLE 9-3

A weightless rectangular beam spans 1 m and is loaded with a vertical downward force $P = 80$ kN at midspan, Fig. 9-3(a). Find the principal stresses at points K, L, M, L', and K', and the stresses acting on an inclined plane of $\theta = +30°$ for the element L'.

SOLUTION

At section KK' a shear of 40 kN and a bending moment of 10 kN·m are necessary to maintain the equilibrium of the segment of the beam. These quantities with their proper sense are in Fig. 9-3(c).

The principal stress at points K and K' follows directly by applying the flexure formula, Eq. 5-1. Since the shearing stresses are distributed parabolically across the cross section of a rectangular beam, no shearing stresses act on these elements, Figs. 9-3(d) and (h).

$$\sigma_{K \text{ or } K'} = \frac{Mc}{I} = \frac{M}{S} = \frac{6M}{bh^2} = \frac{6(10)}{(0.04)(0.3)^2} = 16.67 \times 10^3 \text{ kPa}$$

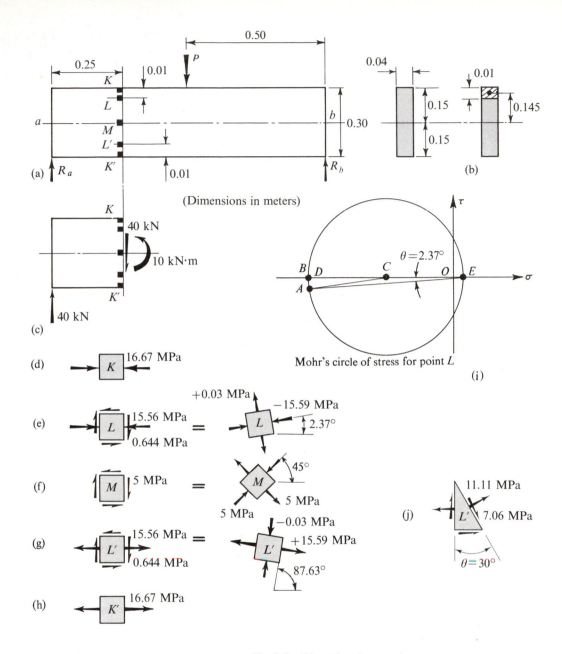

(Dimensions in meters)

Mohr's circle of stress for point L

(i)

Fig. 9-3. (Dimensions in meters)

The normal stresses acting on the elements L and L' shown in the first sketches of Figs. 9-3(e) and (g) are obtained by direct proportion from the normal stresses acting on the elements K and K' (or Eq. 5-1(a) could be used directly). The shearing stresses acting on both these elements are alike. Their sense on the right face of the elements agrees with the sense of the shear at the section KK' in Fig. 9-3(c). The magnitude of these shearing stresses is obtained by applying Eq. 6-6, $\tau = VA_{fghj}\bar{y}/(It)$. For use in this equation, the area A_{fghj} with the corresponding $\bar{y}$ is shown shaded in Fig. 9-3(b).

$$\sigma_{L \text{ or } L'} = \frac{0.14}{0.15} \times \sigma_K = \pm 15.56 \times 10^3 \text{ kPa}$$

$$\tau_{L \text{ or } L'} = \frac{VA_{fghj}\bar{y}}{It} = \frac{(40)(0.04)(0.01)(0.145)}{(1/12)(0.04)(0.3)^3(0.04)} = 644 \text{ kPa}$$

To obtain the principal stresses at L, Mohr's circle of stress is used. Its construction is indicated in Fig. 9-3(i), and the results obtained are shown in the second sketch of Fig. 9-3(e). Note the invariance of the sum of the normal stresses, i.e., $\sigma_x + \sigma_y = \sigma_1 + \sigma_2$ or $-15.56 + 0 = -15.59 + 0.03$. A similar solution for the principal stresses at point L' yields the results indicated in the second sketch of Fig. 9-3(g).

Point M lies on the neutral axis of the beam, hence no flexural stress acts on the corresponding element shown in the first sketch of Fig. 9-3(f). The shearing stress on the right-hand face of the element at M acts in the same direction as the internal shear at the section KK'. Its magnitude can be obtained by applying Eq. 6-6, or directly by using Eq. 6-7, i.e.,

$$\tau_{\max} = \frac{3}{2}\frac{V}{A} = \frac{1.5(40)}{0.04(0.3)} = 5 \times 10^3 \text{ kPa}$$

The *pure* shearing stress transformed into the principal stresses according to Art. 8-6 is shown in the second sketch of Fig. 9-3(f).

It is significant to further examine qualitatively the results obtained. For this purpose the computed principal stresses *acting on the corresponding planes* are shown in Figs. 9-4(a) and (b). By examining Fig. 9-4(a), the characteristic behavior of the algebraically larger (tensile) principal stress at a section of a rectangular beam can be seen. This stress progressively diminishes in its magnitude from a maximum value at K' to zero at K. At the same time the corresponding directions of σ_1 gradually turn through 90°. A similar observation can be made regarding the algebraically smaller (compressive) principal stress σ_2 shown in Fig. 9-4(b).

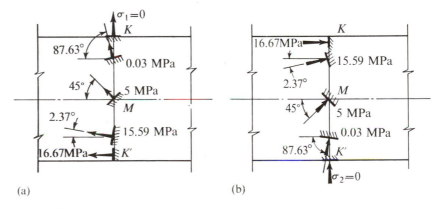

(a)   (b)

Fig. 9-4. (a) Behavior of the algebraically larger principal stress σ_1.
(b) Behavior of the algebraically smaller principal stress σ_2

To find the stresses acting on a plane of $\theta = +30°$ through point L', a direct application of Eqs. 8-1 and 8-2 *using the stresses shown in the first sketch of Fig. 9-3(g)* and $2\theta = 60°$ is made.

$$\sigma_\theta = \frac{+15.56}{2} + \frac{+15.56}{2}\cos 60° + (-0.644)\sin 60° = +11.11 \text{ MPa}$$

$$\tau_\theta = \frac{-15.56}{2}\sin 60° + (-0.644)\cos 60° = -7.06 \text{ MPa}$$

These results are shown in Fig. 9-3(j). The sense of the shearing stress τ_θ is opposite to that shown in Fig. 8-4(b), since the computed quantity is negative. The "wedge technique" explained in Example 8-1 or the Mohr's circle method as explained in Art. 8-8 can be used to obtain the same results.

*9-3. MEMBERS IN A STATE OF TWO-DIMENSIONAL STRESS

Within the scope of the formulas developed in this text, bodies in a state of two-dimensional stress can be studied as was done in the preceding example. A great many points in a stressed body may be investigated for the magnitude and direction of the principal stresses. Then, to study the general behavior of the stresses, selected points can be interconnected to give a visual interpretation of the various aspects of the computed data. For example, the points of algebraically equal principal stresses, regardless of their sense, when connected, provide a "map" of *stress contours.* Any point lying on a stress contour has a principal stress of the same algebraic magnitude.

Similarly, the points at which the directions of the minimum principal stresses form a *constant angle* with the *x*-axis can be connected.* Moreover, since the principal stresses are mutually perpendicular, the direction of the maximum principal stresses through the same points also forms a constant angle with the *x*-axis. The line so connected is a locus of points along which the principal stresses have *parallel directions.* This line is called an *isoclinic line.* The adjective isoclinic is derived from two Greek words, *isos* meaning equal and *klino* meaning slope or incline. Three isoclinic lines can be found by inspection in a rectangular prismatic beam subjected to transverse load acting normal to its axis. The lines corresponding to the upper and lower boundaries of a beam form two isoclinic lines as, at the boundary, the flexural stresses are the principal stresses and act parallel to the boundaries.† On the other hand, the flexural stress is zero at the neutral axis, and only there do pure shearing stresses exist. These pure shearing stresses transform

---

*These angles are usually measured in a counterclockwise direction from the *x*-axis to the nearest line of action of the minimum principal stress. These angles vary from 0° to 90°. An alternative method, amounting to the same thing, consists of measuring angles counterclockwise from a vertical line to the line of action of the algebraically larger principal stress.

†If the bending moments are positive, the line corresponding to the upper boundary of a beam forms a 0° isoclinic line, while the lower boundary corresponds to a 90° isoclinic.

into principal stresses, all of which act at an angle of 45° with the axis of the beam. Hence, another isoclinic line (the 45° isoclinic) is located on the axis of the beam. The other isoclinic lines are curved and are more difficult to determine.

Another set of curves can be drawn for a stressed body for which the magnitude and the sense of the principal stresses are known at a great many points. A curve whose tangent is changing in direction to conform with the direction of the principal stresses is called a *principal stress trajectory* or isostatic line. Like the isoclinic lines, the principal stress trajectories *do not* connect the points of equal stresses, but rather indicate the directions of the principal stresses. Since the principal stresses at any point are mutually perpendicular, the principal stress trajectories for the two principal stresses form a family of orthogonal (mutually perpendicular) curves.* An example of stress trajectories for a rectangular beam loaded with a concentrated force at the midspan is shown in Fig. 9-5. The principal stress trajectories corresponding to the tensile stresses are shown in the figure by solid lines; those for the compressive stresses are shown dashed. The trajectory pattern (not shown) is severely disturbed at the supports and at the point of application of the load P.

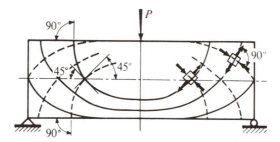

Fig. 9-5. Principal stress trajectories for a rectangular beam

*9-4. THE PHOTOELASTIC METHOD OF STRESS ANALYSIS

The state of stress in any two-dimensional stress problem can be expressed in terms of the stress contours, the isoclinic lines, and the principal stress trajectories discussed in the preceding article. Moreover, it is significant that the application of the same forces in the same manner to any two geometrically similar bodies made from *different elastic materials* causes the same stress distribution. The stress distribution is unaffected† by the elastic constants of a material. Therefore, to determine stresses experimentally,

---

*A somewhat analogous situation is found in fluid mechanics where in "two-dimensional" fluid flow problems the *streamlines* and the *equipotential lines* form an orthogonal system of curves—the *flow net*.

†For this to be true, strictly speaking, the bodies must be simply connected, i.e., without interior holes.

instead of finding the stresses in an actual member, the test specimen is prepared from any material suitable for the type of test to be performed. Glass, celluloid, and particularly certain grades of *Bakelite* have the required optical properties for photoelastic work. More recently, additional materials such as polyurethane rubber, Plexiglas, epoxy resin, and Columbia resin (CR-39) have been used. In a stressed specimen made from one of these materials, the principal stresses *temporarily change the optical properties* of the material. This change in the optical properties can be detected and related to the principal stresses that cause it. The experimental and analytical technique necessary for the analysis of problems in this manner is known as *the photoelastic method of stress analysis*. Only a brief outline of this method will be given here,* commencing with some remarks on light.

Light travels through any given medium in a straight line at a constant velocity. For the purposes at hand, the behavior of light may be explained by considering its single ray as a series of chaotic waves that travel in a number of planes containing the ray. By restricting the vibration of the waves to a single plane, a *plane polarized light* is obtained. This is done by passing the light through a *polarizer*, which may be a suitable pile of plates, a Nicol prism, or a commercially manufactured "Polaroid" element. The

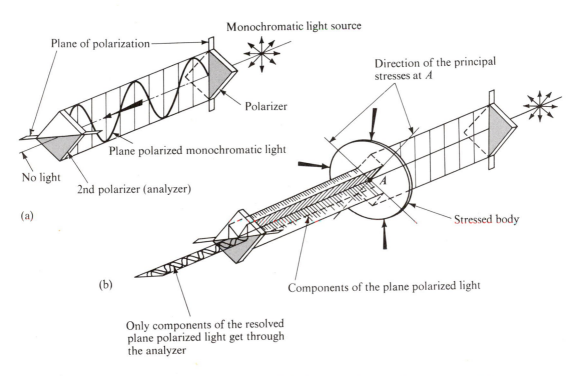

Fig. 9-6. The photoelastic method of stress analysis

*For more details see M. M. Frocht, *Photoelasticity*, vols. I and II, New York: Wiley, 1941 and 1948.

plane of the polarizer through which the transverse vibrations of the light are allowed to pass is called the *plane of polarization*. A schematic diagram of the foregoing definitions is shown in Fig. 9-6(a), where a second polarizer, called the *analyzer*, is also shown. Note that if the planes of polarization of the two polarizers are at right angles to each other, *no light* gets through the analyzer. This arrangement of the analyzer with respect to the polarizer is termed *crossed*.

If the light source* used is monochromatic, i.e., of one color, the transverse vibrations of the plane polarized light are regular, as the wave length in a given medium for any one color is constant. When propagating through the same medium, these transverse vibrations are described by a sinusoidal wave of constant amplitude and frequency.

By inserting an annealed specimen made from a suitable transparent material between the polarizer and the analyzer in the arrangement shown in Fig. 9-6(a), no new phenomenon is observed. However, by stressing the specimen, the optical properties of the material change, and two phenomena take place:†

1. At each point of the stressed body the polarized light wave is *resolved* into two mutually perpendicular components lying in the planes of the principal stresses occurring at that point.
2. The linear velocity of each of the components of the light wave is *retarded* through the stressed specimen in direct proportion to the related principal stress.

These facts are the basis of the photoelastic method of stress analysis. A schematic representation of the behavior of a monochromatic plane polarized wave as it passes through a stressed body and an analyzer is shown in Fig. 9-6(b). In the stressed specimen, a plane polarized wave is resolved into two components whose planes coincide with the planes of the principal stresses, as at point *A*. These components of the sinusoidal vibration leave the specimen with the same frequency but are *out of phase*. The latter effect is caused by the different amount of retardation of the light in the two principal planes of stress. Finally, the light waves emerging from the analyzer are again brought into the same plane, since only certain components of the polarized light can go through the analyzer. The two monochromatic light waves that leave the analyzer vibrate out of phase in the same plane with the same frequency. Their *phase difference*, which is directly proportional to the difference in the principal stresses at a point such as *A* of the stressed body, provides several possibilities that may be observed on a screen placed after the analyzer. If the two waves are out of phase by a full

*Mercury-vapor lamps are commonly used for this purpose.

†The first phenomenon stated was discovered by Sir David Brewster in 1816. The quantitative relation was established by G. Wertheim in 1854. The modern development of photoelasticity and its engineering applications probably owes most to the two professors from England, E. G. Coker and L. N. G. Filon, whose treatise on this subject was first published in 1930.

wave length of the light used, they reinforce each other, and the brightest light is seen on the screen. For other conditions, some interference takes place between the two light waves. *A complete elmination of the light occurs if the amplitudes of the two light waves are equal and are out of phase by one-half wave length or its odd integer multiple.* Therefore, since an infinite number of points in the stressed body affect the plane polarized light in a manner analogous to the point *A*, alternate bright and dark bands become apparent on a screen. The dark bands are called *fringes*.

The greater the difference in the principal stresses, the greater the phase difference between the two light waves emerging from the analyzer. Hence, if forces are *gradually* applied to a specimen until the principal stresses differ sufficiently to cause a phase difference of one-half wave length between the two light waves at some points, the first fringe appears on the screen. Then as the magnitude of the applied forces is increased, the first fringe shifts to a new position and another *"higher order" fringe* makes its appearance on the screen. The second fringe corresponds to the principal stresses that cause a phase difference of $1\frac{1}{2}$ wave lengths. This process can be continued as long as the specimen behaves elastically, and more and more fringes appear on the screen. A photograph with several fringes for a rectangular beam loaded at midspan is shown in Fig. 9-7. In a separate experiment

Fig. 9-7. Fringe photograph of a rectangular beam

fringes can be calibrated with a bar in tension or a beam in pure flexure. The stresses for these simple members can be accurately computed. It is necessary to make the calibration specimens from the same material as the specimen to be investigated. With the calibration data, complex members subjected to complicated loading may be investigated. For each fringe order, the difference of the principal stresses, $\sigma_1 - \sigma_2$, is known from calibration; hence, fringes represent a "map" of the difference in the principal stresses. According to Eq. 8-8, the difference of the principal stresses divided by two is equal to the maximum shearing stress; therefore the fringes also represent the loci of the principal shearing stresses.

A fringe photograph of a stressed body and calibration data are sufficient for determining the magnitude of the maximum or principal shearing stresses. The principal stress at any point of the *unloaded* boundary can also be obtained. At a free boundary, one of the principal stresses that acts normal to the boundary must be zero, and the fringe order is directly related to the other principal stress. Additional experimental work must be performed to determine normal stresses away from the boundaries.

One method of completing the problem consists of obtaining some *very accurate measurements of the change in thickness of the stressed specimen* at a number of points. These measurements, which may be designated by Δt, where t is the thickness of the specimen, are related to the principal stresses; i.e., from the generalized Hooke's law (Eq. 2-6) with $\sigma_z = 0$ one obtains

$$\Delta t = -\nu\left(\frac{\sigma_1 + \sigma_2}{E}\right)t \qquad \text{or} \qquad \sigma_1 + \sigma_2 = -\frac{E}{\nu t}\Delta t$$

Then, from an additional experiment on the same material in simple tension where $\sigma_1 \neq 0$ and $\sigma_2 = 0$, a new calibration chart can be prepared that gives the sum of the principal stresses versus Δt. From the information obtained from these two experiments, a "map" of the *sum* of the principal stresses for the specimen investigated can be prepared. By superposing this "map" with the "map" of the *differences* of the principal stresses obtained from the fringe photograph, the magnitudes of the principal stresses at any point of the stressed specimen can be determined.

Additional information must be found in the picture of fringes to determine the *direction* of the principal stresses. This information is given by the isoclinic line. This is a black line corresponding to the locus of the points where the direction of one of the principal stresses in the stressed body coincides with the plane of the polarized light leaving the polarizer. Rays passing through these points in the stressed specimen are not resolved and are blacked out by the analyzer. By rotating the polarizer into several known positions and maintaining the analyzer crossed, the isoclinic lines can be determined. These lines may be difficult to distinguish from the fringes, as both appear simultaneously on the screen. One method of differentiating the isoclinic lines from the fringes uses white instead of monochromatic light. Using white light makes isoclinics appear black, but the fringes are colored and contain all the visible spectral colors of the white light. On the other hand, to eliminate the isoclinic lines, which are undesirable for fringe photographs, two quarter-wave plates may be inserted into the optical system. Quarter-wave plates resolve the plane polarized light into two mutually perpendicular components; one of these is a quarter wave out of phase with the other. Combination of these components results in a "circularly polarized light." One of these plates is placed between the polarizer and the specimen and the other between the specimen and the analyzer. The fringe photograph shown in Fig. 9-7 was obtained by using this method.

With the aid of analytical methods, a sequence of isoclinic lines and fringe photographs are sufficient to solve the photoelastic problem without finding the sum of the principal stresses experimentally. These procedures are very detailed and laborious, and the reader is referred to books on photo-elasticity for further information.

The photoelastic method of stress analysis is very versatile and has been used to solve numerous problems. Nearly all solutions for the stress-concentration factors have been established by photoelasticity. The inaccuracy of the elementary formulas of mechanics of materials at concentrated forces is clearly brought out by the fringe photographs. For example, in Fig. 9-7, according to the elementary formulas, the fringes in the upper half of the beam should be like those in the lower half. Also note the local disturbance of the stresses at the supports in the same photograph.

The photoelastic method is best adapted to two-dimensional stress problems. Three-dimensional problems have also been analyzed by specialized techniques. The extension of the method to inelastic or plastic problems remains for the present unsolved. The Moiré fringe method is another optical procedure that has met with much success. This method will not be described here.*

9-5. THIN-WALLED PRESSURE VESSELS

The above investigations for the principal stresses in several stressed members illustrated numerous instances where biaxial stresses occur. On the other hand, all of the preceding philosophy of allowable stresses was based on the simple tension or torsion test. Therefore, before the study of the design of members is undertaken, it is important to reach some conclusions regarding the effect of biaxial stresses on the resistance to failure of various materials. Since the answer to this question is found from experiments on thin-walled pressure vessels, a method for analyzing them must be developed first. Attention will be confined to two types of these vessels, the cylindrical pressure vessel and the spherical.† Both of these types of pressure vessels are exceedingly important in industry; hence this topic is in itself of great practical importance.

The walls of an ideal thin-walled pressure vessel act as a membrane, i.e., no bending of the walls takes place. A sphere is an ideal shape for a closed pressure vessel if the contents are of negligible weight. A cylindrical vessel is also good with the exception of the junctures with the ends, a matter to be commented on in more detail later.

The analysis of pressure vessels will begin by considering a cylindrical pressure vessel such as a boiler, as shown in Fig. 9-8(a). A segment is isolated from this vessel by passing two planes perpendicular to the axis of

*See, for example, P. S. Theocaris, "Moiré Fringes: A Powerful Measuring Device," *Applied Mechanics Review*, May 1962, vol. 15.

†For a more general discussion of membrane behavior of axisymmetrical shells of revolution see E. P. Popov, *Introduction to Mechanics of Solids*, Englewood Cliffs, N.J.: Prentice-Hall, 1968.

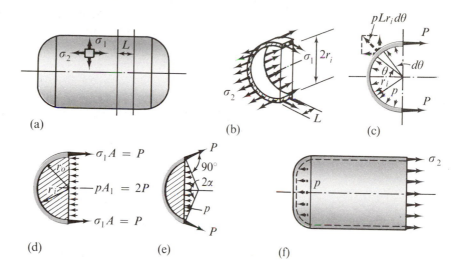

Fig. 9-8. Diagrams for analysis of cylindrical pressure vessels

the cylinder and one additional longitudinal plane *through* the same axis, shown in Fig. 9-8(b). The conditions of symmetry exclude the existence of any shearing stresses in the planes of the sections, as shearing stresses would cause an incompatible distortion of the tube. Therefore, the stresses that can exist on the sections of the cylinder can only be the normal stresses σ_1 and σ_2 shown in Fig. 9-8(b). These stresses are the *principal stresses*. These stresses, multiplied by the respective areas on which they act, maintain the element of the cylinder in equilibrium against the internal pressure.

Let the internal pressure in excess of the external pressure be p psi or Pa (gage pressure), and let the internal radius of the cylinder be r_i. Then the *force* on an infinitesimal area $Lr_i d\theta$ (where $d\theta$ is an infinitesimal angle) of the cylinder caused by the internal pressure acting *normal* thereto is $pLr_i d\theta$, Fig. 9-8(c). The component of this force acting in the horizontal direction is $(pLr_i \, d\theta) \cos\theta$; hence the total resisting force of $2P$ acting on the cylindrical segment is

$$2P = 2 \int_0^{\pi/2} pLr_i \cos d\theta = 2pr_i L$$

Again from symmetry, half of this total force is resisted at the top cut through the cylinder and the other half is resisted at the bottom. The normal stresses σ_2 acting in a direction parallel to the axis of the cylinder do not enter into the above integration.

Instead of obtaining the force $2P$ caused by the internal pressure by integration, as above, a simpler equivalent procedure is available. From an alternate point of view, the two forces P resist the force developed by the internal pressure p, which acts perpendicular to the *projected area* A_1 of the cylindrical segment onto the diametral plane, Fig. 9-8(d). This area in Fig.

9-8(b) is $2r_iL$, hence $2P = A_1 p = 2r_i Lp$. This force is resisted by the forces developed in the material in the longitudinal cuts, and since the outside radius of the cylinder is r_o, the area of *both* longitudinal cuts is $2A = 2L(r_o - r_i)$. Moreover, if the *average* normal stress acting on the longitudinal cut is σ_1, the force resisted by the walls of the cylinder is $2L(r_o - r_i)\sigma_1$. Equating the two forces, $2r_i Lp = 2L(r_o - r_i)\sigma_1$.

Since $r_o - r_i$ is equal to t, the thickness of the cylinder wall, the last expression simplifies to

$$\sigma_1 = \frac{pr_i}{t} \qquad (9\text{-}2)$$

The normal stress given by Eq. 9-2 is often referred to as the *circumferential* or the *hoop stress*. Equation 9-2 is valid only for thin-walled cylinders, as it gives the *average* stress in the hoop. However, as will be shown in Chapter 16, the wall thickness can reach one-tenth of the internal radius and the error in applying Eq. 9-2 will still be small. Since Eq. 9-2 is used primarily for *thin*-walled vessels where $r_i \approx r_o$, the subscript for the radius is usually omitted.

Equation 9-2 can also be derived by passing two longitudinal sections as shown in Fig. 9-8(e). In this treatment, the forces P in the hoop must be considered acting tangentially to the cylinder. The horizontal components of the forces P maintain the horizontal component of the internal pressure in a state of static equilibrium.

The other normal stress σ_2 acting in a cylindrical pressure vessel acts *longitudinally*, Fig. 9-8(b), and it is determined by solving a simple axial force problem. By passing a section through the vessel perpendicular to its axis, a free body as shown in Fig. 9-8(f) is obtained. The force developed by the internal pressure is $p\pi r_i^2$, and the force developed by the longitudinal stress σ_2 in the walls is $\sigma_2(\pi r_o^2 - \pi r_i^2)$. Equating these two forces and solving for σ_2,

$$p\pi r_i^2 = \sigma_2(\pi r_o^2 - \pi r_i^2)$$

$$\sigma_2 = \frac{pr_i^2}{r_o^2 - r_i^2} = \frac{pr_i^2}{(r_o + r_i)(r_o - r_i)}$$

However, $r_o - r_i = t$, the thickness of the cylindrical wall, and since this development is restricted to *thin*-walled vessels, $r_o \approx r_i \approx r$; hence

$$\sigma_2 = \frac{pr}{2t} \qquad (9\text{-}3)$$

Note that for *cylindrical* pressure vessels $\sigma_2 \approx \sigma_1/2$.

An analogous method of analysis can be used to derive an expression for *thin*-walled spherical pressure vessels. By passing a section through the center of the sphere of Fig. 9-9(a), a hemisphere shown in Fig. 9-9(b) is isolated. By using the same notations as above, an equation identical to Eq. 9-3 can be derived. However, for a sphere, *any section that passes through*

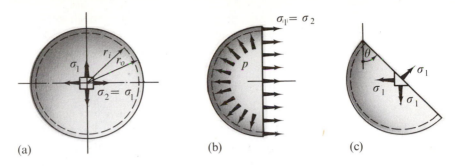

(a)   (b)   (c)

Fig. 9-9. Spherical pressure vessel

the center of the sphere yields the same result. Equal principal stresses act on the elements of the sphere whatever the inclination of the element's side, Fig. 9-9(c). Hence, for thin-walled spherical pressure vessels,

$$\sigma_1 = \sigma_2 = \frac{pr}{2t} \tag{9-3a}$$

*9-6. REMARKS ON THIN-WALLED PRESSURE VESSELS

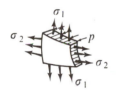

Fig. 9-10. An element of a thin-walled pressure vessel considered to be in a state of biaxial stress

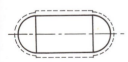

Fig. 9-11. Dashed lines show the tendency (exaggerated), for the cylinder and the ends to expand a different amount under the action of internal pressure.

The state of stress for an element of a thin-walled pressure vessel as given by Eqs. 9-2, 9-3, and 9-3(a) is considered biaxial, although the internal pressure acting on the wall causes a local compressive stress equal to this pressure. Actually a state of triaxial stress exists on the inside of the vessel, Fig. 9-10. However, for thin-walled pressure vessels this latter stress is much smaller than σ_1 and σ_2 and for this reason is generally ignored. The significant stresses acting on the properly oriented elements for cylindrical and spherical pressure vessels are shown in Figs. 9-8(a) and 9-9(a), respectively.

For the same internal pressure, diameter, and wall-thickness, the maximum stress in a spherical vessel is approximately one-half the maximum stress occurring in a cylindrical vessel. Also note that Mohr's circle of stress for the main stresses in a *spherical* vessel degenerates to a point. This means that regardless of the inclination of the plane in the element investigated, the normal stress remains constant and no shearing stresses exist.* The same conclusion is reached by passing an arbitrary cutting plane through the sphere's center, as was done in Fig. 9-9(c).

A discontinuity of the membrane action occurs at the juncture of the cylindrical portion of a pressure vessel with the ends. Under the action of the internal pressure, the cylinder tends to expand as shown by the dashed lines in Fig. 9-11, while the ends tend to expand a different amount, owing to differences in stress. This incompatibility of deformations causes local

---

*By looking on an infinitesimal element *from the top edge*, a new orientation of the element is obtained. For this new orientation, a maximum shearing stress of $\sigma_1/2$ may be seen to exist in the material.

bending and shearing stresses in the neighborhood of the joint, since there must be physical continuity between the ends and the cylindrical wall. For this reason, properly curved ends must be used for pressure vessels. Flat ends are very undesirable.*

A majority of pressure vessels are manufactured from separate curved sheets that are joined. A common method of accomplishing this is to arc-weld the abutting material as shown in Fig. 9-12. Grooves into which the welding metal is deposited are prepared in a number of ways, depending on the thickness of the plates. The so-called single-V butt joint is shown in Fig. 9-12. Other types of butt welds are also used. Their nomenclature depends on the preparation of the groove. For example, if V-grooves are made from both sides, as is done for thicker plates, the weld is called a double-V butt joint. Other terms are single-bevel butt joint, double-bevel butt joint, single-U butt joint, etc. The calculations for the welds are made by assigning an allowable tensile stress to the weld, which is assumed to be of the *same depth as the thickness of the plate*. The allowable stresses are usually expressed as a certain percentage of the strength of the original solid plate of the parent material. This percentage factor varies greatly, depending on the workmanship. For ordinary work, a 20% reduction in the allowable stress for the weld compared to the solid plate may be used. For this factor the efficiency of the joint is said to be 80%. On high grade work, some of the specifications allow 100% efficiency for the welded joint.

In conclusion it must be emphasized that the formulas derived for thin-walled pressure vessels in the preceding article should be used only for cases of *internal pressure*. If a vessel is to be designed for external pressure, as in the case of a vacuum tank or a submarine, *instability* (buckling) of the walls may occur, and stress calculations based on the above formulas are meaningless.

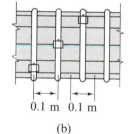

The application of Eqs. 9-2, 9-3, and 9-3a is so direct that no illustrative examples are solved. However, a similar method of analysis is applicable to a related problem, which will now be analyzed.

EXAMPLE 9-4

Determine the pressure p that can be carried by the wooden stave pipe shown in Fig. 9-13. Assume that the wooden staves are of adequate size to span between 10 mm round steel hoops with upset ends† spaced 0.1 m apart. Assume the allowable stress in the steel hoops at 165 MPa, and ignore the effect of crushing of the wood by the hoops.

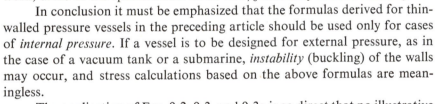

(a)   (b)

Fig. 9-13

*The ASME Unfired Pressure Vessel Code gives practical information on the design of ends; the necessary theory is beyond the scope of this text. In spite of this limitation, the elementary formulas for thin-walled cylinders developed here are suitable in the majority of cases.

†The diameter of the rod at the ends is enlarged by forging, i.e., "upset," in order to maintain the nominal rod diameter at the root of the threads.

SOLUTION

This problem may be analyzed in the same manner as that of the cylindrical pressure vessel. Since the hoops are 0.1 m apart, each hoop resists the force developed by the pressure along one-tenth linear meter of the pipe, Fig. 9-13(b). Then, considering a segment of the pipe 0.1 m long similar to the one shown in Fig. 9-8(b), the internal force tending to burst the pipe in this distance is $2r_iLp = 2(0.5)0.1p = 0.1p$ N. This force is resisted by *two* cross-sectional areas of the hoop. The area of each 10 mm diameter hoop is 78.5 mm² = 78.5×10^{-6} m²; hence it is good for $(78.5 \times 10^{-6})(165 \times 10^6)$ = 12 950 N. By equating the bursting force to the two resisting forces in the hoop, the allowable pressure in the pipe is found, i.e., $0.1p = 2(12\,950)$ and $p = 259\,000$ Pa = 259 kPa.

9-7. FAILURE THEORIES: PRELIMINARY REMARKS

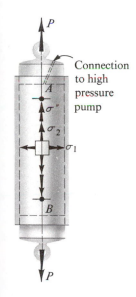

Fig. 9-14. Arrangement for obtaining controlled ratios of the principal stresses

Most of the information on yielding and fracture of materials under the action of biaxial stresses comes from experiments on thin-walled cylinders. A typical arrangement for such experiments is shown in Fig. 9-14. The ends of a thin-walled cylinder of the material being investigated are closed by substantial caps. This forms the hollow interior of a cylindrical pressure vessel. By pressurizing the available space until yielding or bursting point, the elements of the wall are subjected to biaxial stresses of a constant ratio $\sigma_1/\sigma_2 = 2$. By applying an additional tensile force P to the caps, the σ_2 stress is increased to any predetermined amount $\sigma_2 + \sigma''$. By applying a compressive force, the σ_2 stress can be minimized or eliminated. Actual compressive stress in the longitudinal direction is undesirable, as the tube may buckle. By maintaining a fixed ratio between the principal stresses until the failure point is reached, the desired data on a material are obtained. Analogous experiments with tubes simultaneously subjected to torque, axial force, and pressure are also used. An interpretation of these data, together with all other related experimental evidence, including the simple tension tests, permits a formulation of theories of failure for various materials subjected to combined stresses.

Unfortunately, at this date the quantitative criteria for yielding and fracture of materials under multiaxial states of stress are incomplete. A number of questions remain unsettled and are a part of an active area of materials research. As yet no complete answer can be given by any one theory, and there exist several failure or strength theories that are applicable to specific groups of materials. The two widely accepted criteria for the onset of inelastic behavior for ductile materials under combined stresses will be discussed first. This is followed by the presentation of a fracture criterion for brittle materials. It must be emphasized that, in classifying the materials in this manner, strictly speaking, one refers to the brittle or ductile state of the material as this characteristic is greatly affected by temperature as well as by the state of stress itself. More complete answers to such questions are beyond the scope of this book.

9-8. MAXIMUM SHEARING STRESS THEORY

The maximum shearing stress theory,* or simply the maximum shear theory, results from the observation that, in a ductile material, slipping occurs during yielding along critically oriented planes. This suggests that the maximum shearing stress plays the key role, and it is assumed that yielding of the material depends only on the maximum shearing stress that is attained within an element. Therefore, whenever a certain critical value τ_{cr} is reached, yielding in an element commences.† For a given material this value is set equal to the shearing stress at yield in simple tension or compression. Thus, according to Eq. 8-6, if $\sigma_x = \pm \sigma_1 \neq 0$, and $\sigma_y = \tau_{xy} = 0$,

$$\tau_{max} \equiv \tau_{cr} = \left| \pm \frac{\sigma_1}{2} \right| = \frac{\sigma_{yp}}{2} \tag{9-4}$$

which means that if σ_{yp} is the yield-point stress found, for example, in a simple tension test, the corresponding maximum shearing stress is half as large. This conclusion also follows easily from Mohr's circle of stress.

To apply the maximum shearing stress criterion to a given biaxial state of stress, first the maximum shearing stress is determined, and then equated to τ_{max} given by Eq. 9-4. Expressing the maximum shearing stresses for a given state of stress in terms of the principal stresses, then canceling the common 2 in the denominators leads to the following yield criterion:

$$|\sigma_1| \leq \sigma_{yp} \quad \text{and} \quad |\sigma_2| \leq \sigma_{yp} \tag{9-5}$$

and
$$|\sigma_1 - \sigma_2| \leq \sigma_{yp} \tag{9-6}$$

Equations 9-5 apply only when σ_1 and σ_2 are of the *same sign*, and Eq. 9-6 applies only when σ_1 and σ_2 are of the *opposite sign*. In a biaxial case, if the principal stresses have the same sign, the maximum shearing stresses are found by viewing the element along the axis of the smaller stress (see Fig. 8-13); for principal stresses of the opposite sign, the largest shearing stress is given by $(\sigma_1 - \sigma_2)/2$.

For impending yield, the inequalities of Eqs. 9-5 and 9-6 are replaced by equalities. By considering σ_1 and σ_2 as the coordinates of a point in this σ_1-σ_2 stress space, these equations define the limiting hexagon of Fig. 9-15. Stresses falling within the hexagon indicate that no yielding of the material

---

*This theory appears to have been originally proposed by C. A. Coulomb in 1773. In 1868 H. Tresca presented the results of his work on the flow of metals under great pressures to the French Academy. Now this theory often bears his name.

†In single crystals slip occurs along preferential planes and in preferential directions. In studies of this phenomenon the effective component of the shearing stress causing slip must be carefully determined. Here it is assumed that because of the random orientation of numerous crystals the material has isotropic properties, and so by determining τ_{max} one finds the critical shearing stress.

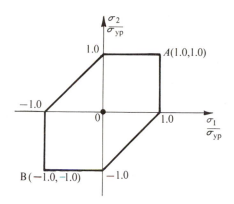

Fig. 9-15. Yield criterion based on maximum shearing stress

has occurred, i.e., the material behaves elastically. The state of stress corresponding to the points falling on the hexagon shows that the material is yielding. No points can lie outside the hexagon.

Note that, according to the maximum shear theory, if hydrostatic tensile or compressive stresses are added, i.e., stresses such that $\sigma_1' = \sigma_2' = \sigma_3'$, no change in the material response is predicted. Adding these stresses merely shifts the Mohr's circles of stress, such as in Fig. 8-13, along the σ axis and τ_{max} remains the same. This matter will be commented upon further in the next article.

The yield criterion just derived is often referred to as the *Tresca yield condition* and is one of the widely used laws of plasticity.

9-9. MAXIMUM DISTORTION ENERGY THEORY

Another widely accepted criterion of yielding for ductile, isotropic materials is based on the energy concepts.* In this approach the total elastic energy is divided into two parts: one associated with the volumetric changes of the material, and the other causing shearing distortions. By equating the shearing distortion energy at yield point in simple tension to that under combined stress, the yield criterion for combined stress is established.

It can be shown† that the yield condition for an ideally plastic material under a triaxial state of stress may be obtained in terms of the principal stresses as

$$(\sigma_1 - \sigma_2)^2 + (\sigma_2 - \sigma_3)^2 + (\sigma_3 - \sigma_1)^2 = 2\sigma_{yp}^2 \qquad (9\text{-}7)$$

For plane stress $\sigma_3 = 0$, and Eq. 9-7 in dimensionless form becomes

$$\left(\frac{\sigma_1}{\sigma_{yp}}\right)^2 - \left(\frac{\sigma_1}{\sigma_{yp}}\frac{\sigma_2}{\sigma_{yp}}\right) + \left(\frac{\sigma_2}{\sigma_{yp}}\right)^2 = 1 \qquad (9\text{-}8)$$

This is an equation of an ellipse, a plot of which is shown in Fig. 9-16. Any stress falling within the ellipse indicates that the material behaves elastically. Points on the ellipse indicate that the material is yielding. This is the same interpretation as that given earlier for Fig. 9-15. On unloading, the material behaves elastically.

*The first attempt to use the total energy as the criterion of yielding was made by E. Beltrami of Italy in 1885. In its present form the theory was proposed by M. T. Huber of Poland in 1904 and was further developed and explained by R. von Mises (1913) and H. Hencky (1925), both of Germany and the United States.

†See, for example, E. P. Popov, *Introduction to Mechanics of Solids*, Englewood Cliffs, N. J.: Prentice-Hall, 1968.

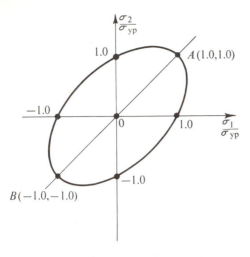

Fig. 9-16. Yield criterion based on maximum distortion energy

It is important to note that this theory does not predict changes in the material response when hydrostatic tensile or compressive stresses are added. This follows from the fact that, since only differences of the stresses are involved in Eq. 9-7, adding a constant stress to each does not alter the yield condition. For this reason in the three-dimensional stress space, the yield surface becomes a cylinder with an axis having all three direction cosines equal to $1/\sqrt{3}$. Such a cylinder is shown in Fig. 9-17. The ellipse in Fig. 9-16 is simply the intersection of this cylinder with the σ_1-σ_2 plane. It can be shown also that the yield surface for the maximum shearing stress criterion is a hexagon that fits into the tube, Fig. 9-17.

The yield condition expressed by Eq. 9-7 can be shown to be another stress invariant. It is also a continuous function. These features make the use of this law of plastic yielding for combined stresses particularly attractive from the theoretical point of view. This widely used law is often referred to as the *Huber-Hencky-Mises* or simply the *von Mises yield condition.**

9-10. MAXIMUM NORMAL STRESS THEORY

The maximum normal stress theory or simply the maximum stress theory† asserts that failure or fracture of a material occurs when the maximum normal stress at a point reaches a critical value regardless of the other stresses. Only the largest principal stress must be determined to apply this criterion. The critical value of stress σ_{ult} is usually determined in a tensile experiment, where the failure of a specimen is defined to be either excessively large elongation or fracture. Usually the latter is implied.

Experimental evidence indicates that this theory applies well to brittle materials in all ranges of stresses providing a tensile principal stress exists. Failure is characterized by the separation, or the cleavage, fracture. This mechanism of failure differs drastically from the ductile fracture, which is

---

*In the past this condition has been also frequently referred to as the *octahedral shearing stress theory*. See A. Nadai, *Theory of Flow and Fracture of Solids*, New York: McGraw-Hill, 1950, p. 104, or F. B. Seely and J. O. Smith, *Advanced Mechanics of Materials* (2nd ed.), New York: Wiley, 1952, p. 61.

†This theory is generally credited to W. J. M. Rankine, an eminent British educator (1820–72). An analogous theory based on the maximum strain, rather than stress, being the basic criterion of failure was proposed by the great French elastician, B. de Saint Venant (1797–1886). Experimental evidence does not corroborate the latter approach.

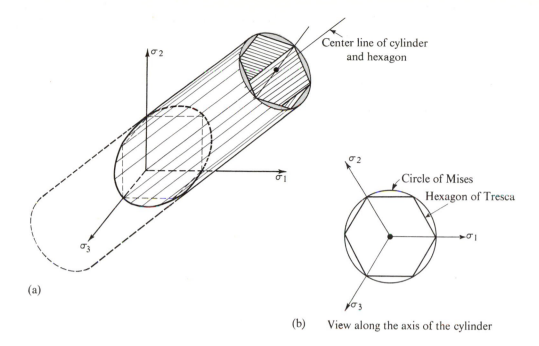

(a)

Center line of cylinder and hexagon

Circle of Mises

Hexagon of Tresca

(b) View along the axis of the cylinder

Fig. 9-17. Yield surfaces for three-dimensional state of stress

accompanied by large deformations due to slip along the planes of maximum shearing stress.

The maximum stress theory can be interpreted on graphs as can other theories. This is done in Fig. 9-18. Failure occurs if points fall on the surface. Unlike the previous theories, this stress criterion gives a bounded surface of the stress space.

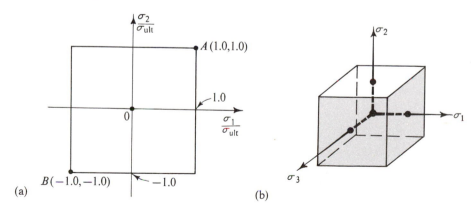

(a)

(b)

Fig. 9-18. Fracture envelopes bssed on maximum stress criterion

9-11. COMPARISON OF THEORIES; OTHER THEORIES

Comparison of some classical experimental results with the yield and fracture theories presented above is shown in Fig. 9-19.* Note the particularly good agreement between the maximum distortion energy theory and experimental results for ductile materials. However, the maximum normal stress theory appears to be best for brittle materials and can be unsafe for ductile materials.

All the theories for uniaxial stress agree since the simple tension test is the standard of comparison. Therefore, if one of the principal stresses at a point is large in comparison with the other, all theories give practically the same results. The discrepancy between the theories is greatest in the second and fourth quadrants, when both principal stresses are numerically equal.

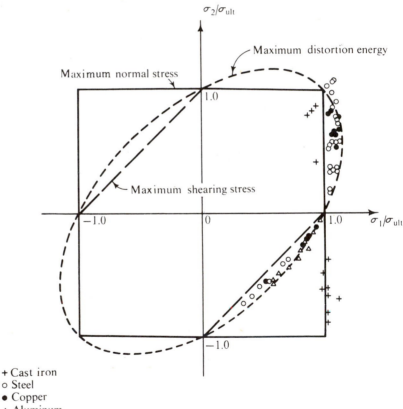

+ Cast iron
o Steel
● Copper
▵ Aluminum

Fig. 9-19. Comparison of yield and fracture criteria with test data

*The experimental points shown on this figure are based on classical experiments by several investigators. The figure is adapted from a compilation made by G. Murphy, *Advanced Mechanics of Materials*, New York: McGraw-Hill, 1964, p. 83.

In the development of the theories discussed above it has been assumed that the properties of material in tension and compression are alike—the plots shown in several of the preceding figures have two axes of symmetry. On the other hand, it is known that some materials such as rocks, cast iron, concrete, and soils, have drastically different properties depending on the sense of the applied stress. An early modification of the maximum shear theory by C. Duguet in 1885 to achieve better agreement with experiment is shown in Fig. 9-20(a). This modification recognizes the high strength of

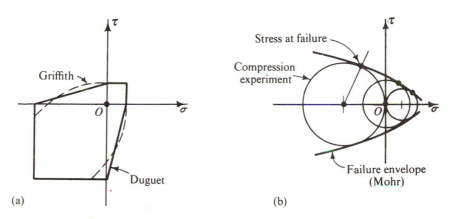

Fig. 9-20. Possible fracture criteria

some materials when subjected to biaxial compression. A. A. Griffith,* in a sense, refined the explanation for the above observation by introducing the idea of surface energy at microscopic cracks and showing the greater seriousness of tensile stresses compared with compressive ones with respect to failure. According to this theory an existing crack will rapidly propagate if the available elastic strain energy release rate is greater than the increase in the surface energy of the crack. The original Griffith concept has been considerably expanded by G. R. Irwin.†

Otto Mohr, in addition to showing the construction of the stress circle bearing his name, suggested another approach for predicting failure of a material. Different experiments such as one in simple tension, one in pure shear, and one in compression are performed first; see Fig. 9-20(b). Then an envelope to these circles defines the failure envelope. Circles drawn tangent to this envelope give the condition of failure at the point of tangency. This approach finds favor in soil mechanics.

*A. A. Griffith, "The Phenomena of Rupture and Flow of Solids," *Philosophical Transactions of the Royal Society of London*, Series A, 1920, vol. 221, pp. 163–98.

†G. R. Irwin, "Fracture Mechanics," *Proceedings*, *First Symposium on Naval Structural Mechanics*, Long Island City, N.Y.: Pergamon Press, 1958, p. 557. Also see *A Symposium on Fracture Toughness Testing and Its Applications*, American Society for Testing and Materials Special Technical Publication No. 381, American Society for Testing and Materials and National Aeronautics and Space Administration, 1965.

Instead of studying the response of materials on stress-space plots as in Fig. 9-20, the stress invariant $(\sigma_x + \sigma_y + \sigma_z)$ and the stress given by Eq. 9-7 may be used as the coordinate axes. Useful fracture criteria have been established from studies based on this approach.*

Sometimes the yield and fracture criteria discussed above are inconvenient to apply. In such cases, interaction curves such as in Fig. 7-6 can be used to advantage. Experimentally determined curves of this type, unless complicated by a local or buckling phenomenon, are equivalent to the strength criteria discussed here.

In the design of members in the next chapter, departures will be made from strict adherence to the yield and fracture criteria established here, although unquestionably these theories provide the rational basis for design.

PROBLEMS FOR SOLUTION

9-1. A 50 mm square steel bar is subjected to an axial tensile force. If the maximum shearing stress caused by this force is 80 000 kN/m², what is the magnitude of the applied force?

9-2. A concrete cylinder tested in a vertical position failed at a compressive stress of 30 MPa. The failure occurred on a plane of 30° with the vertical. On a clear sketch show the normal and the shearing stresses which acted on the plane of failure. *Ans:* $\tau = 13.0$ MN/m².

9-3. A 50 mm by 100 mm wooden post has the grain of the wood running at an angle of 25° with its axis. If for this wood the allowable shearing stress in the direction parallel to the grain is 600 kN/m², what is the allowable axial force for this post controlled by the shearing stress? *Ans:* 7.83 kN.

9-4. A simple beam 50 mm wide by 120 mm high spans 1.50 m and supports a uniformly distributed load of 80 kN/m including its own weight Determine the principal stresses and their directions at points A, B, C, D, and E at the section shown in the figure. *Ans:* $\sigma_C = \pm 5$ MPa.

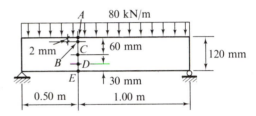

PROB. 9 – 4

9-5. A very short *I*-beam cantilever is loaded as shown in the figure. Find the principal stresses and their direction at points *A*, *B*, and *C*. Point *B* is in the web at the juncture with the flange. Neglect the weight of the beam and ignore the

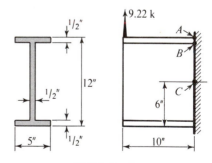

PROB. 9 – 5

*B. Bresler and K. Pister, "Failure of Plain Concrete Under Combined Stresses," *Transactions of the American Society of Civil Engineers*, 1957, vol. 122, p. 1049.

effect of stress concentrations. *I* around the neutral axis for the whole section is 221 in.⁴ Use the accurate formula to determine the shearing stresses. *Ans:* At *A*, 0, −2.50 ksi; at *B*, +0.51 ksi, −2.81 ksi; at *C*, ±1.84 ksi.

9-6. A 100 mm by 500 mm rectangular wooden beam supports a 40 kN load as in the figure. At section *a–a* the grain of the wood makes an angle of 20° with the axis of the beam. Find the shearing stress along the grain of the wood at points *A* and *B* caused by the applied concentrated force. *Ans:* $\tau_A = 835$ kN/m².

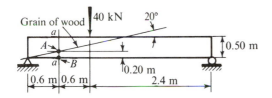

Grain of wood

PROB. 9 – 6

9-7. A cast iron beam is loaded as shown in the figure. Determine the principal stresses at the three points *A*, *B*, and *C* caused by the applied force. The moment of inertia of the cross-sectional area around the neutral axis is 316.2 in.⁴ *Ans:* At *A*, 0, −2,220 psi; at *B*, +37 psi, −1,117 psi; at *C*, +1,960, 0.

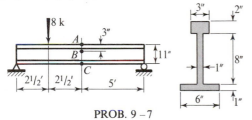

PROB. 9 –7

9-8. At a certain point in a masonry structure, the state of stress caused by the applied forces will be as shown in the sketch. The stone of which the structure is made is stratified and is weak in shear along planes parallel with the plane *A–A*. Is this state of stress permissible? Assume that the allowable stresses in the stone in any direction are 225 psi in tension and 2,000 psi in compression and that the allowable shearing stress parallel with the *A–A* plane is 325 psi. *Ans:* Shear governs, 352 psi.

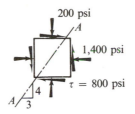

PROB. 9 – 8

9-9. After the erection of a heavy structure, it is estimated that the state of stress in the rock foundation will be essentially two-dimensional and as shown in the figure. If the rock is stratified, the strata making an angle of 30° with the vertical, is the anticipated state of stress permissible? Assume that the static coefficient of friction of rock on rock is 0.50, and along the planes of stratification cohesion amounts to 85 kN/m².

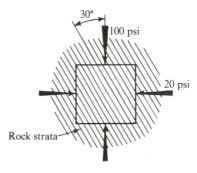

Rock strata

PROB. 9 – 9

9-10. A 0.10 m square bent bar is loaded as shown in the figure. Determine the state of stress at a point lying on the axis of this bar at section

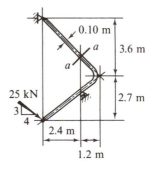

PROB. 9 – 10

a-a caused by the applied inclined force. Show the results on an infinitesimal element. Principal stresses are not required.

9-11. The maximum shearing stress at point *A* in a beam, see figure, is 120 psi. Determine the magnitude of the force *P*. Assume the beam to be weightless.

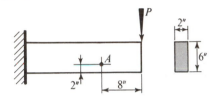

PROB. 9 – 11

9-12. The bracket shown is loaded by a concentrated force *P*, which produces axial compression and bending but no torsion. (a) Set up an element showing the state of stress at point *A* in terms of the applied force. (b) If the horizontal (longitudinal) strain at *A* is 2×10^{-4} in./in. and $E = 30 \times 10^6$ psi, what is the magnitude of the load *P*? For this member $I = 136$ in.^{4} around the centroidal axis. *Ans:* (b) 20.8 k.

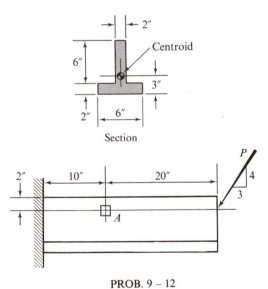

PROB. 9 – 12

9-13. At a point *A* on the upstream surface of a dam the water pressure is −2 MPa. A measured compressive stress parallel to the surface is −3

MPa. Calculate the stresses σ_x, σ_y, and τ_{xy} and indicate their direction on the element shown. *Ans:* $\sigma_x = 2.36$ MPa.

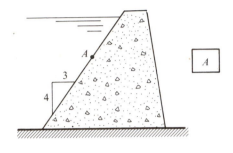

PROB. 9 – 13

9-14. A special hoist is loaded with a 15-kip load suspended by a cable as in the figure. Determine the state of stress at point *A* caused by this load. Show the results on an element with horizontal and vertical faces. *I* of the cross-sectional area around the neutral axis is 165 in.^{4} *Ans:* −4.66 ksi, 1.17 ksi.

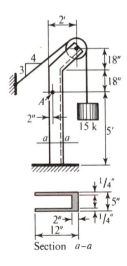

Section *a–a*

PROB. 9 – 14

9-15. By applying a vertical force *P* the toggle clamp, as shown in the figure, exerts a force of 1,000 lb on a cylindrical object. The movable jaw slides in a guide which prevents its upward movement. (a) Determine the magnitude of the applied vertical force *P*, and the downward force com-

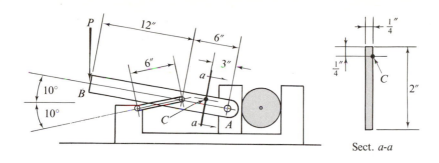

Sect. *a-a*

PROB. 9 – 15

ponent developed at hinge *A*, (b) determine the stresses due to axial trust, transverse shear, and bending moment acting on an element at point *C* of section *a–a*, (c) draw an element at point *C* with sides parallel and perpendicular to the axis of member *BA* and show the stresses acting on the element, (d) using Mohr's circle determine the largest principal stress and the maximum shearing stress at *C*. *Ans:* (a) 117 lb, 58.9 lb., (d) 1,250 psi, 663 psi.

9-16. A vertical T-beam is loaded as shown in the figure. Compute the normal and shearing stresses acting on an element at point *A* caused by the applied loading. Make a sketch of the element and indicate on it the directions of the computed stresses. The cross-sectional area of the beam is 10 in.², and *I* around the neutral axis is 20.8 in.⁴ *Ans:* −3,150 psi, 115 psi.

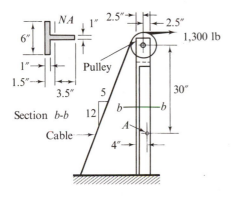

PROB. 9 – 16

9-17. A short rod of 2 in. diameter is subjected to an axial force of 6π kips and a torque of

2π kip-in. Determine the maximum (principal) shearing stresses, and show the results on a properly oriented element. *Ans.* 5 ksi, $18\frac{1}{2}°$.

9-18. A shaft of 2 in. diameter is subjected to an axial tensile force of 12π kips. What torque may be applied to this shaft in addition to this axial force without exceeding the maximum (principal) shearing stress of 10,000 psi? *Ans.* 4π kip-in.

9-19. A section of a 40 mm diameter shaft is simultaneously transmitting a torque of 314 N·m and a bending moment of 314 N·m. Determine the magnitude of the principal shearing stress.

*9-20. A shaft of 2 in. diameter is simultaneously subjected to a torque and a pure bending moment. It is known that at every section of the shaft the largest tensile principal stress caused by the applied loading is 24,000 psi, and that at the same point the largest tensile stress caused by bending, which is 18,000 psi, occurs. Determine the applied bending moment and the applied torque. *Ans.* $M = 4.5\pi$ kip-in., $T = 6\pi$ kip-in.

9-21. A $\frac{1}{2}$ in. diameter drill bit ($A = 0.196$ in.², $I = 0.00306$ in.⁴, $J = 0.00612$ in.⁴) is inserted into a chuck as shown in the figure. During the drilling operation an axial force $P = 3.92$ kips and a torque $T = 10\pi/128$ k-in. act on the bit. If a horizontal force of 35.7 lb. is accidentally applied to the plate being drilled, what is the magnitude of the largest principal stress which develops at the top of the drill bit? Determine the critically stressed point on the drill by inspection. *Ans:* −40 ksi.

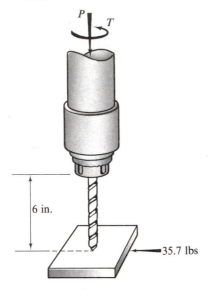

P

T

6 in.

35.7 lbs

PROB. 9 – 21

9-22. A solid circular shaft is loaded as shown in the figure. At section *ABCD* the stresses due to the 10 kN force, and the weight of the shaft and round drum are found to be as follows: maximum bending stress is 40 MN/m², maximum torsional stress is 30 MN/m², and maximum shearing stress due to *V* is 6 MN/m². (a) Set up elements at points *A*, *B*, *C*, and *D* and indicate the magnitudes and directions of the stresses acting on them. In each case state from which direction the element is observed. (b) Using Mohr's circle, find directions and magnitudes of the principal stresses and of the maximum shearing stress at point *A*. *Ans:* (b) $\tau = 36.1$ MPa.

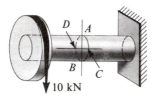

D   A

B   C

10 kN

PROB. 9 – 22

9-23. A circular bar of 2 in. diameter with a rectangular block attached at its free end is suspended as shown in the figure. Also a horizontal

force is applied eccentrically to the block as shown. Analysis of the stresses at section *ABCD* gives the following results: max bending stress is 1,000 psi, max torsional stress is 300 psi, max shearing stress due to *V* is 400 psi, and direct axial stress is 200 psi. (a) Set up an element at point *A* and indicate the magnitudes and directions of the stresses acting on it (the top edge of the element to coincide with section *ABCD*). (b) Using Mohr's circle, find the direction and the magnitude of the maximum (principal) shearing stresses and the associated normal stresses at point *A*. *Ans.* 141 psi, +100 psi, $22\frac{1}{2}°$.

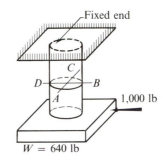

Fixed end

C

D   B

A

1,000 lb

W = 640 lb

PROB. 9 – 23

9-24. A machine bracket is loaded with an inclined force of 4.44 kN as in the figure. Find the principal stresses at point *A*. Show the results on a properly oriented element. Neglect the weight of the member. *Ans:* −17.3 MPa.

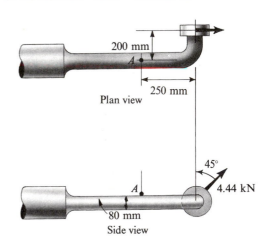

200 mm

A

250 mm

Plan view

45°

A

4.44 kN

80 mm

Side view

PROB. 9 – 24

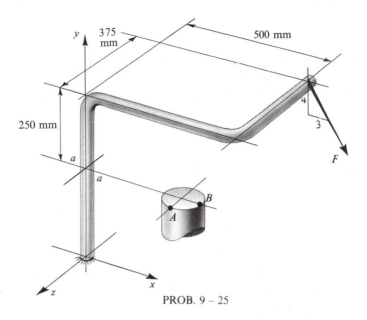

PROB. 9 – 25

9-25. A 50 mm diameter rod is subjected at its free end to an inclined force $F = 225\pi$ N as in the figure. (The force F in plan view acts in the direction of the x axis.) Determine the magnitude and directions of the stresses due to F on the elements A and B at section a–a. Show the results on elements clearly related to the points on the rod. Principal stresses are not required.

9-26. A bent rectangular bar is subjected to an inclined force of 3 000 N as shown in the figure. The cross-section of the bar is 12 mm by 12 mm. (a) Determine the state of stress at point A caused by the applied force and show the results on an element. (b) Find the maximum principal stress. *Ans:* (a) +333 MPa, 578 MPa.

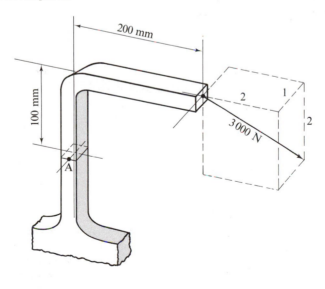

PROB. 9 – 26

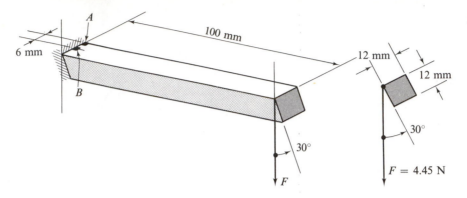

PROB. 9 – 27

9-27. A horizontal 12 mm by 12 mm rectangular bar 100 mm long is attached at one end to a rigid support. Two of the bar's sides form an angle of 30° with the vertical as shown in the figure. By means of an attachment (not shown) a vertical force $F = 4.45$ N is applied acting through a corner of the bar. (See figure.) Calculate the stress at points A and B caused by the applied force F. Neglect stress concentrations. Show the results on the elements viewed from the top. Stress transformations to obtain principal stresses are not required.

9-28. A 2 in. by 2 in. square bar is attached to a rigid support as shown in the figure. Calculate

the state of stress at point A caused by the applied force P of 50 lb. applied to the crank. Neglect stress concentrations and view element A from outside. *Ans:* −212 psi, 167 psi.

9-29. A solid triangular sign is rigidly attached to a 4 in. standard-weight steel pipe as shown in the figure. What principal stresses occur at point A if a wind having an intensity of 30 lb/ft² is blowing on the sign from the reader's side? Neglect the weight of the pipe and sign and the effect of the wind on the pipe itself. Disregard wind suction on the leeward side. *Ans:* +20.4 ksi, −0.2 ksi.

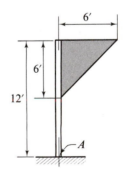

PROB. 9 – 29

9-30. A 400-lb sign is supported by a $2\frac{1}{2}$-in. standard-weight steel pipe as in the figure. The maximum horizontal wind force acting on this sign is estimated to be 90 lb. Determine the state of stress caused by this loading at points A and B at the built-in end. Principal stresses are not

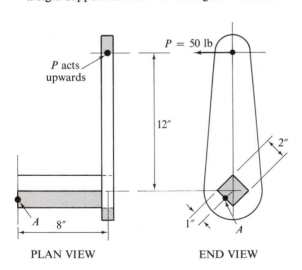

PLAN VIEW   END VIEW

PROB. 9 – 28

required. Indicate results on sketches of elements cut out from the pipe at these points. These elements are to be viewed from outside the pipe. *Ans:* For *A:* 13,295 psi, 1,422 psi; for *B:* 9,912 psi, 1,523 psi.

PROB. 9 – 30

9-31. Approximately what is the bursting pressure for a cold drawn seamless steel tubing of 60 mm outside diameter with 2 mm wall-thickness? The ultimate strength of steel is 380 MN/m².

9-32. A "penstock," i.e., a pipe for conveying water to a hydroelectric turbine, operates at a head of 90 m. If the diameter of the penstock is 0.75 m and the allowable stress 50 MPa, what wall-thickness is required? (The allowable stress is set low to provide for corrosion.)

9-33. A tank of butt-welded construction for the storage of gasoline is to be 40 ft in diameter and 16 ft high. Select the plate thickness for the bottom row of plates. Allow 20 ksi for steel in tension and assume the efficiency of welds at 80%. Add approximately $\frac{1}{8}$ in. to the computed wall thickness to compensate for corrosion. Neglect

local stresses at the juncture of the vertical walls with the bottom. (Specific gravity of the gasoline to be stored is 0.721.) *Ans:* $\frac{1}{4}$ in.

9-34. A steel spherical pressure vessel having an inside diameter of 650 mm is made of two halves welded together. If the efficiency of the welded joint is 75% and the allowable stress in tension for the steel used is 70 000 kN/m², what should the thickness of the wall of this vessel be in order to sustain a pressure of 2 MPa?

9-35. A cylindrical pressure vessel of 2.5 m diameter with walls 12 mm thick operates at 1.5 MPa internal pressure. If the plates are butt-welded on a 30° helical spiral (see figure), determine the stresses acting normal and tangential to the weld. *Ans:* 97.7 MPa, 33.8 MPa.

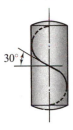

PROB. 9 – 35

9-36. For an industrial laboratory a pilot unit is to employ a pressure vessel of the dimensions shown in the figure. The vessel will operate at an internal pressure of 0.7 MPa. If for this unit 20 bolts are to be used on a 650 mm bolt circle diameter, what size bolts are required? Set the allow-

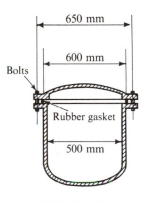

PROB. 9 – 36

able stress in tension for the bolts at 125 MN/m²; however, assume that at the root of the bolt threads the stress concentration factor is $2\frac{1}{2}$.

9-37. An air-chamber for a pump, the sectional side view of which is shown in the figure, consists of two pieces. Compute the number of $\frac{3}{4}$ in. bolts (*net* area 0.302 in.²) required to attach the chamber to the cylinder at plane *A–A*. The allowable tensile stress in the bolts is 6 ksi, and the water and air pressure is 200 psi.

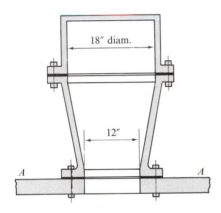

18″ diam.

12″

A   *A*

PROB. 9 – 37

9-38. A water tank made of wood staves is 5 m in diameter and 4 m high. Specify the spacing of 30 mm by 6 mm steel hoops if the allowable tensile stress for steel is set at 90 MPa. Use uniform hoop spacing within each meter of the tank's height. *Ans:* 0.17 m, 0.22 m, etc.

9-39. A piece of 250 mm diameter tubing of 4 mm wall thickness was closed off at the ends as shown in Fig. 9-14. Then this assembly was put into a testing machine and subjected simultaneously to an axial pull *P* and an internal pressure of 2 MPa. What was the magnitude of the applied force *P* if the gage points *A* and *B*, initially precisely 200 mm apart, were found to be 200.04 mm apart after all of the forces were applied? $E = 200$ GPa and $v = 0.25$.

9-40. A cylindrical pressure vessel of 120 in. diameter *outside*, used for processing rubber, is 36 ft long. If the cylindrical portion of the vessel is made from 1 in. thick steel plate and the vessel

operates at 120 psi internal pressure, determine the total elongation of the circumference and the increase in the diameter's dimension caused by the operating pressure. $E = 29 \times 10^6$ psi and $v = 0.3$ *Ans:* 0.0778 in., 0.0247 in.

9-41. A boiler made of $\frac{1}{2}$ in. steel plate is 40 in. in diameter and 8 ft long. It is subjected to an internal pressure of 500 psi. How much will the thickness of the plate change due to this pressure? $E = 30 \times 10^6$ psi and $v = 0.25$. *Ans:* −0.000125 in.

9-42. A compressed air tank is hoisted off the ground by a crane as shown in the figure. The tank is 10 ft long and 2 ft in diameter and has a wall thickness of $\frac{1}{4}$ in. The pressure in the tank is 50 psi. Assume the tank and its contents weigh 65 lb per lineal foot. Determine the principal stresses at points *A* and *B*, and show the results on infinitesimal elements. *Ans:* At *A*: 2,400 psi, 1,045 psi; at *B*, 2,400 psi, 1,165 psi.

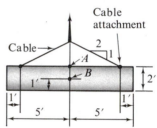

Cable attachment

Cable→

2

1

A

B

2′

1′

1′

5′   5′

PROB. 9 – 42

9-43. A steel pressure vessel 20 in. in diameter and of 0.25-in. wall thickness acts also as an eccentrically loaded cantilever as in the figure. If the internal pressure is 250 psi and the applied weight $W = 31.4$ kips, determine the state of

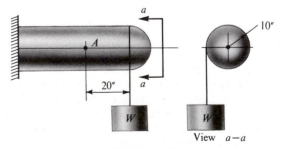

a

10″

A

20″

a

W

W

View *a–a*

PROB. 9 – 43

stress at point *A*. Show the results on an infinitesimal element. Principal stresses are not required. Neglect the weight of the vessel. *Ans:* $\sigma_x = 5$ ksi, $\sigma_y = 10$ ksi, $\tau = 6$ ksi.

9-44. A tank open at the top and closed at the bottom has a diameter of 20 in., a wall-thickness of 0.425 in., and is 14.4 ft long as shown in the figure. This tank is filled to the top with liquid mercury which weighs 850 lb/ft³. If this tank is lifted off the ground by means of a cable and is simultaneously subjected to equal and opposite torques of 81,000 in.-lb at the top and bottom, what is the magnitude and direction of the principal stresses at point *A*, which is 1.44 ft from the top of the tank? Neglect the weight of the tank. Approximately, $I = 1,350$ in.⁴ and $J = 2,700$ in.⁴ *Ans:* $+1,100$ psi, $+100$ psi, $18° 26.5'$.

Section

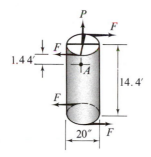

PROB. 9 – 44

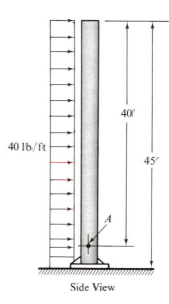
Side View

PROB. 9 – 45

9-45. A fractionating column, 45 ft. long, is made of a 12 in. inside diameter standard steel pipe weighing 49.56 lb/ft. (See Table 8 of the Appendix.) This pipe is operating in a vertical position as indicated on the sketch. If this pipe is internally pressurized to 600 psi, and is subjected to a wind load of 40 lb/ft of height, what is the state of stress at point *A*? Clearly show your calculated stresses on an isolated element; principal stresses need not be found. *Ans:* 9.90 ksi, 4.81 ksi; 0.22 ksi.

9-46. In a certain research investigation on the creep of lead, it was necessary to control the state of stress for the element of a tube. In one such case, a long cylindrical tube with closed ends was pressurized and simultaneously subjected to a torque. The tube was 100 mm in outside diameter with 6 mm walls. What were the principal stresses at the outside surface of the wall of the cylinder if the chamber was pressurized to 1.5 MPa and the externally applied torque was 200 N·m? *Ans:* 12.0 MPa, 4.2 MPa.

10 Design of Members by Strength Criteria

10-1. INTRODUCTION

The selection or design of a member depends on its strength, or its stiffness (deflection), or its stability. Any one of these criteria may govern the size of a member. However, only the *strength* requirement of statically determinate members based on the assumption of elastic behavior of the materials will be considered in this chapter. Thus, the main objective of this chapter is to establish simple and rapid procedures which can be used in practical design problems for selecting a member of adequate strength. Several formulas developed in the earlier chapters and the information on principal stresses and failure theories discussed in the preceding chapter form the basis for the design of members by strength criteria. Study of this chapter will show that in some cases the usual design procedures are cruder than the available theoretical knowledge, but such procedures are usually on the safe side.

This chapter contains only a brief review of the design of axially loaded and torsion members, as this subject was discussed earlier. However, a few additional comments are necessary at this time, since the preceding study of stress transformation at a point gives a more complete picture of the internal stresses. This will be followed by a detailed development of rapid design criteria for beams.

10-2. DESIGN OF AXIALLY LOADED MEMBERS

Axially loaded tensile members and short compression members* are designed by using Eq. 1-1a, i.e., $A = P/\sigma_{\text{allow}}$. The *critical* section for an axially loaded member occurs at a section of minimum cross sectional area, where the stress is a maximum. If an abrupt discontinuity in the cross sectional area is imposed by the design requirements, the use of Eq. 2-11, $\sigma_{\max} = KP/A$, is appropriate. The use of the latter formula is necessary in

---

*Slender compression members are discussed in Chapter 13.

311

the design of machine parts to account for the local stress concentrations where fatigue failure may occur.

Besides the normal stresses, given by the above equations, shearing stresses act on inclined planes even in a state of uniaxial stress. Hence, if a material is weak in shearing strength in comparison to its strength in tension or compression, it will fail along planes approximating the planes of the maximum shearing stress. For example, concrete or cast-iron members in uniaxial compression and duralumin members in uniaxial tension fail on planes inclined to the direction of the load.

Regardless of the type of failure that may actually take place, the allowable stress for design of axially loaded members is customarily based on the *normal* stress. This design procedure is consistent. The maximum normal stress that a material can withstand at the failure point is directly related to the *ultimate* strength of the material. Hence, although the actual break may occur on an inclined plane, the maximum normal stress can be considered as the ultimate normal stress.

10-3. DESIGN OF TORSION MEMBERS

The pertinent formulas for the design of torsion members were established in Chapter 3. For circular shafts, the solution of Eq. 3-5, $J/c = T/\tau_{max}$, at a critical section gives the required parameter, J/c, to provide a member of adequate strength. As shafts are mainly used as parts of machines, Eq. 3-3b, $\tau_{max} = KTc/J$, should be used in most cases. Equation 3-3b, with the stress-concentration factor K, takes care of the high local shearing stress at the changes of the cross-sectional area.

Most torsion members are designed by selecting an *allowable shearing stress*, which is substituted for τ_{max} in Eq. 3-5 or 3-3b. This amounts to a direct use of the maximum shear theory of failure. However, it is well to bear in mind that a state of pure shearing stress, which occurs in torsion, can be transformed into the principal stresses.* In some materials, failure may be caused by one of these principal stresses. For example, a member made of cast iron, a material strong in compression but weaker in tension than in shear, fails in tension.

10-4. DESIGN CRITERIA FOR PRISMATIC BEAMS

If a beam is subjected to *pure bending*, its fibers are in a state of uniaxial stress. If, further, a beam is prismatic, i.e., of a constant cross-sectional area and shape, the critical section occurs at the section of the greatest bending moment. By assigning an allowable stress, the section modulus of such a beam can be determined using Eq. 5-5, $S = M/\sigma_{max}$. Then, once the required

*By rotating an element through 45°, tension-compression stresses are found which are numerically equal to the shearing stresses, Art. 8-6.

section modulus is known, a beam of correct proportions can be selected. However, if a beam resists shear in addition to bending, its design becomes more involved.

Consider the prismatic rectangular beam of Example 9-3 at a section 0.25 m from the left support, where the beam transmits a bending moment *and* a shear, Fig. 10-1(a). The principal stresses at points K, L, M, L', and K' at this section were found before and are reproduced in Fig. 10-1(b). If this section were the critical section, it is seen that the design of this beam, based on the maximum normal stress theory, would be governed by the stresses *at the extreme fibers* as no other stresses exceed these stresses. For a prismatic beam, these stresses depend only on the magnitude of the bending moment and are largest at a section where the maximum bending moment occurs. Therefore, in ordinary design it is *not* necessary to perform the combined stress analysis for interior points. In the example considered, the maximum bending moment is at the middle of the span. The foregoing may be generalized into a basic rule for the design of beams: *A critical section for a prismatic beam carrying transverse force acting normal to its axis occurs where the bending moment reaches its absolute maximum* value.*

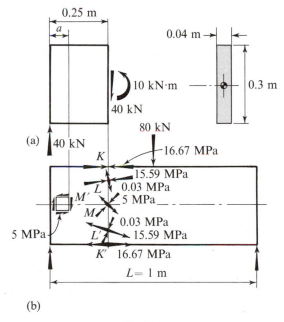

Fig. 10-1

---

*For cross-sections without two axes of symmetry, such as T-beams, made from material which has different properties in tension than in compression, the *largest* moments of *both* senses (positive or negative) must be examined. Under some circumstances, a smaller bending moment of one sense may cause a more critical stress than a larger moment of another sense. The section at which the extreme fiber stress of either sign in relation to the respective allowable stress is highest is the critical section.

The above criterion for the design of prismatic beams is incomplete, as attention was specifically directed to the stresses caused by the moment. In some cases, the shearing stresses caused by the shear at a section may control the design. In the example considered, Figs. 10-1(a) and (b), the magnitude of the shear remains constant at every section through the beam. At a small distance a from the right support, the maximum shear is still 40 kN, while the bending moment, $40a$ kN·m, is small. The maximum shearing stress at the neutral axis corresponding to $V = 40$ kN is the same at point M′ as it is at point M.* Therefore, since in a general problem the bending stresses may be small, they may not control the selection of a beam, and *another critical section for any prismatic beam occurs where the shear is a maximum.* In applying this criterion it is customary to work directly with the maximum shearing stress that may be obtained from Eq. 6-6, $\tau = VQ/(It)$, and not transform τ_{max} so found into the principal stresses. For rectangular and I-beams, the maximum shearing stress given by Eq. 6-6 reduces to Eqs. 6-7 and 6-9, $\tau_{max} = (3/2)(V/A)$ and $(\tau_{max})_{approx} = V/A_{web}$, respectively.

Whether the section where the bending moment is a maximum or the section where the shear is a maximum governs the selection of a prismatic beam depends on the loading and the material used. For most materials, the allowable shearing stress is less than the allowable bending stress. For example, for steel the ratio between these allowable stresses is approximately 0.6, while for some woods it may be as low as 1/15.† Regardless of these ratios of stresses, *the bending stresses usually control the selection of a beam.* Only in beams spanning a short distance does shear control the design. For small lengths of beams, the applied forces and reactions have small moment arms, and the required resisting bending moments are small. On the other hand, the shearing forces may be large if the applied forces are large.

The two criteria for the design of beams are accurate if the two critical sections are in different locations. However, in some instances the maximum bending moment and the maximum shear occur at the *same* section through the beam. In such situations, sometimes higher combined stresses than σ_{max} and τ_{max}, as given by Eqs. 5-5 and 6-6, may exist at the interior points. For example, consider an I-beam of negligible weight that carries a force P at the middle of the span, Fig. 10-2(a). The maximum bending moment occurs at the midspan. Except for sign, the shear is the same on either side of the applied force. At a section just to the right or just to the left of the applied force, the maximum moment *and* the maximum shear occur simultaneously. A section just to the left of P, with the corresponding system of forces acting on it, is shown in Fig. 10-2(b). For this section, it can be shown that the stresses at the extreme fibers are 2.50 ksi, while the principal stresses at the juncture

---

*At point M, the maximum shearing stresses are shown transformed into the principal stresses.

†Wood is very weak in shearing strength *parallel* to its grain.

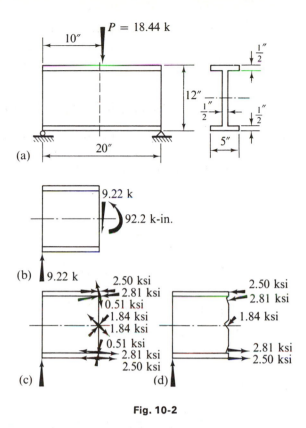

Fig. 10-2

of the web with the flanges, neglecting stress concentrations, are ± 2.81 ksi and ± 0.51 ksi, acting as shown in Figs. 10-2(c) and (d). As usual, local disturbance of stresses in the neighborhood of the applied force P is neglected. From this example it is seen that the maximum normal stress does not always occur at the extreme fibers. Nevertheless, only the extreme fiber stresses and the shearing stresses at the neutral axis are investigated in ordinary design. In the design codes, the allowable stresses are presumably set low enough so that an adequate factor of safety remains, even if the higher combined stresses are disregarded. Also note that, for the same applied force, by increasing the span, the flexural stresses rapidly increase, while the shearing stresses remain constant. In most cases, the flexural stresses are dominant, and the extreme fiber stress is the maximum normal stress. Only for very short beams and unusual arrangements need the combined stress analysis be performed.

From the above discussion it is seen that, for the design of prismatic beams, the critical sections must be determined in every problem, as the design is entirely based on the stresses developed at these sections. The critical sections are best located with the aid of shear and bending-moment diagrams. The required values of $M_{\max}$ and $V_{\max}$ can be determined easily from such diagrams. The construction of these diagrams has been already treated in Chapter 4. However, the importance of these diagrams is so great that an alternative procedure for constructing them rapidly will be discussed next. The procedure to be developed is self-checking.

10-5. SHEAR DIAGRAMS BY SUMMATION

To construct shear diagrams by an alternate procedure, the summation method, certain fundamental relations must be established. Consider an element, Fig. 10-3, isolated from a beam by two adjoining sections taken

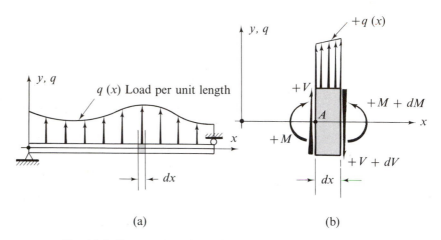

(a)   (b)

Fig. 10-3. Beam and an element cut out from it by two adjoining sections dx apart

perpendicular to the axis of the beam, a distance of dx apart, as was done in Art. 6-2. As before, all forces acting on this element are shown with a positive sense. The positive distributed load* q acts upward and causes an increment in the shear from left to right. The shear and moment change along the beam, so on the right-hand face of the element these quantities are denoted by writing $V + dV$ and $M + dM$.

Writing the summation of the vertical forces and setting it equal to zero for equilibrium,

$$V + q\,dx - (V + dV) = 0 \qquad \text{or} \qquad \frac{dV}{dx} = q \qquad (10\text{-}1)$$

which means that the rate of change of shear along the beam is equal to the applied force in a unit length. Transposing and integrating gives

$$V = \int_o^x q\,dx + C_1 \qquad (10\text{-}2)$$

*A variation of this load within an infinitesimal distance is permissible. The proof of this is analogous to that given in a footnote to Art. 6-2.

By assigning definite limits to this integral, it is seen that the shear at a section is simply an integral (i.e., a sum) of the vertical forces along the beam from the left end of the beam *to the section in question* plus a constant of integration C_1. This constant is equal to the shear on the left-hand end. Between any two definite sections of a beam, the shear changes by the amount of the vertical force included *between* these sections. If no force occurs between any two sections, no change in shear takes place. If a concentrated force comes into the summation, a discontinuity or a "jump" in the value of the shear occurs. The continuous summation process remains valid nevertheless, since a concentrated force may be thought of as being a distributed force extending for an infinitesimal distance along the beam.

On the basis of the above reasoning, a shear diagram can be established by the summation process. For this purpose, *the reactions must always be determined first*. Then the vertical components of forces and *reactions* are successively summed *from the left end* of the beam to preserve the mathematical sign convention for shear adopted in Fig. 4-15. The shear at a section is equal to the sum of *all* vertical forces up to that section.

When the shear diagram is constructed from the load diagram by the summation process, two important observations can be made regarding its shape. First, the sense of the applied load determines the sign of the slope of the shear diagram. If the applied load acts upward, the slope of the shear diagram is positive, and vice versa. Second, this slope is equal to the corresponding applied load. For example, consider a segment of a beam with a uniformly distributed downward load w_o and known shears at both ends as shown in Fig. 10-4(a). Since here the applied load w_o is *negative* and *uniformly distributed*, i.e., $q = -w_o =$ constant, the slope of the shear diagram exhibits the same characteristics. Alternatively, the linearly varying load acting upward on a beam segment with known shears at the ends, shown in Fig. 10-4(b), gives rise to a differently shaped shear diagram. Near the left end of this segment the locally applied *upward* load q_1 is *smaller* than the corresponding one q_2 near the right end. Therefore, the *positive* slope of the shear diagram on the left is *smaller* than it is on the right, and the shear diagram is concave upward.

Do not fail to note that *a mere systematic consecutive summation of the vertical components of the forces is all that is necessary to obtain the shear diagram*. When the consecutive summation process is used, the diagram must close at the right-hand end of the beam, since no shear acts through the beam just beyond the last vertical force or reaction. *The fact that the diagram closes offers an important check on the arithmetical calculations.* This check should never be ignored. It permits one to obtain solutions independently with almost complete assurance of being correct. The semigraphical procedure of integration outlined above is very convenient in practical problems. It is the basis for sketching qualitative shear diagrams rapidly.

From the physical point of view, the shear sign convention is not completely consistent. Whenever beams are analyzed, a shear diagram drawn

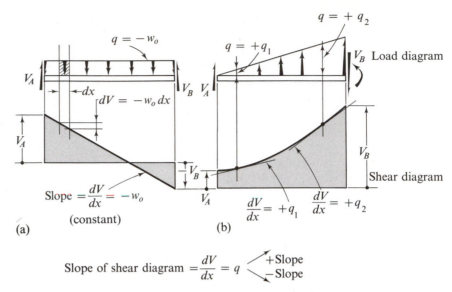

$$\text{Slope of shear diagram} = \frac{dV}{dx} = q \begin{cases} \nearrow +\text{Slope} \\ \searrow -\text{Slope} \end{cases}$$

Fig. 10-4. The relation between the load and the shear diagrams

from one side of the beam is opposite in sign to a diagram constructed by looking at the same beam from the other side. The reader should verify this statement on some simple cases, such as a cantilever with a concentrated force at the end, and a simply supported beam with a concentrated force in the middle. For design purposes the sign of the shear is usually unimportant.

10-6. MOMENT DIAGRAMS BY SUMMATION

To formulate the summation procedure for establishing moment diagrams, again the element shown in Fig. 10-3 must be considered. By taking the moment of forces around A and setting it equal to zero for equilibrium, an expression formerly derived in Art. 6-2 is obtained.

$$\frac{dM}{dx} = V \tag{6-1}$$

This equation states that the rate of change of the bending moment along the beam is equal to the shear. By a fundamental theorem of calculus, Eq. 6-1 also implies that the *maximum or minimum moment occurs at a point where the shear is zero*, since the derivative is then zero. This usually occurs at a point where the shear changes sign.

Transposing and integrating gives

$$M = \int_o^x V \, dx + C_2 \tag{10-3}$$

which is analogous to Eq. 10-2 developed for the construction of shear diagrams. The meaning of the term $V\,dx$ is shown graphically by the shaded areas of the shear diagrams in Fig. 10-5. The summation of these areas between definite sections through a beam corresponds to an evaluation of the above definite integral. If the ends of a beam are on rollers, pin-ended or free, the starting and the terminal moments are zero. If the end is built-in (fixed against rotation), in statically determinate beams the end moment is known from the reaction calculations. If the fixed end of a beam is on the left, this moment with the proper* sign is the *initial constant of integration* C_2.

By proceeding *continuously along the beam from the left-hand end* and summing up the particular areas of the shear diagram with due regard to the sign of the shear, the moment diagram is obtained. This process of deriving the moment diagram from the shear diagram by summation is exactly the same as that employed earlier to go from loading to shear diagrams. *The change in moment in a given segment of a beam is equal to the*

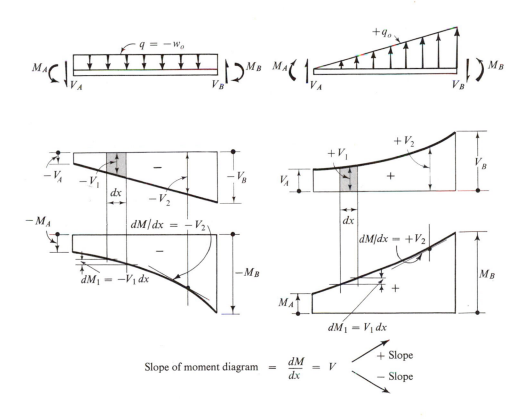

$$\text{Slope of moment diagram} = \frac{dM}{dx} = V$$

Fig. 10-5. The relation between the shear and the moment diagrams

*Bending moments carry signs according to the convention adopted in Fig. 4-17. Moments which cause *compression* in the top fibers of the beam are positive.

area of the corresponding shear diagram. Qualitatively the shape of a moment diagram can be easily established from the slopes at some selected points along the beam. These slopes have the same sign and magnitude as the corresponding shears on the shear diagram, since $dM/dx = V$. Alternatively, the change of moment $dM = V\,dx$ can be studied along the beam. Examples are shown in Fig. 10-5. According to these principles, variable shears cause nonlinear variation of the moment. A constant shear produces a uniform change in the bending moment, resulting in a straight line in the moment diagram. If no shear occurs along a certain portion of a beam, *no change in moment* takes place.

In a bending-moment diagram obtained by summation, *at the right-hand end* of the beam, an invaluable check on the work is available again. *The terminal conditions for the moment must be satisfied.* If the end is free or pinned, the computed sum must equal zero. If the end is built-in, the end moment computed by summation equals the one calculated initially for the reaction. These are the "boundary conditions" and must always be satisfied.

EXAMPLE 10-1

Construct shear and moment diagrams for the symmetrically loaded beam shown in Fig. 10-6(a).

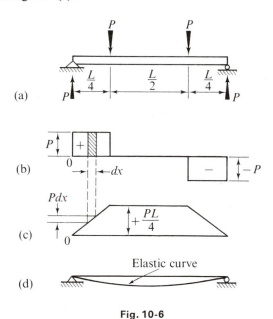

Fig. 10-6

SOLUTION

The reactions are each equal to P. To obtain the shear diagram, Fig. 10-6(b), the summation of forces is started from the left end. The left reaction acts *up* so an ordinate on the shear diagram at this force equal to P is plotted *up*.

Since there are no other forces until the quarter point, *no change in the magnitude of the shear ordinate is made until that point*. Then a downward force P brings the ordinate back to the base line, and this zero ordinate remains until the next downward force P is reached where the shear changes to $-P$. At the right end the upward reaction closes the diagram and provides a check on the work. This shear diagram is *antisymmetrical*.

The moment diagram, Fig. 10-6(c), is obtained by summing up the area of the shear diagram. As the beam is simply supported, the moment at the left end is zero. The sum of the positive portion of the shear diagram *increases at a constant rate* along the beam until the quarter point, where the moment reaches a magnitude of $+PL/4$. This moment remains constant in the middle half of the beam. *No change in the moment can be made in this zone* as there is no corresponding shear area.

Beyond the second force, the moment decreases by $-P\,dx$ in *every dx*. Hence the moment diagram in this zone has a constant, negative slope. Since the positive and the negative areas of the shear diagram are equal, at the right end the moment is zero. This is as it should be, since the right end is on a roller. Thus a check on the work is obtained. This moment diagram is *symmetrical*.

EXAMPLE 10-2

Construct shear and bending-moment diagrams for the beam loaded as shown in Fig. 10-7(a).

SOLUTION

Reactions must be calculated first, and, before proceeding further, the inclined force is resolved into its horizontal and vertical components. The horizontal reaction at A is 30 kips and acts to the right. From $\sum M_A = 0$, the vertical reaction at B is found to be 37.5 kips (check this). Similarly, the reaction at A is 27.5 kips. The sum of the vertical reaction components is 65 kips and equals the sum of the vertical forces.

With reactions known, the summation of forces is begun from the left end of the beam to obtain the shear diagram, Fig. 10-7(b). At first, the downward distributed load accumulates at a rapid rate. Then, as the load intensity decreases, for an equal increment of distance along the beam a smaller change in shear occurs. Hence the shear diagram in the zone CA is a curved line, which is concave up. The total downward force from C to A is 15 kips, and this is the negative ordinate of the shear diagram, *just to the left of the support A*. At A, the *upward* reaction of 27.5 kips moves the ordinate of the shear diagram to $+12.5$ kips. This value of the shear applies to a section through the beam *just to the right* of the support A. The abrupt *change* in the shear at A is equal to the reaction, but this total does not represent the shear through the beam.

No forces are applied to the beam between A and D, hence there is no change in the value of the shear. At D, the 40 kip downward component of the concentrated force drops the value of the shear to -27.5 kips. Similarly, the value of the shear is raised to $+10$ kips at B. Since between E and F the uniformly distributed load acts downward, a decrease in shear takes place at a constant rate of one kip per foot. Thus at F the shear is zero, which serves as the final check.

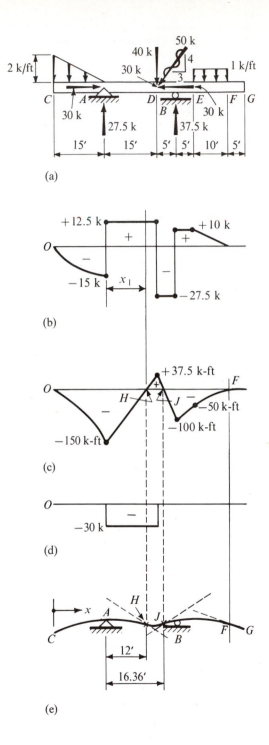

(a)

(b)

(c)

(d)

(e)

Fig. 10-7

To construct the moment diagram shown in Fig. 10-7(c) by the summation method, areas of the shear diagram in Fig. 10-7(b) must be continuously summed from the left end. In the segment CA, at first less area is contributed to the sum in a distance dx than a little further along, so a line that is concave down appears in the moment diagram. The moment at A is equal to the area of the shear diagram in the segment CA. This area is enclosed by a curved line, and it may be determined by integration,* since the shear along this segment may be expressed analytically. This procedure often is tedious, and instead of using it, the bending moment at A may be obtained from the fundamental definition of a moment at a section. By passing a section through A and isolating the segment CA, the moment at A is found. The other areas of the shear diagram in this example are easily determined. Due attention must be paid to the signs of these areas. It is convenient to arrange the work in tabular form. At the right end of the beam, the customary check is obtained.

$$M_A\ldots\ldots \quad -\tfrac{1}{2}(15)2(10) = -150.0 \text{ kip-ft} \qquad \text{(moment around } A\text{)}$$
$$+12.5(15) = +187.5 \qquad \text{(shear area } A \text{ to } D\text{)}$$
$$M_D\ldots\ldots\ldots\ldots\ldots\ldots \quad + 37.5 \text{ kip-ft}$$
$$-27.5(5) = -137.5 \qquad \text{(shear area } D \text{ to } B\text{)}$$
$$M_B\ldots\ldots\ldots\ldots\ldots \quad -100.0 \text{ kip-ft}$$
$$+10(5) = + 50.0 \qquad \text{(shear area } B \text{ to } E\text{)}$$
$$M_E\ldots\ldots\ldots\ldots\ldots \quad - 50.0 \text{ kip-ft}$$
$$+\tfrac{1}{2}(10)10 = + 50.0 \qquad \text{(shear area } E \text{ to } F\text{)}$$
$$M_F\ldots\ldots\ldots\ldots\ldots \quad 0.0 \text{ kip-ft} \quad (\textit{check})$$

The diagram for the axial force is shown in Fig. 10-7(d) (See also Art. 4-8). This compressive force acts in the segment AD of the beam.

*10-7. FURTHER REMARKS ON THE CONSTRUCTION OF SHEAR AND MOMENT DIAGRAMS

In the derivation for moment diagrams by summation of shear-diagram areas, no *external concentrated moment* acting on the infinitesimal element in Fig. 10-3 was included, yet such a moment may actually be applied. Hence the summation process derived applies only to the point of application of an external moment. *At a section just beyond an externally applied moment, a different bending moment is required to maintain the segment of a beam in equilibrium.* For example, in Fig. 10-8 an external clockwise moment M_A is acting on the element of the beam at A. Then, if the internal clockwise moment on the left is M_O, for equilibrium of the element, the resisting counter-clockwise moment

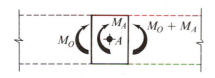

Fig. 10-8. An external concentrated moment acting on an element of a beam

---

*In this case, the shear curve is a second-degree parabola with vertex on a vertical line through A. For areas enclosed by various curves see Table 2 of the Appendix.

on the right is $M_O + M_A$. Situations with other sense of moments may be similarly analyzed. At the point of the externally applied moment, a discontinuity or a "jump" equal to the concentrated moment appears in the moment diagram. Hence, in applying the summation process, due regard must be given the concentrated moments as their effect is not included in the shear-diagram area summation process. The summation process may be applied up to the point of application of a concentrated moment. At this point a vertical "jump" equal to the external moment must be made in the diagram. The direction of this vertical "jump" in the diagram depends upon the sense of the concentrated moment and is best determined with the aid of a sketch analogous to Fig. 10-8. After the discontinuity in the moment diagram is passed, the summation process of the shear-diagram areas may be continued over the remainder of the beam.

EXAMPLE 10-3

Construct the bending-moment diagram for the horizontal beam loaded as shown in Fig. 10-9(a).

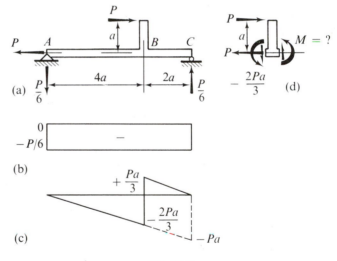

Fig. 10-9

SOLUTION

By taking moments about either end of the beam, the vertical reactions are found to be $P/6$. At A the reaction acts down, at C it acts up. From $\sum F_x = 0$ it is known that at A a horizontal reaction equal to P acts to the left. The shear diagram is drawn next, Fig. 10-9(b). It has a constant negative ordinate for the *whole* length of the beam. After this, by using the summation process, the moment diagram shown in Fig. 10-9(c) is constructed. The moment at the left end of the beam is zero, since the support is pinned. The total change in moment from A to B is given by the area of the shear diagram between these sections and equals $-2Pa/3$. The moment diagram in the zone AB has a constant negative slope. For further analysis, an element is isolated from the beam as shown in Fig. 10-9(d). The moment on the left-hand side of this element is *known to be* $-2Pa/3$, and the concentrated moment caused by the applied force P about the neutral axis of the beam is Pa, hence for equilibrium, on the right side of the element the moment must be $+Pa/3$. At B an upward "jump" of $+Pa$ is made in the moment diagram, and just to the right of B the ordinate is $+Pa/3$. Beyond point B, the summation of the shear diagram area is continued. The area between B and C is equal to $-Pa/3$. This value *closes* the moment diagram at the right end of the beam, and thus the boundary conditions are satisfied. Note that the lines in the moment diagram which are inclined downward to the right are parallel. This follows froms the fact that the shear everywhere along the beam is negative and constant.

EXAMPLE 10-4

Construct shear and moment diagrams for the member shown in Fig. 10-10(a). Neglect the weight of the beam.

SOLUTION

In this case, unlike all cases considered so far, definite dimensions are assigned for the *depth* of the beam. The beam, for simplicity, is assumed to be rectangular in its cross-sectional area, consequently the *neutral axis* lies 0.08 m below the top of the beam. Note carefully that this beam is not supported at the neutral axis.

A free-body diagram of the beam with the applied force resolved into components is shown in Fig. 10-10(b). Reactions are computed in the usual

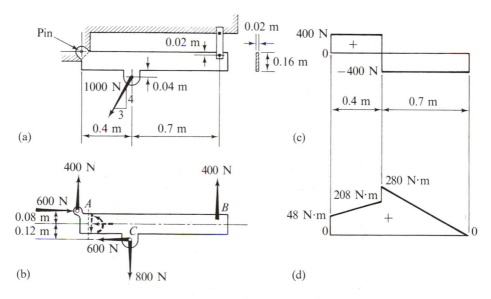

Fig. 10-10

manner. Moreover, since the shear diagram is concerned only with the vertical forces, it is easily constructed and is shown in Fig. 10-10(c).

In constructing the moment diagram shown in Fig. 10-10(d), particular care must be exercised. As was emphasized in Chapter 4, the bending moments may always be determined by considering a segment of a beam, and they are most conveniently computed by taking moments of external forces *around a point on the neutral axis of the beam.* Thus, by passing a section just to the right of *A* and considering the left-hand segment, it can be seen that a positive moment of 48 N·m is resisted by the beam at this end. Hence the plot of the moment diagram must *start* with an ordinate of $+48$ N·m. The other point on the beam where a concentrated moment occurs is *C*. Here the horizontal component of the applied force induces a clockwise moment of $600(0.12) = 72$ N·m around the neutral axis. Just to the right of *C* this moment must be resisted by an additional positive moment. This causes a discontinuity in the moment diagram. The summation process of the shear-diagram areas applies for the segments of the beam where no external moments are applied. The necessary calculations are carried out below in tabular form.

$$M_A \ldots\ldots\ldots\ldots \quad \begin{array}{l} +600(0.08) = +\ 48 \text{ N·m} \\ +400(0.40) = +160 \end{array} \qquad \text{(shear area } A \text{ to } C\text{)}$$

$$\text{Moment just to left of } C \ldots\ldots = +208 \text{ N·m}$$
$$+600(0.12) = +\ 72 \qquad \text{(external moment at } C\text{)}$$

$$\text{Moment just to right of } C \ldots = +280 \text{ kN·m}$$
$$-400(0.70) = -280 \qquad \text{(shear area } C \text{ to } B\text{)}$$

$$M_B \ldots\ldots\ldots\ldots\ldots\ldots\ldots = \quad 0 \ (check)$$

Note that in solving this problem the forces were considered *wherever they actually act on the beam.* The investigation for shear and moments at a section of a beam determines what the beam is actually experiencing. At times this differs from the procedure of determining reactions where the actual framing or configuration of a member is not important.

Occasionally *hinges* or *pinned joints* are introduced into beams. A hinge is capable of transmitting only horizontal and vertical forces. *No moment can be transmitted at a hinged joint.* Therefore the point where a hinge occurs is a particularly convenient location for "separation" of the structure into parts for purposes of computing the reactions. This process is illustrated in Fig. 10-11. Each part of the beam so separated is treated independently. Each hinge provides an extra axis around which moments may be taken to determine reactions. The introduction of a hinge or hinges into a continuous beam in many cases makes the system statically determinate. The introduction of a hinge into a determinate beam results in a beam that is not stable. Note that the reaction at the hinge for one beam acts in an *opposite direction* on the other beam.

In engineering practice it is also common to find several members *rigidly joined* to form a structure. Such a structure may be treated by the methods already discussed, if it can be separated into statically determinate individual beams. To illustrate, consider the structure shown in Fig. 10-12(a).

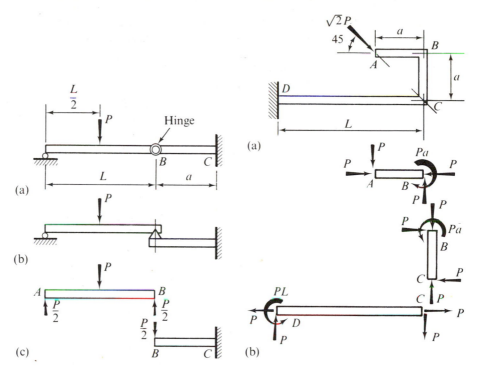

(a)

(b)

(c)

Fig. 10-11. Structures "separated" at hinges to determine the reactions by statics

(a)

(b)

Fig. 10-12. A statically determinate frame separated into individual beams

Beginning at point A, the portions of the structure AB, BC, and CD can be successively isolated as free bodies, and the system of forces at each of the cut sections may be determined. The reader should verify these forces shown in Fig. 10-12(b). Thence, shear and moment diagrams can be constructed for each part, although the sign convention adopted in this text becomes ambiguous for vertical and inclined members. However, if the direction of these quantities is understood by the analyst in a physical sense, no particular difficulty should be encountered in stress analysis or design.

Finally, it must be emphasized that if a moment or a shear is needed at a *particular* section through any member, *the basic method of sections may always be used.* For inclined members, the shear acts *normal to the axis of the beam.*

*10-8. MOMENT DIAGRAM AND THE ELASTIC CURVE

As stated in Art. 4-9, a positive moment causes a beam to deform concave upwards or to "retain water," and vice versa. Hence the shape of the deflected axis of a beam can be *definitely* established from the *sign* of the moment diagram. The trace of this axis of a loaded beam in a deflected position is known as the *elastic curve*. It is customary to show the elastic curve on a sketch where the actual small deflections tolerated in practice are greatly *exaggerated*. A sketch of the elastic curve clarifies the physical action of a beam. Moreover, it forms the most useful basis for quantitative calculations of beam deflections to be discussed in the next chapter. Some of the preceding examples for which bending-moment diagrams were constructed will be used to illustrate the physical action of a beam.

An inspection of Fig. 10-6(c) shows that the bending moment throughout the length of the beam is *positive*. Accordingly, the elastic curve shown in Fig. 10-6(d) is *concave up at every point*. In future work, definiteness regarding the direction of curvature will be essential. The ends of the beam are assumed to rest on immovable supports.

In a more complex moment diagram, Fig. 10-7(c), zones of positive and negative moment occur. Corresponding to the zones of negative moment, a *definite* curvature of the elastic curve that is concave down takes place, Fig. 10-7(e). On the other hand, for the zone *HJ* where the positive moment occurs, the concavity of the elastic curve is upward. Where curves join, as at *H* and *J*, there are lines which are *tangent* to the two joining curves since the beam is physically *continuous*. Also note that the free end *FG* of the beam is tangent to the elastic curve at *F*. There is no curvature in *FG*, since the moment is zero in that segment of the beam.

The point of transition on the elastic curve into reverse curvature is called the *point of inflection* or contraflexure. At this point the moment changes its sign, and the beam is not called upon to resist any moment. This fact often makes these points a desirable place for a field connection of large members, and their location is calculated. A procedure for determining points of inflection will be illustrated in Example 10-5, which follows a summary of the above discussion.

The important process of establishing the elastic curve qualitatively may be summarized as follows:

1. Draw a bending-moment diagram.
2. Sketch the elastic curve, corresponding to the signs of moments without reference to the supports, on the moment diagram.
3. If the beam is on two supports, "bodily lift" the curve so drawn and "set it" on the supports, and, if it is a cantilever, the end of the curve is tangent to the built-in end.

EXAMPLE 10-5

Find the location of the inflection points for the beam analyzed in Example 10-2, Fig. 10-7(a).

SOLUTION

By definition, an inflection point corresponds to a point on a beam where the bending moment is zero. Hence, an inflection point may be located by setting up an algebraic expression for the moment in a beam for the segment where such a point is anticipated, and solving this relation equated to zero. By measuring x from the end C of the beam, Fig. 10-7(e), the bending moment for the segment AD of the beam is $M = -\frac{1}{2}(15)(2)(x-5) + (27.5)(x-15)$. By simplifying and setting this expression equal to zero, a solution for x is obtained.

$$M = 12.5x - 337.5 = 0 \qquad x = 27 \text{ ft}$$

Therefore, the inflection point occuring in the segment AD of the beam is $27 - 15 = 12$ ft from the support A.

Similarly, by writing an algebraic expression for the bending moment for the segment DB and setting it equal to zero, the location of the inflection point J is found.

$$M = -\frac{1}{2}(15)(2)(x-5) + 27.5(x-15) - 40(x-30) = 0$$

where $x = 31.36$ ft, hence the distance $AJ = 16.36$ ft.

Often a more convenient method for finding the inflection points consists of utilizing the known relations between the shear and moment diagrams. Thus, since the moment at A is -150 kip-ft, the point of zero moment occurs when the positive portion of the shear-diagram area from A to H equals this moment, i.e., $-150 + 12.5x_1 = 0$. Hence the distance $AH = 150/12.5 = 12$ ft as before.

Similarly, beginning with a known positive moment of $+37.5$ kip-ft at D, the second inflection point is known to occur when a portion of the negative shear-diagram area between D and J reduces this value to zero. Hence, the distance $DJ = 37.5/27.5 = 1.36$ ft, or the distance $AJ = 15 + 1.36 = 16.36$ ft, Fig. 10-7(e), as before.

10-9. DESIGN OF PRISMATIC BEAMS

The design of a prismatic member is controlled by the maximum stresses developed at the *critical sections*. One critical section occurs where the bending moment is a maximum, the other where the shear is a maximum. To determine the location of these critical sections, shear and moment diagrams are very useful.* The values of maximum moment and shear may be easily found from these diagrams. *The absolute† maximum value of the moment is used in design*, whether positive or negative. Likewise, the absolute maximum shear ordinate is the significant ordinate. For example, consider a simple beam with a concentrated load, as shown in Fig. 10-13. The shear diagram, neglecting the weight of the beam, is shown in Fig. 10-13(a) as it is ordinarily constructed by assuming the applied force concentrated at a point.

*With experience construction of complete diagrams may be eliminated. After reactions are computed, and a section where $V = 0$ or changes sign is determined, the maximum moment corresponding to this section may be found by using the method of sections. For simple loadings, various handbooks give formulas for the maximum shear and moment.

†This is not always true for materials which have different properties in tension and compression.

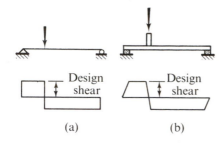

(a)　　　　　　　　(b)

Fig. 10-13. Determination of a design ordinate from the shear diagram

The shear diagram as it more nearly exists is shown in Fig. 10-13(b). Here an allowance is made for the width of the applied force and reactions, assuming them to be uniformly distributed. The assumption of concentrated forces merely straightens the oblique shear lines. In either case, the design shear value is the greater of the positive or negative ordinates and is not the full value of the applied force.

The allowable stresses to be used in design are prescribed by various authorities. In most cases the designer must follow a code depending on the location of the installation. In different codes even for the same material and the same use the allowable stresses differ. The allowable stresses in bending and in shear are different, with the allowable shearing stresses usually being lower.

Sometimes the design of beams is based on their ultimate (plastic) moment capacity. (See Art. 5-8 and Eq. 5-10, which defines the plastic section modulus of a section.) In such problems the assumed design loads are multiplied by a load factor, which defines the ultimate load the beam must carry. For compact, statically determinate beams AISC Code (1970) specifies a factor of 1.70. This means that the collapse of a beam would occur after the design loads are increased by a factor of 1.70. Therefore, the load factor is analogous to the factor of safety in elastic stress analysis. Plastic or limit analysis of beams will be considered further in Chapter 12.

In elastic design, after the critical values of moment and shear are determined and the allowable stresses are selected, the beam is usually first designed to resist a maximum moment using Eq. 5-5 or 5-1 ($\sigma_{max} = M/S$ or $\sigma_{max} = Mc/I$). Then the beam is *checked* for shearing stress. As most beams are governed by flexural stresses, this procedure is convenient. However, in some cases, particularly in timber (and concrete) design, the shearing stress frequently controls the dimensions of the cross section.

The method used in computing the shearing stress depends on the type of beam cross section. For rectangular sections, the maximum shearing stress is 1.5 times the average stress, Eq. 6-7. For wide flange and I-beams, the total allowable vertical shear is taken as the area of the *web* multiplied by an allowable shearing stress, Eq. 6-9. For other cases, Eq. 6-6, $\tau = VQ/(It)$, is used.

Usually there are several types or sizes of commercially available members that may be used for a given beam. Unless specific size limitations are placed on the beam, the lightest member is used for economy. The procedure of selecting a member is a trial-and-error process.

It should also be noted that some beams must be selected on the basis of allowable deflections. This topic will be treated in the next chapter.

EXAMPLE 10-6

Select a Douglas fir beam of rectangular cross section to carry two concentrated forces as shown in Fig. 10-14(a). The allowable stress in bending is 8 MPa, in shear 0.7 MPa, and in bearing perpendicular to the grain of the wood 1.4 MPa.

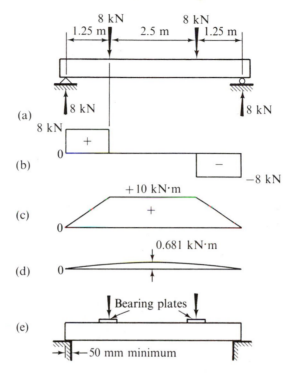

Fig. 10-14

SOLUTION

Shear and moment diagrams for the applied forces are prepared first and are shown respectively in Figs. 10-14(b) and (c). From Fig. 10-14(c) it is seen that $M_{max} = 10$ kN·m. From Eq. 5-5,

$$S = \frac{M}{\sigma_{allow}} = \frac{10 \,(\text{kN·m})}{8\,000 \,(\text{kN/m}^2)} = 1.25 \times 10^{-3} \text{ m}^3$$

By *arbitrarily assuming* that the depth h of the beam is to be two times greater than its width b, from Eq. 5-6,

$$S = \frac{bh^2}{6} = \frac{h^3}{12} = 1.25 \times 10^{-3}$$

hence $h = 0.25$ m and $b = 0.12$ m.

Let a *surfaced* beam 0.14 m by 0.24 m having a section modulus $S = 1.34 \times 10^{-3}$ m³ be used to fulfil this requirement. For this beam, from Eq. 6-7,

$$\tau_{max} = \frac{3V}{2A} = \frac{3(8)}{2(0.14)(0.24)} = 357 \text{ kPa} = 0.357 \text{ MPa}$$

This stress is well within the allowable limit. Hence, the beam is satisfactory.

Note that other proportions of the beam can be used, and a more direct method of design is to find a beam of size corresponding to the wanted section modulus directly from a table similar to Table 10, which gives properties of standard dressed sections in the conventional units.

The above analysis was made without regard for the beam's own weight, which initially was unknown. (Experienced designers usually make an allowance for the weight of the beam at the outset.) However, this may be accounted for now. Assuming that wood weighs 6.5 kN/m³, the beam selected weighs 0.218 kN per linear meter. This uniformly distributed load causes a parabolic bending-moment diagram, shown in Fig. 10-14(d), where the maximum ordinate is $w_oL^2/8 = 0.218(5)^2/8 = 0.681$ kN·m (see Example 4-6). This bending-moment diagram should be added to the moment diagram caused by the applied forces. Inspection of these diagrams shows that the maximum bending moment due to both causes is $0.681 + 10 = 10.68$ kN·m. Hence, the required section modulus actually is

$$S = \frac{M}{\sigma_{allow}} = \frac{10.68}{8\,000} = 1.34 \times 10^{-3} \text{ m}^3.$$

The surfaced 0.14 m by 0.24 m beam already selected provides the required S.

In actual construction, beams are not supported as is shown in Fig. 10-14(a). Wood can be crushed by the supports or the applied concentrated forces. For this reason an adequate bearing area must be provided at the supports and at the applied forces. Assuming that both reactions and the applied forces are 8 kN each, i.e., by neglecting the weight of the beam, it is found that the required bearing area at each concentrated force, by Eq. 1-1, is

$$A = \frac{P}{\sigma_{allow}} = \frac{8}{1\,400} = 0.57 \times 10^{-2} \text{ m}^2$$

These areas can be provided by conservatively specifying that the beam's ends rest on 50 mm by 140 mm (0.7×10^{-2} m²) pads, while at the concentrated forces 80 mm by 80 mm (0.64×10^{-2} m²) steel washers can be used.

EXAMPLE 10-7

Select an I-beam or a wide-flange steel beam to support the load shown in Fig. 10-15(a). Given, $\sigma_{allow} = 24{,}000$ psi, $\tau_{allow} = 14{,}500$ psi.

SOLUTION

The shear and the bending-moment diagrams for the loaded beam are shown in Figs. 10-15(b) and (c), respectively. The maximum moment is

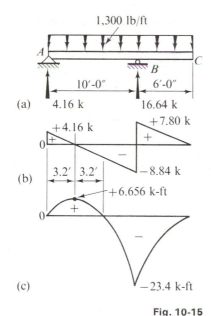

(a) 4.16 k   16.64 k

1,300 lb/ft

A B C

10'-0"   6'-0"

+4.16 k   +7.80 k

(b) 3.2' | 3.2'   −8.84 k

+6.656 k-ft

(c)   −23.4 k-ft

Fig. 10-15

23.4 kip-ft. From Eq. 5-5

$$S = \frac{(23.4)12}{24} = 11.6 \text{ in.}^3$$

Examination of Tables 3 and 4 of the Appendix shows that this requirement for the section modulus is met by a 7-in. S beam weighing 20.0 lb per foot. ($S = 12.1$ in.^{3}) However, lighter members, such as an 8-in. S beam weighing 18.4 lb per foot ($S = 14.4$ in.^{3}) and an 8-in. wide-flange section weighing 17 lb per foot ($S = 14.1$ in.^{3}) can also be used. For weight economy the W 8 × 17 section will be used. The weight of this beam is very small in comparison with the applied load and so is neglected.

From Fig. 10-15(b), $V_{max} = 8.84$ kips. Hence, from Eq. 6-9,

$$(\tau_{max})_{approx} = \frac{V}{A_{web}} = \frac{8,840}{(0.23)8} = 4,800 \text{ psi.}$$

This stress is within the allowable value, and the beam selected is satisfactory.

At the supports or concentrated loads, S- and wide-flange beams should be checked for crippling of the webs. This phenomenon is illustrated at the bottom of Fig. 10-16(a). Crippling of the webs is more critical for members with thin webs than direct bearing of the flanges, which may be investigated as in the preceding problem. To preclude crippling, a design rule is specified by the AISC. It states that the direct stress on area, $(a + k)t$ at the ends or $(a_1 + 2k)t$ at the interior points, must not exceed $0.75\sigma_{yp}$. In these expressions, a and a_1 are the respective lengths of bearing of the applied forces at exterior or interior portions of a beam, Fig. 10-16(b), t is the thickness of the web, and k is the distance from the outer face of the flange to the toe of the web fillet. The values of k and t are tabulated in manufacturers' catalogues.

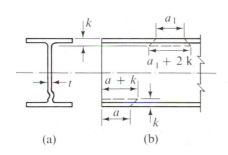

(a)   (b)

Fig. 10-16

For the above problem, assuming $\sigma_{yp} = 36$ ksi, the *minimum* widths of the supports, according to the above rule, are as follows:

At support A: $27(a + k)t = 4.16$

or $27(a + \frac{5}{8})(0.23) = 4.16$ $a = 0.04$ in.

At support B: $27(a_1 + 2k)t = 16.64$

or $27(a_1 + \frac{5}{4})(0.23) = 16.64$ $a_1 = 1.43$ in.

In Chapter 14, another type of connection for beams to other members by small auxiliary angles is fully treated.

The preceding two examples illustrate the design of beams whose cross sections have two axes of symmetry. In both cases the bending moments controlled the design, and, since this is usually true, it is significant to note which members are efficient in flexure. A concentration of as much material as possible away from the neutral axis results in the best sections for resisting flexure, Fig. 10-17(a). Material concentrated near the outside fibers works at a high stress. For this reason, I-type sections, which approximate this requirement, are widely used in practice.

The above statements apply for materials having nearly equal properties in tension and compression. If this is not the case, a deliberate shift of the neutral axis from the mid-height position is desirable. This accounts for the wide use of T and channel sections for cast-iron beams (see Example 5-5).

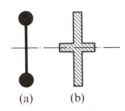

Fig. 10-17. (a) Efficient and (b) inefficient sections in flexure

Finally, two other items warrant particular attention in the design of beams. In many cases, the loads for which a beam is designed are transient in character. They may be placed on the beam all at once, piecemeal, or in *different locations*. The loads, which are not a part of the "dead weight" of the structure itself, are called *live loads*. Live loads must be so placed as to cause the highest possible stresses in a beam. In many cases the placement may be determined by inspection. For example, in a simple beam with a single moving load, the placement of the load at midspan causes the largest bending moment, while placing the same load at the support causes the greatest shear. For most building work, the live load, which supposedly provides for the most severe expected loading condition, is specified in building codes on the basis of so many pounds per square foot of floor area. Multiplying this live load by the spacing of parallel beams gives the *uniformly distributed live load* per unit length of the beam. For design purposes, this load is added to the dead weight of construction. Situations where the applied force is delivered to a beam with a shock or impacts are discussed in Chapter 15.

The second item pertains to *lateral instability* of beams. The beam's flanges, if not held, may be so narrow in relation to the span that a beam may buckle sideways and collapse. The qualitative aspect of this problem was discussed in Art. 5-2. Special formulas applicable to these cases are given in Art. 13-13.

*10-10. DESIGN OF NONPRISMATIC BEAMS

It should be apparent from the preceding discussion that the selection of a prismatic beam is based only on the stresses at the critical sections. At all other sections through the beam the stresses will be below the allowable level. Therefore the potential capacity of a given material is not fully utilized. This situation may be improved by designing a beam of variable cross section, i.e., by making the beam nonprismatic. Since flexural stresses control the design of most beams, as has been shown, the cross sections may everywhere be made just strong enough to resist the corresponding moment. Such beams are called *beams of constant strength*. Shear governs the design at sections through these beams where the bending moment is small.

EXAMPLE 10-8

Design a cantilever of constant strength for resisting a concentrated force applied at the end. Neglect the beam's own weight.

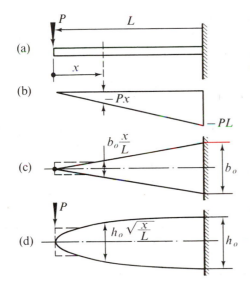

(a)

(b)

(c)

(d)

Fig. 10-18

SOLUTION

A cantilever with a concentrated force applied at the end is shown in Fig. 10-18(a); the corresponding moment diagram is plotted in Fig. 10-18(b). Basing the design on the bending moment, the required section modulus at an arbitrary section is given by Eq. 5-5:

$$S = \frac{M}{\sigma_{\text{allow}}} = \frac{Px}{\sigma_{\text{allow}}}$$

A great many cross-sectional areas satisfy this requirement; so first, it will be assumed that the beam will be of rectangular cross section and of *constant height h*. The section modulus for this beam is given by Eq. 5-6 as $bh^2/6 = S$, hence

$$\frac{bh^2}{6} = \frac{Px}{\sigma_{\text{allow}}} \quad \text{or} \quad b = \left[\frac{6P}{h^2 \sigma_{\text{allow}}}\right] x = \frac{b_o}{L} x$$

where the bracketed expression is a constant and is set equal to b_o/L, so that when $x = L$ the width is b_o. A beam of constant strength with a constant depth in a plan view looks like the wedge* shown in Fig. 10-18(c). Near the free end this wedge must be modified to be of adequate strength to resist the shearing force $V = P$.

If the width or breadth b of the beam is constant,

$$\frac{bh^2}{6} = \frac{Px}{\sigma_{\text{allow}}} \quad \text{or} \quad h = \sqrt{\frac{6Px}{b\sigma_{\text{allow}}}} = h_o \sqrt{\frac{x}{L}}$$

This expression indicates that a cantilever of constant width loaded at the end is also of constant strength if its height varies parabolically from the free end, Fig. 10-18(d).

---

*Since this beam is not of constant cross-sectional area, the use of the elementary flexure formula is not entirely correct. When the angle included by the sides of the wedge is small, little error is involved. As this angle becomes large, the error may be considerable. An exact solution shows that when the total included angle is 40° the solution is in error by nearly 10%.

Beams of constant strength are used in leaf springs and in many machine parts that are cast or forged. In structural work, an approximation to a beam of constant strength is frequently made. For example, the moment diagram for the beam loaded as shown in Fig. 10-19(a) is given by the lines AB and BC in Fig. 10-19(b). By selecting a beam of flexural capacity equal only to M_1, the middle portion of the beam is overstressed. However, cover plates can be provided near the middle of the beam to boost the flexural capacity of the composite beam to the required value of the maximum moment. For the case shown, the cover plates must extend at least over the length DE of the beam, and in practice they are made somewhat longer.

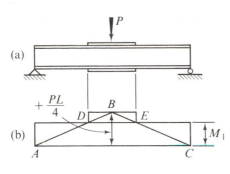

Fig. 10-19

*10-11. DESIGN OF COMPLEX MEMBERS

In many instances the design of complex members cannot be carried out in a routine manner as was done in the preceding problems. Sometimes the size of a member must be *assumed* and a complete stress analysis performed at sections where the stresses appear critical. Designs of this type may require several revisions and much labor. Even experimental methods of stress analysis must be occasionally used since elementary formulas may not be sufficiently accurate. In accurate analyses of manufactured machine parts, the failure theories discussed in Chapter 9 are frequently used.

As a last example in this chapter, a transmission shaft problem will be analyzed. A direct analytical procedure is possible in this problem, which is of great importance in the design of power equipment.

EXAMPLE 10-9

Select the size of a solid steel shaft to drive the two sprockets shown in Fig. 10-20(a). These sprockets drive $1\frac{3}{4}$ in. pitch roller chains* as shown in Figs. 10-20(b) and (c). Pitch diameters of the sprockets shown in the figures are from a manufacturer's catalogue. A 20 hp speed-reducer unit is coupled directly to the shaft and drives it at 63 rpm. At each sprocket 10 hp is taken off. Assume the maximum shear theory of failure, and let $\tau_{\text{allow}} = 6,000$ psi.

SOLUTION

According to Eq. 3-6a the torque delivered to the shaft segment CD is $T = 63,000(\text{Hp}/N) = [(63,000)20]/63 = 20,000$ in.-lb. Hence the torques T_1 and T_2 delivered to the sprockets are $T/2 = 10,000$ in.-lb *each*. Since the chains are arranged as shown in Figs. 10-20(b) and (c), the pull in the chain at sprocket B is $P_1 = T_1/(D_1/2) = 10,000/(10.632/2) = 1,880$ lb. Similarly,

*Similar sprockets and roller chains are commonly used on bicycles.

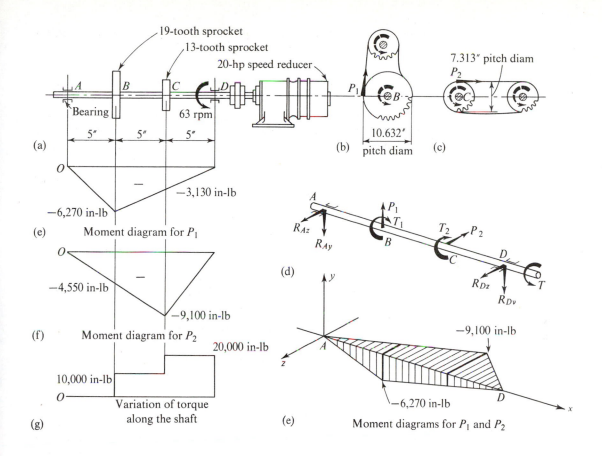

(a)

(b)

(c)

(e) Moment diagram for P_1

−6,270 in-lb
−3,130 in-lb

(f) Moment diagram for P_2

−4,550 in-lb
−9,100 in-lb

(g) Variation of torque along the shaft

20,000 in-lb
10,000 in-lb

(d)

(e) Moment diagrams for P_1 and P_2

−9,100 in-lb
−6,270 in-lb

Fig. 10-20

$P_2 = 10,000/(7.313/2) = 2,730$ lb. The pull P_1 on the chain *is equivalent* to a torque T_1 and a vertical force at B as shown in Fig. 10-20(d). At C the force P_2 acts horizontally and exerts a torque T_2. A complete free-body diagram for the shaft AD is shown in Fig. 10-20(d).

It is seen from the free-body diagram of the shaft that this shaft is simultaneously subjected to bending and torque. These effects on the member are best studied with the aid of appropriate diagrams, which are shown in Figs. 10-20(e), (f), and (g). Next, note that although bending takes place in two planes, a *vectorial resultant of the moments* may be used in the flexure formula, since the beam has a circular cross section.

Bearing the last statement in mind, it will be seen that the general Eq. 8-6, which gives the principal shearing stress at the surface of the shaft, reduces in this problem of bending and torsion to

$$\tau_{\max} = \sqrt{\left(\frac{\sigma_{\text{bending}}}{2}\right)^2 + \tau_{\text{torsion}}^2}$$

or

$$\tau_{\max} = \sqrt{\left(\frac{Mc}{2I}\right)^2 + \left(\frac{Tc}{J}\right)^2}$$

However, since for a circular cross section, $J = 2I$ (Eq. 5-4), $J = \pi d^4/32$ (Eq. 3-2), and $c = d/2$, the last expression simplifies to

$$\tau_{max} = \frac{16}{\pi d^3}\sqrt{M^2 + T^2}$$

Whence, by assigning the allowable shearing stress to τ_{max}, a design formula, based on the maximum shear theory* of failure, for a shaft subjected to bending and torsion is obtained as

$$d = \sqrt[3]{\frac{16}{\pi \tau_{allow}}\sqrt{M^2 + T^2}} \qquad (10\text{-}4)$$

This formula may be used to select the diameter of a shaft simultaneously subjected to bending and torque. In the problem investigated, a few trials should convince the reader that the $\sqrt{M^2 + T^2}$ is largest at the sprocket C; hence, the critical section is at C. Thus,

$$
\begin{aligned}
M^2 + T^2 &= (M_{vert})^2 + (M_{horiz})^2 + T^2 \\
&= (3{,}130)^2 + (9{,}100)^2 + (20{,}000)^2 \\
&= 492{,}600{,}000 \text{ in.}^2\text{-lb}^2
\end{aligned}
$$

$$d = \sqrt[3]{\frac{16}{6{,}000\pi}\sqrt{492{,}600{,}000}} = 2.68 \text{ in.}$$

A $2\frac{11}{16}$ in. diameter shaft, which is a commercial size, should be used.

The effect of shock load on the shaft has been neglected in the foregoing analysis. For some equipment, where its operation is jerky, this condition requires special consideration. The initially assumed allowable stress presumably allows for keyways and fatigue of the material.

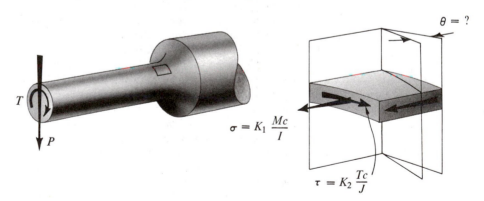

Fig. 10-21. Analysis of a shaft with stress concentrations

*See Problem 10-82 for the formula based on the maximum stress theory of failure.

Although Eq. 10-4 and similar ones based on other failure criteria are widely used in practice, the reader is cautioned in applying them. In many machines, shaft diameters change abruptly, giving rise to stress concentrations. In stress analysis this requires the use of stress concentration factors in bending, which are usually different from those in torsion. Therefore, the problem must be analyzed by considering the actual stresses at the critical section. (See Fig. 10-21.) Then an appropriate procedure, such as Mohr's circle of stress, must be used to determine the significant stress, depending on the selected fracture criteria.

PROBLEMS FOR SOLUTION

10-1 through 10-48. For the beams and frames loaded in one plane as shown in the figures, neglecting the weight of the members, solve the following variations:

A. Without formal computations, sketch shear and moment diagrams directly below a diagram of the given loaded member.

B. Same as A, and, in addition, show the shape of the elastic curve.

C. Plot shear, moment, and, wherever significant, axial force diagrams for the main horizontal members. Determine all critical ordinates. For Probs. 10-36, 10-45, 10-46 and, 10-48 these diagrams must be made for the entire structure.

D. Same as C, and, in addition, determine the points of inflection and show the shape of the elastic curve.

E. Same as C, and, in addition, select the proper size beams of constant cross-sectional dimensions using the elastic approach. The type of beam to be selected is identified in the upper left-hand corner of each figure as:

■ for wooden beams, for which the allowable bending (or normal) and shearing stresses are 8 400 kN/m² and 700 kN/m², respectively, and their width is one-half of their depth.

Ⓢ for standard steel I-shaped beams, and Ⓦ for wide-flange steel beams, for both of which the allowable bending and shearing stresses are 22 ksi and 14.5 ksi, respectively;

● for solid round shafts, for which the allowable bending and shearing stresses are 90 MPa and 50 MPa, respectively.

Assume all beams are laterally braced. Neglect stress-concentrations and the effect of small holes on strength. *Ans:* All shear and moment diagrams must close. For some cases the largest moment is given in parenthesis by the figures in the units of the problem.

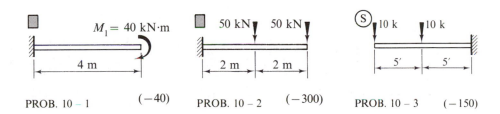

PROB. 10 – 1 (−40)   PROB. 10 – 2 (−300)   PROB. 10 – 3 (−150)

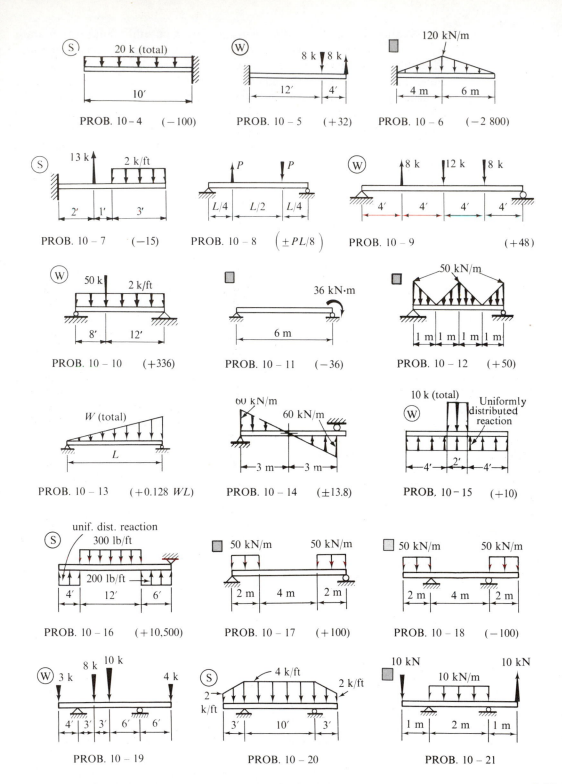

PROB. 10 – 4   (−100)

PROB. 10 – 5   (+32)

PROB. 10 – 6   (−2 800)

PROB. 10 – 7   (−15)

PROB. 10 – 8 $\left(\pm PL/8\right)$

PROB. 10 – 9   (+48)

PROB. 10 – 10   (+336)

PROB. 10 – 11   (−36)

PROB. 10 – 12   (+50)

PROB. 10 – 13   (+0.128 WL)

PROB. 10 – 14   (±13.8)

PROB. 10 – 15   (+10)

PROB. 10 – 16   (+10,500)

PROB. 10 – 17   (+100)

PROB. 10 – 18   (−100)

PROB. 10 – 19

PROB. 10 – 20

PROB. 10 – 21

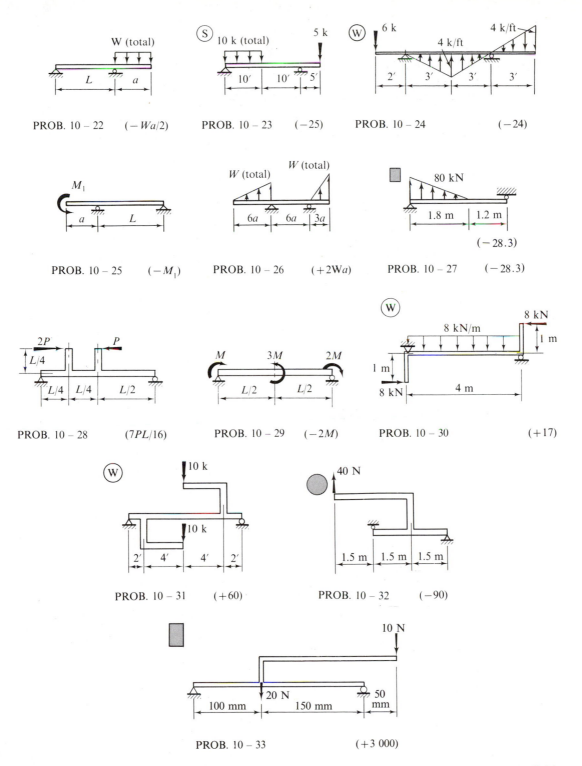

PROB. 10 – 22 $(-Wa/2)$

PROB. 10 – 23 (-25)

PROB. 10 – 24 (-24)

PROB. 10 – 25 $(-M_1)$

PROB. 10 – 26 $(+2Wa)$

(-28.3)

PROB. 10 – 27 (-28.3)

PROB. 10 – 28 $(7PL/16)$

PROB. 10 – 29 $(-2M)$

PROB. 10 – 30 $(+17)$

PROB. 10 – 31 $(+60)$

PROB. 10 – 32 (-90)

PROB. 10 – 33 $(+3\,000)$

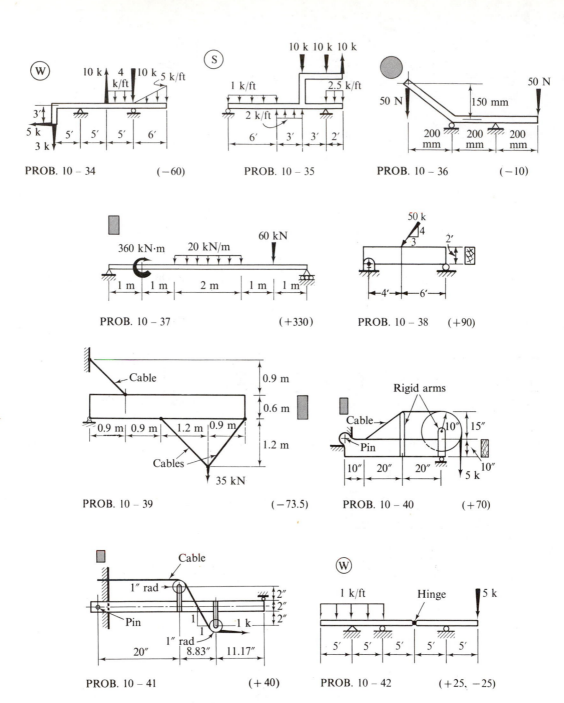

PROB. 10 – 34   (−60)

PROB. 10 – 35

PROB. 10 – 36   (−10)

PROB. 10 – 37   (+330)

PROB. 10 – 38   (+90)

PROB. 10 – 39   (−73.5)

PROB. 10 – 40   (+70)

PROB. 10 – 41   (+40)

PROB. 10 – 42   (+25, −25)

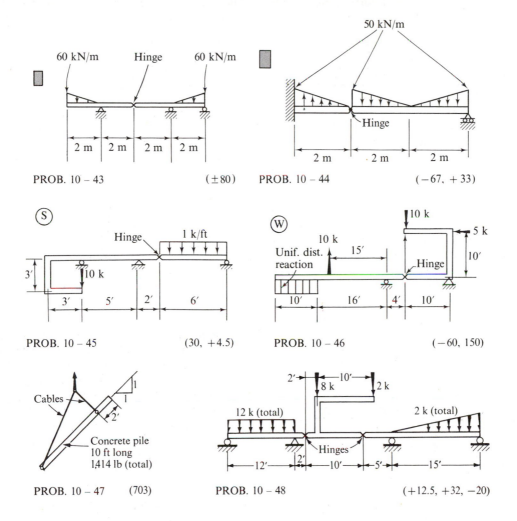

PROB. 10 – 43   (±80)

PROB. 10 – 44   (−67, +33)

PROB. 10 – 45   (30, +4.5)

PROB. 10 – 46   (−60, 150)

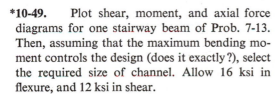

PROB. 10 – 47   (703)

PROB. 10 – 48   (+12.5, +32, −20)

*10-49. Plot shear, moment, and axial force diagrams for one stairway beam of Prob. 7-13. Then, assuming that the maximum bending moment controls the design (does it exactly?), select the required size of channel. Allow 16 ksi in flexure, and 12 ksi in shear.

10-50. Plot shear, moment, and axial force diagrams for the jib-crane loaded as shown in the figure. Neglect the weight of the beam. *Ans:* 128.6 k-in.

10-51. A small narrow barge is loaded as shown in the figure. Plot shear and moment dia-

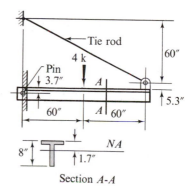

Section *A-A*

PROB. 10 – 50

grams for the applied loading. *Ans: −10ᵏ* (maximum), +50 k-ft (maximum).

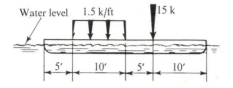

PROB. 10 – 51

10-52. The load distribution for a small, single-engine airplane in flight may be idealized as shown in the figure. In this diagram the vector *A* represents the weight of the engine, *B* the uniformly distributed cabin weight, *C* the weight of the aft fuselage, and *D* the forces from the tail control surfaces. The upward forces *E* are developed by the two longerons from the wings. Using this data construct plausible, qualitative shear and moment diagrams for the fuselage.

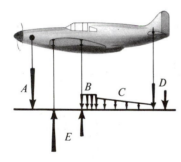

PROB. 10 – 52

For problems **10-53 through 10-55** statically indeterminate beams are loaded as shown in the figures. By the methods of analysis for indeterminate structures, certain quantities, given below, were computed which make the beams statically determinate. Plot shear and moment diagrams for these beams. Indicate all critical values.

10-53. The reaction at *A* is 13 kips (upward).

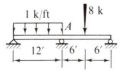

PROB. 10 – 53

10-54. The moment over the support *B* is −400 kN·m.

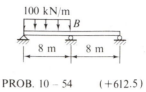

PROB. 10 – 54   (+612.5)

10-55. The moment at the concentrated force is +80 kip-ft.

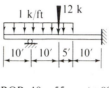

PROB. 10 – 55   (+80)

10-56 through *10-61. The moment diagrams for beams supported at *A* and *B* are as shown in the figures. How are these beams loaded? All curved lines represent parabolas, i.e., plots of equations of the second degree. (*Hint:* construction of shear diagrams aids the solution.) *Ans:* Reaction at *A* in parentheses by the figure.

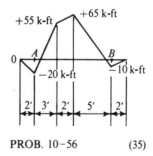

PROB. 10-56   (35)

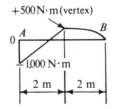

PROB. 10 – 57   (750)

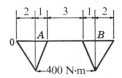

(All dimensions in m)

PROB. 10 – 58   (600)

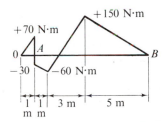

PROB. 10 – 59   (12)

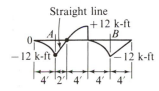

PROB. 10 – 60   (−100)

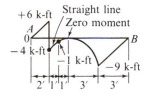

PROB. 10 – 61   (3)

10-62. Rework Problem 10-58 considering the right-hand side of the moment diagram positive.

10-63. A 2 in. by 4 in. (actual size) wooden beam is used in the device illustrated in the figure. What pull may be safely applied to the cables? The allowable bending stress is 1,500 psi, and the allowable shearing stress is 100 psi. Assume that the cables, yokes, etc., are adequate. *Ans:* 267 lb.

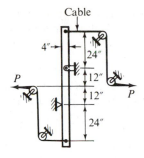

PROB. 10 – 63

10-64. A 4-in.-by-6-in. (actual size) wooden beam is to be symmetrically loaded with two equal loads P as in the figure. Determine the position of these loads and their magnitude when a bending stress of 1,600 psi and a shearing stress of 100 psi are just reached. Neglect the weight of the beam. *Ans:* 1,600 lb, 24 in.

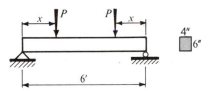

PROB. 10 – 64

10-65. Laminated beams made up from 25 mm by 100 mm boards are to be loaded as shown in the figure. If the allowable bending stress in the wood is 13 MPa, and the shearing strength at the glued joints is 350 kPa, determine: (a) the number of laminates required to carry the bending moment; (b) the number of laminates required to carry the shear force. The number of laminates selected in each case must be a multiple of two. Neglect the weight of the beam.

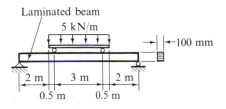

PROB. 10 – 65

10-66. A portion of the floor-framing plan for an office building is shown in the figure. Wooden joists spanning 12 ft are spaced 16 in. apart and support a wooden floor above and a plastered ceiling below. Assume that the floor may be loaded by the occupants everywhere by as much as 75 lb per square foot of floor area (live load). Assume further that floor, joists, and ceiling weigh 25 lb per square foot of the floor area (dead load). (a) Determine the depth required for standard commercial joists nominally 2 in. thick. For wood the allowable bending stress is 1,200 psi and the shearing stress is 100 psi. (b) Select the size required for the steel beam A. Since the joists delivering the load to this beam are spaced closely, assume that the beam is loaded by a uniformly distributed load. The allowable stresses for steel are 20,000 psi and 13,000 psi for bending and shear, respectively. Use a W or an S beam, whichever is lighter. Neglect the width of the column. *Ans:* (a) 2 in. × 10 in. (nominal), (b) W 14 × 30.

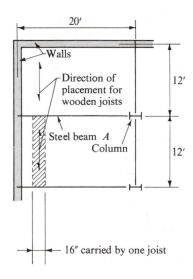

PROB. 10-66

10-67. A bay of an apartment house floor is framed as shown in the figure. Determine the required size of minimum weight for the steel beam "A". Assume that the floor may be loaded everywhere as much as 75 lb/ft² of floor area (live load). Assume further that the weight of the hardwood flooring, structural concrete slab, plas-

tered ceiling below, the weight of the steel beam being selected, etc., also amounts to approximately 75 lb/ft² of floor area (dead load). Use the allowable stresses given in part (b) of Prob. 10-66. *Ans:* S 10 × 25.4.

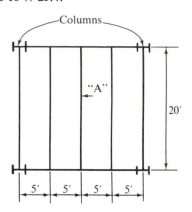

PROB. 10 – 67

10-68. Select the required cross section for a rectangular wooden beam that is to carry a load of $w_o = 20$ kN per meter, including its own weight, for the span in the figure. The allowable bending stress is 9 MPa and the allowable shearing stress is 1 MPa. The beam is to be twice as high as it is wide. Let $a = 3$ m and $b = 1.5$ m.

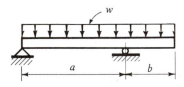

PROB. 10-68

10-69. A T beam is supported in the same manner as the beam in the preceding problem; however, $a = 5$ m and $b = 2.5$ m. The cross-sectional dimensions of the T are in the figure; the

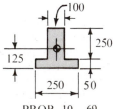

PROB. 10 – 69

moment of inertia around the centroidal axis I is 320×10^6 mm⁴. If the allowable stresses are 8 400 kN/m² in bending and 700 kN/m² in shear, what is the largest load w_o in kN per meter that this beam can carry? All dimensions given in the figure are in mm. *Ans: 4.68 kN/m.*

10-70. A box beam is fabricated from two pieces of $\frac{3}{4}$-in. plywood and two $4\frac{1}{2}$-in.-by-3-in. (actual size) solid wood pieces as shown in the cross-sectional view. If this beam is to be used to carry a concentrated force in the middle of a simple span, (a) what may the magnitude of the maximum applied load P be; (b) how long may the span be; and (c) what size bearing plate should be provided under the concentrated force? Neglect the weight of the beam and assume that there is no danger of lateral buckling. The allowable stresses are: 1,500 psi in bending, 120 psi for shear in plywood, 60 psi for shear in the glued joint, and 400 psi in bearing perpendicular to the grain. *Ans: 3,210 lb, 244 in., 8 in.²*

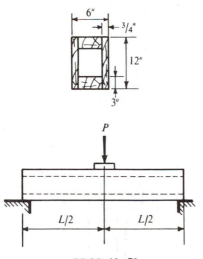

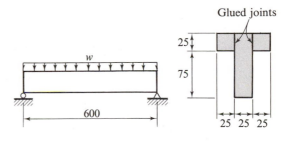

PROB. 10-70

10-71. Determine the size required for an I-shaped beam rail of an overhead traveling crane of 4-ton capacity. The beam is to be attached to the wall at one end and hung from a bracket as in the figure. Assume pinned connection at the wall, and in computations neglect the weight of the beam. Let the allowable bending stress be 12 ksi and the allowable shearing stress be 7 ksi. *Ans: S 15 × 42.9.*

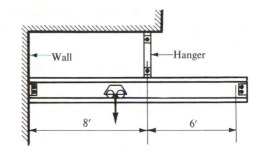

PROB. 10 – 71

10-72. What is the largest allowable load that may be placed at the center of a $4\frac{1}{2}$ ft simple span beam made of a W 10 × 49 section if the allowable bending stress is 24 ksi and the allowable shearing stress is 14 ksi? *Ans: 95.2 kips.*

10-73. A plastic beam is to be made from two 20 mm by 60 mm pieces to span 600 mm and to carry an intermittently applied, uniformly distributed load w. The pieces can be arranged in two alternative ways as in the figure. The allowable stresses are 4 MPa in flexure, 600 kPa shear in plastic, and 400 kPa shear in glue. Which arrangement of pieces should be used, and what load w may be applied?

PROB. 10 – 73

10-74. Find the maximum allowable load w, including own weight, for the plastic beam shown. The allowable stresses are: 3 500 kN/m² for

PROB. 10 – 74 (2)

flexure, 700 kN/m² for shear in plastic, 350 kN/m² for shear in glue. All dimensions given in the figure are in mm. *Ans:* 3.97 kN/m.

10-75.   A 150 mm by 300 mm wooden beam is loaded as shown in the figure. Neglect the weight of the beam. Determine the allowable magnitude of the forces P if $\sigma_{\text{allow}} = 10\,000$ kN/m² and $\tau_{\text{allow}} = 800$ kN/m². *Ans:* 11.25 kN.

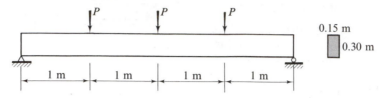

PROB. 10 – 75

10-76.   The allowable bending stress for the beam shown in the figure is $\pm 8\,500$ kN/m². This stress is exceeded if the 30 kN load is applied alone. (a) Determine the minimum value of the load P in order not to exceed the allowable stress. (b) What is the maximum shearing stress in the fully loaded beam. *Ans:* (a) 13.1 kN

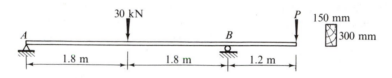

PROB. 10 – 76

10-77.   A four-wheel car running on rails is to be used in light industrial service. When loaded this car will weigh a total of 40 kN. If the bearings are located with respect to the rails as in the figure, what size round axle should be used? Assume the allowable bending stress to be 80 MN/m² and the allowable shearing stress to be 40 MN/m². *Ans:* 50.3 mm.

it is very difficult to determine the magnitudes of the loads that will act on a structure or a machine part. Satisfactory performance in an existing installation may provide the basis for extrapola-

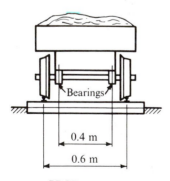

PROB. 10 – 77

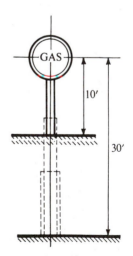

PROB. 10 – 78

10-78.   In many engineering design problems

tion. With this in mind, suppose that a certain sign, such as shown in the figure, has performed satisfactorily on a 4-in. standard steel pipe when its centroid was 10 ft above the ground. What should the size of pipe be if the sign were raised to 30 ft above the ground. Assume that the wind pressure on the sign at the greater height will be 50 per cent greater than it was in the original installation. Vary the size of the pipe along the length as required; however, for ease in fabrication, the successive pipe segments must fit into each other. In arranging the pipe segments also give some thought to aesthetic considerations. For simplicity in calculations, neglect the weight of the pipes and the wind pressure on the pipes themselves.

10-79. Design a cantilever beam of constant strength for resisting a uniformly distributed load. Assume that the width of the beam is constant.

10-80. An S 10 × 25.4 beam is coverplated with two $\frac{1}{2}$ in. by 6 in. plates as shown in Fig. 10-19(a) (I of the composite section is 287.1 in.^{4}), and it spans 20 ft. (a) What concentrated force may be applied at the center of the span if the allowable stress in bending is 16,000 psi? (b) For the above load, where are the theoretical points beyond which the cover plates need not extend? Neglect the weight of the beam, and assume that the beam is braced laterally. *Ans:* (a) 13.9 kips. (b) 4.73 ft from ends.

10-81. The middle part of a simple beam of 2.5 m total length is 150 mm wide by 300 mm deep; the end fifths are 150 mm wide by 200 mm deep. (See figure.) Determine the safe, uniformly distributed load this beam can carry if the allowable bending stress is 10 MPa and the allowable shearing stress is 1 MPa. Neglect

stress concentrations at the change in cross section. *Ans:* 16 kN/m.

10-82. (a) Show that the larger principal stress for a circular shaft simultaneously subjected to a torque and a bending moment is $\sigma_1 = (c/J)(M + \sqrt{M^2 + T^2})$.
(b) Show that the design formula for shafts, on the basis of the maximum stress theory, is

$$d = \sqrt[3]{\frac{16}{\pi\sigma_{allow}}(M + \sqrt{M^2 + T^2})}.$$

10-83. At a critical section a solid circular shaft transmits a torque of 40 kN·m and a bending moment of 10 kN·m. Determine the size of shaft required so that the maximum (principal) shearing stress would not exceed 50 MPa.

10-84. The head shaft of an inclined bucket elevator is arranged as in the figure. It is driven at A at 11 rpm and requires 60 hp for steady operation. Assuming that one-half of the delivered horsepower is used at each sprocket, determine the size of shaft required so that the maximum shearing stress would not exceed 6,000 psi. The assigned stress allows for keyways.

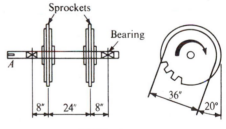

PROB. 10-84

10-85. A shaft is fitted with pulleys as shown in the figure. The end bearings are self-aligning, i.e., they do not introduce moment into the shaft

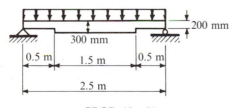

PROB. 10 – 81

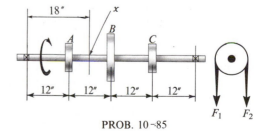

PROB. 10-85

at the supports. Pulley *B* is the driving pulley. Pulley *A* and *C* are the driven pulleys and take off 9,000 in-lb and 3,000 in-lb of torque, respectively. The resultant of the pulls at each pulley is 400 lb acting downward. Determine the size of shaft required so that the principal shearing stress would not exceed 6,000 psi. *Ans:* 2.24 in.

10-86. If the shaft in Prob. 10-85 is 2 in. in diameter, what is the magnitude and direction of the principal stresses at *X*? *Ans:* −13,200 psi, 2,500 psi, 23.5°.

10-87. Two pulleys of 4π in. radius are attached to a 2 in. diameter solid shaft which is supported by the bearings as shown in the figure. If the maximum principal shearing stress is limited to 5 ksi, what is the largest magnitude which the forces *F* can assume? The direct shearing stress caused by *V* need not be considered. *Ans:* 500 lb.

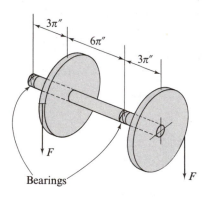

PROB. 10 – 87

10-88. A low-speed shaft is acted upon by an ecentrically applied load *P* caused by a force developed between the gears. Determine the allowable magnitude of the force *P* on the basis of the maximum shearing stress theory if τ_{allow} = 6,500 psi. The small diameter of the overhung shaft is 3 in. Consider the critical section to be where the shaft changes diameter, and that *M* = 3*P* in-lb and *T* = 6*P* in-lb. Note that since the diameter size changes abruptly, the following stress concentration factors must be considered: K_1 = 1.6 in bending, and K_2 = 1.2 in torsion. *Ans:* 4,000 lb.

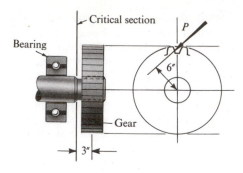

PROB. 10 – 88

10-89. Neglecting the weight of the beam and stress concentrations at the change in cross section, find the largest bending stress for the beam loaded as shown in the figure. All dimensions in the figure are in mm.

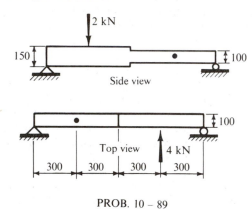

PROB. 10 – 89

10-90. Find the maximum stress in the machine part *AB* loaded as shown in the figure.

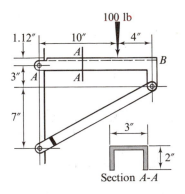

PROB. 10 – 90

The part AB is made from $\frac{1}{8}$ in. hot-rolled steel plate. Assume that the joints are properly designed. Neglect the weight of the parts. The I at section AA around the neutral axis is 0.355 in.^{4} *Ans:* 2,260 psi.

10-91. A 4 m long beam is loaded as shown in the figure. The two applied forces act perpendicularly to the long axis of the beam and are inclined 30° with the vertical. If these forces act through the centroid of the cross-sectional area, find the location and magnitude of the maximum bending stress. Neglect the weight of the beam. *Ans:* ± 172 MN/m^2.

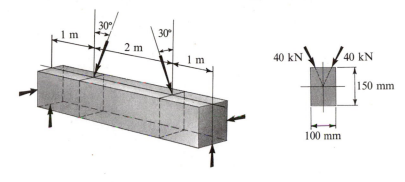

PROB. 10 – 91

11 Deflection of Beams

11-1. INTRODUCTION

Under the action of applied forces the axis of a beam deflects from its initial position. Accurate values for beam deflections are sought in many practical cases. Elements of machines must be sufficiently rigid to prevent misalignment and to maintain dimensional accuracy under load. In buildings, the floor beams cannot deflect excessively to obviate the undesirable psychological effect on the occupants and to minimize or prevent distress in brittle finish materials. Likewise, information on deformation characteristics of members is essential in the study of vibrations of machines as well as of stationary and flight structures.

Basic differential equations for the deflection of beams will be developed in this chapter. Solution of these equations is illustrated in detail. Only deflections caused by forces acting perpendicularly to the axis of a beam are considered. Situations in which axial forces occur simultaneously are discussed in Chapter 13.

The basic theory developed in this chapter is limited to deflections which are small in relation to span length. An idea of the accuracy involved may be gained by noting, for example, that there is approximately a 1% error from the exact solution, if deflections of a simple span are on the order of one-twentieth of its length. By doubling the deflection to one-tenth of the span length, which ordinarily would be considered an intolerably large deflection, the error is raised to approximately 4%. As stiff flexural members are required in most engineering applications, this limitation of the theory is not serious. For clarity, however, the deflections of beams will be shown greatly exaggerated on all diagrams.

Only deflections caused by bending are considered in this chapter. Those due to shear are discussed in Chapter 15 (see especially Example 15-6).

Both the elastic and the inelastic deflections of beams are considered in this chapter. However, since calculations for inelastic deflections of beams are tedious to perform, illustrations are drawn principally from elastic cases.

As the solution of some statically indeterminate elastic beam problems presents no additional mathematical difficulties in comparison with the determinate cases, the solution of such problems is discussed in this chapter. A more complete treatment of statically indeterminate structural systems is given in the next chapter.

After deriving the basic differential equations for beam deflections and exhibiting the boundary conditions, the remainder of the chapter is devoted to two methods for obtaining deflections of a straight beam: direct integration procedures, which are useful if the complete elastic curve needs to be determined; and the so-called *moment-area method*, which is especially convenient if the deflection of only a few points on a beam are of interest.

11-2. STRAIN-CURVATURE AND MOMENT-CURVATURE RELATIONS

To develop the theory of beam deflection, the geometry or kinematics of deformation of a beam element must be considered. The fundamental kinematic hypothesis that plane sections remain plane during deformation, first introduced in Art. 5-3, provides the basis for the theory. This treatment neglects shear deformation of a beam. Fortunately the deflections due to shear usually are very small. (See Example 15-6.)

A segment of an initially straight beam is shown in a deformed state in Fig. 11-1(a). This diagram is analogous to Fig. 5-2, used in establishing the stress distribution in beams due to bending. The deflected axis of the beam,

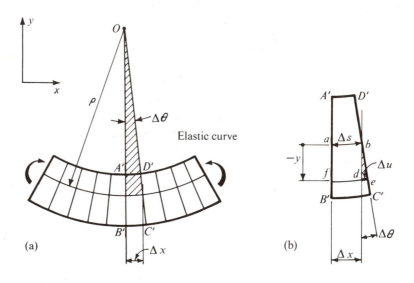

Fig. 11-1. Deformation of a beam segment in bending

i.e., its *elastic curve*, is shown bent into a radius ρ. The center of curvature O for the radius of any element can be found by extending to intersection any two adjoining sections such as $A'B'$ and $D'C'$. For the present it will be assumed that bending is taking place around one of the principal axes of the cross section.

In the enlarged view of the element $A'B'C'D'$ in Fig. 11-1(b), it can be seen that in a bent beam the included angle between two adjoining sections is $\Delta\theta$. If the distance y from the neutral surface to the strained fibers is measured in the usual manner as being positive upwards, the deformation Δu of any fiber can be expressed as

$$\Delta u = -y\,\Delta\theta \qquad (11\text{-}1)$$

For negative y's this yields elongation, which is consistent with the deformation shown in the figure.

The fibers lying in the curved neutral surface of the deformed beam, characterized in Fig. 11-1(b) by the fiber ab, are not strained at all. Therefore the arc length Δs corresponds to the initial length of all fibers between sections $A'B'$ and $D'C'$. Bearing this in mind, upon dividing Eq. 11-1 by Δs, one can form the following relations:

$$\lim_{\Delta s \to 0} \frac{\Delta u}{\Delta s} = -y \lim_{\Delta s \to 0} \frac{\Delta\theta}{\Delta s} \qquad \text{or} \qquad \frac{du}{ds} = -y\frac{d\theta}{ds} \qquad (11\text{-}2)$$

One can recognize that du/ds is the linear strain in a beam fiber at a distance y from the neutral axis. Hence

$$du/ds = \varepsilon \qquad (11\text{-}3)$$

The term $d\theta/ds$ in Eq. 11-2 has a clear geometrical meaning. With the aid of Fig. 11-1(a) it is seen that, since $\Delta s = \rho\,\Delta\theta$,

$$\lim_{\Delta s \to 0} \frac{\Delta\theta}{\Delta s} = \frac{d\theta}{ds} = \frac{1}{\rho} = \kappa \qquad (11\text{-}4)$$

which is the definition of *curvature** κ (kappa).

On the above basis, upon substituting Eqs. 11-3 and 11-4 into Eq. 11-2, one may express the fundamental relation between curvature of the elastic curve and the linear strain as

$$\frac{1}{\rho} = \kappa = -\frac{\varepsilon}{y} \qquad (11\text{-}5)$$

It is important to note that as no material properties were used in deriving Eq. 11-5, this relation can be used for inelastic as well as for elastic problems.

*Note that both θ and s must increase in the same direction.

In the latter case, it is expedient to note that, since $\varepsilon = \varepsilon_x = \sigma_x/E$, and $\sigma_x = -My/I$,

$$\frac{1}{\rho} = \frac{M}{EI} \tag{11-6}$$

This equation relates the bending moment M at a given section of an elastic beam having a moment of inertia I around the neutral axis to the curvature $1/\rho$ of the elastic curve.

EXAMPLE 11-1

For cutting metal a band saw $\frac{1}{2}$ in. wide and 0.025 in. thick runs over two pulleys of 16-in. diameter. What maximum bending stress is developed in the saw as it goes over a pulley? Let $E = 30 \times 10^6$ psi.

SOLUTION

In this application the material must behave elastically. As the thin saw blade goes over the pulley, it conforms to the radius of the pulley; hence, $\rho \approx 8$ in.

Using Eq. 5-1a, $\sigma = -My/I$, together with Eq. 11-6, (after some minor simplifications), yields a generally useful relation:

$$\sigma = -Ey/\rho \tag{11-7}$$

With $y = \pm c$, the maximum bending stress in the saw is determined:

$$\sigma_{max} = \frac{Ec}{\rho} = \frac{(30)10^6(0.0125)}{8} = 46{,}800 \text{ psi}$$

The high stress developed in the band saw necessitates superior materials for this application.

*11-3. THE GOVERNING DIFFERENTIAL EQUATION FOR DEFLECTION OF ELASTIC BEAMS

In texts on analytic geometry it is shown that in Cartesian coordinates curvature of a line is defined as

$$\frac{1}{\rho} = \frac{\dfrac{d^2v}{dx^2}}{\left[1 + \left(\dfrac{dv}{dx}\right)^2\right]^{3/2}} = \frac{v''}{[1 + (v')^2]^{3/2}} \tag{11-8}$$

where x and v are the coordinates of a point on a curve. In terms of the problem being considered, the distance x locates a point on the elastic curve of a deflected beam, and v gives the deflection of the same point from its initial position.

If Eq. 11-8 were substituted into Eq. 11-5 or 11-6, the exact differential equation of the elastic curve would result. In general, the solution of such an equation is very difficult to achieve. However, since, the deflections tolerated in the vast majority of engineering structures are very small, the slope dv/dx of the elastic curve is also very small. Therefore the square of the slope v' is a negligible quantity in comparison with unity, and Eq. 11-8 simplifies to

$$\frac{1}{\rho} \approx \frac{d^2v}{dx^2} \qquad (11\text{-}9)$$

On this basis the governing differential equation for the deflection of an elastic beam* follows from Eq. 11-6 and is

$$\frac{d^2v}{dx^2} = \frac{M}{EI} \qquad (11\text{-}10)$$

where it is understood that $M = M_{zz}$, and $I = I_{zz}$.

Note that in Eq. 11-10 the xyz coordinate system is employed to locate the material points in a beam for calculating the moment of inertia I. On the other hand, in the planar problem, it is the xv system of axes that is used to locate points on the elastic curve.

The positive direction of the v axis is taken to have the same sense as that of the positive y axis and the positive direction of the applied load q, Fig. 11-2. Note especially that if the positive slope dv/dx of the elastic curve becomes more positive as x increases, the curvature $1/\rho \approx d^2v/dx^2$ is positive. This sense of curvature agrees with the induced curvature caused by the applied positive moments M. For this reason the signs are positive on both sides of Eq. 11-10.

In some texts the positive direction for deflection v is taken downward with the x-axis directed to the right. For such a choice of coordinates the positive curvature is concave downwards. Whereas, if the usual sense for positive moments is retained, Fig. 11-2(a), the corresponding curvature of the bent beam is concave upwards. Therefore, since the curvature induced by the positive moments M is opposite to that associated with the positive curvature of the elastic curve, one has

$$\frac{d^2v}{dx^2} = -\frac{M}{EI} \qquad (11\text{-}11)$$

In this text only Eq. 11-10 is employed.

It is important and interesting to note that for the elastic curve, at the level of accuracy of Eq. 11-10, one has $ds = dx$. This follows from the fact that, as before, the square of the slope dv/dx is negligibly small compared with unity, and

$$ds = \sqrt{dx^2 + dv^2} = \sqrt{1 + (v')^2}\, dx \approx dx \qquad (11\text{-}12)$$

*The equation of the elastic curve was formulated by James Bernoulli, a Swiss mathematician, in 1694. Leonhard Euler (1707–83) greatly extended its applications.

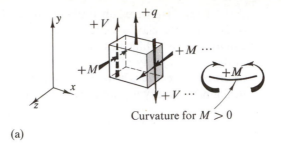

(a)

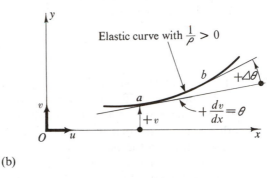

(b)

Fig. 11-2. Moment and its relation to curvature

Thus, in the small deflection theory, no difference in length is said to exist between the initial length of the beam axis and the arc of the elastic curve. Stated alternatively, there is no horizontal displacement of the points lying on the neutral surface, i.e., at $y = 0$.

11-4. ALTERNATIVE DIFFERENTIAL EQUATIONS OF ELASTIC BEAMS

The differential relations among the applied loads, shear, and moment (Eqs. 6-1 and 10-1) can be combined with Eq. 11-10 to yield the following useful sequence of equations:

$$v = \text{deflection of the elastic curve}$$

$$\theta = \frac{dv}{dx} = v' = \text{slope of the elastic curve}$$

$$M = EI\frac{d^2v}{dx^2} = EIv''$$

$$V = \frac{dM}{dx} = \frac{d}{dx}\left(EI\frac{d^2v}{dx^2}\right) = (EIv'')'$$

$$q = \frac{dV}{dx} = \frac{d^2}{dx^2}\left(EI\frac{d^2v}{dx^2}\right) = (EIv'')''$$

(11-13)

In applying these relations, the sign convention shown in Fig. 11-2 must be adhered to strictly. For beams with constant flexural rigidity EI, Eq. 11-13 simplifies into three alternative equations for determining the deflection of a loaded beam:

$$EI\frac{d^2v}{dx^2} = M(x) \tag{11-14}$$

$$EI\frac{d^3v}{dx^3} = V(x) \tag{11-15}$$

$$EI\frac{d^4v}{dx^4} = q(x) \tag{11-16}$$

The choice of equation for a given case depends on the ease with which an expression for load, shear, or moment can be formulated. Fewer constants of integration are needed in the lower-order equations.

11-5. BOUNDARY CONDITIONS

For the solution of beam deflection problems, in addition to the differential equations, boundary conditions must be prescribed. Several types of homogeneous boundary conditions are as follows:

1. *Clamped or fixed support:* In this case the displacement v and the slope dv/dx must vanish. Hence at the end considered, where $x = a$,

$$v(a) = 0, \qquad v'(a) = 0 \tag{11-17a}$$

2. *Roller or pinned support:* At the end considered, no deflection v nor moment M can exist. Hence

$$v(a) = 0, \qquad M(a) = EIv''(a) = 0 \tag{11-17b}$$

Here the physically evident condition for M is related to the derivative of v with respect to x from Eq. 11-14.

3. *Free end:* Such an end is free of moment and shear. Hence

$$M(a) = EIv''(a) = 0, \qquad V(a) = (EIv'')'_{x=a} = 0 \tag{11-7c}$$

4. *Guided support:* In this case free vertical movement is permitted, but the rotation of the end is prevented. The support is not capable of resisting any shear. Therefore

$$v'(a) = 0, \qquad V(a) = (EIv'')'_{x=a} = 0 \tag{11-17d}$$

The same boundary conditions for constant EI are summarized in Fig. 11-3. Note the two basically different types of boundary conditions. Some pertain to the force quantities and are said to be *static boundary conditions.* Others describe geometrical or deformational behavior of an end; these are *kinematic boundary conditions.*

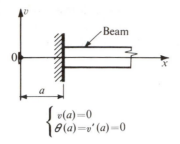

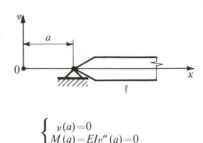

$$\begin{cases} v(a)=0 \\ \theta(a)=v'(a)=0 \end{cases}$$

$$\begin{cases} v(a)=0 \\ M(a)=EIv''(a)=0 \end{cases}$$

(a) Clamped support

(b) Simple support

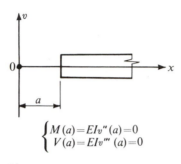

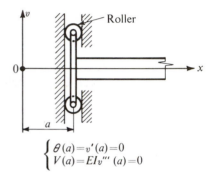

$$\begin{cases} M(a)=EIv''(a)=0 \\ V(a)=EIv'''(a)=0 \end{cases}$$

$$\begin{cases} \theta(a)=v'(a)=0 \\ V(a)=EIv'''(a)=0 \end{cases}$$

(c) Free end

(d) Guided support

Fig. 11-3. Homogeneous boundary conditions for beams with constant *EI*. In (a) both conditions are *kinematic*; in (c) both are *static*; in (b) and (d), conditions are mixed.

Nonhomogeneous boundary conditions, where a given shear, moment, rotation, or displacement is prescribed at the boundary, also occur in applications. In such cases the zeros in the appropriate Eqs. 11-17a through 11-17d are replaced by the specified quantity.

In some solutions the physical requirements of continuity of the elastic curve must be brought in to supplement the boundary conditions. This means that at any juncture of two zones of a beam the deflection and the tangent to the elastic curve must be the same, regardless of the direction from

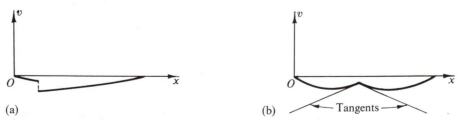

(a)

(b) Tangents

Fig. 11-4. Impossible situations in a continuous elastic curve.

which the common point is approached. The situations in Fig. 11-4 are impossible. The requirements of force and moment equilibrium are contained implicitly in the conditions of continuity.

11-6. SOLUTION OF BEAM DEFLECTION PROBLEMS BY DIRECT INTEGRATION

As a general example of calculating beam deflection, consider Eq. 11-16, $EIv^{iv} = q(x)$. By successively integrating this expression four times, the formal solution for v is obtained. Thus

$$EIv^{iv} = EI\frac{d^4v}{dx^4} = EI\frac{d}{dx}(v''') = q(x)$$

$$EIv''' = \int_0^x q\, dx + C_1$$

$$EIv'' = \int_0^x dx \int_0^x q\, dx + C_1x + C_2 \qquad (11\text{-}18)$$

$$EIv' = \int_0^x dx \int_0^x dx \int_0^x q\, dx + C_1x^2/2 + C_2x + C_3$$

$$EIv = \int_0^x dx \int_0^x dx \int_0^x dx \int_0^x q\, dx + C_1x^3/3! + C_2x^2/2! + C_3x + C_4$$

If, instead, one started with Eq. 11-14, $EIv'' = M(x)$, after two integrations the solution is

$$EIv = \int_0^x dx \int_0^x M\, dx + C_3x + C_4 \qquad (11\text{-}19)$$

In both equations the constants C_1, C_2, C_3, and C_4, corresponding to the homogeneous solution of the differential equations, must be determined from the conditions at the boundaries. The constants C_1 and C_2 were encountered in Chapter 10 in the solution of the differential equations of equilibrium (Arts. 10-5 and 10-6). In Eq. 11-19 the constants C_1 and C_2 are incorporated into the expression of M. The constants C_1, C_2, $C_3/(EI)$, and $C_4/(EI)$, respectively, are usually* the initial values of V, M, θ, and v at the origin.

The first term on the right hand of the last part of Eq. 11-18 and the corresponding one in Eq. 11-19 are the particular solutions of the respective differential equations. The one in Eq. 11-18 is especially interesting as it depends only on the loading condition of the beam. This term remains the same regardless of the prescribed boundary conditions. The latter are brought into the problem from the homogeneous solution of the differential equation.

If the loading, shear, and moment functions are continuous and the flexural rigidity EI is constant, the evaluation of the particular integrals is

---

*In certain cases where transcendental functions are used, these constants do not have this meaning. Basically, the whole function, which includes the constants of integration, must satisfy the conditions at the boundary.

very direct. When discontinuities occur, solutions can be found for each segment of a beam in which the functions are continuous; the complete solution is then achieved by enforcing continuity conditions at the common boundaries of the beam segments. Alternatively, graphical or numerical procedures* of successive integrations can be used very effectively in the solution of practical problems.

The procedures discussed above are quite general and are applicable to both statically determinate and indeterminate elastic beams. In the next four examples, however, alternative solutions for determinate cases only will be illustrated. In one of the examples the case of a variable I is treated. Statically indeterminate beams will be considered in the following article.

EXAMPLE 11-2

A bending moment M_1 is applied at the free end of a cantilever of length L and of constant flexural rigidity EI, Fig. 11-5(a). Find the equation of the elastic curve.

SOLUTION

The boundary conditions are recorded near the figure from inspection of the conditions at the ends. At $x = L$, $M(L) = +M_1$, a nonhomogenoeus condition.

From a free-body diagram of Fig. 11-5(b), it can be observed that the bending moment is $+M_1$ throughout the beam. By applying Eq. 11-14, integrating successively, and making use of the boundary conditions, one obtains the solution for v:

$$EI\frac{d^2v}{dx^2} = M = M_1$$

$$EI\frac{dv}{dx} = M_1 x + C_3$$

But $\theta(0) = 0$; hence at $x = 0$ one has $EIv'(0) = C_3 = 0$ and

$$EI\frac{dv}{dx} = M_1 x$$

$$EIv = \tfrac{1}{2}M_1 x^2 + C_4$$

But $v(0) = 0$; hence $EIv(0) = C_4 = 0$ and

$$v = M_1 x^2/(2EI) \qquad (11\text{-}20)$$

The positive sign of the result indicates that the deflection due to M_1 is upward. The largest value of v occurs at $x = L$. The slope of the elastic curve at the free end is $+M_1 L/(EI)$ radians.

---

*Such procedures are of great importance in complicated problems. For example, see N. M. Newmark, "Numerical Procedure for Computing Deflections, Moments, and Buckling Loads," *Trans. ASCE*, vol. 108, (1943), 1161.

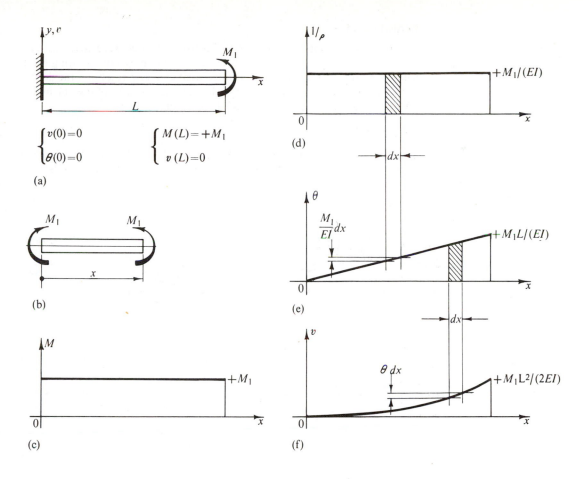

$$\begin{cases} v(0)=0 \\ \theta(0)=0 \end{cases} \qquad \begin{cases} M(L)=+M_1 \\ v(L)=0 \end{cases}$$

(a)

(b)

(c)

(d)

(e)

(f)

$+M_1/(EI)$

$\dfrac{M_1}{EI}dx$

$+M_1L/(EI)$

$\theta\,dx$

$+M_1L^2/(2EI)$

$+M_1$

Fig. 11-5

Equation 11-20 shows that the elastic curve is a parabola. However, every element of the beam experiences equal moments and deforms alike. Therefore the elastic curve should be a part of a circle. The inconsistency results from the use of an approximate relation for the curvature $1/\rho$. It can be shown that the error committed is in the ratio of $(\rho - v)^3$ to ρ^3. As the deflection v is much smaller than ρ, the error is not serious.

It is important to associate the above successive integration procedure with a graphical solution or interpretation. This is shown in the sequence of Figs. 11-5(c) through (f). First the conventional moment diagram is shown. Then from Eqs. 11-9 and 11-10, $1/\rho \approx d^2v/dx^2 = M/(EI)$, the curvature diagram is plotted in Fig. 11-5(d). For the elastic case this is simply a plot of $M/(EI)$. By integrating the curvature diagram one obtains the θ diagram. In the next integration the elastic curve is obtained. In this problem since the beam is fixed at the origin, the conditions $\theta(0) = 0$, and $v(0) = 0$ are used in constructing the diagrams. This graphical approach or its numerical equivalents are very useful in the solution of problems with variable EI.

EXAMPLE 11-3

A simple beam supports a uniformly distributed downward load w_o. The flexural rigidity EI is constant. Find the elastic curve by the following three methods: (a) Use the second-order differential equation to obtain the deflection of the beam. (b) Use the fourth-order equation instead of the one in (a). (c) Illustrate a graphical solution of the problem.

SOLUTION

Case (a). A diagram of the beam together with the implied boundary conditions is in Fig. 11-6(a). The expression for M for use in the second-order

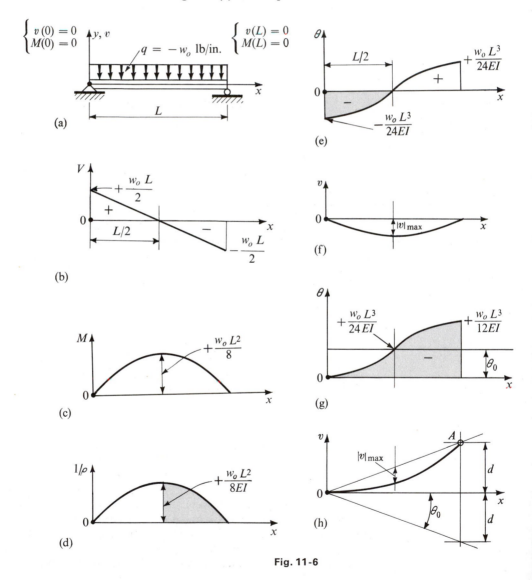

Fig. 11-6

differential equation has been found in Example 4-6. From Fig. 4-20

$$M = \frac{w_o L x}{2} - \frac{w_o x^2}{2}$$

Substituting this relation into Eq. 11-14, integrating it twice in succession, and making use of the boundary conditions, one finds the equation of the elastic curve:

$$EI\frac{d^2v}{dx^2} = M = \frac{w_o L x}{2} - \frac{w_o x^2}{2}$$

$$EI\frac{dv}{dx} = \frac{w_o L x^2}{4} - \frac{w_o x^3}{6} + C_3$$

$$EIv = \frac{w_o L x^3}{12} - \frac{w_o x^4}{24} + C_3 x + C_4$$

But $v(0) = 0$; hence $EIv(0) = 0 = C_4$; and, since $v(L) = 0$,

$$EIv(L) = 0 = \frac{w_o L^4}{24} + C_3 L \qquad \text{and} \qquad C_3 = -\frac{w_o L^3}{24}$$

$$v = -\frac{w_o}{24EI}(L^3 x - 2L x^3 + x^4) \qquad \qquad (11\text{-}21)$$

By virtue of symmetry, the largest deflection occurs at $x = L/2$. On substituting this value of x into Eq. 11-21, one obtains

$$|v|_{max} = 5w_o L^4/(384EI) \qquad \qquad (11\text{-}22)$$

The condition of symmetry could also have been used to determine the constant C_3. As it is known that $v'(L/2) = 0$, one has

$$EIv'(L/2) = \frac{w_o L(L/2)^2}{4} - \frac{w_o(L/2)^3}{6} + C_3 = 0$$

and, as before, $C_3 = -(1/24)w_o L^3$.

Case (b). Application of Eq. 11-16 to the solution of this problem is direct. The constants are found from the boundary conditions.

$$EI\frac{d^4v}{dx^4} = q = -w_o$$

$$EI\frac{d^3v}{dx^3} = -w_o x + C_1$$

$$EI\frac{d^2v}{dx^2} = -\frac{w_o x^2}{2} + C_1 x + C_2$$

But $M(0) = 0$; hence $EIv''(0) = 0 = C_2$; and, since $M(L) = 0$,

$$EIv''(L) = 0 = -\frac{w_o L^2}{2} + C_1 L \qquad \text{or} \qquad C_1 = \frac{w_o L}{2}$$

hence

$$EI\frac{d^2v}{dx^2} = \frac{w_o L x}{2} - \frac{w_o x^2}{2}$$

The remainder of the problem is the same as in Case (*a*). In this approach no preliminary calculation of reactions is required. As will be shown later, this is advantageous in some statically indeterminate problems.

Case (*c*). The steps needed for a graphical solution of the complete problem are in Figs. 11-6(b) through (f). In Figs. 11-6(b) and (c) the conventional shear and moment diagrams are shown. The curvature diagram is obtained by plotting $M/(EI)$, as in Fig. 11-6(d).

Since, by virtue of symmetry, the slope to the elastic curve at $x = L/2$ is horizontal, $\theta(L/2) = 0$. Therefore, the construction of the θ diagram can be begun from the center. In this procedure, the right ordinate in Fig. 11-6(e) must equal the shaded area of Fig. 11-6(d), and vice versa. By summing the θ diagram, one finds the elastic deflection v. The shaded area of Fig. 11-6(e) is equal numerically to the maximum deflection. In the above, the condition of symmetry was employed. A generally applicable procedure follows.

After the curvature diagram is established as in Fig. 11-6(d), the θ diagram can be constructed with an assumed initial value of θ at the origin. For example, let $\theta(0) = 0$ and sum the curvature diagram to obtain the θ diagram, Fig. 11-6(g). Note that the shape of the curve so found is identical to that of Fig. 11-6(e). Summing the area of the θ diagram gives the elastic curve. In Fig. 11-6(h) this curve extends from O to A. This violates the boundary condition at A, where the deflection must be zero. Correct deflections are given, however, by measuring them vertically from a straight line passing through O and A. This inclined line corrects the deflection ordinates caused by the incorrectly assumed $\theta(0)$. In fact, after constructing Fig. 11-6(h), one knows that $\theta(0) = -d/L = -w_o L^3/(24EI)$. When this value of $\theta(0)$ is used, the problem reverts to the preceding solution (Figs. 11-6(e) and (f)). In Fig. 11-6(h) inclined measurements have no meaning, The procedure described is applicable for beams with overhangs. In such cases the base line for measuring deflections must pass through the support points.

EXAMPLE 11-4

A simple beam supports a concentrated downward force P at a distance a from the left support, Fig. 11-7(a). The flexural rigidity EI is constant. Find the equation of the elastic curve by successive integration.

SOLUTION

The solution will be obtained using the second-order differential equation. The reactions and boundary conditions are noted in Fig. 11-7(a). The moment diagram plotted in Fig. 11-7(b) clearly shows that a discontinuity at $x = a$ exists in $M(x)$, requiring two different functions for it. At first the solution proceeds independently for each segment of the beam.

For segment *AD*:	For segment *DB*:
$\dfrac{d^2v}{dx^2} = \dfrac{M}{EI} = \dfrac{Pb}{EIL}x$	$\dfrac{d^2v}{dx^2} = \dfrac{M}{EI} = \dfrac{Pa}{EI} - \dfrac{Pa}{EIL}x$
$\dfrac{dv}{dx} = \dfrac{Pb}{EIL}\dfrac{x^2}{2} + A_1$	$\dfrac{dv}{dx} = \dfrac{Pa}{EI}x - \dfrac{Pa}{EIL}\dfrac{x^2}{2} + B_1$
$v = \dfrac{Pb}{EIL}\dfrac{x^3}{6} + A_1 x + A_2$	$v = \dfrac{Pa}{EI}\dfrac{x^2}{2} - \dfrac{Pa}{EIL}\dfrac{x^3}{6} + B_1 x + B_2$

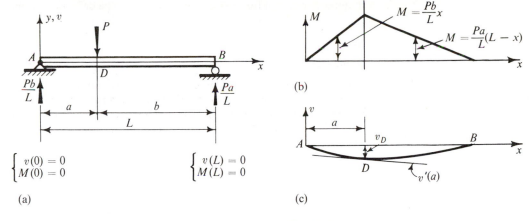

$$\begin{cases} v(0) = 0 \\ M(0) = 0 \end{cases}$$

$$\begin{cases} v(L) = 0 \\ M'(L) = 0 \end{cases}$$

(a)

(b)

(c)

Fig. 11-7

To determine the four constants $A_1, A_2, B_1,$ and B_2, two boundary and two continuity conditions must be used.

For segment AD:

$$v(0) = 0 = A_2$$

For segment DB:

$$v(L) = 0 = \frac{PaL^2}{3EI} + B_1L + B_2$$

Equating deflections for both segments at $x = a$:

$$v_D = v(a) = \frac{Pa^3b}{6EIL} + A_1a = \frac{Pa^3}{2EI} - \frac{Pa^4}{6EIL} + B_1a + B_2$$

Equating slopes for both segments at $x = a$:

$$\theta_D = v'(a) = \frac{Pa^2b}{2EIL} + A_1 = \frac{Pa^2}{EI} - \frac{Pa^3}{2EIL} + B_1$$

Upon solving the four equations simultaneously, one finds

$$A_1 = -\frac{Pb}{6EIL}(L^2 - b^2) \qquad A_2 = 0$$

$$B_1 = -\frac{Pa}{6EIL}(2L^2 + a^2) \qquad B_2 = \frac{Pa^3}{6EI}$$

With these constants, for example, the elastic curve for the left segment AD of the beam becomes

$$v = [(Pb/(6EIL)][x^3 - (L^2 - b^2)x] \qquad (11\text{-}23)$$

The largest deflection occurs in the longer segment of the beam. If $a > b$, the point of maximum deflection is at $x = \sqrt{a(a + 2b)/3}$, which follows from setting the expression for the slope equal to zero. The deflection at this point is

$$|v|_{max} = \frac{Pb(L^2 - b^2)^{3/2}}{9\sqrt{3}\,EIL} \qquad (11\text{-}24)$$

Usually the deflection at the center of the span is very nearly equal to the numerically largest deflection. Such a deflection is much simpler to determine, which recommends its use. If the force P is applied at the middle of the span, i.e. $a = b = L/2$, it can be shown by direct substitution into Eq. 11-23 or 11-24 that at $x = L/2$

$$|v|_{max} = PL^3/(48EI) \qquad (11\text{-}25)$$

EXAMPLE 11-5

A simply supported beam 5 m long is loaded with a 20 N downward force at a point 4 m from the left support, Fig. 11-8(a). The moment of inertia of the cross section of the beam is $4I_1$ for the segment AB, and I_1 for the remainder of the beam. Determine the elastic curve.

SOLUTION

A similar problem was solved in the preceding example. Another useful technique will be illustrated here which is convenient in some complicated problems where different M/EI expressions are applicable to several segments of the beam.* This method consists of selecting an origin at one end of the beam and carrying out successive integrations until expressions for θ and v are obtained for the first segment. The values of θ and v are then determined at the end of the first segment. Due to continuity conditions, these become the initial constants in the integrations carried out for the next segment. This process is repeated until the far end of the beam is reached, then the boundary conditions are imposed to determine the remaining unknown constants. A new origin is used at every juncture of the segments, and all x's are taken to be positive in the same direction.

For segment AB: $0 < x < 4$

$$M = 4x \qquad \text{and} \qquad EI = 4EI_1$$

$$\frac{d^2v}{dx^2} = \frac{M}{EI} = \frac{x}{EI_1}$$

$$\theta = \frac{dv}{dx} = \frac{x^2}{2EI_1} + A_1$$

$$v = \frac{x^3}{6EI_1} + A_1x + A_2$$

At $x = 0$: $v(0) = v_A = 0$, and $\theta(0) = \theta_A$. Hence, $A_1 = \theta_A$ and $A_2 = 0$.

---

*Singularity functions may be used to set up and solve the differential equations for many beam deflection problems. See, for example, E. P. Popov, *Introduction to Mechanics of Solids*, Englewood Cliffs, N.J.: Prentice-Hall, 1968.

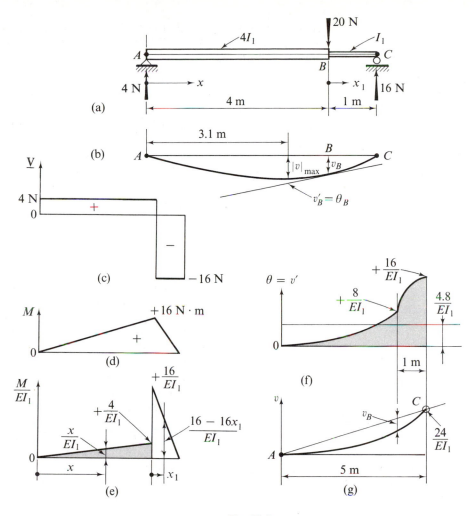

Fig. 11-8

At the end of segment AB:

$$\theta(4) = \theta_B = \frac{8}{EI_1} + \theta_A \quad \text{and} \quad v(4) = v_B = \frac{32}{3EI_1} + 4\theta_A$$

For segment BC: $0 < x_1 < 1$

$$M = 4(4 + x_1) - 20x_1 = 16 - 16x_1 \quad \text{and} \quad EI = EI_1$$

$$\frac{d^2v}{dx_1^2} = \frac{16}{EI_1} - \frac{16x_1}{EI_1}$$

$$\theta = \frac{dv}{dx_1} = \frac{16x_1}{EI_1} - \frac{8x_1^2}{EI_1} + A_3$$

$$v = \frac{8x_1^2}{EI_1} - \frac{8x_1^3}{3EI_1} + A_3x_1 + A_4$$

At $x_1 = 0$: $v(0) = v_B$ and $\theta(0) = \theta_B$. Hence, $A_4 = v_B = 32/(3EI_1) + 4\theta_A$, and $A_3 = \theta_B = 8/(EI_1) + \theta_A$. The expressions for θ and v in segment BC are then obtained as

$$\theta = \frac{16x_1}{EI_1} - \frac{8x_1^2}{EI_1} + \frac{8}{EI_1} + \theta_A$$

$$v = \frac{8x_1^2}{EI_1} - \frac{8x_1^3}{3EI_1} + \frac{8x_1}{EI_1} + \theta_A x_1 + \frac{32}{3EI_1} + 4\theta_A$$

Finally, the boundary condition at C is applied to determine the value of θ_A. At $x_1 = 1$: $v(1) = v_c = 0$

$$0 = \frac{8}{EI_1} - \frac{8}{3EI_1} + \frac{8}{EI_1} + \theta_A + \frac{32}{3EI_1} + 4\theta_A$$

and

$$\theta_A = -\frac{4.8}{EI_1}$$

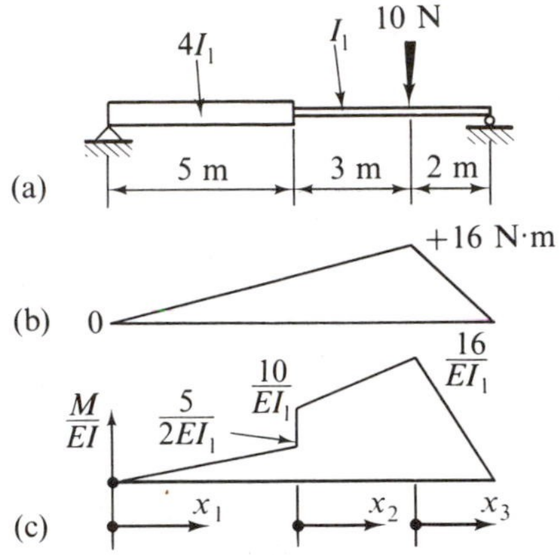

(a)

(b)

(c)

Fig. 11-9. Multiple selection of the origins of x for a discontinuous $M/(EI)$ function

Substituting this value of θ_A into the respective expressions for θ and v, equations for these quantities can be obtained for either segment. For example, the equation for the slope in segment AB is $\theta = x^2/(2EI_1) - 24/(5EI_1)$. Upon setting this quantity equal to zero, x is found to be 3.1 m. The maximum deflection occurs at this value of x, and $|v|_{max} = 9.95/(EI_1)$. Characteristically, the deflection at the center of the span (at $x = 2.5$ m) is nearly the same, being $9.4/(EI_1)$.

A self-explanatory graphical procedure is shown in Figs. 11-8(d) through (g). Variations in I cause virtually no complications in the graphical solution, a great advantage in complex problems. Multiple origins can be used as shown in Fig. 11-9 to simplify the numerical work as in the present example.

11-7. STATICALLY INDETERMINATE ELASTIC BEAM PROBLEMS

In a large and important class of beam problems, reactions cannot be determined using the conventional procedures of statics. For example, for the beam shown in Fig. 11-10(a), four reaction components are unknown. The three vertical components cannot be found from equations of static equilibrium. Further examination of Fig. 11-10(a) shows that any one of the vertical reactions can be removed and the beam would remain in equilibrium. Therefore any one of these reactions may be said to be superflous, or redundant, for maintaining equilibrium. Problems with extra or redundant reactive forces and/or moments are called (*externally*) *statically indeterminate*.

When the number of unknown reactions exceeds by one that which can be determined by statics, the member is said to be indeterminate to the *first degree*. As the number of unknowns increases, the degree of indeterminacy also increases. For example, by providing one more support than

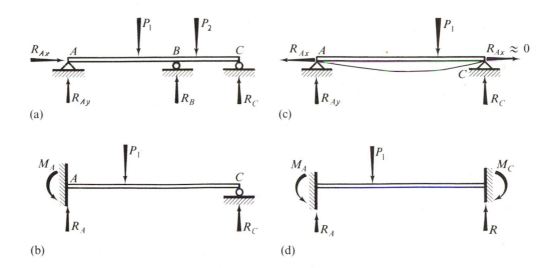

Fig. 11-10. Illustrations of statical indeterminacy of beams. In (a) and (b), the beams are indeterminate to the first degree. If it is assumed that the horizontal components of the reactions are negligible, the beam in (c) is determinate and in (d) indeterminate to the second degree.

shown for the beam in Fig. 11-10(a), the beam would become indeterminate to the second degree. The beam of Fig. 11-10(b) is indeterminate to the first degree since either M_A or R_C can be considered redundant.

As $ds \approx dx$, for small deflections according to Eq. 11-12 no significant axial strain can develop in a transversely loaded beam.* Therefore, the horizontal components of the reactions in situations with immovable supports, such as shown in Figs. 11-10(c) and (d), are negligible. On this basis, the beam shown in Fig. 11-10(c), with pins at both ends, is a determinate beam. The beam of Fig. 11-10(d) is indeterminate to the second degree.

To determine the elastic deflection of statically indeterminate beams, the procedure of solving the differential equations is practically the same as that discussed above for determinate beams. The only difference is that kinematic boundary conditions replace some of the static ones. As the degree of indeterminacy increases, as in continuous members, the number of simultaneous equations for determining the constants increases. In such problems the number of constants to be found is no longer limited to a maximum of four.

EXAMPLE 11-6

Find the equation of the elastic curve for the uniformly loaded, two-span continuous beam shown in Fig. 11-11(a). The EI is constant.

*The horizontal force becomes important in thin plates. See S. Timoshenko and S. Woinowsky-Krieger, *Theory of Plates and Shells* (2nd ed.), New York: McGraw-Hill, 1959, p. 6.

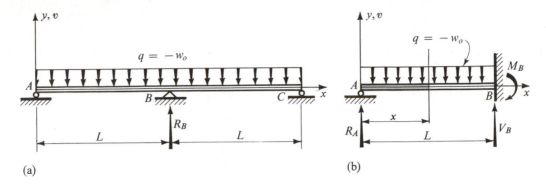

Fig. 11-11

SOLUTION

It is possible to obtain the solution to this problem in a manner similar to that adopted in Example 11-5, by carrying out the successive integrations for one segment at a time (proceeding from A to B and then B to C), and imposing the continuity requirements and boundary conditions. However, the given problem is symmetrical around the support at B. Therefore the tangent to the elastic curve at B is horizontal, and an equivalent problem involving one-half of the original beam shown in Fig. 11-11(b) can be analyzed instead. This new problem can be solved using the fourth-order differential equation with the following four boundary conditions:

$$v_A = 0, \qquad M_A = EIv''(0) = 0, \qquad v_B = 0, \qquad \text{and} \quad v'_B = 0$$

Alternatively, on designating the unknown reaction at A as R_A, one may state the bending moment within the span as

$$M = R_A x - w_o x^2/2$$

By substituting this relation into Eq. 11-14, integrating it twice, and making use of three of the kinematic boundary conditions stated above, one finds the unknown constants R_A, C_3, and C_4.

Case (a):   Fourth-order differential equation solution.

$$EI\frac{d^4v}{dx^4} = q = -w_o$$

$$EI\frac{d^3v}{dx^3} = -w_o x + C_1$$

$$EI\frac{d^2v}{dx^2} = -\frac{w_o x^2}{2} + C_1 x + C_2$$

$$EI\frac{dv}{dx} = -\frac{w_o x^3}{6} + C_1\frac{x^2}{2} + C_2 x + C_3$$

$$EI v = -\frac{w_o x^4}{24} + C_1\frac{x^3}{6} + C_2\frac{x^2}{2} + C_3 x + C_4$$

Boundary conditions:

$$v(0) = v_A = 0 = C_4$$

$$v''(0) = v_A'' = 0 = C_2$$

$$v'(L) = v_B' = 0 = -\frac{w_o L^3}{6} + C_1 \frac{L^2}{2} + C_3$$

$$v(L) = v_B = 0 = -\frac{w_o L^4}{24} + C_1 \frac{L^3}{6} + C_3 L$$

Solving the last two equations simultaneously,

$$C_1 = \frac{3 w_o L}{8} \quad \text{and} \quad C_3 = -\frac{w_o L^3}{48}$$

Substituting the values of these constants into the equation for the elastic curve, one obtains

$$EI\, v = -(w_o/48)(2x^4 - 3Lx^3 + L^3 x)$$

for the span AB. The elastic curve for the span BC follows from the symmetry that exists around the support B.

Case (b):   Second-order differential equation solution.

$$EI\frac{d^2 v}{dx^2} = R_A x - \frac{w_o x^2}{2}$$

$$EI\frac{dv}{dx} = R_A \frac{x^2}{2} - \frac{w_o x^3}{6} + C_3$$

$$EI\, v = R_A \frac{x^3}{6} - \frac{w_o x^4}{24} + C_3 x + C_4$$

Boundary conditions:

$$v(0) = v_A = 0 = C_4$$

$$v'(L) = v_B' = 0 = R_A \frac{L^2}{2} - \frac{w_o L^3}{6} + C_3$$

$$v(L) = v_B = 0 = R_A \frac{L^3}{6} - \frac{w_o L^4}{24} + C_3 L$$

Once again solving the last two equations,

$$R_A = \frac{3 w_o L}{8} \quad \text{and} \quad C_3 = -\frac{w_o L^3}{48}$$

which, upon substitution into the equation for the elastic curve, leads to

$$EI\, v = -(w_o/48)(2x^4 - 3Lx^3 + L^3 x)$$

which is the same as the equation obtained earlier using the fourth-order differential equation.

11-8. REMARKS ON THE ELASTIC DEFLECTION OF BEAMS

The integration procedures discussed above for obtaining the elastic deflections of loaded beams are generally applicable. The reader must realize, however, that numerous problems with different loadings have been solved and are readily available.* Nearly all the tabulated solutions are made for simple loading conditions. Therefore in practice the deflections of beams subjected to several or complicated loading conditions are usually synthesized from the simpler loadings, using the principle of superposition. For example, the problem in Fig. 11-12 can be separated into three different cases as shown. The algebraic sum of the three separate deflections caused by the separate loads for the same point gives the total deflection.

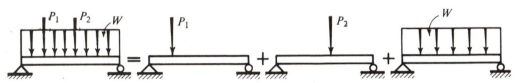

Fig. 11-12. Resolution of a complex problem into several simpler problems in computing deflections

Note that in applying superposition, the solution of Example 11-4 for a concentrated force P at an arbitrary location may be used for determining deflections of beams with the same boundary conditions for any loading. For distributed loads, P must be replaced by $q\,dx$ and integrated over the loaded range.

The procedures discussed above for determining elastic deflection of straight beams can be extended to structural systems consisting of several flexural members. For example, consider the simple frame shown in Fig. 11-13(a), for which the deflection of point C due to the applied force P is sought. The deflection of the vertical leg BC alone can be found by treating it as a cantilever fixed at B. However, due to the applied load, joint B deflects and rotates. This is determined by studying the behavior of the member AB.

A free-body diagram for the member AB is in Fig. 11-13(b). This member is seen to resist the axial force P and a moment $M_1 = Pa$. Usually the effect of the axial force P on deflections due to bending can be neglected.† The axial elongation of a member usually is also very small in comparison with the bending deflections. Therefore the problem here can be reduced to that of determining the deflection and rotation of B caused by an end moment M_1.

*See any civil or mechanical engineering handbook.

†Recall discussion in connection with Fig. 7-1 and see Chapter 13 on beam columns.

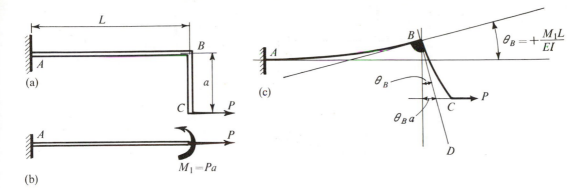

Fig. 11-13. A method of analyzing deflections of frames

This has been done in ·Example 11-2; from it the angle θ_B is noted on Fig. 11-13(c). By multiplying this angle θ_B by the length a of the vertical member, the deflection of point C due to rotation of joint B is found. The cantilever deflection of the member BC when treated alone is augmented by the amount $\theta_B a$. The vertical deflection of C is equal to the vertical deflection of point B.

In interpreting the shape of deformed structures such as shown in Fig. 11-13(c) it must be kept clearly in mind that the deformations are greatly exaggerated. In the small deformation theory discussed here, the cosines of all small angles such as θ_B are taken to equal unity. Both the deflections and the rotations of the elastic curve are small.

Beams with overhangs can also be analyzed conveniently using the concept of superposition in the manner described above. For example, the portion of a beam between the supports, as AB in Fig. 11-14(a), is isolated*

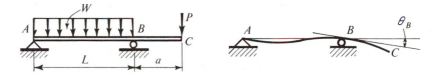

Fig. 11-14. A method of analyzing deflections of an overhang

and rotation of the tangent at B is found. The remainder of the problem is analogous to the case discussed before.

Approximations similar to those just discussed are also made in composite structures. In Fig. 11-15(a), for example, a simple beam rests on a rigid support at one end and on a yielding support with a spring constant† k at the

---

*The effect of the overhang on the beam segment AB must be included by introducing at the support B a bending moment $-Pa$.

†See Art. 7-9, where k is defined as a force necessary to deflect a spring by a unit distance. More generally, a beam or any other member may be considered as a "spring" and k may be defined in the same manner at the point of load application.

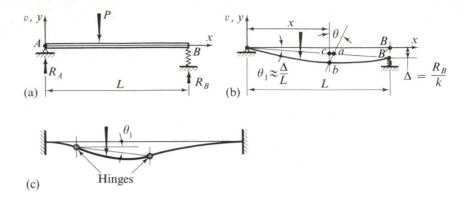

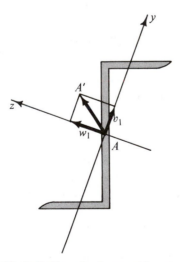

Fig. 11-15. Deflections in composite structure

other end. If R_B is the reaction at B, the support B settles $\Delta = R_B/k$, Fig. 11-15(b). A rigid beam would assume the direction of the line AB' making an angle $\theta_1 = \tan^{-1}(\Delta/L) \approx \Delta/L$ radians with the horizontal line. For an elastic beam, the elastic curve between A and B' may be found in the usual manner. However, since the ordinates such as ab, Fig. 11-15(b), make a very small angle θ with the vertical, $ab \approx cb$. Hence the deflection of a point such as b is very nearly $\theta_1 x + cb$. Deflections of beams in situations where hinges are introduced, Fig. 11-15(c), are treated similarly. For these, the tangent to the adjoining elastic curves is *not continuous* across a hinge.

*11-9. ELASTIC DEFLECTION OF BEAMS IN UNSYMMETRICAL BENDING

In the preceding discussion the deflection of beams was assumed to take place in one plane. More precisely, the foregoing theory applies to deflections of beams when the applied moments act around one of the principal axes of the cross section and when deflection takes place in a plane normal to such an axis. Unless this is the case, another moment develops tending to bend the beam around the other principal axis (see Art. 5-7). If unsymmetrical bending occurs, the elastic deflection problem can be solved by superposition. The elastic curve in the plane containing one of the principal axes is determined by considering only the effect of the components of forces acting parallel to this axis. The elastic curve in the plane containing the other principal axis is found similarly. A vectorial addition of the deflections so found at a particular point of a beam gives the total displacement of the beam at that point.

Fig. 11-16. Deflection of a beam subjected to unsymmetrical bending

For example, if a certain beam made of a Z section is subjected to unsymmetrical bending and the deflection at a particular point is v_1 in the y direction, and w_1 in the z direction, Fig. 11-16, the total deflection is $v_1 + w_1$, i.e., the distance AA'. Such deflections, without torsion, occur only if the applied forces pass through the shear center (see Art. 6-7).

*11-10. INELASTIC DEFLECTION OF BEAMS

All the preceding solutions for beam deflections apply only if the material behaves elastically. This limitation is the result of introducing Hooke's law into the strain-curvature relation, Eq. 11-5, to yield the moment-curvature equation, Eq. 11-6. The subsequent procedures of approximating the curvature as d^2v/dx^2 and the integration schemes have nothing to do with the material properties.

If attention is limited to statically determinate beams, the bending moments throughout a member can be determined regardless of the material properties of the beam. Then, if a relationship between the bending moment and curvature is available for a given cross section, the curvature diagram or function for the given beam can be established. Upon two successive integrations of the curvature relation, with adjustments for the boundary conditions, the inelastic deflection of a given beam can be found. This will be illustrated in the next two examples.

Superposition does not apply in inelastic problems since deflections are not linearly related to the applied forces. As a consequence of this, time-consuming trial-and-error solutions are often required to calculate deflections in the indeterminate beams. The bending moments depend on the reactions, and the latter depend on the nonlinear response of the beam to deformations. This will not be pursued in this book. An approach for the plastic strength analysis of statically indeterminate beams will be given, in the next chapter however.

EXAMPLE 11-7

Determine and plot the moment-curvature relationship for an elastic-ideally plastic rectangular beam.

SOLUTION

In a rectangular elastic-plastic beam at y_0, where the juncture of the elastic and plastic zones occurs, the linear strain $\varepsilon_x = \pm\varepsilon_{yp}$ see Fig. 5-17. Therefore, according to Eq. 11-5, with the curvature $1/\rho = \kappa$,

$$\frac{1}{\rho} = \kappa = -\frac{\varepsilon_{yp}}{y_0} \quad \text{and} \quad \kappa_{yp} = -\frac{\varepsilon_{yp}}{h/2}$$

where the last expression gives the curvature of the member at impending yielding when $y_o = h/2$. From the above relations

$$\frac{y_o}{h/2} = \frac{\kappa_{yp}}{\kappa}$$

On substituting this expression into Eq. 5-11, one obtains the required moment-curvature relationship:

$$M = M_p \left[1 - \frac{1}{3} \left(\frac{y_0}{h/2} \right)^2 \right] = \frac{3}{2} M_{yp} \left[1 - \frac{1}{3} \left(\frac{\kappa_{yp}}{\kappa} \right)^2 \right] \qquad (11\text{-}26)$$

This function is plotted in Fig. 11-17. Note how rapidly it approaches the asymptote. At curvature just double that of the impending yielding, eleven-twelfths or 91.6% of the ultimate plastic moment M_p is already reached. At this point the middle half of the beam remains elastic.

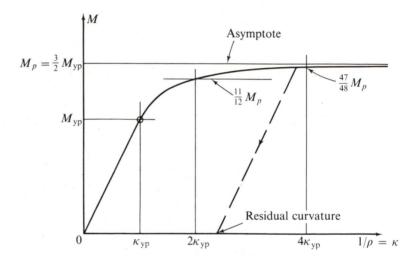

Fig. 11-17. Moment-curvature relation for a rectangular beam

On releasing an applied moment the beam rebounds elastically as shown in the figure. On this basis residual curvature can be determined.

The reader should recall that the ratio of M_p to M_{yp} varies for different cross sections.

EXAMPLE 11-8

A 3 in. wide mild-steel cantilever beam has the other dimensions as shown in Fig. 11-18(a). Determine the tip deflection caused by applying the two loads of 5 kips each. Assume $E = 30 \times 10^3$ ksi, and $\sigma_{yp} = \pm 40$ ksi.

SOLUTION

The moment diagram is in Fig. 11-18(b). From $\sigma_{max} = Mc/I$ it is found that the largest stress in the beam segment ab is 24.4 ksi, which indicates

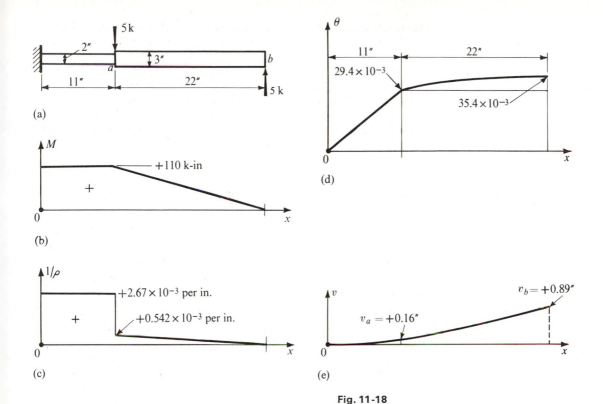

(a)

(b)

(c)

(d)

(e)

Fig. 11-18

elastic behavior. An analogous calculation for the shallow section of the beam gives a stress of 55 ksi, which is not possible as the material yields at 40 ksi.

A check of the ultimate capacity for the 2-in. deep section based on Eq. 5-9 gives

$$M_p = M_{ult} = \sigma_{yp}\frac{bh^2}{4} = \frac{40 \times 3 \times 2^2}{4} = 120 \text{ k-in.}$$

This calculation shows that although the beam yields partially, it can carry the applied moment. The applied moment is $\frac{11}{12}M_p$. According to the results found in the preceding example, this means that the curvature in the 2 in. deep section of the beam is twice that at the beginning of yielding. Therefore the curvature in the 11 in. segment of the beam adjoining the support is

$$\frac{1}{\rho} = 2\kappa_{yp} = 2\frac{\varepsilon_{yp}}{h/2} = 2\frac{\sigma_{yp}}{Eh/2} = \frac{2 \times 40}{30 \times 10^3 \times 1} = 2.67 \times 10^{-3} \text{ per in.}$$

The maximum curvature for segment ab is

$$\frac{1}{\rho} = \frac{M_{max}}{EI} = \frac{\sigma_{max}}{Ec} = \frac{24.4}{30 \times 10^3 \times 1.5} = 0.542 \times 10^{-3} \text{ per in.}$$

These data on curvatures are plotted in Fig. 11-18(c). On integrating this twice with $\theta(0) = 0$ and $v(0) = 0$, the deflected curve, Fig. 11-18(e), is obtained. The tip deflection is 0.89 in. upward.

If the applied loads were released, the beam would rebound elastically. This amounts to 0.64 in. at the tip, and a residual tip deflection of 0.25 in. would remain. The residual curvature would be confined to the 2 in. deep segment of the beam.

If the end load were applied alone, the beam would collapse. Super-position cannot be used to solve this problem.

*11-11. INTRODUCTION TO THE MOMENT-AREA METHOD

In numerous engineering applications where deflections of beams must be determined, the loading is complex, and the cross-sectional areas of the beam vary. This is the usual situation in shafts of machines, where gradual or stepwise variations in the shaft diameter are made to accommodate rotors, bearings, collars, retainers, etc. Likewise, haunched or tapered beams are frequently employed in aircraft as well as in bridge construction. By inter-preting semigraphically the mathematical operations of solving the governing differential equation, an effective procedure for obtaining deflections in com-plicated situations has been developed. Using this alternative procedure, one finds that problems with load discontinuities and arbitrary variations of inertia of the cross-sectional area of a beam cause no complications and require only a little more arithmetical work for their solution. The solution of such problems is the objective in the following articles on the moment-area method.*

The method to be developed usually is used to obtain only the displace-ment and rotation at a single point on a beam. It may be used to determine the equation of the elastic curve, but no advantage is gained in comparison with the direct solution of the differential equation. Often, however, it is the deflection and/or the angular rotation of the elastic curve, or both, at a particular point of a beam that are of greatest interest in the solution of prac-tical problems.

The method of moment areas is just an alternative method for solving the deflection problem. It possesses the same approximations and limitations discussed earlier in connection with the solution of the differential equation of the elastic curve. By applying it, one determines only the deflection due to the flexure of the beam; deflection due to shear is neglected. Here the applica-tion of the method will be limited to statically determinate beams. Statically indeterminate situations will be considered in the next chapter.

*The development of the moment-area method for finding deflections of beams is due to Charles E. Greene, of the Univeristy of Michigan, who taught it to his classes in 1873. Somewhat earlier, in 1868, Otto Mohr, of Dresden, Germany, developed a similar method which appears to have been unknown to Professor Greene.

*11-12. DERIVATION OF THE MOMENT-AREA THEOREMS

The necessary theorems are based on the geometry of the elastic curve and the associated $M/(EI)$ diagram. Boundary conditions do not enter into the derivation of the theorems since the theorems are based only on the interpretation of definite integrals. As will be shown later, further geometrical considerations are necessary to solve a complete problem.

For deriving the theorems, Eq. 11-10, $d^2v/dx^2 = M/(EI)$, can be rewritten in the following alternative forms:

$$\frac{d^2v}{dx^2} = \frac{d}{dx}\left(\frac{dv}{dx}\right) = \frac{d\theta}{dx} = \frac{M}{EI} \quad \text{or} \quad d\theta = \frac{M}{EI}dx \qquad (11\text{-}27)$$

From Fig. 11-19(a), the quantity $[M/(EI)]dx$ corresponds to an infinitesimal area of the $M/(EI)$ diagram. According to Eq. 11-27 this area is equal to the change in angle between two adjoining tangents. The contribution of an angle change in one element to the deformation of the elastic curve is shown in Fig. 11-19(b).

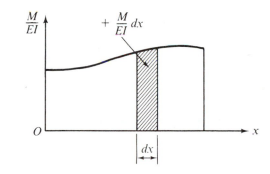

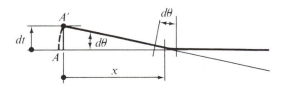

Fig. 11-19. Interpretation of a small angle change in an element

If the small angle change $d\theta$ for an element is multiplied by a distance x from an arbitrary origin to the same element, a vertical distance dt is obtained, see Fig. 11-19(b). As only small deflections are considered, no distinction between the arc AA' and the vertical distance dt need be made. Based on this geometrical reasoning, one has

$$dt = x\, d\theta = \frac{M}{EI}x\, dx \qquad (11\text{-}28)$$

Formally integrating Eqs. 11-27 and 11-28 between any two points such as A and B on a beam (see Fig. 11-20), yields the two moment-area theorems. The first is

$$\int_A^B d\theta = \theta_B - \theta_A = \Delta\theta_{BA} = \int_A^B \frac{M}{EI} dx \qquad (11\text{-}29)$$

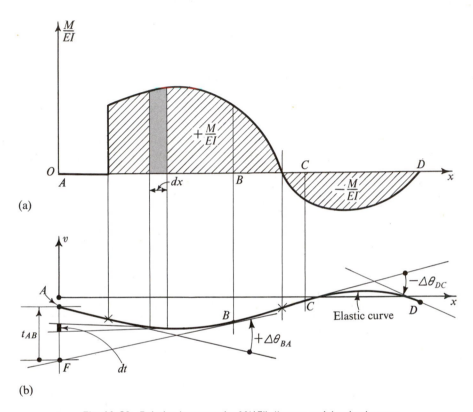

Fig. 11-20. Relation between the $M/(EI)$ diagram and the elastic curve

This states that the change in angle measured in radians between the two tangents at any two points A and B on the elastic curve is equal to the $M/(EI)$ area bounded by the ordinates through A and B. Therefore, if the slope of the elastic curve at one point, as at A, is known, the slope at another point on the right, as at B, can be determined:

$$\theta_B = \theta_A + \Delta\theta_{BA} \qquad (11\text{-}30)$$

The first theorem shows that a numerical evaluation of the $M/(EI)$ area bounded between the ordinates through any two points on the elastic curve gives the angular rotation between the corresponding tangents. In performing this summation, areas corresponding to the positive bending moments are taken positive, and those corresponding to the negative moments are taken negative. If the sum of the areas between any two points such as A and B is positive, the tangent on the right rotates in the counterclockwise direction; if negative, the tangent on the right rotates in a clockwise direction (see Fig. 11-20(b)). If the net area is zero, the two tangents are parallel.

The quantity dt in Fig. 11-20(b) is due to the effect of curvature of an element. By summing this effect for all elements from A to B, the vertical distance AF is obtained. Geometrically this distance represents the displacement or deviation of a point A from a tangent to the elastic curve at B. Henceforth, it will be termed the *tangential deviation* of a point A from a tangent at B and will be designated t_{AB}. The foregoing, in mathematical form, gives the second moment-area theorem:

$$t_{AB} = \int_A^B d\theta\, x = \int_A^B \frac{M}{EI} x\, dx \qquad (11\text{-}31)$$

This states that the tangential deviation of a point A on the elastic curve from a tangent through another point B also on the elastic curve is equal to the statical (or first) moment of the bounded section of the $M/(EI)$ diagram around a vertical line through A. In most cases, the tangential deviation is not in itself the desired deflection of a beam.

Using the definition of the center of gravity of an area, one may for convenience restate Eq. 11-31 in numerical applications in a simpler form as

$$t_{AB} = \Phi \bar{x} \qquad (11\text{-}32)$$

where Φ is the total area of the $M/(EI)$ diagram between the two points considered and $\bar{x}$ is the horizontal distance to the centroid of this area from A.

By analogous reasoning the deviation of a point B from a tangent through A is

$$t_{BA} = \Phi \bar{x}_1 \qquad (11\text{-}33)$$

where the same $M/(EI)$ area is used, but $\bar{x}_1$ is measured from the vertical line through point B, Fig. 11-21. Note carefully the order of the subscript letters for t in these two equations. The point whose deviation is being determined is written first.

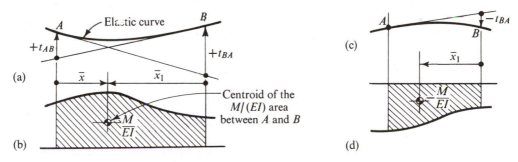

Fig. 11-21. The meaning of signs for tangential deviation

In the above equations, the distances $\bar{x}$ or $\bar{x}_1$ are always taken positive, and as E and I are also positive quantities, the sign of the tangential deviation depends on the sign of the bending moments. A positive value for the tangential deviation indicates that a given point lies above a tangent to the elastic curve drawn through the other point and vice versa, Fig. 11-21.

The above two theorems are applicable between any two points on a continuous elastic curve of any beam for any loading. They apply between and beyond the reactions for overhanging and continuous beams. However, it must be emphasized that only relative rotation of the tangents and only tangential deviations are obtained directly. A further consideration of the geometry of the elastic curve at the supports to include the boundary conditions is necessary in every case to determine deflections. This will be illustrated in the examples that follow.

In applying the moment-area method a carefully prepared sketch of the elastic curve is always necessary. Since no deflection is possible at a pinned or a roller support, the elastic curve is drawn passing through such supports. At a fixed support, neither displacement nor rotation of the tangent to the elastic curve is permitted, so the elastic curve must be drawn tangent to the direction of the unloaded axis of the beam. In preparing a sketch of the elastic curve in the above manner, it is customary to exaggerate the anticipated deflections. On such a sketch the deflection of a point on a beam is usually referred to as being above or below its initial position, without much emphasis on the signs. To aid in the application of the method, useful properties of areas enclosed by curves and centroids are assembled in Table 2 of the Appendix.

EXAMPLE 11-9

Consider an aluminum cantilever beam 1.6 m long with a 10 kN force applied 0.4 m from the free end, as shown in Fig. 11-22(a). For a distance

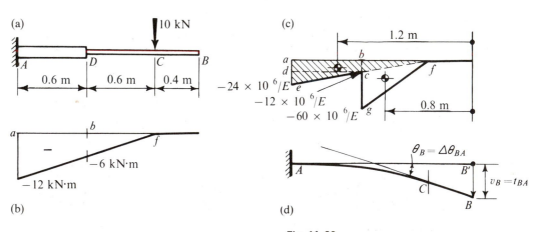

Fig. 11-22

of 0.6 m from the fixed end, the beam is of greater depth than it is beyond, having $I_1 = 5 \times 10^{-4}$ m⁴. For the remaining 1 m of the beam, $I_2 = 1 \times 10^{-4}$ m⁴. Find the deflection and the angular rotation of the free end. Neglect the weight of the beam, and assume E for aluminum at 70×10^9 N/m².

SOLUTION

The bending-moment diagram is in Fig. 11-22(b). By dividing all ordinates of the M diagram by EI, the $M/(EI)$ diagram in Fig. 11-22(c) is obtained.* Two ordinates appear at point D. One, $-12 \times 10^6/E$, is applicable just to the left of D; the other, $-60 \times 10^6/E$, applies just to the right of D. Since the bending moment is negative from A to C, the elastic curve throughout this distance is concave down, Fig. 11-22(d). At the fixed support A, the elastic curve must start out tangent to the initial direction AB' of the unloaded beam. The unloaded straight segment CB of the beam is tangent to the elastic curve at C.

After the foregoing preparatory steps, from the geometry of the sketch of the elastic curve it may be seen that the distance BB' represents the desired deflection of the free end. However, BB' is also the tangential deviation of the point B from the tangent at A. Therefore the second moment-area theorem may be used to obtain t_{BA}, which in this special case represents the deflection of the free end. Also, from the geometry of the elastic curve it is seen that the angle included between the lines BC and AB' is the angular rotation of the segment CB. This angle is the same as the one included between the tangents to the elastic curve at the points A and B, and the first moment-area theorem may be used to compute this quantity.

It is convenient to extend the line ec in Fig. 11-22(c) to the point f for computing the area of the $M/(EI)$ diagram. This gives two triangles, the areas of which are easily calculated.†
The area of triangle afe:

$$\Phi_1 = -\frac{1}{2}\frac{(1.2)(24 \times 10^6)}{E} = -14.4 \times \frac{10^6}{E}$$

The area of triangle fcg:

$$\Phi_2 = -\frac{1}{2}\frac{(0.6)(48 \times 10^6)}{E} = -14.4 \times \frac{10^6}{E}$$

$$\theta_B = \Delta\theta_{BA} = \int_A^B \frac{M}{EI}\,dx = \Phi_1 + \Phi_2 = -\frac{28.8 \times 10^6}{70 \times 10^9}$$

$$= -0.411 \times 10^{-3} \text{ radians}$$

$$v_B = t_{BA} = \Phi_1\bar{x}_1 + \Phi_2\bar{x}_2 = \frac{-14.4 \times 10^6}{E}(1.2) + \frac{-14.4 \times 10^6}{E}(0.8)$$

$$= -0.411 \times 10^{-3} \text{ m} = -0.411 \text{ mm}$$

*In constructing this diagram kilonewtons (kN) are changed to newtons (N).

†A little ingenuity in such cases saves arithmetical work. Of course it is perfectly correct in this example to use two triangular areas dce and bfg, and a rectangle $abcd$.

Note the numerical smallness of both the above values. The negative sign of $\Delta\theta$ indicates clockwise rotation of the tangent at B in relation to the tangent at A. The negative sign of t_{BA} means that point B is below a tangent through A.

EXAMPLE 11-10

Find the deflection due to the concentrated force P applied as shown in Fig. 11-23(a) at the center of a simply supported beam. The flexural rigidity EI is constant.

SOLUTION

The bending-moment diagram is in Fig. 11-23(b). Since EI is constant, the $M/(EI)$ diagram need not be made, as the areas of the bending-moment diagram divided by EI give the necessary quantities for use in the moment-area theorems. The elastic curve is in Fig. 11-23(c). It is concave upward throughout its length as the bending moments are positive. This curve must pass through the points of the support at A and B.

It is apparent from the sketch of the elastic curve that the desired quantity is represented by the distance CC'. Moreover, from purely geo-metrical or kinematic considerations, $CC' = C'C'' - C''C$, where the distance $C''C$ is measured from a tangent to the elastic curve passing through the point of support B. However, since the deviation of a support point from a tangent to the elastic curve at the other support may always be computed by the second moment-area theorem, a distance such as $C'C''$ may be found by proportion from the geometry of the figure. In this case, t_{AB} follows by

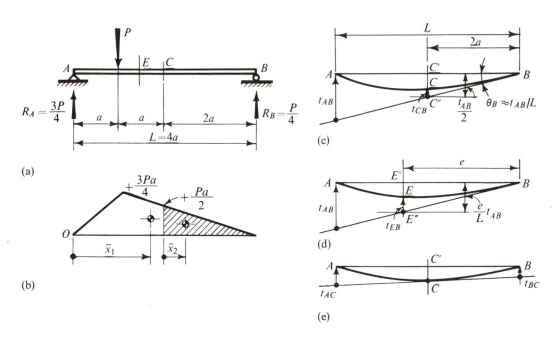

(a)

(b)

(c)

(d)

(e)

Fig. 11-23

taking the whole $M/(EI)$ area between A and B and multiplying it* by its $\bar{x}$ measured from a vertical through A, hence $C'C'' = \frac{1}{2}t_{AB}$. By another application of the second theorem, t_{CB}, which is equal to $C''C$, is determined. For this case, the $M/(EI)$ area is shaded in Fig. 11-23(b), and, for it, the $\bar{x}$ is measured from C. Since the right reaction is $P/4$ and the distance $CB = 2a$, the maximum ordinate for the shaded triangle is $+Pa/2$.

$$v_C = C'C'' - C''C = (t_{AB}/2) - t_{CB}$$

$$t_{AB} = \Phi_1 \bar{x}_1 = \frac{1}{EI}\left(\frac{4a}{2}\frac{3Pa}{4}\right)\frac{(a + 4a)}{3} = +\frac{5Pa^3}{2EI}$$

$$t_{CB} = \Phi_2 \bar{x}_2 = \frac{1}{EI}\left(\frac{2a}{2}\frac{Pa}{2}\right)\frac{(2a)}{3} = +\frac{Pa^3}{3EI}$$

$$v_C = \frac{t_{AB}}{2} - t_{CB} = \frac{5Pa^3}{4EI} - \frac{Pa^3}{3EI} = \frac{11Pa^3}{12EI}$$

The positive signs of t_{AB} and t_{CB} indicate that points A and C lie above the tangent through B. As may be seen from Fig. 11-23(c), the deflection at the center of the beam is in a downward direction.

The slope of the elastic curve at C can be found from the slope at one of the ends and from Eq. 11-30. For point B on the right

$$\theta_B = \theta_C + \Delta\theta_{BC} \quad \text{or} \quad \theta_C = \theta_B - \Delta\theta_{BC}$$

$$\theta_C = \frac{t_{AB}}{L} - \Phi_2 = \frac{5Pa^2}{8EI} - \frac{Pa^2}{2EI} = \frac{Pa^2}{8EI} \quad \text{(radians counterclockwise)}$$

The above procedure for finding the deflection of a point on the elastic curve is generally applicable. For example, if the deflection of the point E, Fig. 11-23(d), at a distance e from B is wanted, the solution may be formulated as

$$v_E = E'E'' - E''E = (e/L)t_{AB} - t_{EB}$$

By locating the point E at a variable distance x from one of the supports, the equation of the elastic curve be obtained.

To simplify the arithmetical work, some care in selecting the tangent at a support must be exercised. Thus, although $v_C = t_{BA}/2 - t_{CA}$ (not shown in the diagram), this solution would involve the use of the unshaded portion of the bending-moment diagram to obtain t_{CA}, which is more tedious.

ALTERNATE SOLUTION

The solution of the foregoing problem may be based on a different geometrical concept. This is illustrated in Fig. 11-23(e), where a tangent to the elastic curve is drawn at C. Then, since the distances AC and CB are equal,

$$v_C = CC' = (t_{AC} + t_{BC})/2$$

*See Table 2 of the Appendix for the centroid of the whole triangular area. Alternatively, by treating the whole $M/(EI)$ area as two triangles,

$$t_{AB} = \frac{1}{EI}\left(\frac{a}{2}\frac{3Pa}{4}\right)\frac{2a}{3} + \frac{1}{EI}\left(\frac{3a}{2}\frac{3Pa}{4}\right)\left(a + \frac{3a}{3}\right) = +\frac{5Pa^3}{2EI}$$

i.e., the distance CC' is an average of t_{AC} and t_{BC}. The tangential deviation t_{AC} is obtained by taking the first moment of the unshaded $M/(EI)$ area in Fig. 11-23(b) about A, and t_{BC} is given by the first moment of the shaded $M/(EI)$ area about B. The numerical details of this solution are left for completion by the reader. This procedure is usually longer than the first.

Note particularly that if the elastic curve is not symmetrical, the tangent at the center of the beam is not horizontal.

EXAMPLE 11-11

For a prismatic beam loaded as in the preceding example, find the maximum deflection caused by the applied force P, Fig. 11-24(a).

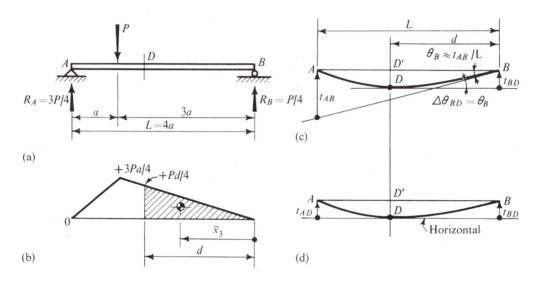

Fig. 11-24

SOLUTION

The bending-moment diagram and the elastic curve are in Figs. 11-24(b) and (c), respectively. The elastic curve is concave up throughout its length, and the maximum deflection occurs where the tangent to the elastic curve is horizontal. This point of tangency is designated in the figure by D and is located by the unknown horizontal distance d measured from the right support B. Then, by drawing a tangent to the elastic curve through point B at the support, one sees that $\Delta\theta_{BD} = \theta_B$ since the line passing through the supports is horizontal. However, the slope θ_B of the elastic curve at B may be determined by obtaining t_{AB} and dividing it by the length of the span. On the other hand, by using the first moment-area theorem, $\Delta\theta_{BD}$ may be expressed in terms of the shaded area in Fig. 11-24(b). Equating $\Delta\theta_{BD}$ to θ_B and solving for d locates the horizontal tangent at D. Then, again from geometrical considerations, it is seen that the maximum deflection represented by DD' is equal to the tangential deviation of B from a horizontal tangent through D, i.e., t_{BD}.

$$t_{AB} = \Phi_1 \bar{x}_1 = +\frac{5Pa^3}{2EI} \qquad \text{(see Example 11-10)}$$

$$\theta_B = \frac{t_{AB}}{L} = \frac{t_{AB}}{4a} = \frac{5Pa^2}{8EI}$$

$$\Delta\theta_{BD} = \frac{1}{EI}\left(\frac{d}{2}\frac{Pd}{4}\right) = \frac{Pd^2}{8EI} \qquad \text{(area between } D \text{ and } B\text{)}$$

Since $\theta_B = \theta_D + \Delta\theta_{BD}$ and it is required that $\theta_D = 0$,

$$\Delta\theta_{BD} = \theta_B, \qquad \frac{Pd^2}{8EI} = \frac{5Pa^2}{8EI} \qquad \text{hence } d = \sqrt{5}\,a$$

$$v_{\max} = v_D = DD' = t_{BD} = \Phi_3 \bar{x}_3$$

$$= \frac{1}{EI}\left(\frac{d}{2}\frac{Pd}{4}\right)\frac{2d}{3} = \frac{5\sqrt{5}\,Pa^3}{12EI} = \frac{11.2Pa^3}{12EI}$$

After the distance d is found, the maximum deflection may also be obtained, as $v_{\max} = t_{AD}$, or $v_{\max} = (d/L)t_{AB} - t_{DB}$ (not shown). Also note that using the condition $t_{AD} = t_{BD}$, Fig. 11-24(d), an equation may be set up for d.

It should be apparent from the above solution that it is easier to calculate the deflection at the center of the beam, which was illustrated in Example 11-10, than to determine the maximum deflection. Yet, by examining the end results, one sees that numerically the two deflections differ little: $v_{\text{center}} = 11Pa^3/(12EI)$ as opposed to $v_{\max} = 11.2Pa^3/(12EI)$. For this reason, in many practical problems of simply supported beams where all the applied forces act in the same direction, it is often sufficiently accurate to calculate the deflection at the center instead of attempting to obtain the true maximum.

EXAMPLE 11-12

In a simply supported beam, find the maximum deflection and rotation of the elastic curve at the ends caused by the application of a uniformly distributed load of w_o lb per foot, Fig. 11-25(a). The flexural rigidity EI is constant.

SOLUTION

The bending-moment diagram is in Fig. 11-25(b). As established in Example 4-6, it is a second-degree parabola with a maximum value at the vertex of $w_o L^2/8$. The elastic curve passing through the points of the support A and B is shown in Fig. 11-25(c).

In this case, the $M/(EI)$ diagram is symmetrical about a vertical line passing through the center. Therefore the elastic curve must be symmetrical, and the tangent to this curve at the center of the beam is horizontal. From the figure, it is seen that $\Delta\theta_{BC}$ is equal to θ_B, and the rotation of the end B is equal to one-half the area* of the whole $M/(EI)$ diagram. The distance CC' is the desired deflection, and from the geometry of the figure it is seen to be equal to t_{BC} (or t_{AC}, not shown).

---

*See Table 2 of the Appendix for a formula giving an area enclosed by a parabola as well as for $\bar{x}$.

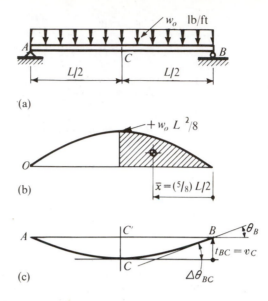

(a)

(b)

$\bar{x} = (^5/_8)\, L/2$

(c)

θ_B

$t_{BC} = v_C$

$\Delta\theta_{BC}$

Fig. 11-25

$$\Phi = \frac{1}{EI}\left(\frac{2}{3}\,\frac{L}{2}\,\frac{w_o L^2}{8}\right) = \frac{w_o L^3}{24EI}$$

$$\theta_B = \Delta\theta_{BC} = \Phi = +\frac{w_o L^3}{24EI}$$

$$v_C = v_{\max} = t_{BC} = \Phi\bar{x} = \frac{w_o L^3}{24EI}\,\frac{5L}{16} = \frac{5w_o L^4}{384EI}$$

The value of the deflection agrees with Eq. 11-22, which expresses the same quantity derived by the integration method. Since the point B is above the tangent through C, the sign of v_C is positive.

EXAMPLE 11-13

Find the deflection of the free end A of the beam shown in Fig. 11-26(a) caused by the applied forces. The EI is constant.

SOLUTION

The bending-moment diagram for the applied forces is in Fig. 11-26(b). The bending moment changes sign at $a/2$ from the left support. At this point an inflection in the elastic curve takes place. Corresponding to the positive moment, the curve is concave up, and vice versa. The elastic curve is so drawn and passes over the supports at B and C, Fig. 11-26(c). To begin, the inclination of the tangent to the elastic curve at the support B is determined by finding t_{CB} as the statical moment of the areas with the proper signs of the $M/(EI)$ diagram between the verticals through C and B about C.

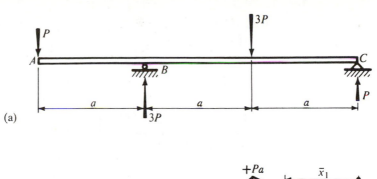

(a)

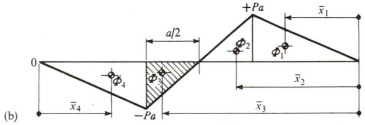

(b)

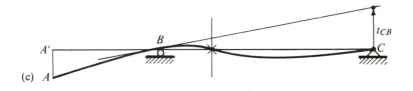

(c)

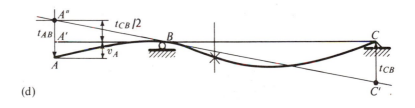

(d)

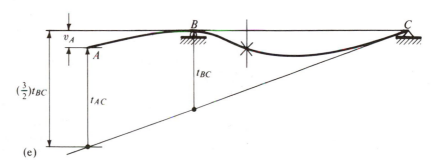

(e)

Fig. 11-26.

$$t_{CB} = \Phi_1 \bar{x}_1 + \Phi_2 \bar{x}_2 + \Phi_3 \bar{x}_3$$

$$= \frac{1}{EI}\left[\frac{a}{2}(+Pa)\frac{2a}{3} + \frac{1}{2}\frac{a}{2}(+Pa)\left(a + \frac{1}{3}\frac{a}{2}\right) + \frac{1}{2}\frac{a}{2}(-Pa)\left(\frac{3a}{2} + \frac{2}{3}\frac{a}{2}\right)\right]$$

$$= +\frac{Pa^3}{6EI}$$

The positive sign of t_{CB} indicates that the point C is above the tangent through B. Hence a corrected sketch of the elastic curve is made, Fig. 11-26(d), where it is seen that the deflection sought is given by the distance AA' and is equal to $AA'' - A'A''$. Further, since the triangles $A'A''B$ and $CC'B$ are similar, the distance $A'A'' = t_{CB}/2$. On the other hand, the distance AA'' is the deviation of the point A from the tangent to the elastic curve at the support B. Hence

$$v_A = AA' = AA'' - A'A'' = t_{AB} - t_{CB}/2$$

$$t_{AB} = \frac{1}{EI}(\Phi_4 \bar{x}_4) = \frac{1}{EI}\left[\frac{a}{2}(-Pa)\frac{2a}{3}\right] = -\frac{Pa^3}{3EI}$$

where the negative sign means that point A is below the tangent through B. This sign is not used henceforth, as the geometry of the elastic curve indicates the direction of the actual displacements. Thus the deflection of point A below the line passing through the supports is

$$v_A = \frac{Pa^3}{3EI} - \frac{1}{2}\frac{Pa^3}{6EI} = \frac{Pa^3}{4EI}$$

This example illustrates the necessity of watching the signs of the quantities computed in the applications of the moment-area method, although usually less difficulty is encountered than in the above example. For instance, if the deflection of the end A is established by first finding the rotation of the elastic curve at C, no ambiguity in the direction of tangents occurs. This scheme of analysis is shown in Fig. 11-26(e), where $v_A = \frac{3}{2}t_{BC} - t_{AC}$.

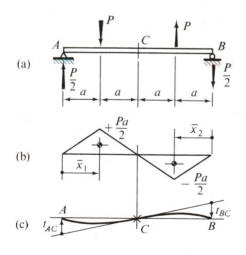

Fig. 11-27

EXAMPLE 11-14

A simple beam supports two equal and opposite forces P at the quarter points, Fig. 11-27(a). Find the deflection of the beam at the middle of the span. The EI is constant.

SOLUTION

The bending-moment diagram and elastic curve with a tangent at C are shown in Figs. 11-27(b) and (c), respectively. Then, since the statical moments of the positive and negative areas of the bending-moment diagram around A and B, respectively, are numerically equal, i.e., $|t_{AC}| = |t_{BC}|$, the deflection of the beam at the center of the span is *zero*. The elastic curve in this case is *anti*symmetrical. Noting this, much work may be avoided in obtaining the deflections at the *center of* *the span.* The deflection of any other point on the elastic curve must be found in the usual manner.

The foregoing examples illustrate the manner in which the moment-area method can be used to obtain the deflection of any statically determinate beam. No matter how complex the $M/(EI)$ diagrams may become, the above procedures are applicable. In practice, any $M/(EI)$ diagram whatsoever may be approximated by a number of rectangles and triangles. It is also possible to introduce concentrated angle changes at hinges to account for discontinuities in the directions of the tangents at such points. The magnitudes of the concentrations can be found from kinematic requirements.*

For complicated loading conditions, deflections of elastic beams determined by the moment-area method are often best found by superposition. In this manner the areas of the separate $M/(EI)$ diagrams may become simple geometrical shapes. In the next chapter, superposition will be used in solving statically indeterminate problems.

The method described here can be used very effectively in determining the inelastic deflection of beams, providing the $M/(EI)$ diagrams are replaced by the appropriate curvature diagrams.

PROBLEMS FOR SOLUTION

11-1. A 2 mm by 6 mm steel strip 3142 mm long is clamped at one end as shown in the figure. What is the required end moment to force the strip to touch the wall? What would the maximum stress be when the strip is in the bent condition? Let $E = 200$ GPa. *Ans:* 200 MPa.

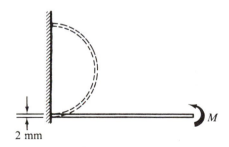

2 mm

PROB. 11 – 1

11-2. A round aluminum bar of 6 mm diameter is bent into a circular ring having a mean diameter of 3 m. What is the maximum stress in the bar? $E = 7 \times 10^7$ kN/m². *Ans:* 140 MPa.

11-3. What will the radius of curvature of a W 8 × 17 beam bent around the *X-X* axis be if the stress in the extreme fibers is 20 ksi? $E = 29 \times 10^6$ psi. *Ans:* 483 ft.

11-4. If the equation of the elastic curve for a simply supported beam of length L having a constant EI is $v = (k/360EI)(-3x^5 + 10x^3L^2 - 7xL^4)$, how is the beam loaded?

11-5. An elastic beam of constant EI and of length L has the deflected shape $EI\,v(x) = M_0(x^3 - x^2L)/(4L)$. (a) Determine the loading

---

*For a systematic treatment of more complex problems, see for example A. C. Scordelis and C. M. Smith, "An Analytical Procedure for Calculating Truss Displacements," *Proceedings of the American Society of Civil Engineers*, paper no. 732, July 1955, vol. 732.

and support conditions. (b) Plot the shear and moment diagrams for the beam and sketch the deflected shape.

11-6.   Rework Example 11-2 by taking the origin of the coordinate system at the free end. *Ans:* $v = (M_1/2EI)(x^2 - 2Lx + L^2)$.

*11-7.   Using the exact differential equation, Eq. 11-8, show that the equation of the elastic curve in Example 11-2 is $x^2 + (v - \rho)^2 = \rho^2$, where ρ is a constant. (*Hint: let* $dv/dx = \tan\theta$ and integrate.)

11-8 through 11-19.   For the beams loaded as shown in the figures, determine the equations of the elastic curves using either the second-order differential equation or the fourth-order equation, as directed. In all cases, *EI* is constant. *Ans:* See Table 11 in the Appendix for some of the answers, and

Prob. 11-10:
$$EIv = -W(x^5 - 5L^4x + 4L^5)/(60L^2)$$
Prob. 11-11:
$$EIv = kL^3x^2/6 - kL^2x^3/12 + kx^5/120$$
Prob. 11-12:
$$EIv = -k(L/\pi)^4 \sin \pi x/L$$
Prob. 11-13:
$$EIv = kL^3x^3/72 - kx^6/360 - kL^5x/90$$
Prob. 11-17:
$$EIv = A[Lx^3/24 - x^5/(60L) - 5L^3x/192]$$

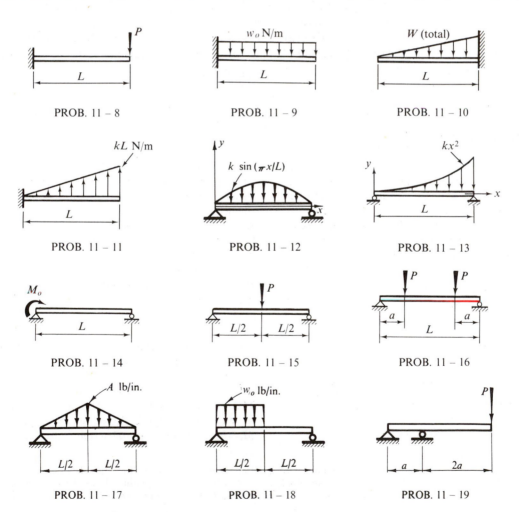

PROB. 11 – 8   PROB. 11 – 9   PROB. 11 – 10

PROB. 11 – 11   PROB. 11 – 12   PROB. 11 – 13

PROB. 11 – 14   PROB. 11 – 15   PROB. 11 – 16

PROB. 11 – 17   PROB. 11 – 18   PROB. 11 – 19

11-20. If a cantilever is loaded as shown in Prob. 11-8 and its width *and* flexural strength are constant, see Fig. 10-18(d), what is the equation of the elastic curve? Neglect the increase in depth at the end of the beam for the shear force. *Ans:* $EI_o v = 2PL^2 x - \frac{4}{3} PL^{3/2} x^{3/2} - \frac{2}{3} PL^3$, where I_o is the moment of inertia of the cross-section at the fixed end.

11-21. If in Prob. 11-15 the cross-sectional area of the beam is constant, and the left half of the span is made of steel ($E = 30 \times 10^6$ psi) and the right half is made of aluminum ($E = 10 \times 10^6$ psi), determine the equation of the elastic curve.

11-22. Rework Example 11-4 using the procedure of Example 11-5.

11-23. Determine the equations of the elastic curve for the beam loaded as shown in Fig. 11-9.

11-24. Compute the maximum flexural stress and the maximum deflection for an S 15 × 42.9 simply supported beam spanning 20 ft loaded with a 10 kip concentrated downward force in the middle of the span and a uniformly distributed gravity load, including the weight of the beam, of 1 kip per ft. $E = 29 \times 10^6$ psi. Use the formulas given in the Appendix and the method of superposition. *Ans:* 20.1 ksi, −0.500 in.

11-25. Using a semigraphical procedure such as shown in Figs. 11-6 and 11-8, find the deflection of the beam at midspan, see figure. Let $EI = 23,200$ lb-in.² Neglect the effect of the axial force on the deflection. *Ans:* 0.0905 in.

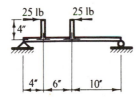

PROB. 11 – 25

11-26. A W 10 × 49 cantilever ($I = 113.6 \times 10^6$ mm⁴) supports a concentrated force as shown in the figure. Calculate the deflections caused by the applied force (a) at the applied force, (b) at the free end. Use the formulas given in the Appendix. $E = 200 \times 10^6$ kN/m². (*Hint:* the slope of the beam between the applied force and the free end is constant.) *Ans:* 7.49 mm.

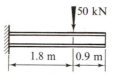

PROB. 11 – 26

11-27. A W 8 × 40 beam is loaded as shown in the figure. Using the equations given in the Appendix and the method of superposition, calculate the deflection at the center of the span. $E = 29 \times 10^6$ psi.

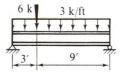

PROB. 11 – 27

11-28. The maximum deflection for a simple beam spanning 24 ft and carrying a uniformly distributed load of 100 kips (total) is limited to 0.5 in. (a) Specify the required W beam. (b) Determine the maximum fiber stress. Let $E = 29 \times 10^6$ psi.

11-29 through 11-31. Establish the equations of the elastic curves for the statically indeterminate beams shown in the figures. All beams have a constant EI.

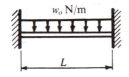

PROB. 11 – 29

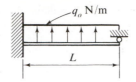

PROB. 11 – 30

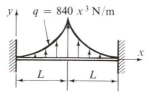

PROB. 11 – 31

11-32. Rework Prob. 11-30 if the applied load $q = \sin x$. Let $L = \pi$. *Ans:* $EIv = \sin x - x^3/(2\pi^2) + 3\,x^2/(2\pi) - x$.

11-33. For the beam loaded as shown in the figure, after obtaining the equation of the elastic curve, (a) determine the ratio of the moment at the built-in end to the applied moment M_a; (b) determine the rotation of the free end. The EI is constant. *Ans:* $-\frac{1}{2}$; $-M_aL/(4EI)$.

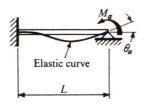

PROB. 11 – 33

11-34. One end of an elastic beam is displaced an amount Δ relative to the other end as shown in the figure. No rotation of the ends is permitted to occur. Derive the expression for the elastic

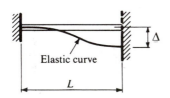

PROB. 11 – 34

curve, and plot the shear and moment diagrams. The EI is constant.

11-35. A S 3 × 5.7 cantilever is 42 in. long. What is its "spring constant" for the downward application of a force at the end? $E = 29 \times 10^6$ psi. *Ans:* 2.94 k/in.

11-36. A W 10 × 33 beam is loaded as shown in the figure. A 1 in. round steel rod 8 ft long provides the anchorage for the right-hand end. Determine the deflection of the left end caused by the applied concentrated force. $E = 29 \times 10^6$ psi.

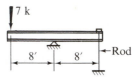

PROB. 11 – 36

11-37. The data for a beam loaded as shown in Fig. 11-14 are: $w_0 = 30$ kN/m, $P = 25$ kN, $L = 3$ m, and $a = 1.2$ m. If the beam is made from a W 8 × 20 section ($I = 29 \times 10^6$ mm⁴), what is the deflection of the free end C caused by the applied loads? $E = 200 \times 10^6$ kN/m².

11-38. A small frame is loaded as shown in the figure. The cross-sectional areas of the vertical and the horizontal parts of the frame are equal. Determine the vertical deflection of the end at the applied force. Neglect the effect of the axial force on deflection and neglect shortening of the column due to the same force. Express the result in terms of P, L, E and I. (*Hint:* synthesize the solution from the results of Example 11-2 and the appropriate formula from the Appendix.) *Ans:* 13 $PL^3/(192EI)$.

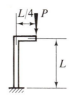

PROB. 11 – 38

11-39. Determine the deflection of the end of an S 8 × 18.4 cantilever $6\frac{1}{2}$ ft long due to an inclined force of 1,200 lb applied at the end as shown in the figure. Let $E = 29 \times 10^3$ ksi. *Ans:* 0.454 in., 0.111 in.

PROB. 11 – 39

11-40. The M/EI diagram for a simple beam is as shown in the figure. Draw a qualitative elastic curve and compute t_{AB} and $\Delta\theta_{AB}$. *Ans:* 878,000/EI in., 6,000/EI.

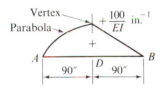

PROB. 11 – 40

11-41. If a positive M/EI diagram for a simple beam were as shown in the figure, what would the statical moment of the shaded area around a line through X give? Illustrate on a sketch.

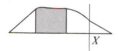

PROB. 11 – 41

11-42. An aluminum bar 40 mm wide and 30 mm deep is supported and loaded as shown in the figure. Calculate the angles (in radians) between the tangent to the elastic curve at A and the corresponding tangents at B, C, and D caused by

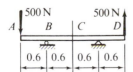

PROB. 11 – 42

the applied loads. Let $E = 7 \times 10^7$ kN/m². All dimensions shown in the figure are in meters.

11-43. If an additional downward force of 10 kN is applied at D to the beam of Example 11-9, what will be the deflection of the end B?

11-44. Using the moment-area method, find the maximum deflection for the beam of Prob. 11-10.

11-45 through 11-59. Using the moment-area method for the members loaded as shown in the figures, determine the deflection and the slope of the elastic curve at points A. Specify whether the deflection is upward or downward. If the size of the member is not given, assume that EI is constant over the entire length. Neglect the weight of the members. Wherever needed, assume $E = 29 \times 10^6$ psi or 200 GPa. Wherever the answer is expressed in terms of EI, no adjustment for units need be made. *Ans:* The deflection sought is noted in the lower right-hand corner.

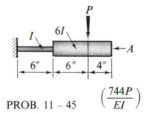

PROB. 11 – 45 $\left(\dfrac{744P}{EI}\right)$

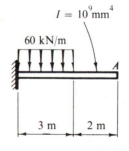

PROB. 11 – 46 (5.74)

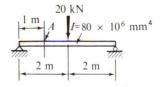

PROB. 11 – 47

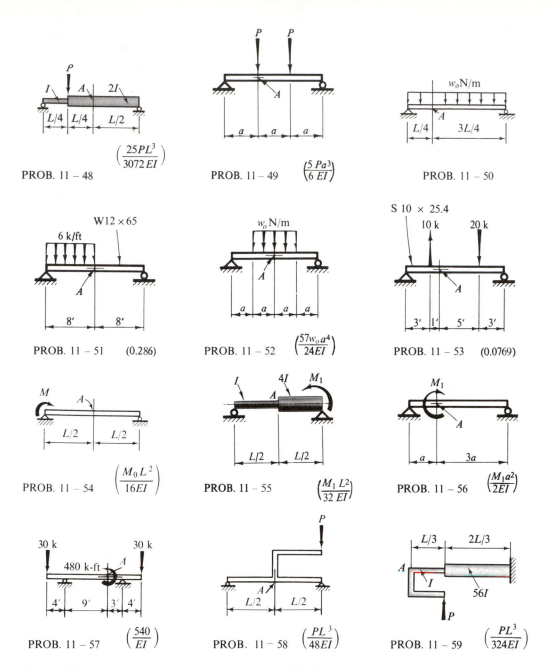

$$\left(\frac{25PL^3}{3072\,EI}\right)$$

PROB. 11 – 48

PROB. 11 – 49 $\quad \left(\dfrac{5\,Pa^3}{6\,EI}\right)$

PROB. 11 – 50

PROB. 11 – 51 (0.286)

PROB. 11 – 52 $\quad \left(\dfrac{57w_oa^4}{24EI}\right)$

PROB. 11 – 53 (0.0769)

PROB. 11 – 54 $\quad \left(\dfrac{M_0L^2}{16EI}\right)$

PROB. 11 – 55 $\quad \left(\dfrac{M_1L^2}{32\,EI}\right)$

PROB. 11 – 56 $\quad \left(\dfrac{M_1a^2}{2EI}\right)$

PROB. 11 – 57 $\quad \left(\dfrac{540}{EI}\right)$

PROB. 11 – 58 $\quad \left(\dfrac{PL^3}{48EI}\right)$

PROB. 11 – 59 $\quad \left(\dfrac{PL^3}{324EI}\right)$

11-60. Using the moment-area method, establish the equation of the elastic curve for the beam of Prob. 11-8.

11-61. Using the moment-area method, establish the equation of the elastic curve for the beam of Prob. 11-15.

*11-62. Using the moment-area method, establish the equation of the elastic curve for the beam of Prob. 11-50.

11-63. Using the moment-area method, determine the maximum deflection for the beam of Prob. 11-54.

11-64. A beam of variable I is loaded as shown in the figure. (a) Determine the deflection at the center of the span caused by the two concentrated forces. (b) Locate the point at which the maximum deflection occurs. (The magnitude of this deflection need not be found.) Assume that E and I are given and have appropriate units. *Ans:* (a) $123/EI$, (b) 1.85 m from B.

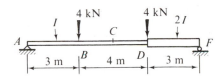

PROB. 11 – 64

11-65. Determine the maximum deflection of the beam in Prob. 11-49.

11-66. Determine the maximum deflection of the beam in Prob. 11-52.

11-67 through 11-72. Using the moment-area method for the members loaded as shown in the figures, determine the location and magnitude of the maximum deflection *between* the supports. Disregard the effect of axial forces on deflections wherever this condition occurs. Other conditions are the same as in Probs. 11-45 through 11-59. *Ans:* Lower right-hand corner of a figure.

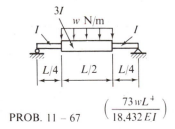

PROB. 11 – 67 $\left(\dfrac{73\,wL^4}{18{,}432\,EI}\right)$

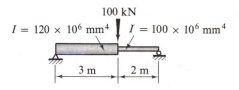

PROB. 11 – 68

$EI = 1{,}800$ lb-in.^{2}

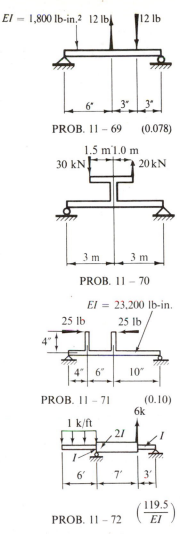

PROB. 11 – 69   (0.078)

PROB. 11 – 70

$EI = 23{,}200$ lb-in.

PROB. 11 – 71   (0.10)

PROB. 11 – 72 $\left(\dfrac{119.5}{EI}\right)$

11-73. For the beam of Prob. 11-69, determine (a) the deflection at the center of the span and (b) the deflection at the point of inflection. *Ans:* (a) 0.075 in., (b) 0.047 in.

11-74. Rework Example 11-13 using the procedure shown in Fig. 11-26(e).

11-75 through 11-80. Using the moment-area method, determine the deflection and slope of the elastic curve of the overhang at A for the beams loaded as shown. Other conditions are the same as in Probs. 11-67 through 11-72. *Ans:* Deflection in lower right-hand corner of each figure.

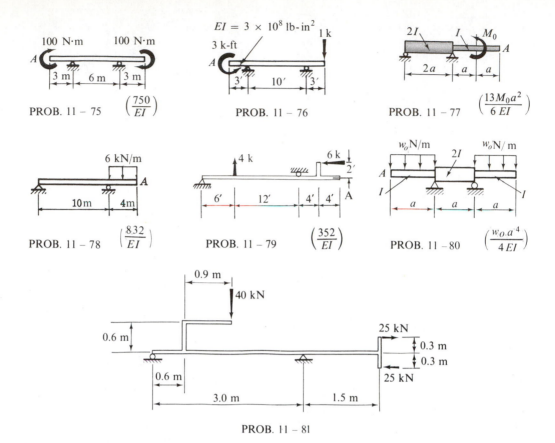

PROB. 11 – 75 $\left(\dfrac{750}{EI}\right)$

PROB. 11 – 76

PROB. 11 – 77 $\left(\dfrac{13M_0a^2}{6\,EI}\right)$

PROB. 11 – 78 $\left(\dfrac{8.32}{EI}\right)$

PROB. 11 – 79 $\left(\dfrac{352}{EI}\right)$

PROB. 11 – 80 $\left(\dfrac{w_0 \cdot a^4}{4\,EI}\right)$

PROB. 11 – 81

11-81. Find the location and magnitude of the maximum *upward* deflection of the beam shown in the figure. Express results in terms of EI. *Ans:* 0.824 m, 5.09/(EI).

11-82. Determine the maximum *upward* deflection for the overhang of a beam loaded as shown in the figure. The E and I are constant. *Part Ans:* At 2.93 ft from right support.

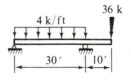

PROB. 11 – 82

11-83. A structure is formed by joining a simple beam to a cantilever beam with a hinge as shown in the figure. If a 10 N force is applied at the center of the simple span, determine the deflection caused by this force at A. Use the moment-area method. The EI is constant over the entire structure.

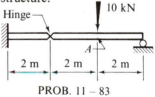

PROB. 11 – 83

11-84. Two beams, both of the same, constant, flexural rigidity EI, are connected by a hinge as shown in the figure. Find the deflection at A caused by the applied force P. Use the moment-area method. *Ans:* $Pa^3/(3EI)$.

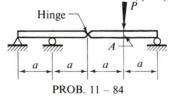

PROB. 11 – 84

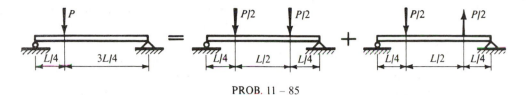

PROB. 11 – 85

11-85. Determine the deflection at the midspan of a simple beam, loaded as shown in the figure, by solving the two separate problems indicated and superposing the results. Use the moment-area method. The EI is constant. *Ans:* $11PL^3/(768EI)$.

11-86. Consider the structure loaded with the force P as shown in Fig. 11-13(a). What uniformly distributed load w_o must be applied to the horizontal beam AB such that the horizontal displacement of point C will return to its unloaded condition?

11-87. A steel beam is supported and loaded as shown in the figure. The force $P = 10$ kips, and $M_A = 1,200$ k-in. With the horizontal line AC as a reference, determine the slope and the vertical deflection at end C. For this beam $E = 30 \times 10^6$ psi and $I = 1,000$ in.[4] The spring at B, when isolated, requires a force of 20 kips to shorten it 1 in. Use any method you wish to compute the required slope and deflection.

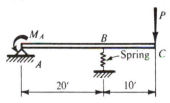

PROB. 11 – 87

11-88. A 1-in. square bar of a linearly elastic-plastic material is to be wrapped around a round

PROB. 11 – 88

mandrel as shown in the figure. (a) What mandrel diameter D is required so that the outer thirds of the cross sections become plastic, i.e., the elastic core is $\frac{1}{3}$ in. deep by 1 in. wide? Assume the material to be initially stress-free with $\sigma_{yp} = 40$ ksi. Let $E = 30 \times 10^6$ psi. The pitch of the helix angle is so small that only the bending of the bar in a plane need be considered. (b) What will be the diameter of the coil after the release of the forces used in forming it? Stated alternatively, determine the coil diameter after the elastic spring-back. *Ans:* (b) 482 in.

11-89. A rectangular, weightless, simple beam of linearly elastic-plastic material is loaded in the middle by the force P as shown in the figure. (a) Determine the magnitude of the force P that would cause the plastic zone to penetrate $\frac{1}{4}$ of the beam depth from each side. (b) For the above loading condition sketch the moment-curvature diagram clearly showing it for the plastic zone.

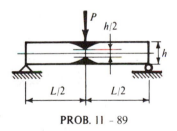

PROB. 11 – 89

11-90 and 11-91. Using the moment-area method, determine the deflection at the center of the span for the beams loaded as shown in the figures. In Prob. 11-91, $I_1 = 10 \times 10^6$ mm⁴, $I_2 = 20 \times 10^6$ mm⁴, $E = 10 \times 10^7$ kN/m². *Ans:* Prob. 11-90; $4Pa^3/(3EI)$.

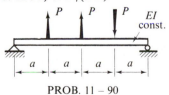

PROB. 11 – 90

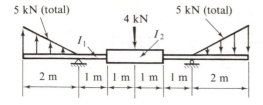

PROB. 11 – 91

11-92. Beam *AB* is subjected to an end moment at *A* and an unknown concentrated moment M_C as shown in the figure. Using the moment-area method, determine the magnitude of the bending moment M_C so that the deflection at point *B* will be equal to zero. The *EI* is constant. *Ans:* 23.5 kN·m.

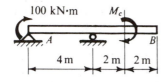

PROB. 11 – 92

11-93. The beam shown in the figure has a constant $EI = 3,600 \times 10^6$ lb-in.² Determine the distance *a* such that the deflection at *A* would be 0.25 in. if the end were subjected to a concentrated moment $M_0 = 15$ k-ft.

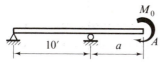

PROB. 11 – 93

11-94. A light pointer is attached only at *A* to a 6-in.-by-6-in. (actual) wooden beam as shown in the figure. Determine the position of the end of the pointer after a concentrated force of 1,200 lb is applied. Let $E = 1.2 \times 10^6$ psi. *Ans:* 0.036 in.

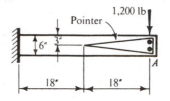

PROB. 11 – 94

11-95. What must the ratio of the loads *W* and *P* be for the beam loaded as shown in the figure so that the elastic curve shall be horizontal at the supports? The *EI* is constant. *Ans:* $\frac{3}{8}$.

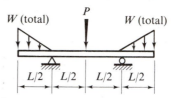

PROB. 11 – 95

11-96. The beam *ABCD* is initially horizontal. A load *P* is then applied at *C* as shown in the figure. It is desired to place a vertical force at *B* to bring the position of the beam at *B* back to the original level *ABCD*. What force is required at *B*? *Ans:* 7P/8.

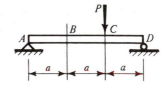

PROB. 11 – 96

12 Statically Indeterminate Problems

12-1. INTRODUCTION

The simplest problems in the mechanics of materials are externally statically determinate. In such cases the reactions and the internal system of stress resultants at a section can be determined from statics without considering deformations. In the preceding chapter, only on introducing the differential equations for the deflection of beams did methods for the analysis of statically indeterminate elastic beams become possible. In this chapter, procedures for the solution of statically indeterminate problems will be extended to include additional situations.

In the next two articles of this chapter, procedures for the analysis of statically indeterminate systems, which are applicable to both linear and nonlinear material response, will be discussed. It will be shown how the equations of static equilibrium can be supplemented by additional equations based on considerations of the geometry of deformation. The additional equations required will be formulated using displacement compatibility conditions. In the inelastic analysis of the indeterminate systems, such procedures can become very complex.

The principle of superposition is then used to obtain general methods, which are very effective for the solution of highly indeterminate problems involving linear elastic materials. These are commonly referred to as the *force method* (or flexibility method)* and the *displacement method* (or stiffness method). Further, a method of analysis is also developed for indeterminate beams by applying the moment-area technique in conjunction with the force method. Some attention is given to the *three-moment equation*, which is a useful recurrence formula for the analysis of continuous elastic beams.

A procedure for determining the limit or the ultimate carrying capacity of determinate as well as indeterminate beams made of ductile materials is treated at the end of this chapter.

*Also called the *method of consistent deformations*.

12-2. A GENERAL APPROACH

In all statically indeterminate problems the equations of static equilibrium remain valid. These equations are necessary but not sufficient to solve the indeterminate problems. The supplementary equations are established from considerations of the geometry of deformation. In structural systems, of physical necessity, certain elements or parts must deflect together, twist together, expand together, etc., or remain stationary. Formulating such observations quantitatively provides one with the required additional equations. For example, a statement of a common displacement of several members of a joint can give the required relation. Such kinematic equations are independent of the mechanical properties of materials and thus are not limited to the linear elastic response.

The necessary procedures for determining the linear deformation of axially loaded rods, the angular twist of shafts, and the deflection of beams were developed earlier. Here, except for designating the forces acting on such members as unknowns by some appropriate algebraic symbols, the same procedures apply. As before, the smallness of the deformations in comparison with the linear dimensions of the body is tacitly assumed.

Several examples illustrating the method of supplementing the equilibrium equations with the displacement relationships follow.

EXAMPLE 12-1

A stepped bar is built in at both ends onto immovable supports, Fig. 12-1(a). The left part of the bar has a cross-sectional area A_1; the area of the right part is A_2. (a) If the material of the bar is elastic with an elastic modulus E, what are the reactions R_1 and R_2 caused by the application of an axial force P at the point of discontinuity of the section? (b) If $A_1 = 6 \times 10^{-4}$ m^2, $A_2 = 12 \times 10^{-4}$ m^2, $a = 0.75$ m, $b = 0.50$ m, and the material is linearly elastic-perfectly plastic as shown in Fig. 12-1(d), determine the displacement u_1 of the step as a function of the applied force P. Let $E = 200 \times 10^9$ Pa.

SOLUTION

Case (a). The point on the rod where the force P is applied deflects the same amount whether the right or the left part of the bar is considered. By separating the bar at P, the two free-body diagrams in Figs. 12-1(b) and (c) are obtained. The left part of the rod is subjected throughout its length to a tensile force R_1 and elongates an amount u_1. The right part contracts u_2 under the action of a compressive force R_2. Of physical necessity, the absolute values of the two deflections must be the same:

From statics: $\qquad\qquad R_1 + R_2 = P$
From geometry:* $\qquad\qquad |u_1| = |u_2|$

---

*By considering elongation of the bar positive and contraction negative, alternatively one has $u_1 + u_2 = 0$, which means that the deformation of the bar from end to end is zero.

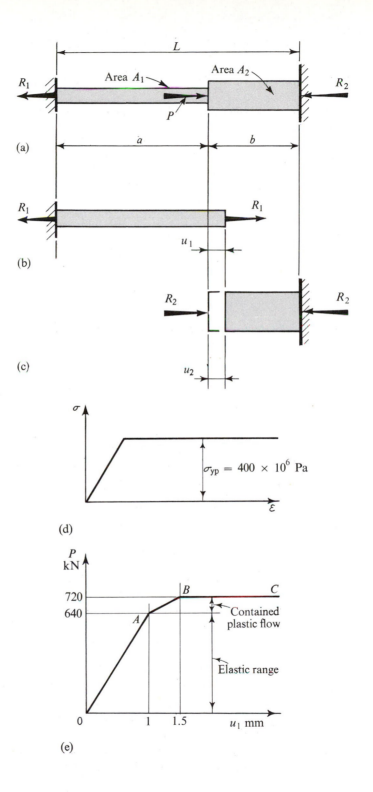

(a)

(b)

(c)

(d)

(e)

Fig. 12-1

405

On applying Eq. 2-4, $\Delta = u = PL/(AE)$, the above relation yields

$$\frac{R_1 a}{A_1 E} = \frac{R_2 b}{A_2 E}$$

Solving the two equations simultaneously gives

$$R_1 = \frac{P}{1 + aA_2/(bA_1)} \quad \text{and} \quad R_2 = \frac{P}{1 + bA_1/(aA_2)} \quad (12\text{-}1)$$

Case (b). By direct substitution of the given data into Eq. 12-1,

$$R_1 = \frac{P}{1 + 0.75(12 \times 10^{-4})/[0.5(6 \times 10^{-4})]} = \frac{P}{4} \quad \text{and} \quad R_2 = \frac{3P}{4}$$

Hence the normal stresses are

$$\sigma_1 = R_1/A_1 = 10^4 P/24$$

and
$$\sigma_2 = R_2/A_2 = 10^4 P/16 \quad \text{(compression)}$$

As $|\sigma_2| > \sigma_1$, the load at impending yield is found by setting $|\sigma_2| = 400 \times 10^6$ Pa. At this load the right part of the bar just reaches yield, and the strain attains the magnitude of $\varepsilon_{yp} = \sigma_{yp}/E$. Therefore

$$P_{yp} = 16 \times 10^{-4}\sigma_{yp} = 640 \times 10^3 \text{ N} = 640 \text{ kN}$$

and
$$|u_2| = |u_1| = \varepsilon_{yp}b = 0.001 \text{ m} = 1 \text{ mm}$$

These quantities locate point A in Fig. 12-1(e).

On increasing P above 640 kN, the right part of the bar continues to yield carrying a compressive force $R_2 = \sigma_{yp}A_2 = 480$ kN. At the point of impending yield for the whole bar, the left part just reaches yield. This occurs when $R_1 = \sigma_{yp}A_1 = 240$ kN and the strain in the left part just reaches $\varepsilon_{yp} = \sigma_{yp}/E$. Therefore

$$P = R_1 + R_2 = 720 \text{ kN} \quad \text{and} \quad u_1 = \varepsilon_{yp}a = 0.001\,5 \text{ m} = 1.5 \text{ mm}$$

These quantities locate point B in Fig. 12-1(e). Beyond this point the plastic flow is uncontained, and $P = 720$ kN is the ultimate or limit load of the rod.

Note the simplicity of calculating the limit load, which, however, provides no information on the deflection characteristics of the system. In general, plastic limit analysis is simpler than elastic analysis, which in turn is simpler than tracing the elastic-plastic load-deflection relationship.

Torsional problems may also be statically indeterminate. For example, if a shaft is built-in at both ends and a torque is applied at some intermediate point, the two torques at the ends cannot be found from statics alone. However, analogous to the above example, by separating the shaft at the applied torque and equating the angular rotation of one part of the shaft to the angular rotation of the other part, an auxiliary geometrical equation can be formulated and the problem solved.

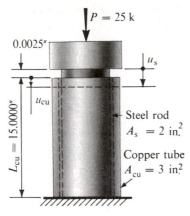

P = 25 k

0.0025"

u_s

u_{cu}

$L_{cu} = 15.0000''$

Steel rod
$A_s = 2$ in.^{2}

Copper tube
$A_{cu} = 3$ in.^{2}

Fig. 12-2

EXAMPLE 12-2

A steel rod 2 in.^{2} in cross-sectional area and 15.0025 in. long is loosely inserted into a copper tube as in Fig. 12-2. The copper tube has a cross-sectional area of 3 in.^{2} and is 15.0000 in. long. If an axial force $P = 25$ kips is applied through a rigid cap, what stresses will develop in the two materials? Assume that the elastic moduli of steel and copper are $E_s = 30 \times 10^6$ psi and $E_{cu} = 17 \times 10^6$ psi, respectively.

SOLUTION

If the applied force P is sufficiently large to close the small gap, a force P_s will be developed in the steel rod and a force P_{cu} in the copper tube. Moreover, upon loading, the steel rod will compress axially u_s, which is as much as the axial deformation u_{cu} of the copper tube plus the initial gap. Hence

From statics: $\qquad\qquad P_s + P_{cu} = 25{,}000$ lb

From geometry: $\qquad\qquad u_s = u_{cu} + 0.0025$

On applying Eq. 2-4, $\Delta = u = PL/(AE)$,

$$\frac{P_s L_s}{A_s E_s} = \frac{P_{cu} L_{cu}}{A_{cu} E_{cu}} + 0.0025$$

or
$$\frac{15.0025}{2(30)10^6} P_s - \frac{15}{3(17)10^6} P_{cu} = 0.0025$$

or
$$P_s - 1.176 P_{cu} = 10{,}000 \text{ lb}$$

Solving these equations simultaneously gives

$$P_{cu} = 6{,}900 \text{ lb} \qquad \text{and} \qquad P_s = 18{,}100 \text{ lb}$$

and dividing these forces by the respective cross-sectional areas gives

$$\sigma_{cu} = 6{,}900/3 = 2{,}300 \text{ psi} \qquad \text{and} \qquad \sigma_s = 18{,}100/2 = 9{,}050 \text{ psi}$$

If either of these stresses were above the proportional limit of its material or if the applied force were too small to close the gap, the above solution would not be valid. Also note that since the deformations considered are small, it is sufficiently accurate to use $L_s = L_{cu}$.

ALTERNATE SOLUTION

The force F necessary to close the gap may be found first, using Eq. 2-4. In developing this force the rod acts as a "spring" and resists a part of the applied force. The remaining force P' causes equal deflections u'_s and u'_{cu} in the two materials.

$$F = \frac{u A_s E_s}{L_s} = \frac{(0.0025)2(30)10^6}{15.0025} = 10{,}000 \text{ lb} = 10 \text{ kips}$$

$$P' = P - F = 25 - 10 = 15 \text{ kips}$$

Then if P'_s is the force resisted by the steel rod, in addition to the force F, and P'_{cu} is the force carried by the copper tube,

From statics: $$P'_s + P'_{cu} = P' = 15$$

From geometry: $$u'_s = u'_{cu} \quad \text{or} \quad \frac{P'_s L_s}{A_s E_s} = \frac{P'_{cu} L_{cu}}{A_{cu} E_{cu}}$$

$$\frac{15.0025}{2(30)10^6} P'_s = \frac{15}{3(17)10^6} P'_{cu}, \qquad P'_{cu} = \frac{17}{20} P'_s$$

By solving the two appropriate equations simultaneously, it is found that $P'_{cu} = 6.9$ kips and $P'_s = 8.1$ kips, or $P_s = P'_s + F = 18.1$ kips.

If $(\sigma_{yp})_s = 40$ ksi and $(\sigma_{yp})_{cu} = 10$ ksi, the limit load for this assembly can be determined as follows:

$$P_{ult} = (\sigma_{yp})_s A_s + (\sigma_{yp})_{cu} A_{cu} = 110 \text{ kips}$$

At the ultimate load both materials yield, therefore the small discrepancy in the initial lengths of the parts is of no consequence.

EXAMPLE 12-3

Three bars of elastic-perfectly plastic material are symmetrically arranged in a plane to form the system shown in Fig. 12-3(a). Investigate the load-deflection characteristics of joint C. The cross-sectional area A of each bar is the same.

SOLUTION

A free-body diagram of joint C is shown in Fig. 12-3(b), from which, for small deformations, an equation of equilibrium is obtained as

$$F_1 + 2F_2 \cos \alpha = P \tag{12-2}$$

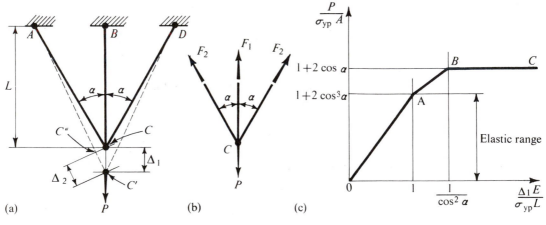

Fig. 12-3

This relation holds true regardless of the material response. The latter, however, depends on the attained magnitude of strain.

The deformed structure is shown in Fig. 12-3(a) by the dashed lines AC', BC', and DC'. The elongation of the bar BC' is Δ_1. The elongation of the inclined bars is Δ_2. For compatible displacements

$$\Delta_2 = \Delta_1 \cos \alpha \qquad (12\text{-}3)$$

where it is assumed that because of the smallness of the deformations being considered, the arc CC'' with the center at A can be replaced by a normal to AC'.

Equation 12-3 applies both in the elastic and the inelastic strain ranges provided the deformation remains small. For the elastic range, by noting that the inclined bars are $L/\cos \alpha$ long, and applying Eq. 2-4 one has

$$\frac{F_2(L/\cos \alpha)}{AE} = \frac{F_1 L}{AE} \cos \alpha, \qquad \text{i.e.,} \qquad F_2 = F_1 \cos^2 \alpha \qquad (12\text{-}4)$$

Solving Eqs. 12-2 and 12-4 simultaneously yields

$$F_1 = \frac{P}{1 + 2\cos^3 \alpha} \qquad \text{and} \qquad F_2 = \frac{P}{1 + 2\cos^3 \alpha} \cos^2 \alpha \qquad (12\text{-}5)$$

From this it is seen that the maximum force and stress occur in the vertical bar. At impending yield, $F_1 = \sigma_{yp} A$ and $\Delta_1 = (\sigma_{yp}/E)L$. With F_1 known, the maximum force P that can be carried elastically follows from Eqs. 12-2 and 12-4. This condition, with $P = \sigma_{yp} A(1 + 2\cos^3 \alpha)$, corresponds to point A in Fig. 12-3(c).

On increasing the force P above the impending yield in the vertical bar, the force $F_1 = \sigma_{yp} A$ remains constant, and Eq. 12-2 alone becomes sufficient for determining the force F_2. The inclined bars behave elastically until their stresses reaches σ_{yp}. This occurs when $F_2 = \sigma_{yp} A$. At impending yield in the inclined bars, using Eq. 12-2, $P = \sigma_{yp} A(1 + 2\cos \alpha)$. This condition corresponds to the limit load for the system.

At impending yield, $\Delta_2 = (\sigma_{yp}/E)(L/\cos \alpha)$. Hence, from Eq. 12-3, $\Delta_1 = (\sigma_{yp}/E)L/\cos^2 \alpha$. This value locates the abscissa for point B in Fig. 12-3(c). Beyond this point uncontained plastic flow takes place.

Note again the simplicity in finding the limit load as one works directly with a statically determinate system.

EXAMPLE 12-4

Two cantilever beams AD and BE of equal flexural rigidity $EI = 24 \times 10^6$ N·m², shown in Fig. 12-4(a), are interconnected by a taut steel rod DC ($E_s = 200 \times 10^9$ Pa). The rod DC is 5 m long and has a cross section of 3×10^{-4} m². Find the deflection of the cantilever AD at D due to a force $P = 50$ kN applied at E.

SOLUTION

By separating the structure at D, the two free-body diagrams in Figs. 12-4(b) and (c) are obtained. In both diagrams the same unknown force X is shown

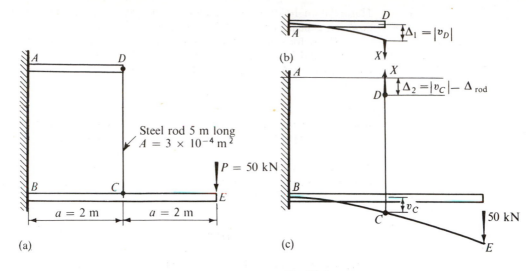

Fig. 12-4

acting (a condition of statics). The deflection of the point D is the same, whether the beam AD at D or the top of the rod DC is considered. The deflection Δ_1 of the point D in Fig. 12-4(b) is caused by X. The deflection Δ_2 of the point D on the rod is equal to the deflection v_C of the beam BE caused by the forces P and X less the elastic stretch of the rod DC.

From statics: $\qquad X_{\text{pull on } AD} = X_{\text{pull on } DC} = X$

From geometry: $\qquad \Delta_1 = \Delta_2 \qquad \text{or} \qquad |v_D| = |v_C| - \Delta_{\text{rod}}$

Beam deflections can be found using any one of the methods discussed in the preceding chapter. From Table 11 of the Appendix, in terms of the notation of this problem, one has

$$v_D = -\frac{Xa^3}{3EI} = -\frac{X \times 2^3}{3 \times 24 \times 10^6} = -\frac{10^{-6}X}{9} \qquad \text{(down)}$$

$$v_{C \text{ due to } X} = +\frac{10^{-6}X}{9} \qquad \text{(up)}$$

$$v_{C \text{ due to } P} = -\frac{P}{6EI}[2(2a)^3 - 3(2a)^2 a + a^3] = -\frac{10^{-3}(125)}{9} \qquad \text{(down)}$$

and using Eq. 2-4

$$\Delta_{\text{rod}} = \frac{XL_{CD}}{A_{CD}E} = \frac{X(5)}{3(10^{-4})200(10^9)} = \frac{10^{-6}X}{12}$$

Then

$$\frac{10^{-6}X}{9} = \frac{10^{-3}(125)}{9} - \frac{10^{-6}X}{9} - \frac{10^{-6}X}{12}$$

and $\qquad\qquad X = +45\,500 \text{ N} = 45.5 \text{ kN}$

and $\qquad v_D = -\dfrac{10^{-6}(45\,500)}{9} = -0.005\,05 \text{ m} = -5.05 \text{ mm} \qquad \text{(down)}$

Note particularly that the deflection of point C is caused by the applied force P at the end of the cantilever as well as by the unknown force X.

12-3. STRESSES CAUSED BY TEMPERATURE

It was possible to disregard the deformations caused by temperature in statically determinate systems since in such situations the members are free to expand or contract. However, in statically indeterminate systems, expansion or contraction of a body can be inhibited or entirely prevented in certain directions. This can cause significant stresses and must be investigated.

The free deformations caused by a change in temperature must be known for the determination of stresses caused by temperature. For a body of length L having a uniform thermal strain, the linear deformation Δ due to a change in temperature of δT degrees is

$$\Delta = \alpha(\delta T)L \qquad (12\text{-}6)$$

where α is the coefficient of thermal expansion.

The solution of indeterminate problems involving temperature deformations follows the concepts discussed in the preceding article. Two examples follow to illustrate some of the details of solution.

EXAMPLE 12-5

A copper tube 12 in. long and having a cross-sectional area of 3 in.^{2} is placed between two very rigid caps made of Invar,* Fig. 12-5(a). Four $\frac{3}{4}$ in. steel bolts are symmetrically arranged parallel to the axis of the tube and are lightly tightened. Find the stress in the tube if the temperature of the assembly is

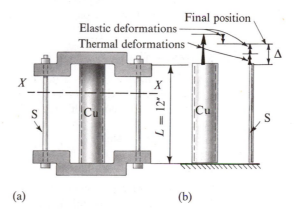

Fig. 12-5

*Invar is a steel alloy which at ordinary temperatures has an $\alpha \approx 0$ and for this reason is used in the best grades of surveyor's tapes and watch springs.

raised from 60°F to 160°F. Let $E_{cu} = 17 \times 10^6$ psi, $E_s = 30 \times 10^6$ psi, $\alpha_{cu} = 0.0000091$ per °F, and $\alpha_s = 0.0000065$ per °F.

SOLUTION

If the copper tube and the steel bolts were free to expand, the axial thermal elongations shown in Fig. 12-5(b) would take place. However, since the axial deformation of the tube must be the same as that of the bolts, the copper tube will be pushed back and the bolts will be pulled out so that the net deformations will be the same. Moreover, as can be established by considering a free body of the assembly above some arbitrary plane such as X-X in Fig. 12-5(a), the compressive force P_{cu} in the copper tube and the tensile force P_s in the steel bolts are equal. Hence

From statics: $\qquad\qquad P_{cu} = P_s = P$

From geometry: $\qquad\qquad \Delta_{cu} = \Delta_s = \Delta$

This kinematic relation, on the basis of Fig. 12-5(b) with the aid of Eqs. 12-6 and 2-4, becomes

$$\alpha_{cu}(\delta T)L_{cu} - \frac{P_{cu}L_{cu}}{A_{cu}E_{cu}} = \alpha_s(\delta T)L_s + \frac{P_sL_s}{A_sE_s}$$

or, since $\delta T = 100°$ and 0.442 in.^{2} is the cross section of one bolt,

$$(0.0000091)100 - \frac{P_{cu}}{3(17)10^6} = (0.0000065)100 + \frac{P_s}{4(0.442)30(10)^6}$$

Solving the two equations simultaneously, $P = 6,750$ lb. Therefore the stress in the copper tube is $\sigma_{cu} = 6,750/3 = 2,250$ psi.

The kinematic expression used above may also be set up on the basis of the following statement: The differential expansion of the two materials due to the change in temperature is accommodated by or is equal to the elastic deformations that take place in the two materials.

EXAMPLE 12-6

A steel bolt having a cross-sectional area $A_1 = 1$ in.^{2} is used to grip two steel washers of total thickness L, each having the cross-sectional area $A_2 = 9$ in.^{2}, Fig. 12-6(a). If the bolt in this assembly is tightened initially so that its stress is 20 ksi, what will be the final stress in this bolt after a force $P = 15$ kips is applied to the assembly?

SOLUTION

A free body corresponding to the initial conditions of the assembly is in Fig. 12-6(b), where I_t is the initial tensile force in the bolt and I_c is the initial compressive force in the washers. From statics, $I_t = I_c$. A free body of the assembly after the force P is applied is shown in Fig. 12-6(c), where X designates the increase in the tensile force in the bolt, and Y is the decrease in the compressive force on the washers due to P. As a result of these forces X and

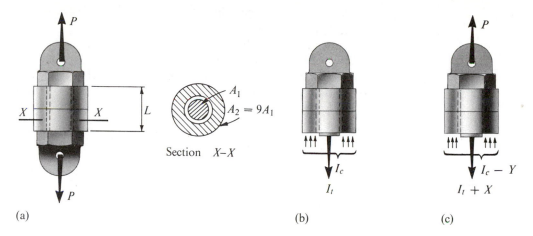

Fig. 12-6

Y, if the adjacent parts remain in contact, the bolt elongates the same amount as the washers expand elastically. Hence the final conditions are the following:

From statics: $\qquad\qquad P + (I_c - Y) = (I_t + X)$

or since $\qquad\qquad\qquad\qquad I_c = I_t$

$$X + Y = P$$

From geometry: $\qquad\qquad \Delta_{\text{bolt}} = \Delta_{\text{washers}}$

On applying Eq. 2-4,

$$\frac{XL}{A_1 E} = \frac{YL}{A_2 E}, \quad \text{i.e.,} \quad Y = \frac{A_2}{A_1} X$$

Solving the two equations simultaneously,

$$X = \frac{P}{1 + (A_2/A_1)} = \frac{P}{1 + 9} = 0.1P = 1{,}500\ \text{lb}$$

Therefore the increase of the stress in the bolt is $X/A_1 = 1{,}500$ psi, and the stress in the bolt after the application of the force P becomes 21,500 psi. This remarkable result indicates that most of the applied force is carried by decreasing the initial compressive force on the assembled washers since $Y = 0.9P$.

The solution is not valid if one of the materials ceases to behave elastically or if the applied force is such that the initial precompression of the assembled parts is destroyed.

Situations approximating the above idealized problem are found in many practical applications. A hot rivet used in the assembly of plates, upon cooling, develops within it enormous tensile stresses. Thoroughly tightened bolts, as in a head of an automobile engine or in a flange of a pressure vessel, have high initial tensile stresses; so do the steel tendons in a prestressed concrete beam. It is crucially important that on applying the working loads, only a small increase occur in the initial tensile stresses.

12-4. ANALYSIS OF INDETERMINATE SYSTEMS BASED ON SUPERPOSITION

Structural systems that experience only small deformations and are composed of linear elastic materials are linear structural systems. The principle of superposition is applicable for such structures and forms the basis for two of the most effective methods for the analysis of indeterminate systems.

In the first of these methods, a statically indeterminate system is reduced initially to one that is determinate by removing redundant (super-fluous) reactions for maintaining static equilibrium. Then these reactions are considered as externally applied loads, and their magnitudes are so adjusted as to satisfy the prescribed deformation conditions at their points of application. Once the redundant reactions are determined, the system is statically deter-minate and can be analyzed for strength or stiffness characteristics by the methods introduced earlier. This widely used method is commonly referred to as the *force method* (or the flexibility method).

In the second method, referred to as the *displacement method* (or the stiffness method), the joint displacements of a structure are treated as the unknowns. The system is first reduced to a series of members whose *joints* are imagined to be completely restrained from any movement. The joints are then released to an extent sufficient to satisfy the force equilibrium conditions at each joint. This method is extremely well-suited for computer coding and, hence, is even more widely used in practice than the force method, especially for the analysis of large-scale structures.

While some of the older classical methods continue to have limited utility, the force and displacement methods are the two modern approaches to the solution of indeterminate structural systems. These methods may also be formulated in a more general context by using the energy principles in structural mechanics. They are then applicable to both linear and nonlinear problems.

*12-5. FORCE METHOD

The first step in the analysis of structural systems using the force method is the determination of the degree of statical indeterminacy, which is the same as the number of redundant reactions as discussed in Art. 11-7. The redundant reactions are temporarily removed to obtain a statically determinate structure which is referred to as the *released* or *primary* structure. Then, since this structure is artificially reduced to statical determinacy, it is possible to find any desired deflection by the methods previously discussed. For example, by removing the redundant reaction* at A from the indeter-

---

*In the analysis of beams, the bending moments at the supports are often treated as redun-dants. In such cases, rotations of tangents at the supports are considered instead of deflec-tions.

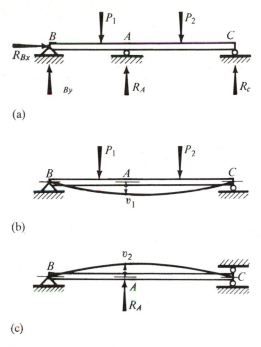

(a)

(b)

(c)

Fig. 12-7. Illustration of the superposition technique for the force method

minate beam shown in Fig. 12-7(a), the deflection v_1 at A, Fig. 12-7(b), can be found. By reapplying the removed redundant reaction R_A to the same determinate beam, Fig. 12-7(c), the deflection v_2, found as a function of R_A, can also be determined. Then, superposing (adding) the two deflections, since $v_1 + v_2 = 0$, one finds a solution for R_A. The effect of this superposition is that, under the action of the applied forces and the redundant reaction, point A does not actually move.

This procedure is often referred to as the *force method*, since the redundant forces are treated as the unknowns in this approach. Note especially that, after the redundant reactions are determined, the problem becomes statically determinate, and further analysis of stresses and deformations proceeds in the usual manner.

EXAMPLE 12-7

Rework Example 12-1 using the force method, Fig. 12-8(a).

SOLUTION

The rod is imagined cut at the left support. Then, in this determinate member, the cut end deflects an amount u_1 because of the applied force P, Fig. 12-8(b). Reapplication of the unknown force R_1, Fig. 12-8(c), causes a deflection u_2. Superposing these deflections in order to have no movement of the left end of

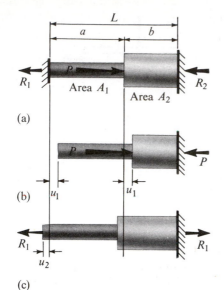

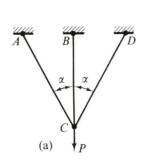

Fig. 12-8

the rod, as required by the conditions of the problem, gives

$$u_1 + u_2 = 0$$

Then, using Eq. 2-4, $\Delta = u = PL/(AE)$, and taking deflections to the right as positive,

$$\frac{Pb}{A_2E} - \left(\frac{R_1 a}{A_1 E} + \frac{R_1 b}{A_2 E}\right) = 0$$

and

$$R_1 = \frac{P}{1 + aA_2/(bA_1)}$$

which is the same result as that obtained in Example 12-1. The right reaction can be found from the condition of statics: $R_1 + R_2 = P$.

EXAMPLE 12-8

Rework Example 12-3 using the force method, Fig. 12-9(a).

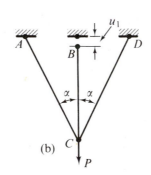

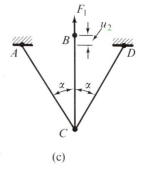

Fig. 12-9

SOLUTION

The bar BC is imagined to be cut at support B and the displacement u_1 with zero stress in the member BC is determined, Fig. 12-9(b). Then, the unknown force F_1 in the bar BC is applied, as shown in Fig. 12-9(c), and causes a deflection u_2. The superposition of these two results, together with the geometrical condition that the deflection of support B is zero, i.e., $u_1 + u_2 = 0$, enables the determination of the force F_1. Using Eq. 2-4, and arbitrarily taking downward deflections as positive,

$$\left[\frac{\left(\dfrac{P}{2\cos\alpha}\right)\left(\dfrac{L}{\cos\alpha}\right)}{AE}\frac{1}{\cos\alpha}\right] - \left[\frac{F_1 L}{AE} + \frac{\left(\dfrac{F_1}{2\cos\alpha}\right)\left(\dfrac{L}{\cos\alpha}\right)}{AE}\frac{1}{\cos\alpha}\right] = 0$$

and

$$F_1 = \frac{P}{1 + 2\cos^3\alpha}$$

which is the same result as that obtained in Example 12-3. The force F_2 in bars AC and DC can then be found from the conditions of statics: $F_1 + 2F_2 \cos \alpha = P$.

The force method is also very convenient in the analysis of elastic statically indeterminate torsional problems. In such cases, the torsion member is imagined cut at one of the supports, and the rotation of the released end is computed. The magnitude of the redundant torque is determined by making it of sufficient magnitude to restore the cut end of the member to its true position.

EXAMPLE 12-9

Plot shear and moment diagrams for a uniformly loaded beam fixed at one end and simply supported at the other, Fig. 12-10(a). The EI is constant.

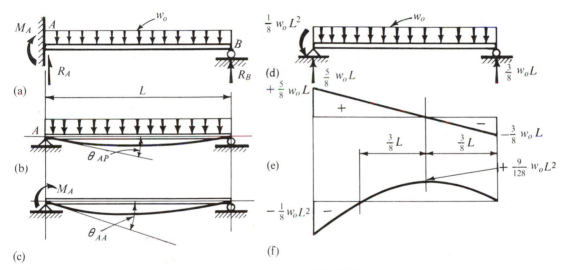

Fig. 12-10

SOLUTION

This beam is indeterminate to the first degree, but it can be reduced to determinacy by removing M_A as in Fig. 12-10(b). A positive moment M_A acting at A on the same structure is shown in Fig. 12-10(c). The rotations at A for the two determinate cases can be found from Table 11 in the Appendix. (Also see Example 11-3.) The requirement of zero rotation at A in the original structure provides the necessary equation for determining M_A.

$$\theta_{AP} = w_o L^3/(24EI) \quad \text{(clockwise)}$$

and

$$\theta_{AA} = M_A L/(3EI) \quad \text{(clockwise)}$$

$$\theta_A = \theta_{AP} + \theta_{AA} = 0$$

Taking clockwise rotations as positive,

$$\frac{w_o L^3}{24EI} + \frac{M_A L}{3EI} = 0 \quad \text{and} \quad M_A = -\frac{w_o L^2}{8}$$

The negative sign of the result indicates that M_A acts in the direction opposite to that assumed. Its correct sense is shown in Fig. 12-10(d).

The remainder of the problem can be solved with the aid of statics. Reactions, shear diagram, and moment diagram are in Figs. 12-10(d), (e), and (f), respectively.

This problem may also be analyzed by treating R_B as the redundant.

As stated earlier, the superposition procedure is applicable to linear systems that are indeterminate to a high degree. In such cases, it is essential to remember that the displacement of every point on a structure reduced to statical determinacy is affected by each reapplied redundant force. As an example, consider the beam in Fig. 12-11(a).

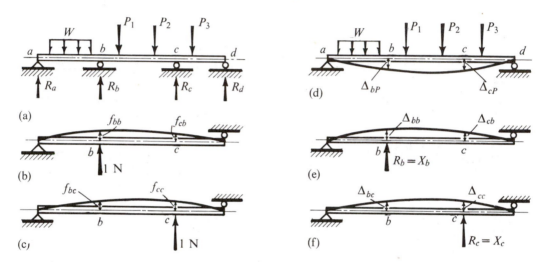

Fig. 12-11. Superposition method for a continuous beam

On removing the redundant reactions* R_b and R_c the beam becomes determinate, and the deflections at b and c can be computed, Fig. 12-11(d). These deflections are designated Δ_{bP} and Δ_{cP}, respectively, where the first subscript indicates the point where the deflection occurs, and the second the cause of the deflection. By reapplying R_b to the same beam, the deflections at b and c due to R_b at b can be found Fig. 12-11(e). These deflections are designated Δ_{bb} and Δ_{cb}, respectively. Similarly, Δ_{bc} and Δ_{cc}, due to R_c, can be established Fig. 12-11(f). Superposing the deflections at each support and setting the sum equal to zero, since points b and c actually do not deflect, one obtains two equations:

$$\Delta_b = \Delta_{bP} + \Delta_{bb} + \Delta_{bc} = 0$$
$$\Delta_c = \Delta_{cP} + \Delta_{cb} + \Delta_{cc} = 0$$

(12-7)

*The choice of redundant reactions is arbitrary.

These can be rewritten in a more meaningful form using *flexibility coefficients* f_{bb}, f_{bc}, f_{cb}, and f_{cc}, which are defined as the deflections shown in Figs. 12-11(b) and (c) due to unit forces applied in the direction of the redundants. Then, since a linear structural system is being considered, the deflection at point b due to the redundants can be expressed as

$$\Delta_{bb} = f_{bb}X_b \quad \text{and} \quad \Delta_{bc} = f_{bc}X_c \tag{12-8}$$

and similarly at point c as

$$\Delta_{cb} = f_{cb}X_b \quad \text{and} \quad \Delta_{cc} = f_{cc}X_c \tag{12-9}$$

where X_b and X_c are dimensionless factors which, on being multiplied by the respective unit forces, acquire the units of the redundant quantities. Using this notation, Eq. 12-7 becomes

$$f_{bb}X_b + f_{bc}X_c + \Delta_{bP} = 0$$
$$f_{cb}X_b + f_{cc}X_c + \Delta_{cP} = 0 \tag{12-10}$$

where the only unknown quantities are X_b and X_c; simultaneous solution of these equations constitutes the solution of the problem.

The canonical form of superposition equations* of the force method for a system with n unknown redundants reads:

$$f_{aa}X_a + f_{ab}X_b + \cdots + f_{an}X_n + \Delta_{aP} = \Delta_a$$
$$f_{ba}X_a + f_{bb}X_b + \cdots + f_{bn}X_n + \Delta_{bP} = \Delta_b \tag{12-11}$$
$$\cdots\cdots\cdots\cdots\cdots\cdots\cdots\cdots\cdots\cdots\cdots$$
$$f_{na}X_a + f_{nb}X_b + \cdots + f_{nn}X_n + \Delta_{nP} = \Delta_n$$

In general, the deflections of various points designated in Eq. 12-11 as $\Delta_a, \Delta_b, \ldots, \Delta_n$ need not necessarily be zero. In these equations the quantities f_{ij}, Δ_{iP}, and Δ_i represent either linear or angular deflections depending on whether they are associated with a force or a couple.

It can be shown† that the matrix of the flexibility coefficients f_{ij} is symmetric, i.e., $f_{ij} = f_{ji}$.

EXAMPLE 12-10

For the fixed-ended beam shown in Fig. 12-12(a), using the force method, find the moments developed at the supports due to the applied force P. The EI is constant.

*Sometimes these are referred to as the Maxwell-Mohr equations.

†See, for example, E. P. Popov, *Introduction to Mechanics of Solids*, Englewood Cliffs, N.J.: Prentice-Hall, 1968, p. 494.

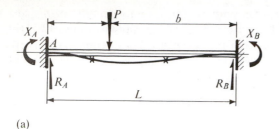

(a)

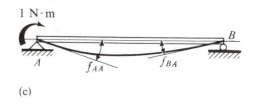

(c)

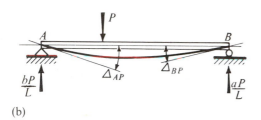

(b)

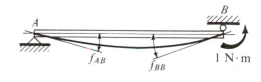

(d)

Fig. 12-12

SOLUTION

This beam, indeterminate to the second degree, is reduced to determinacy by removing the end moments, Fig. 12-12(b). In this determinate beam the rotations of the tangents at the supports due to the applied load can be taken from the solution in Example 11-4. This yields

$$|\Delta_{AP}| = \left|\left(\frac{dv}{dx}\right)_{x=0}\right| = \frac{Pab}{6EIL}(a + 2b)$$

$$|\Delta_{BP}| = \left|\left(\frac{dv}{dx}\right)_{x=L}\right| = \frac{Pab}{6EIL}(b + 2a)$$

The rotations of the beam ends due to the unit couples shown in Figs. 12-12(c) and (d) can be found from Table 11 of the Appendix. With $M_o = 1$, this gives

$$|f_{AA}| = |f_{BB}| = \frac{L}{3EI} \quad \text{and} \quad |f_{AB}| = |f_{BA}| = \frac{L}{6EI}$$

The sense of all of the above rotations is shown in Fig. 12-12. This must be carefully noted in setting up the superposition relations formally stated by Eq. 12-11. In each equation the positive displacements are measured in the direction of the displacement caused by the corresponding redundant quantity. On this basis two equations are obtained:

$$\circlearrowleft + \quad \Delta_A \equiv \theta_A = +\frac{L}{3EI}X_A + \frac{L}{6EI}X_B + \frac{Pab}{6EIL}(a + 2b) = 0$$

$$\circlearrowright + \quad \Delta_B \equiv \theta_B = +\frac{L}{6EI}X_A + \frac{L}{3EI}X_B + \frac{Pab}{6EIL}(b + 2a) = 0$$

Solving these equations simultaneously,

$$X_A = -\frac{Pab^2}{L^2} \quad \text{and} \quad X_B = -\frac{Pa^2b}{L^2}$$

where the negative signs of the bending moments indicate that the assumed directions were chosen incorrectly.

This problem may also be solved by treating R_B and X_B as the redundants since their temporary removal makes the structure determinate. This procedure is particularly convenient to apply if it is specified that one of the supports moves vertically. In such a case the deflection caused by the applied forces and the redundants is equated to the movement of the support.

*12-6. DISPLACEMENT METHOD

In the force method discussed in the preceding article, the redundant forces were assumed to be the unknowns, and a *released* structure was first obtained by removal of the redundants. In the displacement method, on the other hand, the displacement—both linear and angular—of the joints (points of contact of members with supports, points of intersection of two or more members or free ends of projecting members) are taken as the unknowns. The first step in applying this method is to prevent these joint displacements, which are also called *kinematic indeterminants* or *degrees of freedom*. The suppression of these degrees of freedom results in a modified system that is composed of a series of members each of whose end points are restrained from translations and rotations. The calculation of reactions at these artificially restrained ends due to the externally applied loads may be carried out using other methods of analysis such as the force method. However, the results of such calculations are usually available for a variety of loading conditions in tables such as Table 12 of the Appendix. In this type of analysis, counterclockwise moments and upward reactions acting *on either end of a member* are *arbitrarily* taken to be positive. This results in a consistent formulation of superposition equations. **This sign convention differs from the previously used beam sign convention.**

Restraining the support B of a propped cantilever shown in Fig. 12-13(a) reduces the problem to that of a fixed-end beam, and the moment M_1 can be found at B, Fig. 12-13(b). The end-moment M_2 required to rotate the fixed-end B through an angle θ_B can also be determined. Superposing the two moments, and noting that $M_1 + M_2 = 0$ since the actual beam under consideration is simply supported at B, one finds a solution for θ_B, the rotation of the beam at B.

This procedure is often referred to as the *displacement method*, since the joint displacements are treated as the unknowns in this approach. Note especially that, once the joint displacements are determined, all of the support reactions can be

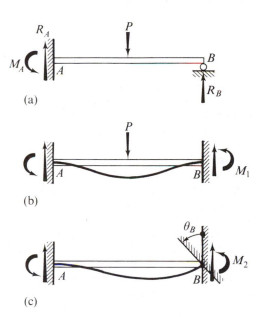

(a)

(b)

(c)

Fig. 12-13

obtained by superposition of the corresponding reactions such as those found in Fig. 12-13(b) and (c). Further analysis of stresses and deformations can then proceed in the usual manner.

EXAMPLE 12-11

Rework Example 12-9 using the displacement method, Fig. 12-14(a).

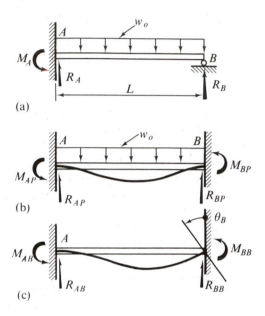

(a)

(b)

(c)

Fig. 12-14

SOLUTION

Assuming that axial deformations are neglected, this beam is kinematically indeterminate to the first degree since the only unknown joint displacement is the rotation of the beam at support B. A fixed-end beam is obtained by restraining the rotation at B, Fig. 12-14(b). The reactions at A and B due to the uniformly distributed load can be found from Table 12 of the Appendix. Here these reactions carry a double subscript; the first identifies their location, the second identifies that they are caused by the applied load, which is called P for generality. The same table can also be consulted to obtain the reactions on the unloaded fixed-end beam whose end B is subjected to a rotation θ_B. Here again the first subscript identifies the position of the reaction, and the second identifies the location of the applied displacement. The requirement of zero moment at B in the original structure provides the necessary condition for determining θ_B.

Using the above notation, and noting that the applied load w_o acts downward (negative), from Table 12 of the Appendix,

$$M_{BP} = -w_o L^2/12, \qquad M_{BB} = +4EI\theta_B/L$$

The equilibrium condition requires that the net moment M_B at the end B be zero, i.e.,

$$M_B = M_{BP} + M_{BB} = 0$$

Hence,
$$-\frac{w_oL^2}{12} + \frac{4EI}{L}\theta_B = 0, \qquad \text{or} \qquad \theta_B = \frac{w_oL^3}{48EI}$$

Then using superposition, and again making use of Table 12,

$$M_A = M_{AP} + M_{AB} = +\frac{w_oL^2}{12} + \frac{2EI}{L}\theta_B = +\frac{w_oL^2}{8} \qquad \text{(counterclockwise)}$$

$$R_A = R_{AP} + R_{AB} = \frac{w_oL}{2} + \frac{6EI}{L^2}\theta_B = +\frac{5}{8}w_oL \qquad \text{(upward)}$$

$$R_B = R_{BP} + R_{BB} = \frac{w_oL}{2} - \frac{6EI}{L^2}\theta_B = +\frac{3}{8}w_oL \qquad \text{(upward)}$$

The signs of all quantities are in accordance with the sign convention shown in Figs. 12-13(b) and (c).

According to the beam sign convention, the moment M_A at A is negative. The reactions and the conventional shear and moment diagrams are as shown in Figs. 12-10(d), (e), and (f), respectively.

EXAMPLE 12-12

Rework Example 12-3 using the displacement method, Fig. 12-15(a).

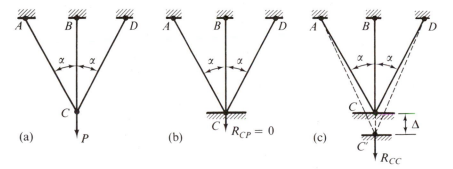

(a)   (b)   (c)

Fig. 12-15

SOLUTION

The unknown joint displacements in this problem are the horizontal and vertical movements of the point C. However, due to the symmetry involved, the horizontal displacement of C is zero, and only one degree of kinematic indeterminacy or degree of freedom remains.

The first step in the solution of this indeterminate system, using the displacement method, is to eliminate this degree of freedom and obtain a system consisting of three restrained bars, Fig. 12-15(b). For such a system, the reaction* R_{CP} at C is zero, since there are no externally applied loads

*The same system of notation is used as in the preceding example.

(other than the load applied directly on the joint C) between the restrained ends on any of the three bars.*

The next step in the solution process is to calculate the force R_{CC} required to subject joint C to a vertical displacement Δ, Fig. 12-15(c). For compatible deformations, $\Delta_{BC} = \Delta$, and $\Delta_{AC} = \Delta_{DC} = \Delta \cos \alpha$. Then, using Eq. 2-4,

$$F_{BC} = \frac{\Delta A E}{L} \qquad \text{(tension)}$$

$$F_{AC} = F_{DC} = \frac{(\Delta \cos \alpha) A E}{L/\cos \alpha} = \frac{\Delta A E}{L} \cos^2 \alpha \qquad \text{(tension)}$$

and
$$R_{CC} = F_{BC} + F_{AC} \cos \alpha + F_{DC} \cos \alpha$$

or
$$R_{CC} = \frac{\Delta A E}{L}(1 + 2 \cos^3 \alpha)$$

Finally, superposing the results of the two steps and imposing the force equilibrium condition in the vertical direction at the joint C, one obtains

$$R_C = R_{CP} + R_{CC} = P$$

hence
$$\frac{\Delta A E}{L}(1 + 2 \cos^3 \alpha) = P$$

or
$$\Delta = \frac{PL}{AE(1 + 2 \cos^3 \alpha)}$$

Substituting this expression for Δ into the preceding equations for the member forces expressed in terms of Δ, one obtains

$$F_{BC} = \frac{P}{1 + 2 \cos^3 \alpha} \qquad \text{and} \qquad F_{AC} = F_{DC} = \frac{P \cos^2 \alpha}{1 + 2 \cos^3 \alpha}$$

which are the same as the results found in Examples 12-3 and 12-8.

It should be noted that if in the above problem there were no symmetry about the vertical axis (either due to lack of symmetry in the structure itself or due to application of the load P at C other than in a vertical direction), a horizontal displacement would also have to be imposed on the joint C. Two force equilibrium equations, in the horizontal and vertical directions, must then be set up and solved simultaneously for the horizontal and vertical displacements of C. Further, it is interesting to note that the addition of any number of bars (Fig. 12-16), does not increase the kinematic indeterminacy of this structure, and the maximum number of simultaneous equations to be solved remains at two for such a system. In the force method, on the other hand, the number of statical redundants increases, and the number of simultaneous equations equals the number of redundants. However, this does not imply that the displacement method always involves the solution of fewer equations compared to the force method. Consider, for example,

*Suppose an additional downward load P acts on the bar BC at a distance $L/4$ above C. Using Eq. 12-1, one can then obtain $R_{CP} = -3P/4$.

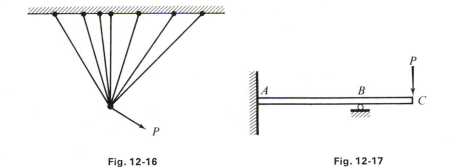

Fig. 12-16

Fig. 12-17

the case of a propped cantilever with an overhang, Fig. 12-17. This beam is statically indeterminate only to the first degree but kinematically indeterminate to the third degree (rotations at B and C, and vertical displacement at C); hence one needs to solve only a single equation in the force method but three simultaneous equations need to be solved in the displacement method solution of this problem.

Consider now the beam shown in Fig. 12-18(a), where the guided support at c allows for vertical displacement but no rotation of the beam at c. The other degree of freedom of this beam is the rotation of its tangent at the support b. This beam is thus kinematically indeterminate to the second degree. Upon restraining these two degrees of freedom, one obtains a system consisting of two fixed-end beams ab and bc, Fig. 12-18(d). The effect of the externally applied loads on these two fixed-end beams is to produce a set of reactive forces at the supports that are referred to as the *fixed-end actions*. Thus, for example, S_{bP} is the fixed-end moment at b, and S_{cP} is the fixed-end

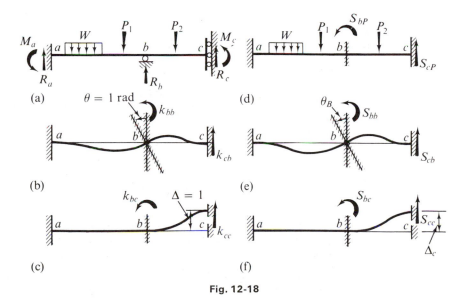

Fig. 12-18

reaction at c. Then the fixed support at b is rotated through an angle θ_b which gives rise to S_{bb} and S_{cb}, as shown in Fig. 12-18(e) at b and c, respectively. Similarly, S_{bc} and S_{cc} are caused by a vertical movement of the fixed support at c through a distance Δ_c, Fig. 12-18(f). Since there is no externally applied moment at b and no vertical force at c in the original structure, the respective resultant forces S_b and S_c at these points are equal to zero. Alternatively, these two forces may be found by superposing (summing) the results of three separate analyses, Figs. 12-18(d) through (f). These considerations lead to two simultaneous equations:

$$S_b = S_{bP} + S_{bb} + S_{bc} = 0$$
$$S_c = S_{cP} + S_{cb} + S_{cc} = 0 \tag{12-12}$$

These can be rewritten in a more meaningful form using *stiffness coefficients* k_{bb}, k_{bc}, k_{cb}, and k_{cc}, which are defined as the fixed-end actions shown in Figs. 12-18(b) and (c) due to the unit displacements (linear or angular) corresponding to the kinematic indeterminants. Then, since a linear system is being considered, the moments at b due to the kinematic indeterminants can be expressed as

$$S_{bb} = k_{bb}X_b \quad \text{and} \quad S_{bc} = k_{bc}X_c \tag{12-13}$$

and similarly the vertical reactions at c as

$$S_{cb} = k_{cb}X_b \quad \text{and} \quad S_{cc} = k_{cc}X_c \tag{12-14}$$

where numerically X_b and X_c are the unknown joint displacements to be determined at b and c, respectively. Using this notation, Eq. (12-12) becomes

$$k_{bb}X_b + k_{bc}X_c + S_{bP} = 0$$
$$k_{cb}X_b + k_{cc}X_c + S_{cP} = 0 \tag{12-15}$$

which can be solved simultaneously for the unknowns X_b and X_c.

The general form of these *force equilibrium* equations for a system having n degrees of kinematic indeterminacy can be written as

$$k_{aa}X_a + k_{ab}X_b + \cdots + k_{an}X_n + S_{aP} = S_a$$
$$k_{ba}X_a + k_{bb}X_b + \cdots + k_{bn}X_n + S_{bP} = S_b \tag{12-16}$$
$$\cdots\cdots\cdots\cdots\cdots\cdots\cdots\cdots\cdots\cdots$$
$$k_{na}X_a + k_{nb}X_b + \cdots + k_{nn}X_n + S_{nP} = S_n$$

The right-side terms S_i in these equations correspond to the *external forces* applied *at* the designated points. In the absence of such forces, which is often the case, these terms are zero. In these equations the quantities k_{ij}, S_{iP}, and S_i represent either forces or moments, depending on whether they are associated with a displacement or a rotation.

It is useful to note that the stiffness coefficients are symmetric*, i.e., $k_{ij} = k_{ji}$.

EXAMPLE 12-13

Calculate the rotations at b and c for the continuous beam of constant EI loaded as shown in Fig. 12-19(a), using the displacement method.

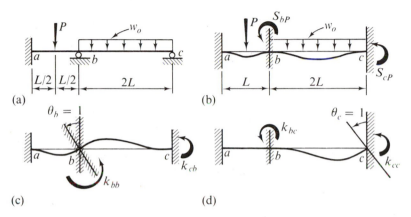

(a)

(b)

(c)

(d)

Fig. 12-19

SOLUTION

The freedom of rotation at supports b and c render this beam kinematically indeterminate to the second degree. By temporarily restraining the supports at b and c against rotation, one obtains a system of two fixed-end beams acted on by the external forces, Fig. 12-19(b). The fixed-end moments S_{bP} and S_{cP} can be obtained using standard formulas such as those given in Table 12 of the Appendix. Thus, noting that fixed-end moments act on the joint at b from both sides,

$$S_{bP} = \left[-\frac{PL}{8}\right]_{ba} + \left[\frac{w_o(2L)^2}{12}\right]_{bc} = -\frac{PL}{8} + \frac{w_oL^2}{3}$$

and

$$S_{cP} = -\frac{w_o(2L)^2}{12} = -\frac{w_oL^2}{3}$$

The *stiffness coefficients* can be calculated by subjecting the temporarily fixed ends b and c, to unit rotations *one at a time*, Figs. 12-19(c) and (d). Again, using formulas in Table 12 of the Appendix and by noting that the two adjoining spans contribute to the stiffness of the joint at b, one has

$$k_{bb} = \left[\frac{4EI}{L}\right]_{ba} + \left[\frac{4EI}{2L}\right]_{bc} = \frac{6EI}{L} \qquad k_{bc} = \left[\frac{2EI}{2L}\right]_{bc} = \frac{EI}{L}$$

$$k_{cb} = \left[\frac{2EI}{2L}\right]_{cb} = \frac{EI}{L} \qquad k_{cc} = \left[\frac{4EI}{2L}\right]_{cc} = \frac{2EI}{L}$$

*See, for example, N. Willems and W. M. Lucas, Jr., *Matrix Analysis for Structural Engineers*, Englewood Cliffs, N.J.: Prentice-Hall, 1968, p. 78.

The sense of all the above fixed-end forces are shown in Fig. 12-19 and must be carefully noted in setting up the force equilibrium superposition equations formally stated in Eq. 12-16. In each equation, positive forces are measured in the direction of the forces caused by the corresponding kinematic unknown. On this basis two equations are obtained:

$$S_b = \frac{6EI}{L} X_b + \frac{EI}{L} X_c - \frac{PL}{8} + \frac{w_o L^2}{3} = 0$$

$$S_c = \frac{EI}{L} X_b + \frac{2EI}{L} X_c - \frac{w_o L^2}{3} = 0$$

Solving these equations simultaneously,

$$X_b = \frac{L^2}{11EI}\left(\frac{P}{4} - w_o L\right) = \theta_b$$

and
$$X_c = \frac{L^2}{11EI}\left(-\frac{P}{8} + \frac{7}{3} w_o L\right) = \theta_c$$

In this problem the joint a is fixed and does not rotate, therefore $\theta_a = X_a = 0$. With this information on the rotation of the joints, the reactions caused by these rotations can be determined for each span with the aid of Table 12 in the Appendix. For each span these reactions are superposed with the fixed-end actions caused by the applied load, Fig. 12-19(b). For example, by defining the moment at the end b for the beam bc as M_{bc}, then

$$M_{bc} = [S_{bP}]_{bc} + \frac{4EI}{(2L)}\theta_b + \frac{2EI}{(2L)}\theta_c = +\frac{4}{11} w_o L^2 + \frac{3}{88} PL$$

Analogous calculations for the moment M_{ba} at the end b for the beam ba shows this moment to be numerically the same as M_{bc}, but of opposite sign. This is as it should be, since in this case no external moment acts at b, and $S_b \equiv M_b = M_{ba} + M_{bc} = 0$.

*12-7. MOMENT-AREA METHODS FOR STATICALLY INDETERMINATE BEAMS

The basic concept underlying the force method discussed in Art. 12-5 is used, together with the moment-area theorems derived in Art. 11-12, for the development of a special technique applicable for the analysis of *indeterminate beams*. This moment-area technique is then generalized in the next two articles to obtain a recurrence formula which is convenient for the analysis of continuous beams.

The application of the moment-area method with the superposition technique for solving indeterminate beam problems can be greatly accelerated, by the following reasoning: Restrained* and continuous beams differ from simply supported beams mainly by the presence of redundant moments at

*Indeterminate beams with one or more ends fixed are called restrained beams.

the supports. Therefore the bending-moment diagrams for these beams may be considered to consist of two independent parts—one part for the moment caused by all of the applied loading on a beam assumed to be simply supported, the other part for the redundant moments. Thus the effect of redundant end moments is superposed on a beam assumed to be simply supported. Physically this notion can be clarified by imagining an indeterminate beam cut through at the supports while the vertical reactions are maintained. The continuity of the elastic curve of the beam is preserved by the redundant moments.

Although the critical ordinates of the bending-moment diagrams caused by the redundant moments are not known, their shape is known. Application of a redundant moment at an end of a simple beam results in a triangular-shaped moment diagram with a maximum at the applied moment and a zero ordinate at the other end. Likewise, when end moments are present at both ends of a simple beam, two triangular moment diagrams superpose into a trapezoidal-shaped diagram. (Verify these statements).

The known and the unknown parts of the bending-moment diagram together give a complete bending-moment diagram. This whole diagram can then be used in applying the moment-area theorems to the continuous elastic curve of a beam. The geometrical conditions of a problem, such as the continuity of the elastic curve at the support or the tangents at built-in ends which cannot rotate, permit a rapid formulation of equations for the unknown values of the redundant moments at the supports.

For beams of variable flexural rigidity, $M/(EI)$ diagrams must be used.

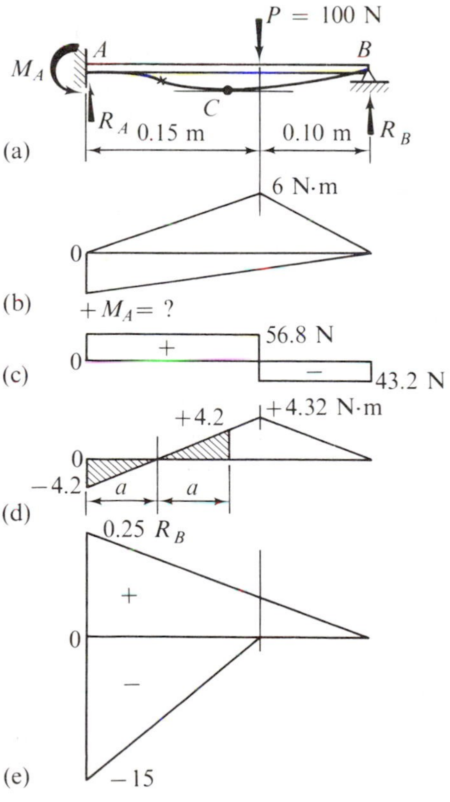

EXAMPLE 12-14

Find the maximum downward deflection of the small aluminum beam shown in Fig. 12-20(a) due to an applied force $P = 100$ N. The beam's constant flexural rigidity $EI = 60$ N·m².

SOLUTION

The solution of this problem consists of two parts. First, a redundant reaction must be determined to establish the numerical values for the bending-moment diagram; then the usual moment-area procedure is applied to find the deflection.

Imagining the beam is released from the redundant end moment, a simple beam-moment diagram is constructed above the base line in Fig. 12-20(b). The moment diagram of known shape due to the unknown

Fig. 12-20

redundant moment M_A is shown on the same diagram below the base line. One assumes M_A to be positive, since in this manner its correct sign is obtained automatically according to the beam convention. The composite diagram represents a *complete* bending-moment diagram.

The tangent at the built-in end remains horizontal after the application of the force P. Hence the geometrical condition is $t_{BA} = 0$. An equation formulated on this basis yields a solution for M_A.* The equations of static equilibrium are used to compute the reactions. The final bending-moment diagram, Fig. 12-20(d), is obtained in the usual manner after the reactions are known. Thus, since $t_{BA} = 0$,

$$\frac{1}{EI}\left[\frac{1}{2}(0.25)(6)\frac{1}{3}(0.25 + 0.10) + \frac{1}{2}(0.25)M_A\frac{2}{3}(0.25)\right] = 0$$

Hence, $M_A = -4.2 \text{ N·m}$

$$\sum M_A = 0 \circlearrowright +, \qquad 100(0.15) - R_B(0.25) - 4.2 = 0, \qquad R_B = 43.2 \text{ N}$$
$$\sum M_B = 0 \circlearrowleft +, \qquad 100(0.10) + 4.2 - R_A(0.25) = 0, \qquad R_A = 56.8 \text{ N}$$
$$\textit{Check:} \qquad\qquad \sum F_y = 0 \uparrow +, \qquad 43.2 + 56.8 - 100 = 0$$

The maximum deflection occurs where the tangent to the elastic curve is horizontal, point C in Fig. 12-20(a). Hence, by noting that the tangent at A is also horizontal and using the first moment-area theorem, point C is located. This occurs when the shaded areas in Fig. 12-20(d) having opposite signs are equal, i.e., at a distance $2a = 2(4.2/56.8) = 0.148$ m from A. The tangential deviation t_{AC} (or t_{CA}) gives the deflection of point C.

$$v_{max} = v_C = t_{AC}$$

$$= \frac{1}{EI}\left[\frac{1}{2} \times 0.074(+4.2)\left(0.074 + \frac{2}{3} \times 0.074\right)\right.$$

$$\left. + \frac{1}{2} \times 0.074(-4.2)\frac{1}{3} \times 0.074\right]$$

$$= (15.36)10^{-3}/(EI) = 0.000\,256 \text{ m} \qquad \text{(down)}$$

ALTERNATE SOLUTION

A rapid solution can also be obtained by plotting the moment diagram by cantilever parts. This is shown in Fig. 12-20(e). Note that one of the ordinates is in terms of the redundant reaction R_B. Again, using the geometrical condition $t_{BA} = 0$, one obtains an equation yielding R_B. Other reactions follow by statics. From $t_{BA} = 0$

$$\frac{1}{EI}\left[\frac{1}{2}(0.25)(+0.25R_B)\frac{2}{3}(0.25) + \frac{1}{2}(0.15)(-15)\left(0.1 + \frac{2}{3} \times 0.15\right)\right] = 0$$

Hence, $R_B = 43.2$ N, up as assumed.

$$\sum M_A = 0 \circlearrowleft +, \qquad M_A + 43.2(0.25) - 100(0.15) = 0, \qquad M_A = 4.2 \text{ N·m}$$

After the combined moment diagram is constructed, Fig. 12-20(d), the remainder of the work is the same as in the preceding solution.

---

*See Table 2 of the Appendix for the centroidal distance of a whole triangle.

EXAMPLE 12-15

Find the moments at the supports for a fixed-end beam loaded with a uniformly distributed load of w_o newtons per unit length, Fig. 12-21(a).

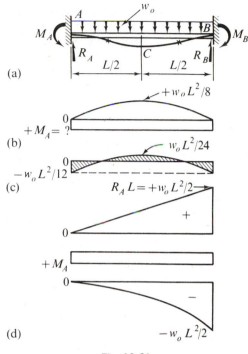

Fig. 12-21

SOLUTION

The moments at the supports are called fixed-end moments, and their determination is of great importance in structural theory. Due to symmetry in this problem, the fixed-end moments are equal, as are the vertical reactions which are $w_oL/2$ each. The moment diagram for this beam considered to be simply supported is a parabola as shown in Fig. 12-21(b), while the fixed-end moments give the rectangular diagram shown in the same figure.

Although this beam is statically indeterminate to the second degree, because of symmetry a single equation based on a geometrical condition is sufficient to yield the redundant moments. From the geometry of the elastic curve, any one of the following conditions may be used: $\Delta\theta_{AB} = 0$,* $t_{BA} = 0$, or $t_{AB} = 0$. From the first condition, $\Delta\theta_{AB} = 0$,

$$\frac{1}{EI}\left[\frac{2}{3}L\left(+\frac{w_oL^2}{8}\right) + L(+M_A)\right] = 0$$

then

$$M_A = M_B = -\frac{w_oL^2}{12}$$

*Also since the tangent at the center of the span is horizontal, $\Delta\theta_{AC} = 0$ and $\Delta\theta_{CB} = 0$.

The composite moment diagram is shown in Fig. 12-21(c). In comparison with the maximum bending moment of a simple beam, a considerable reduction in the magnitude of the critical moments occurs.

ALTERNATE SOLUTION

The moment diagram by cantilever parts is shown in Fig. 12-21(d). Noting that $R_A = R_B = w_o L/2$, and using the same geometrical condition as above, $\Delta \theta_{AB} = 0$, one can verify the former solution as follows:

$$\frac{1}{EI}\left[\frac{1}{2}L\left(+\frac{w_o L^2}{2} \right) + L(+M_A) + \frac{1}{3}L\left(-\frac{w_o L^2}{2} \right) \right] = 0$$

or $$M_A = -\frac{w_o L^2}{12}$$

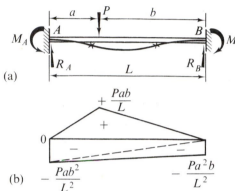

(a)

(b)

Fig. 12-22

EXAMPLE 12-16

Rework Example 12-10 using the moment-area method, Fig. 12-22(a).

SOLUTION

Treating the beam AB as a simple beam, the moment diagram due to P is shown above the base line in Fig. 12-22(b). The fixed-end moments are *not equal* and result in the trapezoidal diagram. Three geometrical conditions for the elastic curve are available to solve this problem which is indeterminate to the second degree:

(a) $\Delta \theta_{AB} = 0$, since the tangents at A and B are parallel.
(b) $t_{BA} = 0$, since the support B does not deviate from a fixed tangent at A.
(c) Similarly, $t_{AB} = 0$.

Any two of the above conditions may be used; arithmetical simplicity of the resulting equations governs the choice. Thus, using condition (a), which is always the simplest, and condition (b), two equations are*

$$\Delta \theta_{AB} = \frac{1}{EI}\left(\frac{1}{2}L\frac{Pab}{L} + \frac{1}{2}LM_A + \frac{1}{2}LM_B \right) = 0$$

or $$M_A + M_B = -\frac{Pab}{L}$$

$$t_{BA} = \frac{1}{EI}\left[\frac{1}{2}L\frac{Pab}{L}\frac{1}{3}(L + b) + \frac{1}{2}LM_A\frac{2}{3}L + \frac{1}{2}LM_B\frac{1}{3}L \right] = 0$$

or $$2M_A + M_B = -\frac{Pab}{L^2}(L + b)$$

Solving the two reduced equations simultaneously gives

$$M_A = -\frac{Pab^2}{L^2} \quad \text{and} \quad M_B = -\frac{Pa^2 b}{L^2}$$

*See Table 2 of the Appendix for the centroidal distance of a whole triangle.

These results agree with those found in Example 12-10, except that the signs of the moments follow the beam convention of signs. These signs resulted from taking M_A and M_B as positive quantities initially.

EXAMPLE 12-17

Plot moment and shear diagrams for a continuous beam loaded as shown in Fig. 12-23(a). The EI is constant for the whole beam.

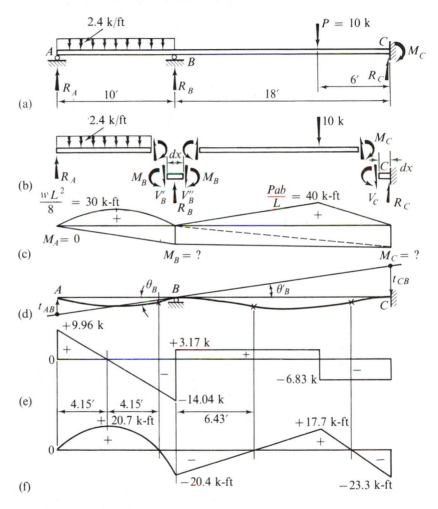

Fig. 12-23

SOLUTION

This beam is statically indeterminate to the second degree. By treating each span as a simple beam with the redundant moments, as in Fig. 12-23(b), the moment diagram of Fig. 12-23(c) is obtained. No end moments exist at A as this end is on a roller. The clue to the solution is contained in two geometrical conditions for the elastic curve for the whole beam, Fig. 12-23(d):

(a) $\theta_B = \theta_B'$. Since the beam is physically continuous, there is a line at the support B that is tangent to the elastic curve in *either* span.

(b) $t_{BC} = 0$, since the support B does not deviate from a fixed tangent at C.

To apply condition (a), t_{AB} and t_{CB} are determined, and, *dividing these quantities by the respective span lengths*, the two angles θ_B and θ_B' are obtained. These angles are equal. However, although t_{CB} is algebraically expressed as a positive quantity, the tangent through point B is *above* point C. Therefore this deviation must be considered negative. Hence, by using condition (a), one equation with the redundant moments is obtained.

$$t_{AB} = \frac{1}{EI}\left[\frac{2}{3}10(+30)\frac{1}{2}10 + \frac{1}{2}10(+M_B)\frac{2}{3}10\right]$$

$$= \frac{1}{EI}\left(1{,}000 + \frac{1}{3}100M_B\right)$$

$$t_{CB} = \frac{1}{EI}\left[\frac{1}{2}18(+40)\frac{1}{3}(18+6) + \frac{1}{2}18(+M_B)\frac{2}{3}18 + \frac{1}{2}18(+M_C)\frac{1}{3}18\right]$$

$$= \frac{1}{EI}(2{,}880 + 108M_B + 54M_C)$$

Since $\theta_B = \theta_B'$, or $\dfrac{t_{AB}}{L_{AB}} = -\dfrac{t_{CB}}{L_{CB}}$,

$$\frac{1}{EI}\left(\frac{1{,}000 + \frac{1}{3}\times 100\ M_B}{10}\right) = -\frac{1}{EI}\left(\frac{2{,}880 + 108M_B + 54M_C}{18}\right)$$

or

$$\frac{28}{3}M_B + 3M_C = -260$$

Using condition (b) for the span BC provides another equation, $t_{BC} = 0$, or

$$\frac{1}{EI}\left[\frac{1}{2}18(+40)\frac{1}{3}(18+12) + \frac{1}{2}18(+M_B)\frac{1}{3}18 + \frac{1}{2}18(+M_C)\frac{2}{3}18\right] = 0$$

or

$$3M_B + 6M_C = -200$$

Solving the two reduced equations simultaneously,

$$M_B = -20.4 \text{ ft-lb} \quad \text{and} \quad M_C = -23.3 \text{ ft-lb}$$

where the signs agree with the convention of signs used for beams. These moments with their proper sense are shown in Fig. 12-23(b).

After the redundant moments M_A and M_B are found, no new techniques are necessary to construct the moment and shear diagrams. *However, particular care must be exercised to include the moments at the supports while computing shears and reactions.* Usually, isolated beams as shown in Fig. 12-23(b) are the most convenient free bodies for determining shears. Reactions follow by adding the shears on the adjoining beams. In units of kips and feet,

For free body *AB:*

$$\Sigma M_B = 0 \; \curvearrowright +, \qquad 2.4(10)5 - 20.4 - 10R_A = 0, \qquad R_A = 9.96 \text{ kips} \uparrow$$

$$\Sigma M_A = 0 \; \curvearrowleft +, \qquad 2.4(10)5 + 20.4 - 10V'_B = 0, \qquad V'_B = 14.04 \text{ kips} \uparrow$$

For free body *BC:*

$$\Sigma M_C = 0 \; \curvearrowright +, \quad 10(6) + 20.4 - 23.3 - 18V''_B = 0, \quad V''_B = 3.17 \text{ kips} \uparrow$$

$$\Sigma M_B = 0 \; \curvearrowleft +, \quad 10(12) - 20.4 + 23.3 - 18V_C = 0, \quad V_C = R_C = 6.83 \text{ kips} \uparrow$$

Check: $R_A + V'_B = 24$ kips $\uparrow$ and $V''_B + R_C = 10$ kips $\uparrow$

From above, $R_B = V'_B + V''_B = 17.21$ kips $\uparrow$.

The complete shear and moment diagrams are shown in Figs. 12-23(e) and (f), respectively.

*12-8. THE THREE-MOMENT EQUATION

Generalizing the procedure used in the preceding example, a recurrence formula, i.e., an equation that can be applied repeatedly for *every two adjoining spans*, can be derived for continuous beams. For any *n* number of spans, $n - 1$ such equations may be written. This gives enough simultaneous equations for the solution of redundant moments over the supports. This recurrence formula is called the *three-moment equation* because three unknown moments appear in it.

Consider a continuous beam, such as shown in Fig. 12-24(a), subjected to any transverse loading. For *any* two adjoining spans, such as *LC* and *CR*, the bending-moment diagram is considered to consist of two parts. The areas A_L and A_R to the left and to the right of center support *C* in Fig. 12-24(b), correspond to the bending-moment diagrams in the respective spans if these spans are treated as being *simply supported*. These moment diagrams depend entirely upon the nature of the known forces applied within each span. The other part of the moment diagram of known shape is due to the unknown moments M_L at the left support, M_C at the center support, and M_R at the right support.

Next, the elastic curve shown in Fig. 12-24(c) must be considered. This curve is continuous for any continuous beam. Hence the angles θ_C and θ'_C, which define, from the respective sides, the inclination of the same tangent to the elastic curve at *C*, are equal. By using the second moment-area theorem to obtain t_{LC} and t_{RC}, these angles are defined as $\theta_C = t_{LC}/L_L$ and $\theta'_C = -t_{RC}/L_R$, where L_L and L_R are span lengths on the left and on the right of *C*, respectively. The negative sign for the second angle is necessary since the tangent from point *C* is *above* the support *R*, whereas a positive deviation of t_{RC} locates a tangent below the same support. Hence, following

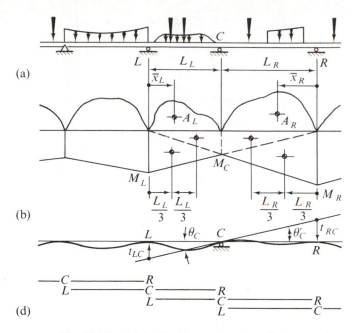

(a)

(b)

(d)

Fig. 12-24. Deriving the three-moment equation

the steps outlined,

$$\theta_C = \theta_C' \quad \text{or} \quad \frac{t_{LC}}{L_L} = -\frac{t_{RC}}{L_R}$$

and

$$\frac{1}{L_L}\frac{1}{EI_L}\left(A_L\bar{x}_L + \frac{1}{2}L_L M_L \frac{1}{3}L_L + \frac{1}{2}L_L M_C \frac{2}{3}L_L\right)$$

$$= -\frac{1}{L_R}\frac{1}{EI_R}\left(A_R\bar{x}_R + \frac{1}{2}L_R M_R \frac{1}{3}L_R + \frac{1}{2}L_R M_C \frac{2}{3}L_R\right)$$

where I_L and I_R are the respective moments of inertia of the cross-sectional area of the beam in the left and the right spans. Throughout each span, I_L and I_R are assumed constant.* The term $\bar{x}_L$ is the distance *from the left support* L to the centroid of the area A_L, and $\bar{x}_R$ is a similar distance for A_R measured *from the right support* R. The terms M_L, M_C, and M_R denote the unknown moments at the supports.

Simplifying the above expression, the three-moment equation† is

$$L_L M_L + 2\left(L_L + \frac{I_L}{I_R}L_R\right)M_C + \frac{I_L}{I_R}L_R M_R = -6A_L\frac{\bar{x}_L}{L_L} - 6A_R\frac{I_L\bar{x}_R}{I_R L_R} \qquad (12\text{-}17)$$

Equation 12-17 applies to continuous beams on unyielding supports, with the beam in each span of constant I. In a particular problem, all terms,

---

*The expressions become very involved if variation of I occurs *within* any one span.
†The three-moment equation was originally derived by E. Clapeyron (a French engineer) in 1857, and sometimes is referred to as Clapeyron's equation.

with the exception of the redundant moments at the supports, are constant. A sufficient number of simultaneous equations for the unknown moments is obtained by successively imagining the supports of the adjoining spans as *L*, *C*, and *R* as shown in Fig. 12-24(d). *However, in these equations the subscripts of the M's must correspond to the actual designation of the supports,* such as *A*, *B*, *C*, etc. Also note that at pinned ends of beams the moments are known to be zero. Likewise, if a continuous beam has an overhang, *the moment at the first support is known from statics.* Fixed supports will be discussed in Example 12-19. For symmetrical beams symmetrically loaded, work can be minimized by noting that moments at symmetrically placed supports are equal.

In deriving the three-moment equation, the moments at the supports were assumed positive. Hence an algebraic solution of simultaneous equations automatically gives the correct sign of moments according to the convention for beams.

*12-9. CONSTANTS FOR SPECIAL LOAD CASES

As a specific example of the evaluation of the constant terms on the right side of the three-moment equation, consider two adjoining spans loaded with the *concentrated forces* P_L and P_R, Fig. 12-25. Considering these spans as simply supported, since the maximum moment in the left span is $+P_L ab/L_L$, and $\bar{x}_L = \frac{1}{3}(L_L + a)$, then

$$-6A_L \frac{\bar{x}_L}{L_L} = -6\left(\frac{L_L}{2}\right)\frac{P_L ab}{L_L}\frac{(L_L + a)}{3L_L} = -P_L ab\left(1 + \frac{a}{L_L}\right) \quad (12\text{-}18)$$

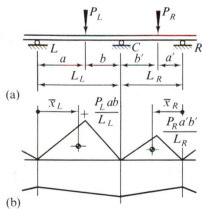

(a)

(b)

Fig. 12-25. Establishing the constants on the right of the three-moment equation for concentrated loads

Similarly, by interchanging the role of the dimensions *a* and *b* in the right-hand span, i.e., *by always measuring a's from the outside support toward the force,*

$$-6A_R \frac{I_L}{I_R}\frac{\bar{x}_R}{L_R} = -P_R a'b'\left(1 + \frac{a'}{L_R}\right)\frac{I_L}{I_R} \quad (12\text{-}19)$$

If a number of concentrated forces occur within a span, the contribution of each one of them to the above constant may be treated separately. Hence, a constant term for the right side of the three-moment equation applicable for any number of concentrated forces applied within the spans is

$$-\sum P_L ab\left(1 + \frac{a}{L_L}\right) - \sum P_R a'b'\left(1 + \frac{a'}{L_R}\right)\frac{I_L}{I_R} \quad (12\text{-}20)$$

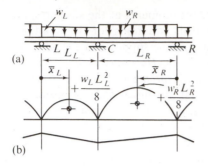

(a)

(b)

Fig. 12-26. Establishing the constants on the right of the three-moment equation for uniformly distributed loads

where the summation sign designates the fact that a separate term appears for *every concentrated force* P_L in the left span, and similarly, *for every* P_R force in the right span. In both cases, *a* or *a'* is the distance from the *outside support* to the particular concentrated force, and *b* or *b'* is the distance to the force from the *center support*. If any one of these forces acts upwards, the term contributed to the constant by such a force is of opposite sign.

The constant for the right side of the three-moment equation, when *uniformly distributed loads* are applied to a beam, is determined similarly. Thus, using the diagram shown in Fig. 12-26,

$$-6A_L\frac{\bar{x}_L}{L_L} = -6\left(\frac{2}{3}L_L\right)\frac{w_LL_L^2}{8}\frac{1}{2}\frac{L_L}{L_L} = -\frac{w_LL_L^3}{4} \qquad (12\text{-}21)$$

and similarly,

$$-6A_R\frac{I_L}{I_R}\frac{\bar{x}_R}{L_R} = -\frac{w_RL_R^3}{4}\frac{I_L}{I_R} \qquad (12\text{-}22)$$

Constants for other types of loading can be determined by using the same procedure as above.

EXAMPLE 12-18

Find the moments at all supports and the reactions at C and D for the continuous beam loaded as shown in Fig. 12-27(a). The flexural rigidity EI is constant.

SOLUTION

By using Eq. 12-17 and treating the span AB as the left span and BC as the right, one equation is written. From statics, the beam convention being used for signs, $M_A = -20(2.5) = -50$ N·m. Equations 12-20 and 12-21 are used to obtain the right terms. The moments of inertia I_R and I_L are equal.

$$6M_A + 2(6 + 10)M_B + 10M_C$$
$$= -\frac{8(6)^3}{4} - 16(7.5)(2.5)\left(1 + \frac{7.5}{10}\right) - 24(5)5\left(1 + \frac{5}{10}\right)$$

Substituting $M_A = -50$ N·m and simplifying gives

$$32M_B + 10M_C = -1\,557$$

Next, Eq. 12-12 is again applied for the spans BC and CD. No constant terms are contributed to the right side of the three-moment equation by the unloaded span CD. At the pinned end, $M_D = 0$.

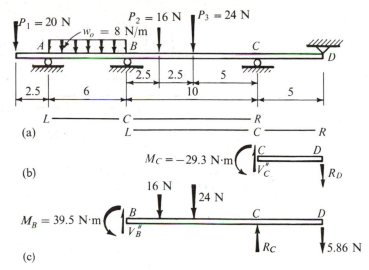

(a)

(b)

$M_C = -29.3 \ \text{N·m}$

(c)

$M_B = 39.5 \ \text{N·m}$

(Dimensions in meters)

Fig. 12-27.

$$10M_B + 2(10 + 5)M_C + 5M_D$$

$$= -16(2.5)(7.5)\left(1 + \frac{2.5}{10}\right) - 24(5)5\left(1 + \frac{5}{10}\right)$$

or
$$10M_B + 30M_C = -1\,275$$

Solving the reduced equations simultaneously gives

$$M_B = -39.5 \ \text{N·m} \qquad \text{and} \qquad M_C = -29.3 \ \text{N·m}$$

Isolating the span CD as in Fig. 12-27(b), one obtains the reaction R_D from statics. Instead of isolating the span BC and computing V'_C to add to V''_C to find R_C, as was done in Example 12-17, the free body shown in Fig. 12-27(c) is used. For free body CD:

$$\Sigma M_C = 0 \circlearrowright +, \qquad 29.3 - 5R_D = 0, \qquad R_D = 5.86 \ \text{N} \downarrow$$

For free body BD:

$$\Sigma M_B = 0 \circlearrowright +,$$
$$16(2.5) + 24(5) - R_C(10) + 5.86(15) - 39.5 = 0, \qquad R_C = 20.84 \ \text{N}$$

EXAMPLE 12-19

Rework Example 12-17 using the three-moment equation, Fig. 12-28.

SOLUTION

No difficulty is encountered in setting up a three-moment equation for the spans AB and BC. This is done in a manner analogous to that in the preceding

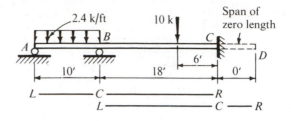

Fig. 12-28

example. Note that an unknown moment does exist at the built-in end, and, since the end A is on a roller, $M_A = 0$.

$$10M_A + 2(10 + 18)M_B + 18M_C = -\frac{2.4(10)^3}{4} - 10(6)12\left(1 + \frac{6}{18}\right)$$

$$56M_B + 18M_C = -1,560$$

To set up the next equation, an artifice is introduced. An imaginary span of zero length is added at the fixed end, and the three-moment equation is set up in the usual manner:

$$18M_B + 2(18 + 0)M_C + 0(M_D) = -10(12)6\left(1 + \frac{12}{18}\right)$$

$$18M_B + 36M_C = -1,200$$

Solving the reduced equations simultaneously,

$$M_B = -20.4 \text{ kip-ft} \quad \text{and} \quad M_C = -23.3 \text{ kip-ft}$$

The remainder of the problem is the same as before.

The use of a zero-length span at the fixed ends of beams is justified by the moment-area procedure. This expedient is equivalent to the requirement of a zero deviation of a support nearest the fixed end from the tangent at the fixed end. For example, if the second of the above reduced equations is divided through by 6, the corresponding equation in Example 12-17 is obtained; there the latter condition was used directly.

*12-10. LIMIT ANALYSIS OF BEAMS

Examples of ultimate or limit load calculations for axially loaded bar systems made of elastic-plastic material were given at the beginning of this chapter (see Examples 12-1 and 12-3). It is important to note that in such problems there are three stages of loading. First, there is the range of linear elastic response (see Figs. 12-1(e) and 12-3(c)). Then a portion of a structure yields as the remainder continues to deform elastically. This is the range of contained plastic flow. Finally, the structure continues to yield at no increase in load. At this stage the plastic deformation of the structure becomes unbounded.* This condition corresponds to the limit load for the structure.

Since the same general behavior is exhibited by elastic-plastic beams, the objective now is to develop a procedure for determining the limit loads for them. By bypassing the earlier stages of loading and going directly to the determination of the limit load, the procedure becomes relatively simple. For background, some of the results previously established will be re-examined.

Typical moment-curvature relationships of elastic-plastic beams having several different cross sections are shown in Fig. 12-29. Such results

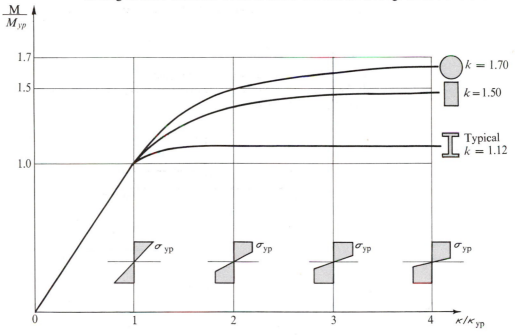

Fig. 12-29. Moment-curvature relations for circular, rectangular, and I cross sections. $M_p/M_{yp} = k$, the shape factor

were established in Example 11-7 for a rectangular beam. (See Fig. 11-17.) Note especially the rapid ascent of the curves toward their respective asymptotes as the cross sections plastify. This means that very soon after exhausting the elastic capacity of a beam, a rather constant moment is both achieved and maintained. This condition is likened to a plastic hinge. In contrast to a frictionless hinge capable of permitting large rotations at no moment, the plastic hinge allows large rotations to occur at a constant moment. This constant moment is approximately M_p, the ultimate or plastic moment for a cross section.

*In reality a structure cannot be permitted to deform excessively.

Using plastic hinges, a sufficient number can be inserted into a structure at the points of maximum moments to create a kinematically admissible collapse mechanism. Such a mechanism, permitting unbounded movement of a system, enables one to determine the ultimate or limit carrying capacity of a beam or of a frame. This approach will now be illustrated by several examples, confining the discussion to beams.

When the limit analysis approach is used for the selection of members, the working loads are multiplied by a load factor larger than unity to obtain the limit loads for which the calculations are performed. This is analogous to the use of the factor of safety in elastic analyses. In structural steel work the term *plastic design* is commonly applied to this approach.

EXAMPLE 12-20

A force P is applied at the middle of a simply supported beam, Fig. 12-30(a). If the beam is made of a ductile material, what is the limit load P_{ult}? Neglect the weight of the beam.

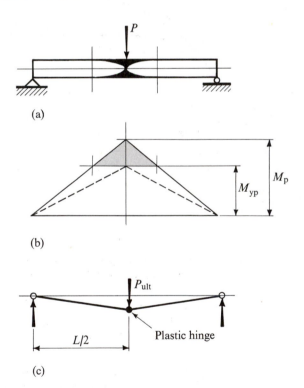

(a)

(b)

(c)

Fig. 12-30

SOLUTION

The shape of the moment diagram is the same regardless of the load magnitude. For any value of P, the maximum moment $M = PL/4$, and if $M \le M_{yp}$, the beam behaves elastically. Once M_{yp} is exceeded, contained yielding of the beam commences and continues until the maximum plastic moment M_p is reached. At that instant a plastic hinge is formed in the middle of the span forming the collapse mechanism shown in Fig. 12-30(c). By setting the plastic moment M_p equal to $PL/4$ with $P = P_{ult}$, one obtains the result sought:

$$P_{ult} = 4M_p/L$$

Note that consideration of the actual plastic zone indicated shaded in Fig. 12-30(a) is unnecessary in this calculation.

EXAMPLE 12-21

A restrained beam of ductile material is loaded as shown in Fig. 12-31(a). Find the limit load P_{ult}. Neglect the weight of the beam.

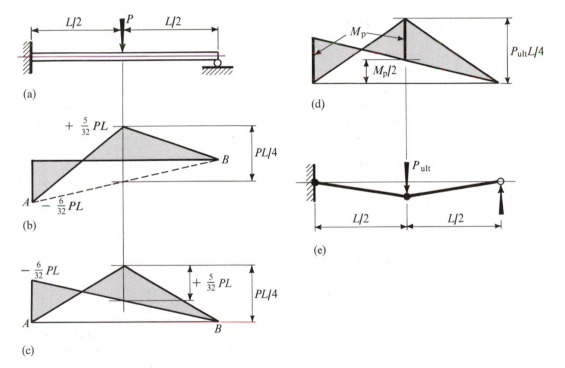

Fig. 12-31

SOLUTION

The results of an elastic analysis are shown in Fig. 12-31(b) in the usual manner. The same results are replotted in Fig. 12-31(c) from a horizontal baseline AB. In both diagrams the value of the moment ordinates are the same, and the shaded portions of the diagrams represent the final results. Note that the auxiliary ordinate $PL/4$ has precisely the value of the maximum moment in a simple beam with a concentrated force in the middle.

By setting the maximum elastic moment equal to M_{yp}, one obtains the load P_{yp} at impending yield:

$$P_{yp} = (16/3)M_{yp}/L$$

When the load is increased above P_{yp}, the moment at the built-in end can reach but cannot exceed M_p. This is also true of the moment at the middle of the span. These limiting conditions are shown in Fig. 12-31(d). The sequence in which M_p's occur is unimportant. In determining the limit load it is necessary to have only a kinematically admissible mechanism. With the two plastic hinges and a roller on the right this condition is assured, Fig. 12-31(e).

From Fig. 12-31(d) it can be seen that by proportions the end moment M_p gives an ordinate of $M_p/2$ at the middle of the span. Therefore, the simple beam ordinate $P_{ult}L/4$ in the middle of the span must be equated to $3M_p/2$ to obtain the limit load. This gives

$$P_{ult} = 6M_p/L$$

Comparing this result with P_{yp}, one has

$$P_{ult} = \frac{9M_p}{8M_{yp}}P_{yp} = \frac{9}{8}kP_{yp}$$

which shows that the increase in P_{ult} over P_{yp} is due to two causes: $M_p > M_{yp}$, and the maximum moments are distributed more advantageously in the plastic case. (Compare the moment diagrams in Figs. 12-31(c) and (d).)

EXAMPLE 12-22

A restrained beam of ductile material carries a uniformly distributed load as shown in Fig. 12-32(a). Find the limit load w_{ult}.

SOLUTION

In this problem two plastic hinges are required to create a collapse mechanism. One of these hinges will be at the built-in end. The location of the hinge associated with the other maximum moment is not immediately known since the moment near the middle of the span changes very gradually. However, one can assume the mechanism to be as shown in Fig. 12-32(c) since this would be in agreement with the moment diagram of Fig. 12-32(b).

For purposes of analysis, the beam with the plastic hinges is separated into two parts as in Figs. 12-32(d) and (e). Then by noting that no shear is possible at C since it is the point of maximum moment on a continuous curve, one may write two equations of static equilibrium:

$$\sum M_A = 0 \circlearrowleft +, \qquad M_p - w_{ult}a^2/2 = 0$$
$$\sum M_B = 0 \circlearrowright +, \qquad 2M_p - w_{ult}(L-a)^2/2 = 0$$

The simultaneous solution of these equations gives $a = (\sqrt{2} - 1)L$, which locates the plastic hinge C. The same equations yield the limit load

$$w_{ult} = 2M_p/a^2 = 2M_p/[(\sqrt{2} - 1)L]^2$$

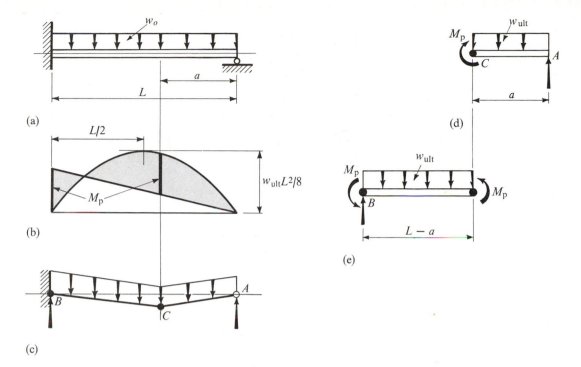

Fig. 12-32

In problems with several concentrated forces applied to a beam, a search for the interior plastic hinge must also be made. The least load for an assumed interior hinge under any one of the loads constitutes the solution of the problem. An equilibrium at any higher load requires moments greater than M_p and is therefore impossible. To determine this may require several trials.

EXAMPLE 12-23

A fixed beam of ductile material supports a uniformly distributed load, Fig. 12-33(a). Determine the limit load w_{ult}.

SOLUTION

According to the elastic analysis (see Example 12-15 and Fig. 12-21(c)) the maximum moments occur at the built-in ends and are equal to $w_o L^2/12$. Therefore

$$M_{yp} = w_{yp} L^2/12 \quad \text{or} \quad w_{yp} = 12 M_{yp}/L^2$$

On increasing the load, plastic hinges develop at the supports. The collapse mechanism is not formed, however, until a plastic hinge also occurs in the middle of the span, Figs. 12-33(b) and (c).

The maximum moment for a simply supported, uniformly loaded beam

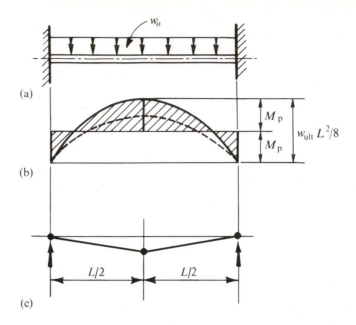

(a)

(b)

M_p

M_p

$w_{ult} L^2/8$

(c)

$L/2$ $L/2$

Fig. 12-33

is $w_o L^2/8$. Therefore, as can be seen from Fig. 12-33(b), to obtain the limit load in a clamped beam, this quantity must be equated to $2M_p$ with $w_o = w_{ult}$. Hence

$$w_{ult}L^2/8 = 2M_p \qquad \text{or} \qquad w_{ult} = 16M_p/L^2$$

Comparing this result with w_{yp}, one has

$$w_{ult} = \frac{4M_p}{3M_{yp}}w_{yp} = \frac{4}{3}kw_{yp}$$

As in Example 12-21, the increase of w_{ult} over w_{yp} depends on the shape factor k and the equalization of the maximum moments.

The analysis of continuous beams proceeds in a manner analogous to the above. Ordinarily the collapse of such beams occurs locally in only one of the spans. For example, for the beam shown in Fig. 12-34(a), a mechanism can form as in Fig. 12-34(b) or 12-34(c), depending on the relative magnitudes of the loads and span lengths. Such problems revert to cases already considered. Local collapse mechanisms for the interior spans require the formation of three hinges resembling those in the last example. The mechanisms for frames can become quite complex; the treatment of such problems is beyond the scope of this text.*

*For further details see P. G. Hodge, *Plastic Analysis of Structures*, New York: McGraw-Hill, 1959.

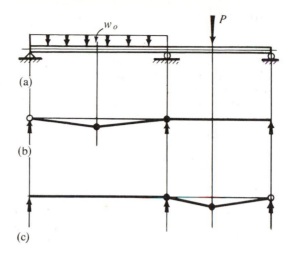

Fig. 12-34

*12-11. CONCLUDING REMARKS

In practice, statically indeterminate members occur in numerous situations. Some of the methods for analyzing these members have been discussed in this chapter. Sometimes the stresses caused by indeterminacy, particularly those due to temperature, are undesirable. More often, however, members are deliberately arranged to be indeterminate as such members are stiffer, which is highly desirable in many cases. A reduction in stresses can also be accomplished. For example, the maximum bending moments in indeterminate beams are usually smaller than the maximum moments in similar determinate beams. This permits selection of smaller members and results in an economy of material.

There are also some disadvantages to using indeterminate members. Some uncertainty always exists as to whether the supports are capable of completely fixing the ends. Likewise, the supports may settle or move in relation to each other. Then the calculated elastic stresses or deflections may be seriously in error. These matters are of little concern in a statically determinate structure. Finally, the method of elastic analysis becomes very involved when the degree of indeterminacy is high. However, this situation has been largely overcome by specialized methods and a wider use of digital computers.

For situations where the applied loads are static in character, and the materials employed are ductile, the plastic method of design offers advantages.

12-1. Find the reactions at the built-in ends for an elastic rod loaded as shown in the figure and draw the axial force diagram. The cross-sectional area of the rod is 2 in.^{2}, $a = 6$ in., and $b = 12$ in. The applied force $P_1 = 20$ k, and $P_2 = 40$ k. *Ans:* -5k on the left.

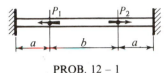

PROB. 12 – 1

12-2. A bar is built-in at both ends to immovable supports as in the preceding problem and carries two forces P_1 and P_2 as shown in the figure. The magnitude of P_2 is twice that of P_1. (a) Assuming linearly elastic behavior determine the reactions and the axial force distribution in the bar. Plot the axial force and the axial deformation diagrams. (b) If $\sigma_{yp} = 400$ MN/m^2, determine the range of the contained plastic flow. Plot a diagram showing the variation in the magnitude of the force P_2 as a function of its displacement. Let the cross-sectional area of the bar be 400 mm^2, and $E = 200\,000$ MN/m^2, For both cases $b = 2a$. *Ans:* (a) $-P_1/4$ on left.

12-3. A material possesses a nonlinear stress-strain relationship given as $\sigma = K\varepsilon^n$, where K and n are material constants. If a rod made of this material and of constant area A is initially fixed at both ends and is then loaded as shown in the figure, how much of the applied force P is carried by the left support? *Ans:* $P/[(a/b)^n + 1]$.

PROB. 12 – 2

12-4. An elastic round shaft of constant cross-sectional area is subjected to a torque T_1 as shown in the figure. If both ends of the shaft are built in, what torques are resisted at each end? *Ans:* $T_A = T_1 b/L$.

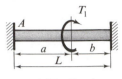

PROB. 12 – 4

12-5. A round bar of constant cross-sectional area is built in at both ends and is subjected to a torque T_1 as shown in the figure of the preceeding problem. (a) Assuming linearly elastic behavior of the material, determine the reactions. Plot the torque $T(x)$ and the angle of twist $\phi(x)$ diagrams. (b) If the bar is 2 in. in diameter, $a = 30$ in., and $b = 20$ in., determine and plot the relationship between the angle of twist ϕ at $x = 30$ in. and the applied torque T_1. Construct this diagram analogously to the one shown in Fig. 12-1(e). Assume the material to be elastic-perfectly plastic with $\tau_{yp} = 20$ ksi, and $G = 12 \times 10^6$ psi.

12-6. A 10 kN weight is to be lifted by means of two rods, each nearly 3 m long as shown in the figure. One rod is made of steel ($E = 200\,000$ MN/m^2), the other of aluminum ($E = 70\,000$ MN/m^2). Each rod has a cross-sectional area of 120 mm^2. Which rod must be made shorter and by how much in order to distribute the load equally between them? Assume elastic behavior.

PROB. 12 – 6

12-7. A rigid platform rests on two aluminum bars ($E = 10^7$ psi) each 10.000 in. long. A third bar made of steel ($E = 30 \times 10^6$ psi) and standing in the middle is 9.995 in. long. (a) What will be the stress in the steel bar if a load P of 100 kips is applied on the platform? (b) How much do the aluminum bars shorten? *Ans:* (a) 15 ksi; (b) 0.01 in.

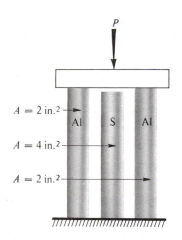

PROB. 12 – 7

12-8. A rod is fixed at *A* and loaded with an axial force *P* as shown in the figure. The material is elastic, perfectly plastic with $E = 200$ GPa and a yield stress of 200 MPa. Prior to loading, a gap of 2 mm exists between the end of the rod and the fixed support *C*. Plot the load-displacement diagram for the load point assuming *P* increases from zero to its ultimate value for the rod. The cross-section from *A* to *B* is 200 mm², and that from *B* to *C* is 100 mm².

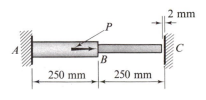

PROB. 12 – 8

12-9. If a load of 1 kip is applied to a rigid bar suspended by three wires as shown in the figure, what force will be resisted by each wire? The outside wires are aluminum ($E = 10^7$ psi). The inside wire is steel ($E = 30 \times 10^6$ psi). Initially

there is no slack in the wire. *Ans:* Each wire carries $\frac{1}{3}$ kip.

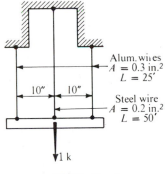

PROB. 12 – 9

12-10. If the bar in Prob. 12-9 is not rigid but has an $I = 0.222$ in.⁴ and, being of steel, an *E* of 30×10^6 psi, what forces will be developed in each wire? *Ans:* 428 lb in middle wire.

12-11. Three springs and a flexible bar are assembled as shown in the figure. If a load of 300 lb is applied to the bar, how is it distributed between the springs? The top spring has a spring constant of 100 k/in., and the bottom ones 50 k/in. each. The *EI* of the bar is 3.6×10^6 lb-in.² *Ans:* 200 lb middle spring.

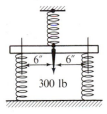

PROB. 12 – 11

12-12. A rigid bar is supported by a pin at *A* and two linearly elastic wires at *B* and *C* as shown in the figure. The area of the wire at *B* is 60 mm², and for the one at *C* is 120 mm². Determine the reactions at *A*, *B*, and *C* caused by the applied load $P = 6$ kN. *Ans:* 4.8 kN.

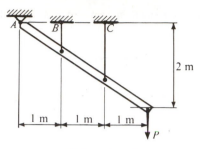

PROB. 12 – 12

12-13. A load $P = 1$ kN is applied to a rigid bar suspended by three wires as shown in the figure. All wires are of equal size and the same material. For each wire $A = 80$ mm², $E = 200$ GPa, and $L = 4$ m. If initially there were no slack in the wires, how will the applied load distribute between the wires? *Ans:* 83.3 N, 333.3 N, 583.3 N.

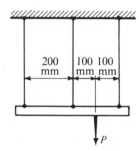

PROB. 12 – 13

12-14. A 30-in. cantilever beam of constant flexural rigidity, $EI = 10^7$ lb-in.², initially has a gap of 0.05 in. between its end and the spring. The spring constant $k = 10$ kips per inch. If a force of 100 lb, shown in the figure, is applied to the end of this cantilever, how much of this force will be carried by the spring? *Ans:* 40 lb.

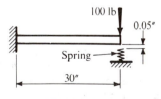

PROB. 12 – 14

12-15. A W 8 × 17 beam is simply supported at the ends A and B and goes over the middle of

two W 8 × 17 cross beams as shown in the figure. When erected, the cross beams just touch the beam AB. What will be the reactions at A and B if a uniformly distributed load of 2 kips per ft is applied to the upper beam. $E = 29 \times 10^6$ psi. *Ans:* 6.74 kips.

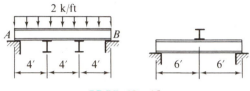

PROB. 12 – 15

12-16. Two identical, horizontal, simply supported beams span 3.6 m each. The beams cross each other at right angles at their respective midspans. When erected, there is a 6 mm gap between the two beams. If a concentrated downward force of 50 kN is applied at mid-span to the upper beam, how much will the lower beam carry? The EI of each beam is 6,000 kN·m².

12-17. The midpoint of a cantilever beam 6 m long rests on the midspan of a simply supported beam 8 m long. Determine the deflection of point A, where the beams meet, which results from the application of a 40 kN load at the end of the cantilever beam. State the answer in terms of EI which is the same and is constant for both beams.

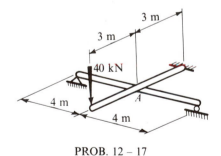

PROB. 12 – 17

12-18. The beam AB in an unloaded condition just touches a spring at midspan as shown in the figure. What is the spring stiffness k that will make the forces in all three supports equal for a uniformly distributed, vertical load w_o. Use Table 11 in the Appendix. *Ans:* 384 $EI/(7L^3)$.

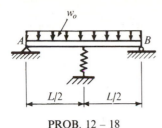

PROB. 12 – 18

12-19.   Two vertical beams 2 m long are connected at their midspans by a taut wire as shown in the figure. The *EI* for the beam on the left is 50 kN·m² and for the beam on the right is 150 kN·m². The cross-sectional area of the wire is 65 mm² and its $E = 70 \times 10^6$ kN/m². Find the stress in the wire after the two 2 kN forces are applied to the beams at midspan. Use the beam deflection formula given in Table 11 of the Appendix. *Ans:* 30 MN/m².

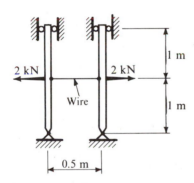

PROB. 12 – 19

12-20.   One end of a W 18 × 50 beam is cast into concrete. It was intended to support the other end with a 1-in.² steel rod 12 ft long as shown in the figure. During the installation, however, the nut on the rod was poorly tightened and in the unloaded condition there is a ½-in. gap

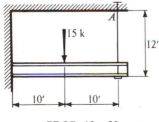

PROB. 12 – 20

between the top of the nut and the bottom of the beam. What tensile force will develop in the rod because a force of 15 kips is applied at the middle of the beam? Let $E = 30 \times 10^6$ psi. *Ans:* 2.03 kips.

12-21.   An aluminum rod 7 in. long, having two different cross-sectional areas, is inserted into a steel link as shown in the figure. If at 60°F no axial force exists in the aluminum rod, what will be the magnitude of this force when the temperature rises to 160°F? $E_a = 10^7$ psi and $\alpha_a = 12.0 \times 10^{-6}$ per °F; $E_s = 30 \times 10^6$ psi and $\alpha_s = 6.5 \times 10^{-6}$. *Ans:* 1,650 lb.

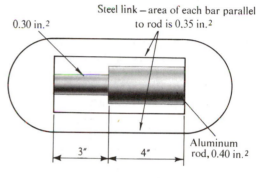

PROB. 12 – 21

12-22.   If the rod of Prob. 12-1, in addition to the application of the forces P_1 and P_2, experiences a drop in temperature of 100°F, what would the reaction on the right be? Let $\alpha = 6.5 \times 10^{-6}$ per °F, and $E = 30 \times 10^6$ psi. *Ans:* +14 k.

12-23.   If in Example 12-6, instead of a bolt, a rivet with no initial tension at 1,600°F is used in the assembly of the washers, what tensile stress will develop in the rivet when the temperature drops to 100°F? Let $E = 30 \times 10^6$ psi, $\sigma_{yp} = 40$ ksi, and $\alpha = 6.5 \times 10^{-6}$ per °F.

12-24.   Determine whether the small pressure vessel of Prob. 9-36 can operate satisfactorily at an internal pressure of 700 kPa. Here this conclusion is to be based only on the behavior of the bolts. For this purpose assume that 20 bolts each having a 200 mm² cross-sectional area at the root of the threads are to be used. The allowable stress for the bolts in tension is considered to be satisfactory at 200 MN/m²; however, at the root of the bolt threads a stress concentration factor of 2.5 is considered necessary. Moreover, for operation under pressure, the bolts will be tightened to develop in each one of them an initial force of 8 kN. *Ans:* 164 MPa.

12-25. A gray cast iron bolt $\frac{3}{4}$ in. in diameter passes through an aluminum tube 3.000 in. long which has an inside diameter of 1 in. and an outside diameter of 2 in. The nut on the end of the bolt is so tightened that at 60°F the stress in the tube becomes 600 psi. Compute the temperature at which the stress in the shank of the bolt will double. Let $E_{CI} = 12 \times 10^6$ psi and $\alpha_{CI} = 0.0000062$ per °F, and $E_{Al} = 10 \times 10^6$ psi and $\alpha_{Al} = 0.0000124$ per °F. *Ans:* 112.8°F.

12-26. A rigid bar, *ABC*, is supported by three tension members as shown in the figure. Members *BB'* and *CC'* are subjected to a drop in temperature of 50°C. Determine the resulting force in member *CC'*. For each tension member $A = 1,200$ mm², $E = 80\,000\,000$ kN/m², and $\alpha = 0.000\,010$ per °C.

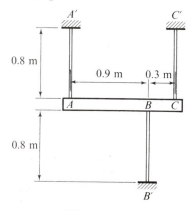

PROB. 12–26

12-27. A steel piano wire 30 in. long is stretched from the middle of an aluminum beam *AB* to a rigid support at *C* as shown in the figure. What is the increase in stress in the wire if the temperature drops 100°F? See Table 11 in the Appendix for the beam deflection formula. The cross-sectional area of the wire is 0.0001 in.²; $E = 30 \times 10^6$ psi. For the aluminum beam $EI = 1,040$ lb-in.² Let $\alpha_s = 6.5 \times 10^{-6}$ per °F, $\alpha_a = 12.9 \times 10^{-6}$ per °F. *Ans:* 6,500 psi.

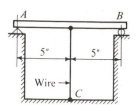

PROB. 12 – 27

12-28. Three steel wires attached to a rigid bar support a load of 300 lb. Initially this load is equally distributed between the three wires. What will each wire carry if the temperature of the right wire raises 84°F? Assume for all wires $E = 30 \times 10^6$ psi, $A = 0.011$ in.², and $\alpha = 6.5 \times 10^{-6}$ per °F. *Ans:* 70 lb, 160 lb, 70 lb.

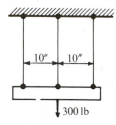

PROB. 12 – 28

12-29. A steel wire 5 m in length with cross-sectional area equal to 160 mm² is stretched tightly between the midpoint of the simple beam and the free end of the cantilever as shown in the figure. Determine the deflection of the end of the cantilever as a result of a temperature drop of 50°C. For steel wire: $E = 200$ GPa, $\alpha = 12 \times 10^{-6}$ per °C. For both beams: $I = 10 \times 10^{-6}$ m⁴ and $E = 10$ GPa.

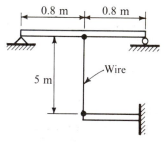

PROB. 12 – 29

12-30. Two steel cantilever beams *AB* and *CD* are connected by a taut steel wire *BC* having a length of 150 in. under initial no-load conditions, see figure. Determine the stress produced in the wire by a 2-kip load applied at C, and a temperature drop of 100°F, in the wire only. For beams AB and CD: $E = 30 \times 10^6$ psi, and $I = 24$ in.⁴ For wire BC: $E = 30 \times 10^6$ psi, $A = 0.1$ in.², and $\alpha = 6.0 \times 10^{-6}$ per °F. *Ans: 11.6 ksi.*

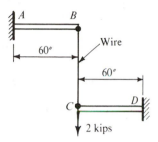

PROB. 12 – 30

12-31. At a given temperature an elastic cantilever just rests against the frictionless plane as shown in the figure. Calculate the maximum bending moment in the beam if the temperature of the beam is raised δT. Neglect the weight of the beam and the effect of axial force on bending deflection. The quantities A, E, I, and α are given. *Ans: $\alpha L \delta T/[1/(AE) + L^2/(3EI)]$.*

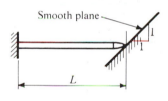

PROB. 12 – 31

12-32. A 1-in. stainless steel square bar 40 in. long lies between two parallel frictionless surfaces. If one side of this initially straight bar is maintained at a temperature 500°F higher than that of the opposite side, what is the deflection of the bar at the center from the chord through the ends? What moments applied at the ends would straighten out the bar? Assume that the temperature varies linearly across the thickness of the bar. Let $\alpha = 10 \times 10^{-6}$ per °F, and $E = 27 \times 10^6$ psi.

12-33. A thin ring is heated in oil 150°C above room temperature. In this condition the ring just

slips on a solid cylinder as shown in the figure. Assuming the cylinder to be completely rigid, (a) determine the hoop stress which develops in the ring upon cooling, and (b) determine what bearing pressure develops between the ring and the cylinder. Let $\alpha = 2 \times 10^{-5}$ per °C, and $E = 7 \times 10^7$ kN/m².

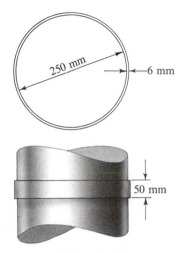

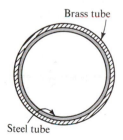

PROB. 12 – 33

12-34. A cylindrical pressure vessel is made by shrinking a brass shell over a mild steel shell. Both cylinders have a wall thickness of $\frac{1}{4}$ in. The nominal diameter of the vessel is 30 in. and is to be used in all calculations involving the diameter. When the brass cylinder is heated 100°F above room temperature, it exactly fits over the steel cylinder which is at room temperature. What is the stress in the brass cylinder when the composite vessels cool to room temperature? For brass $E_b = 16 \times 10^6$ psi, and $\alpha_b = 10.7 \times 10^{-6}$ per °F. For steel $E_s = 30 \times 10^6$ psi, and $\alpha_s = 6.7 \times 10^{-6}$ per °F. *Ans: 11 ksi.*

Brass tube

Steel tube

PROB. 12 – 34

*12-35. A circular rod of elastic material 1 is subjected to a small torsional deformation of θ radians per inch. Then, by a special process, a uniform layer of elastic cladding material 2 is deposited on the surface of the rod. After removal from the process and curing, the initial torque is released. What torsional deformation, in terms of the angle θ, of the composite rod will occur when the torque is released? Assume $G_2 = 10\ G_1$, $t/c = \frac{1}{40}$. *Ans:* $\theta/2$.

PROB. 12 – 35

*12-36. A tube of wall thickness 2 mm and radius 25 mm is attached at the ends by means of rigid flanges to a solid shaft of 25 mm diameter as shown in the figure of Prob. 3-25. (a) Prior to welding, the shaft is subjected to a torque of 200 N·m and maintained in this condition during the welding operation. If both the tube and the shaft are made of steel, what torque will be carried by the tube if the initial 200 N·m torque is released after welding is completed? (b) Same as (a) except the tube is made of aluminum while the shaft is of steel. (c) What additional torque must be applied opposite to the direction of the original 200 N·m torque so that the aluminum tube of (b) is just about to yield? At this stage what is the angle of twist in the 500 mm length of the aluminum tube? $G_{St} = 84\,000$ MN/m²; $G_{Al} = 28\,000$ MN/m²; for aluminum $\tau_{yp} = 150$ MN/m².

12-37. A solid brass, circular shaft is built in at both ends and two torques, $T_1 = 314$ lb-in.

and $T_2 = 628$ lb-in., are applied to it as shown in the figure. Determine the torque at A, and plot the torque and the angle of twist diagrams. Assume that the material behaves linearly elastically with a $G = 5.6 \times 10^6$ psi. The diameter $d_1 = 2.83$ in. and $d_2 = 2.38$ in.

12-38. Suppose that in Example 12-3 the cross-sectional area of each bar is 2 in.², the distance $L = 100$ in., and $\alpha = 30°$. The bars are made of steel with a well-defined yield stress of 40 ksi. Let the elastic modulus $E = 30 \times 10^6$ psi. During manufacture, by mistake, the middle bar was made 0.100 in. too short, i.e., before assembly, the three bars looked as shown in Fig. 12-9(b). (a) What residual stresses develop in the bars as a result of a forced assembly? Assume that no buckling of the bars can occur. (b) On the same graph show load-deflection diagrams analogous to Fig. 12-3(c) for the initially stress-free assembly, and the one with the residual stresses found above.

12-39. An L-shaped steel shaft of 2.125-in. diameter is built in at one end to a rigid wall and is simply supported at the other end as shown in the figure. In plan the bend is 90°. What bending moment will be developed at the built-in end due to the application of a 2,000-lb force at the corner of the shaft? Assume $E = 30 \times 10^6$ psi, $G = 12 \times 10^6$ psi, and for simplicity let $I = 1.00$ in.⁴ and $J = 2.00$ in.⁴ *Ans:* 49,600 lb-in.

2,000 lb

30″   30″

PROB. 12 – 39

12-40. A steel wire 4 m long is stretched from the end of a bent 20 mm diameter circular steel rod ABC to a rigid support D as shown in the figure. Compute the stresses acting on the element A, located at the support on top of the rod, caused by a temperature drop in the wire of 80°C. Do not compute principal stresses. A

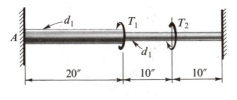

20″   10″   10″

PROB. 12 – 37

wire $= 6.5$ mm². Let $E = 200\,000$ MPa, $G = 84\,000$ MPa, and $\alpha = 12 \times 10^{-6}$ per °C.

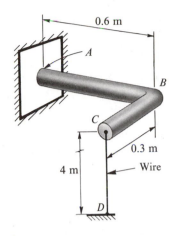

PROB. 12 – 40

12-41. A horizontal thin-walled tube is attached at one end to a rigid support and has a rigid arm clamped to it at its other end. (See figure.) A vertical rod passes through the arm and has an initial tension of 100 lbs. If the nut is tightened so as to move it 0.200 in. along the rod, what total force will develop in the rod? The mean diameter of the tube is 2 in., and its thickness is $1/(5\pi)$. Hence its $A = 0.40$ in.² and $J = 0.40$ in.⁴ The area of the rod is 0.00714 in.² Assume the arm to be infinitely rigid, but for the tube and the rod let $E = 30 \times 10^6$ psi, and $G = 12 \times 10^6$ psi.

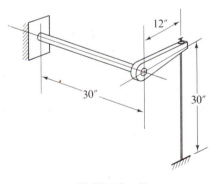

PROB. 12 – 41

12-42. The temperature in a furnace is measured by means of a stainless steel wire placed in it. The wire is fastened to the end of a cantilever beam outside the furnace. The strain measured by the strain gage glued to the outside of the beam is a measure of the temperature. Assuming that the full length of the wire is heated to the furnace temperature, what is the change in furnace temperature if the gage records a change in strain of -100×10^{-6} in. per inch. Assume that the wire has sufficient amount of initial tension to perform as intended. The mechanical properties of the materials are as follows: $\alpha_{ss} = 9.5 \times 10^{-6}$ per °F, $\alpha_a = 12 \times 10^{-6}$ per °F, $E_{ss} = 30 \times 10^6$ psi, $E_a = 10 \times 10^6$ psi, $A_{wire} = 5 \times 10^{-4}$ in.², $I_{beam} = 6.5 \times 10^{-4}$ in.⁴ The depth of the small beam is 0.25 in. *Ans: 96.8°F.*

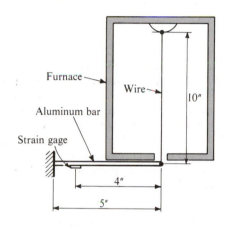

PROB. 12 – 42

12-43. Rework Example 12-10 by treating R_B and X_B as redundants. Employ the superposition equations, Eq. 12-11.

12-44 through 12-46. (a) For the beams loaded as shown in the figures, using the method of superposition, determine the redundant reactions. In all cases EI is constant. (b) For the same beams, plot shear and moment diagrams. *Ans: M_A is given in parentheses by the figures.*

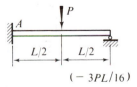

$(-\,3PL/16\,)$

PROB. 12 – 44

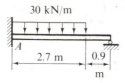

30 kN/m

A

2.7 m | 0.9 m

PROB. 12 – 45

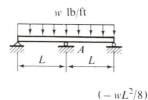

w lb/ft

A

L | L

$(-wL^2/8)$

PROB. 12 – 46

12-47. Using the superposition equations, Eq. 12-11, determine the reactions at the supports caused by the applied load for the beam shown in the figure. Treat the moment and the vertical reaction at b as redundants, and determine the end deflection and rotation of a determinate beam by any of the previously developed methods. If desired, the answer given to Prob. 11-11 may be used. *Ans:* $M_a = -kL^3/30$, $M_b = -kL^3/20$.

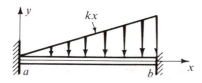

y

kx

a b x

PROB. 12 – 47

12-48. Using the formulas for determinate cases given in Table 11 of the Appendix and superposition, determine the fixed end moments for the beam loaded as shown in the figure.

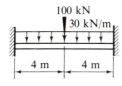

100 kN

30 kN/m

4 m | 4 m

PROB. 12 – 48

12-49. Revise the conditions of Prob. 12-47 by assuming that end b is simply supported. For this new problem, determine the reactions using the displacement method. Make use of the answers given to Prob. 12-47, but exercise care in interpreting the signs.

12-50. Rework Example 12-12 with another force P applied at $\frac{1}{4}$ the length of BC on the central bar BC above C.

12-51. Five steel bars each having the cross-sectional area of 500 mm² are assembled in a symmetrical manner as shown in the figure. Assume that the steel behaves as linearly elastic-plastic material with $E = 200 \times 10^6$ kN/m², and $\sigma_{yp} = 250$ MN/m². Determine and plot the load deflection characteristics of joint A due to an application of a downward force P. Assume that initially all bars are free of stress.

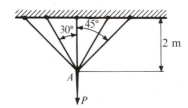

30° 45°

2 m

A

P

PROB. 12 – 51

12-52. Rework Example 12-12 assuming that member AC has twice as large an area as either member BC or DC.

12-53. In Example 12-13, by determining M_{ba} from the rotations, show that $M_{ba} = -M_{bc}$. Similarly show that $M_{cb} = 0$.

12-54. For the continuous beam loaded as shown in the figure, using the displacement method, determine the bending moment acting over the middle support, and plot shear and

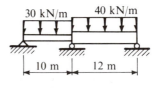

30 kN/m 40 kN/m

10 m | 12 m

PROB. 12 – 54

moment diagrams. The I of the beam in the right span is twice as large as that in the left span. *Ans:* $M_{max} = 490$ kN·m, $M_{min} = -504$ kN·m.

12-55. Rework Prob. 12-54 assuming that the left support is fixed. *Ans:* $M_B = -459$ kN·m.

12-56 and 12-57. For the beams loaded as shown in the figures, using the moment-area method, determine the redundant reactions and plot shear and moment diagrams. In all cases EI is constant. (*Hint:* In Prob. 12-57 treat the reaction on the right as the redundant.) *Ans: M_A is* given in parentheses by the figure.

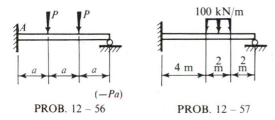

PROB. 12 – 56   PROB. 12 – 57

12-58. (a) Using the moment area method, determine the redundant moment at the built in end for the beam shown in the figure and plot the shear and moment diagrams. Neglect the weight of the beam. (b) Select the depth for a 200 mm wide wooden beam using an allowable bending stress of 8 000 kN/m² and a shearing stress of 1 000 kN/m². (c) Determine the maximum deflection of the beam between the supports and the maximum deflection of the overhang. For wood, E is 10^7 kN/m². *Ans:* (a) ± 45 kN·m.

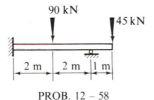

PROB. 12 – 58

12-59. (a) Using the moment area method, determine the redundant moment at the built-in end for the beam shown in the figure, and plot the shear and moment diagrams. Neglect the weight of the beam. (b) Select a W beam using an allowable bending stress of 18,000 psi and a shearing stress of 12,000 psi. (c) Determine the

maximum deflection of the beam between the supports and the maximum deflection of the overhang. Let $E = 29 \times 10^6$ psi. *Ans:* (b) W 16 × 26.

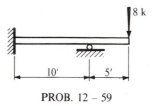

PROB. 12 – 59

12-60. For the beam loaded as shown in the figure, (a) determine the ratio of the moment at the built in end to the applied moment M_A; (b) determine the rotation of the end A. The EI is constant. *Ans:* $-\frac{1}{2}$, $-M_A L/(4EI)$

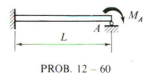

PROB. 12 – 60

12-61. For the beam loaded as shown in the figure, (a) determine the ratio of the moment at the built in end to the applied moment M_A; (b) determine the rotation of the end A.

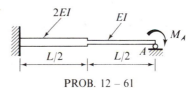

PROB. 12 – 61

12-62. Using the moment area method, show that the maximum deflection of a beam fixed at both ends and carrying a uniformly distributed load is one fifth the maximum deflection of the same beam simply supported. The EI is constant.

12-63 and 12-64. For the beams loaded as shown in the figures, using the moment-area method, (a) determine the fixed-end moments and plot shear and moment diagrams. Neglect the weight of the beams. (b) Express the maximum deflection in terms of the loads, distances, and EI. No adjustment for units need be made. *Ans:* For (b) by the figures.

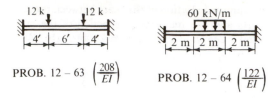

PROB. 12 – 63 $\left(\dfrac{208}{EI}\right)$　　PROB. 12 – 64 $\left(\dfrac{122}{EI}\right)$

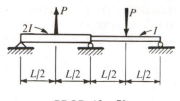

PROB. 12 – 70

12-65.　A W 12 × 36 beam is loaded as shown in the figure. Using the moment-area method and neglecting the weight of the beam, determine (a) the fixed-end moments, (b) the maximum bending stress, (c) the deflection at the midspan. Let $E = 29 \times 10^6$ psi.　*Ans:* (b) 13.9 ksi; (c) −0.133 in.

12-71.　For the continuous beam loaded as shown in the figure, using the moment-area method, determine the bending moment directly over the center suppport. The *EI* is constant. *Ans:* −13.44 k-ft.

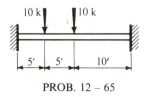

PROB. 12 – 65

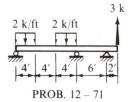

PROB. 12 – 71

12-66 through 12-68.　For beams of constant flexural rigidity loaded as shown in the figures, using the moment-area method, determine the fixed-end moments.

12-72.　Rework Example 12-17 after assuming that both supports *A* and *C* are fixed.　*Ans:* $M_A = -22.2$ k-ft, $M_B = -15.7$ k-ft.

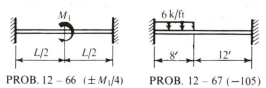

PROB. 12 – 66　$(\pm M_1/4)$　　PROB. 12 – 67 (-105)

12-73.　Rework Example 12-17 after assuming that the right support *C* is pinned.　*Ans:* $M_B = -28.9$ k-ft.

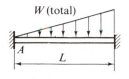

PROB. 12 – 68　$(-WL/15)$

12-74.　A beam of constant flexural rigidity *EI* is built in at both its ends a distance *L* apart. If one of the supports settles vertically downward an amount Δ relative to the other support (without causing any rotation), what moments will be induced at the ends?　*Ans:* $\pm 6\,EI\Delta/L^2$.

12-69.　Rework Prob. 12-54 using the moment-area method.

12-70.　A beam having a variable moment of inertia is loaded as shown in the figure. (a) Using the moment-area method determine the moment over the middle support. (b) Find all reactions.

12-75.　For the continuous beam loaded as shown in the figure, using the moment-area

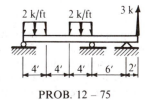

PROB. 12 – 75

method or the three-moment equation, determine the bending moment directly over the center support. The EI is constant. *Ans:* -13.44 k-ft.

12-76. A beam of constant flexural rigidity EI is continuous over four spans of equal length L. Plot shear and moment diagrams for this beam if throughout its length it is loaded with a uniformly distributed load of w_o lb per foot. Use the three-moment equations to determine the moments over the supports. (*Hint:* Take advantage of symmetry.) *Ans:* Moment over the center support is $w_o L^2/14$, the reactions at ends are $11 w_o L/28$ each.

12-77. Rework Example 12-17, Fig. 12-23, after assuming that both supports A and C are fixed. Use the three-moment equation. *Ans:* $M_A = -22.2$ k-ft, $M_B = -15.7$ k-ft.

12-78. A restrained beam of ductile material is loaded with two concentrated forces P as shown in the figure. Determine the limit loads P_{ult}. Neglect the weight of the beam. (*Hint:* The possibility of a plastic hinge must be checked under each load.) *Ans:* $4M_p/L$.

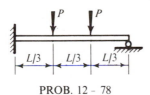

PROB. 12 – 78

12-79. Using the three moment equation, determine the moments over the supports for the beam loaded as shown in the figure. The EI is constant. *Ans:* -60 k-ft, -2.7 k-ft, -27.4 k-ft, and 0.

12-80. Using limit analysis, calculate the value of P that would cause (flexural) collapse of the two-span beam shown. The beam has a rectangular cross section 120 mm wide and 300 mm deep. The yield stress is 15 MPa. Neglect the weight of the beam.

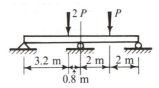

PROB. 12 – 80

12-81. A two-span, continuous, prismatic beam carries a concentrated force P in the middle of one span, and a uniformly distributed load w_o in the other span. Using the plastic method of analysis, determine the ratio of $w_o L$ to P necessary for the (flexural) collapse to occur in both spans simultaneously. Neglect the weight of the beam.

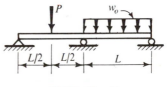

PROB. 12 – 81

12-82. Using the limit analysis approach, select a steel W section for the loading condition shown in the figure. Assume $\sigma_{\text{yp}} = 40$ ksi, a shape factor of 1.10 and use a load factor of 2. *Ans:* W 16 × 26.

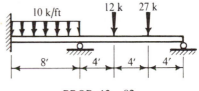

PROB. 12 – 82

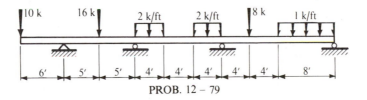

PROB. 12 – 79

12-83. A prismatic "weightless" beam is loaded as shown in the figure. What is the magnitude of the governing maximum plastic moment? *Ans:* $w_o L^2/3$.

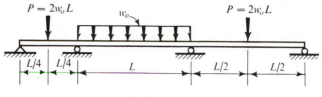

PROB. 12 – 83

13 Columns

13-1. INTRODUCTION

At the beginning of this text it was stated that the selection of structural and machine elements is based on three characteristics: strength, stiffness, and stability. The procedures of stress and deformation analyses were discussed in some detail in the preceding chapters. In all cases hitherto considered, the members were assumed to be in stable equilibrium. But not all structural arrangements are necessarily stable. For example, consider a stick one meter long having the cross-sectional area of an ordinary pencil. If this stick were stood on end, one might conclude that the stress at the base would be equal to the total weight of the stick divided by its cross-sectional area. However, the equilibrium of this stick is precarious. With the smallest imperfection in the stick or the lightest gust of wind the stick would fall down, thus the aforementioned stress calculation is meaningless. This physically obvious example is introduced to accustom the reader to the thought that stability considerations may be primary in some problems.

The chief concern of this chapter will be the analysis of columns, i.e., of compression members of constant cross-sectional area. In common with all instability problems, the column problem is complicated by the fact that different phenomena contribute to the capacity of a column, depending on its length. Thus if a square-ended steel rod of, say, 5 mm diameter were made 10 mm long to act as a "column," no question of instability would enter, and considerable force could be applied to this piece. On the other hand, if the same rod were made a meter long, at a smaller applied force than the one a short piece could carry, the rod could become laterally unstable through sideways buckling and could collapse. An ordinary slender yardstick, if subject to an axial compression, fails in this manner. The consideration of material strength alone is not sufficient to predict the behavior of such a member.

The same phenomenon occurs in numerous other situations where compressive stresses are present. Thin sheets, although fully capable of sustaining tensile loadings, are very poor in transmitting compression. Narrow beams, unbraced laterally, can snap sidewise and collapse under an applied load. Vacuum tanks, as well as submarine hulls, unless properly designed can severely distort under external pressure and can assume shapes that differ drastically from their original geometry. A thin-walled tube can wrinkle like tissue paper when subjected to a torque. For example see Fig. 13-1.* During some stages of firing, the thin casings of missiles are critically loaded in compression. These are crucially important problems of engineering design. Moreover, usually the buckling or wrinkling phenomena observed in loaded members occurs rather suddenly. For this reason many structural instability failures are spectacular and are very dangerous.

(a)

(b)

Fig. 13-1. (a) Typical buckle pattern for thin-walled cylinder in compression; (b) typical buckle pattern for pressurized cylinder in torsion. (Courtesy Dr. L. A. Harris of North American Aviation, Inc.)

A vast number of structural instability problems suggested by the above listing are beyond the scope of this text.† Here the column problem will mainly be considered. The range of column lengths at which instability occurs, and formulas for the carrying capacity of slender columns, will be

*Figures taken from L. A. Harris, H. W. Suer, and W. T. Skene, "Model Investigations of Unstiffened and Stiffened Circular Shells," *Experimental Mechanics*, July 1961, pp. 3 and 5.

†See, for example, S. P. Timoshenko and J. M. Gere, *Theory of Elastic Stability* (2nd ed.), New York: McGraw-Hill, 1961.

established. Through algebraic manipulations, the expressions to be derived for the carrying capacity of columns are made to look like the usual stress formulas. This customary conversion of column formulas leads some engineers to extrapolate the usual strength formulas into the range where the instability phenomenon governs. The history of some engineering structural failures attests to this tragic blunder; bridges, buildings, machines, and airplanes turned to "spaghetti."

At the end of the chapter the problem of lateral instability of beams is briefly discussed.

13-2. STABILITY OF EQUILIBRIUM

A perfectly straight needle balanced on its tip may be said to be in equilibrium. However, the least disturbance or imperfection in its manufacture would make this equilibrium impossible. This kind of equilibrium is said to be unstable, and it is imperative to avoid analogous situations in structural systems.

To clarify the problem further, again consider a vertical rigid bar with a torsional spring of stiffness k at the base as shown in Fig. 13-2(a).* The behavior of such a bar subjected to a vertical force P and a horizontal force F is shown in Fig. 13-2(b) for a large and a small F. The question then arises: How will this system behave if $F = 0$? This is the limiting case, and it corresponds to the investigation of pure buckling.

To answer this question analytically, the system must be deliberately displaced a small (infinitesimal) amount consistent with the boundary conditions. Then, if the restoring forces are greater than the forces tending to upset the system, the system is stable, and vice versa.

The rigid bar shown in Fig. 13-2(a) can experience only rotation as it cannot bend; i.e., the system has one degree of freedom. For an assumed rotation θ the restoring moment is $k\theta$, and, with $F = 0$, the upsetting moment is $PL \sin \theta \approx PL\theta$. Therefore, if

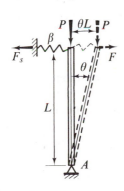

Fig. 13-A

$$k\theta > PL\theta \qquad \text{the system is stable}$$

and if

$$k\theta < PL\theta \qquad \text{the system is unstable}$$

Right at the transition point, $k\theta = PL\theta$, and the equilibrium is neither stable nor unstable but is neutral. The force associated with this condition is the critical or buckling load, which will be designated P_{cr}. For the bar system considered

$$P_{cr} = k/L \qquad (13\text{-}1)$$

*Instead of a torsional spring of stiffness k at the base, one might introduce a linear spring of stiffness $\beta = k/L^2$ in the horizontal direction at the top of the bar, as shown in Fig. 13-A. Then, assuming a *small* rotation θ of the rigid bar, $\sum M_A = 0$ gives $P_{cr}\theta L + FL = F_s L$ Since $F_s = \beta\theta L$, $P_{cr} = (\beta\theta L - F)/\theta$ And, with $F = 0$, since $\beta = k/L^2$, one obtains $P_{cr} = \beta L = k/L$.

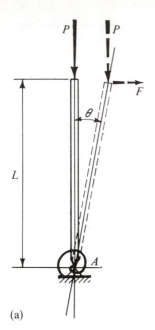

(a)

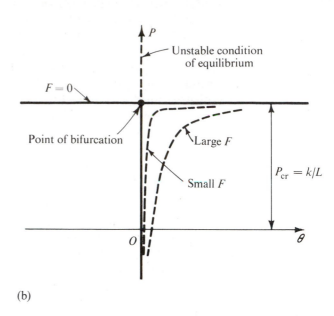

(b)

Fig. 13-2. Buckling behavior of a rigid bar

This condition establishes the inception of buckling. At this force two equilibrium positions are possible, the straight form and a deflected form infinitesimally near it. Since two branches of the solution are thus possible, this is called the *bifurcation point* of the equilibrium solution. For $P > k/L$, the system is unstable. As the solution has been linearized (by assuming θ to be small), there is no possibility of having θ become indefinitely large at P_{cr}. In the sense of large displacements, there is always a point of stable equilibrium at $\theta < \pi$.*

The behavior of perfectly straight, concentrically loaded, elastic columns, i.e., of ideal columns, is highly analogous to the behavior described in the simple example above. From a linearized formulation of the problem, the critical buckling loads can be determined. Examples will be given in the following articles.

The critical loads do not describe the action of the buckling itself. By using an exact differential equation of the elastic curve for large deflections, it is possible† to find equilibrium positions higher than P_{cr} corresponding to the applied force P. The results of such an analysis are illustrated in Fig. 13-3. Note especially that increasing P_{cr} by a mere 1.5 per cent causes a maximum

*See, for example, E. P. Popov, *Introduction to Mechanics of Solids*, Englewood Cliffs, N.J.: Prentice-Hall, 1968.

†See S. P. Timoshenko and J. M. Gere, *Theory of Elastic Stability* (2nd ed.), New York: McGraw-Hill, 1961, p. 76.

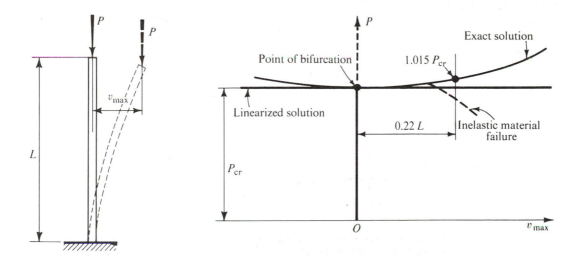

Fig. 13-3. Behavior of an ideal elastic column

sideways deflection of 22 per cent of the column length.* For practical reasons such enormous deflections can seldom be tolerated. Moreover, the material usually cannot resist the induced bending stresses. Therefore, real columns fail inelastically. In the vast majority of engineering applications, P_{cr} represents the ultimate capacity for a straight, concentrically loaded column.

13-3. THE EULER FORMULA FOR A PIN-ENDED COLUMN

At the critical load, a column that is circular or tubular in its cross-sectional area may buckle sideways in any direction. In the more usual case, a compression member does not possess equal flexural rigidity in all directions. The moment of inertia I_{xx} is a maximum around one centroidal axis of the cross-sectional area, Fig. 13-4, a minimum around the other. The significant flexural rigidity EI of a column depends on the *minimum I*, and at the critical load a column buckles either to one side or the other in the *X-X* plane. The use of a minimum *I* in the derivation that follows is understood.

Consider a column with its ends free to rotate around frictionless pins. Such columns, having pinned supports at both ends, are called *pin-ended columns*. In Fig. 13-5(a) the equivalent boundary conditions are shown by means of rounded ends in frictionless supports. *The buckled shape shown is possible only at a critical or Euler load, as prior to this load the column remains straight.* The *smallest* force at which a buckled shape is possible is the critical force.

*The fact that an elastic column continues to carry a load beyond the buckling stage can be demonstrated by applying a force in excess of the buckling load to a flexible bar or plate such as a carpenter's saw.

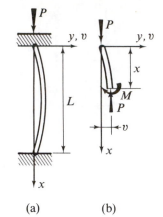

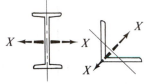

(a)   (b)

Fig. 13-4. Columns buckle in the plane of the major axis of the cross-section.

Fig. 13-5. Pin- or round-ended column used in the derivation of the Euler load.

It is seen from Fig. 13-5(b) that the bending moment at any section is $-Pv$, which conforms to the usual sign convention for beams.* With this value of the bending moment, the differential equation for the elastic curve, Eq. 11-10, *at* the critical load is

$$\frac{d^2v}{dx^2} = \frac{M}{EI} = -\frac{P}{EI}v \qquad (13\text{-}2)$$

by letting $\lambda^2 = P/(EI)$, and transposing, gives

$$\frac{d^2v}{dx^2} + \lambda^2 v = 0 \qquad (13\text{-}3)$$

This is an equation of the same form as the one for simple harmonic motion, and its solution is

$$v = A \sin \lambda x + B \cos \lambda x \qquad (13\text{-}4)$$

where A and B are arbitrary constants that must be determined from the conditions at the boundary. These conditions are

$$v(0) = 0 \quad \text{and} \quad v(L) = 0$$

Hence $v(0) = 0 = A \sin 0 + B \cos 0 \quad \text{or} \quad B = 0$

and $v(L) = 0 = A \sin \lambda L \qquad (13\text{-}5)$

Equation 13-5 can be satisfied by taking $A = 0$; however, as can be seen from Eq. 13-4, this corresponds to a condition of no buckling of the

---

*For the positive direction of the deflection v shown, the bending moment is negative. (Rotate the diagram through 90° counterclockwise.) If the column were deflected in the other direction, the moment would be positive. However, v would be negative. Hence, to make Pv positive, it must likewise be treated as a negative quantity.

column and for an instability problem represents a trivial solution. Alternatively, Eq. 13-5 is also satisfied if the sine term vanishes. This requires that λL equal $n\pi$, where n is an integer. Hence, using the earlier definition of λ as $\sqrt{P/(EI)}$, and solving $\sqrt{P/(EI)}L = n\pi$ for P, which is the critical force since it makes the curved shape possible,

$$P_{cr} = \frac{n^2\pi^2 EI}{L^2} \tag{13-6}$$

However, since the least value of the critical or Euler load is sought, n in Eq. 13-6 must be taken as one. Therefore the *Euler load formula* for a column pin-ended at both ends* is

$$P_{cr} = \frac{\pi^2 EI}{L^2} \tag{13-7}$$

where I is the *least* moment of inertia of the constant cross-sectional area of a column, and L is its length. This case of a column with pinned or round ends at both ends is often referred to as the *fundamental case*.

According to Eq. 13-4, at the critical load, the equation of the elastic curve is

$$v = A \sin \lambda x \tag{13-8}$$

This is the characteristic function or eigenfunction of this problem, and, since n can assume any integral value, there is an infinite number of such functions. In this linearized solution the amplitude A of the buckling mode remains indeterminate. For $n = 1$, the elastic curve is a half-wave sine curve. This shape, together with the modes corresponding to $n = 2$ and $n = 3$, is shown in Fig. 13-6. The higher modes have no physical significance in bucking problems since the least critical load occurs at $n = 1$.

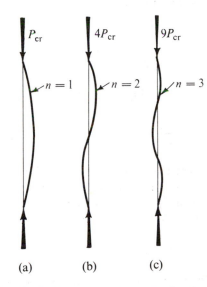

Fig. 13-6. Pin-ended column and its first three buckling modes

*13-4. EULER FORMULAS FOR COLUMNS WITH DIFFERENT END RESTRAINTS

The same procedure as that discussed in the preceding article can be used to determine the critical loads for columns with different boundary conditions. The solutions of these problems are very sensitive to the end restraints. Consider, for example, a column with one end fixed and the other pinned as shown in Fig. 13-7 (where it is shown in a horizontal position for

*This formula was derived by the great mathematician Leonhard Euler in 1757.

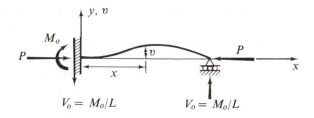

$V_o = M_o/L$ $V_o = M_o/L$

Fig. 13-7

convenience). In this case, the unknown moment at the fixed end and the necessary support reactions to maintain the column in equilibrium must also be taken into account in setting up the differential equation for the elastic curve at the critical load:

$$\frac{d^2v}{dx^2} = \frac{M}{EI} = \frac{-Pv + M_o(1 - x/L)}{EI} \tag{13-9}$$

Letting $\lambda^2 = P/(EI)$ as before, and transposing, gives

$$\frac{d^2v}{dx^2} + \lambda^2 v = \frac{\lambda^2 M_o}{P}\left(1 - \frac{x}{L}\right) \tag{13-9a}$$

The *homogeneous solution* of this differential equation, i.e., when the right side equals zero, is the same as that given in Eq. 13-4. The *particular solution*, due to the nonzero right side, is given by dividing the term on that side by λ^2. The complete solution is then given as

$$v = A \sin \lambda x + B \cos \lambda x + (M_o/P)(1 - x/L) \tag{13-10}$$

where A and B are arbitrary constants and M_o the unknown moment at the fixed end. The three boundary conditions, namely, $v(0) = v(L) = 0$ and $v'(0) = 0$ are then used and, analogously to Eq. 13-5, one obtains the transcendental equation

$$\lambda L = \tan \lambda L \tag{13-11}$$

which must be satisfied for a nontrivial equilibrium shape of the column at the critical load. The smallest root* of Eq. 13-11 is

$$\lambda L = 4.493$$

whence the corresponding critical load for a column fixed at one end and pinned at the other is

$$P_{cr} = 20.19EI/L^2 = 2.05\pi^2 EI/L^2 \tag{13-12}$$

*See S. P. Timoshenko and J. M. Gere, *Theory of Elastic Stability* (2nd ed.), New York: McGraw-Hill, 1961.

In the case of a column fixed at both ends, Fig. 13-8(d), the critical load is

$$P_{cr} = 4\pi^2 EI/L^2 \tag{13-13}$$

The last two equations show that by restraining the ends the critical loads are substantially raised above those in the fundamental case, Eq. 13-7. On the other hand, the critical load for a free-standing column* (Fig. 13-8(b)) with a load at the top is

$$P_{cr} = \pi^2 EI/(4L^2) \tag{13-14}$$

In this extreme case the critical load is only one-fourth of that for the fundamental case.

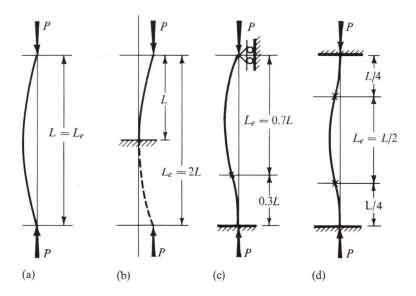

Fig. 13-8. Effective lengths of columns with different restraints

All the above formulas can be made to resemble the fundamental case, providing that the effective column lengths are used instead of the actual column length. This length turns out to be the distance between the inflection points on the elastic curves, or hinges if there are any. The effective column length L_e for the fundamental case is L, but for the cases discussed above it is $0.7L$, $0.5L$, and $2L$, respectively. For a general case, $L_e = KL$, where K is the effective length factor which depends on the end restraints.

In contrast to the classical cases shown in Fig. 13-8, actual compression members are seldom truly pinned or completely fixed against rotation at the ends. Because of the uncertainty regarding the fixity of the ends, columns

*A telephone pole having no external braces and with a heavy transformer at the top is an example.

are often assumed to be pin-ended. With the exception of the case shown in Fig. 13-8(b), where it cannot be used, this procedure is conservative.

The above equations become completely misleading in the inelastic range and must not be used in the form given (see Art. 13-8).

The critical-load formulas for columns are truly remarkable, since *no strength property of the material appears in them*, yet they determine the carrying capacity of a column. The only material property involved is the elastic modulus E, which physically represents the stiffness characteristic of the material.

*13-5. ELASTIC BUCKLING OF COLUMNS USING FOURTH-ORDER DIFFERENTIAL EQUATION

The Euler formulas for the critical loads of columns can also be obtained by solving a fourth-order differential equation of the elastic curve at the critical load and using appropriate boundary conditions, depending on the end restraints. Such an equation can be obtained by considering the equilibrium of an infinitesimal element* as was done in the case of beams (Chapter 11), but here it suffices to note that it may be written in the form

$$\frac{d^4v}{dx^4} + \lambda^2\frac{d^2v}{dx^2} = 0 \qquad (13\text{-}15)$$

where $\lambda^2 = P/(EI)$ as before (note that two successive differentiations of Eq. 13-3 or 13-9a lead to Eq. 13-15). The solution of this fourth-order homogeneous differential equation and several of its derivatives are given as

$$v = C_1 \sin \lambda x + C_2 \cos \lambda x + C_3 x + C_4 \qquad (13\text{-}16a)$$

$$v' = C_1\lambda \cos \lambda x - C_2\lambda \sin \lambda x + C_3 \qquad (13\text{-}16b)$$

$$v'' = -C_1\lambda^2 \sin \lambda x - C_2\lambda^2 \cos \lambda x \qquad (13\text{-}16c)$$

$$v''' = -C_1\lambda^3 \cos \lambda x + C_2\lambda^3 \sin \lambda x \qquad (13\text{-}16d)$$

For a pin-ended column, the boundary conditions are

$$v(0) = 0, \qquad v(L) = 0, \qquad M(0) = EIv''(0) = 0$$

and
$$M(L) = EIv''(L) = 0$$

Using these conditions with Eqs. 13-16(a) and (c), one obtains

$$C_2 \qquad\qquad + C_4 = 0$$
$$C_1 \sin \lambda L + C_2 \cos \lambda L + C_3 L + C_4 = 0$$
$$-C_2\lambda^2 EI \qquad\qquad = 0$$
$$-C_1\lambda^2 EI \sin \lambda L - C_2\lambda^2 EI \cos \lambda L \qquad\qquad = 0$$

*See E. P. Popov, *Introduction to Mechanics of Solids*, Englewood Cliffs, N.J.: Prentice-Hall, 1968, p. 522.

To satisfy this set of equations, C_1, C_2, C_3, and C_4 could be set equal to zero, which, is a trivial solution however. A nontrivial solution demands that the determinant of the coefficients for a set of homogeneous algebraic equations be equal to zero. Therefore, with $\lambda^2\,EI = P$,

$$\begin{vmatrix} 0 & 1 & 0 & 1 \\ \sin \lambda L & \cos \lambda L & L & 1 \\ 0 & -P & 0 & 0 \\ -P \sin \lambda L & -P \cos \lambda L & 0 & 0 \end{vmatrix} = 0$$

The evaluation of this determinant leads to $\sin \lambda L = 0$, which is precisely the same condition as given by Eq. 13-5.

This approach is advantageous in problems with different boundary conditions where the axial force and EI remain constant throughout the length of the column. The method cannot be applied if the axial force extends over only a part of a member.

*13-6. ANALYSIS OF BEAM-COLUMNS

A beam that is acted on by an axial compressive force in addition to transversely applied loads is referred to as a *beam-column*. Detailed treatment of this topic will not be presented in this book,* but a simple case will be examined to illustrate the significant effect of the axial force in such problems. Consider, for example, an elastic beam-column subjected to an axial force P and an upward transverse load F at its midspan, Fig. 13-9(a). The free-body diagram for the deflected beam-column is shown in Fig. 13-9(b).

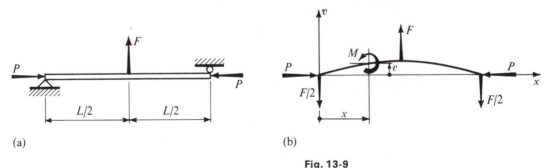

(a)   (b)

Fig. 13-9

This diagram permits the formulation of the total bending moment M, which includes the effect of the axial force P multiplied by the deflection v. The total moment divided by EI could be set equal to the expression for the

*See books on stability :for example, S. P. Timoshenko and J. M. Gere, *Theory of Elastic Stability* (2nd ed.), New York: McGraw-Hill, 1961, and A. Chajes, *Principles of Structural Stability Theory*, Englewood Cliffs, Prentice-Hall, 1974.

exact curvature, Eq. 11-8. However, this curvature will be taken as d^2v/dx^2 customarily; i.e., the expression $M = EIv''$ will be accepted. This yields accurate results only for small deflections and rotations, and the acceptance of this approximation will lead to infinite deflections at the critical loads.

Thus, using the relation $M = EIv''$ and noting that for the left side of the span $M = -(F/2)x - Pv$, one has

$$EIv'' = M = -Pv - (F/2)x \qquad (0 \le x \le L/2)$$

or
$$EIv'' + Pv = -(F/2)x$$

By dividing through by EI and letting $\lambda^2 = P/(EI)$, after some simplification, the governing differential equation becomes

$$\frac{d^2v}{dx^2} + \lambda^2 v = -\frac{\lambda^2 F}{2P}x \qquad (0 \le x \le L/2) \tag{13-17}$$

The homogeneous solution for this differential equation has the well-known form of the one for simple harmonic motion; the particular solution equals the right-hand term divided by λ^2. Therefore, the complete solution is

$$v = C_1 \sin \lambda x + C_2 \cos \lambda x - (F/2P)x \tag{13-18}$$

The constants C_1 and C_2 follow from the boundary condition $v(0) = 0$ and from a condition of symmetry $v'(L/2) = 0$. The first condition gives

$$v(0) = C_2 = 0$$

Since
$$v' = C_1 \lambda \cos \lambda x - C_2 \lambda \sin \lambda x - F/(2P)$$

with C_2 already known to be zero, the second condition gives

$$v'(L/2) = C_1 \lambda \cos \lambda L/2 - F/(2P) = 0$$

or
$$C_1 = F/[2P\lambda \cos (\lambda L/2)]$$

On substituting this constant into Eq. 13-18,

$$v = \frac{F}{2P\lambda} \frac{1}{\cos \lambda L/2} \sin \lambda x - \frac{F}{2P}x \tag{13-19}$$

The maximum deflection occurs at $x = L/2$. Thus, after some simplifications,

$$v_{max} = [F/(2P\lambda)](\tan \lambda L/2 - \lambda L/2) \tag{13-20}$$

From this it can be concluded that the absolute maximum moment, occurring at the midspan, is

$$M_{max} = \left| -\frac{FL}{4} - Pv_{max} \right| = \frac{F}{2\lambda} \tan \frac{\lambda L}{2} \tag{13-21}$$

Note that the expressions given by Eqs. 13-19, 13-20, and 13-21 become infinite if $\lambda L/2$ is a multiple of $\pi/2$ since this makes $\cos \lambda L/2$ equal to zero and $\tan \lambda L/2$ infinite. Stated algebraically this occurs when

$$\frac{\lambda L}{2} = \sqrt{\frac{P}{EI}} \frac{L}{2} = \frac{n\pi}{2} \qquad (13\text{-}22)$$

where n is an integer. Solving this equation for P, one obtains the magnitude of P causing either infinite deflections or bending moment. This corresponds to the condition of the critical axial force P_{cr} for this bar:

$$P_{\text{cr}} = \frac{n^2 \pi^2 EI}{L^2} \qquad (13\text{-}23)$$

For the smallest critical force, the integer $n = 1$. This result is the Euler buckling load that was discussed in Arts. 13-3 and 13-4.

It is significant to note that bending moments in slender members may be substantially increased by the presence of axial compressive forces, Eq. 13-21. When such forces exist, the deflections caused by the transverse loading are magnified, as given by Eqs. 13-19 and 13-20 (for example, verify that, for $\lambda L/2 = \pi/4$, $v_{\max} = FL^3/(36EI) > FL^3/(48EI)$, which is the maximum deflection in the absence of P, i.e., $\lambda = 0$). For tensile forces, on the other hand, the deflections are reduced.

13-7. LIMITATIONS OF THE EULER FORMULAS

The elastic modulus E was used in the derivation of the Euler formulas for columns, therefore, all the reasoning presented earlier is applicable *while the material behavior remains linearly elastic.* To bring out this significant limitation, Eq. 13-7 will be written in a different form. By definition, $I = Ar^2$, where A is the cross-sectional area and r is its *radius of gyration.* Substitution of this relation into Eq. 13-7 gives

$$P_{\text{cr}} = \frac{\pi^2 EI}{L^2} = \frac{\pi^2 E A r^2}{L^2}$$

or
$$\sigma_{\text{cr}} = \frac{P_{\text{cr}}}{A} = \frac{\pi^2 E}{(L/r)^2} \qquad (13\text{-}24)$$

where the *critical stress* σ_{cr} for a column is defined as P_{cr}/A, i.e., as an *average* stress over the cross-sectional area A of a column at the critical load P_{cr}. The length* of the column is L, and r is the *least* radius of gyration of the cross-sectional area, since the original Euler formula is in terms of the minimum I. The ratio L/r of the column length to the *least* radius of gyration

*By using the effective length L_e, the expression becomes general.

is called the column *slenderness ratio*. *No factor of safety is included in the above equation.*

A graphical interpretation of Eq. 13-24 is shown in Fig. 13-10, where the critical column stress is plotted versus the slenderness ratio for three different materials. For each material E is constant, and the resulting curve is a hyperbola. However, since Eq. 13-24 is based on the elastic behavior of a material, σ_{cr} determined by this equation cannot exceed the proportional limit of a material. Therefore the hyperbolas shown in Fig. 13-10 are drawn dashed beyond the individual material's proportional limit, and these portions of the curves *cannot be used*. The necessary modifications of Eq. 13-24 to include inelastic material response will be discussed in the next article.

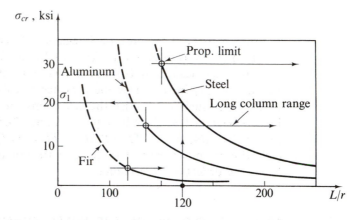

Fig. 13-10. Variation of the critical column stress with the slenderness ratio for three different materials

The useful portions of the hyperbolas do not represent the behavior of one column, but rather the behavior of an infinite number of ideal columns. For example, a particular steel column, say with an $L/r = 120$, may at the most carry a load of $\sigma_1 A$. Also note that σ_{cr} *always decreases with increasing ratios of L/r*. Moreover, note that, a precise definition of a long column is now possible with the aid of these diagrams. Thus, a column is said to be long if the elastic Euler formula applies. The beginning of the long-column range is shown for three materials in Fig. 13-10.

EXAMPLE 13-1

Find the shortest length L for a pin-ended steel column having a cross-sectional area of 60 mm by 100 mm for which the elastic Euler formula applies. Let $E = 200 \times 10^9$ Pa and assume the proportional limit to be at 250 MPa.

SOLUTION

The minimum moment of inertia of the cross-sectional area $I_{min} = (0.100)(0.06)^3/12 = 1.8 \times 10^{-6}$ m⁴. Hence

$$r = r_{min} = \sqrt{\frac{I_{min}}{A}} = \sqrt{\frac{1.8 \times 10^{-6}}{(0.06)(0.100)}} = \sqrt{3} \times 10^{-2}\,\text{m}$$

Then, using Eq. 13-24, $\sigma_{cr} = \pi^2 E/(L/r)^2$. Solving it for the L/r ratio at the proportional limit,

$$\left(\frac{L}{r}\right)^2 = \frac{\pi^2 E}{\sigma_{cr}} = \frac{\pi^2 (200)10^9}{(250)10^6} = 800\pi^2$$

or $\quad \dfrac{L}{r} = 88.9 \quad$ and $\quad L = (88.9)\sqrt{3} \times 10^{-2} = 1.54\,\text{m}$

Therefore, if this column is 1.54 m or more in length, it will buckle elastically as, for such dimensions of the column, the critical stress at buckling will not exceed the proportional limit for the material.

13-8. GENERALIZED EULER BUCKLING-LOAD FORMULAS

A typical compression stress-strain diagram for a specimen that is prevented from buckling is represented in Fig. 13-11(a). In the stress range from 0 to A, the material behaves elastically. If the stress in a column at buckling does not exceed this range, the column buckles elastically. The hyperbola expressed by Eq. 13-24, $\sigma_{cr} = \pi^2 E/(L/r^2)$, is applicable in such a case. This portion of the curve is shown as ST in Fig. 13-11(b). It is important to recognize that this curve does not represent the behavior of one column but rather the behavior of an infinite number of the ideal columns of different lengths. The hyperbola beyond the useful range is shown in the figure by dashed lines.

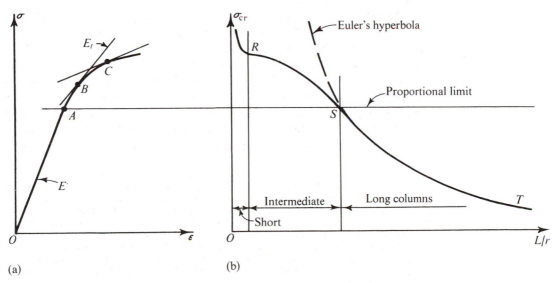

(a)　　　　　　　　　(b)

Fig. 13-11. (a) Compression stress-strain diagram; (b) critical stress in columns versus slenderness ratio

A column with an L/r ratio corresponding to point S in Fig. 13-11(b) is the shortest column of a given material and size that will buckle elastically. A shorter column, having a still smaller L/r ratio, will not buckle at the proportional limit of the material. On the compression stress-strain diagram, Fig. 13-11(a), this means that the stress level in the column has passed point A and has reached some point B perhaps. At this higher stress level, it may be said that a column of different material has been created, in effect, since the stiffness of the material is no longer represented by the elastic modulus. At this point, the material stiffness is given instantaneously. by the tangent to the stress-strain curve, i.e., by the tangent modulus E_t. The column remains stable if its new flexural rigidity E_tI at B is sufficiently great, and it can carry a higher load. As the load is increased, the stress level rises, whereas the tangent modulus decreases. A column of ever "less stiff material" is acting under an increasing load. Substitution of the tangent modulus E_t for the elastic modulus E is then the only modification necessary to make the elastic buckling formulas applicable in the inelastic range. Hence the generalized Euler buckling-load formula, or the tangent modulus formula* becomes

$$\sigma_{cr} = \frac{\pi^2 E_t}{(L/r)^2} \tag{13-25}$$

Since stresses corresponding to the tangent moduli can be obtained from the compression stress-strain diagram, the L/r ratio at which a column will buckle with these values can be obtained from Eq. 13-25. A plot representing this behavior for low and intermediate ratios of L/r is shown in Fig. 13-11(b) by the curve from R to S. Tests on individual columns verify this curve with remarkable accuracy.†

As mentioned earlier, columns that buckle elastically are referred to sometimes as *long columns*. Columns having low L/r ratios exhibiting essentially no buckling phenomena are called *short columns*. At low L/r ratios, ductile materials "squash out" and can carry very large loads.

If the length L in Eq. 13-25 is treated as the effective length of a column, different end conditions can be analyzed. Following this procedure, for comparative purposes plots of critical stress σ_{cr} versus the slenderness ratio L/r for fixed-ended columns and pin-ended ones are shown in Fig. 13-12. It is important to note that the carrying capacity for the two cases is in a ratio

---

*The tangent modulus formula gives the carrying capacity of a column defined at the instant it tends to buckle. As a column deforms further, the fibers on the concave side continue to exhibit approximately the tangent modulus E_t. The fibers on the convex side, however, are relieved of some stress and rebound with the original elastic modulus E. These facts led to the establishment of the so-called *double-modulus theory* of load-carrying capacity of columns. The end results as obtained by this theory do not differ greatly from those obtained by the tangent modulus theory. For further details and significant refinements see F. R. Shanley, "Inelastic Column Theory," *Journal of Aeronautical Sciences*, Vol. 14, no. 5 (May 1947); and F. Bleich, *Buckling Strength of Metal Structures*, (New York: McGraw-Hill, 1952).

†See Bleich, *Buckling Strength of Metal Structures*, p. 20.

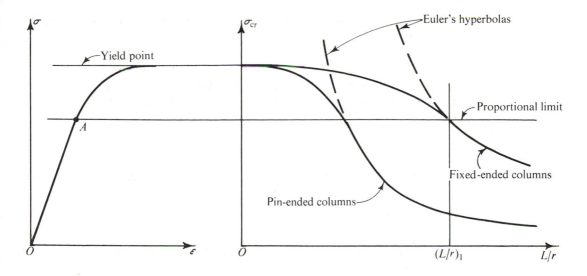

Fig. 13-12. Comparison of the behavior of columns with different end conditions

of 4 to 1 only for columns having the slenderness ratio $(L/r)_1$ or greater. For smaller L/r ratios, progressively less benefit is derived from restraining the ends. At low L/r ratios the curves merge. It makes little difference whether a "short block" is pinned or fixed at the ends as strength rather than buckling determines the behavior.

*13-9. THE SECANT FORMULA

A different method of analysis may be used to determine the capacity of a column than was discussed above. Since no column is perfectly straight nor are the applied forces perfectly concentric, the behavior of real columns may be studied with some statistically determined imperfections or possible misalignments of the applied loads. Then, for the design of an actual column, which is termed "straight," a probable crookedness or an effective load eccentricity may be assigned. Also there are many columns where an eccentric load is deliberately applied. Thus, an eccentrically loaded column can be studied and its capacity determined on the basis of an allowable elastic stress. This does not determine the ultimate capacity of a column.

To analyze the behavior of an eccentrically loaded column, consider the column shown in Fig. 13-13. If the origin of the coordinate axes is taken at the upper force P, the bending moment at any section is $-Pv$, and the differential equation for the elastic curve is the same as for a concentrically loaded column, i.e., as in Eq. 13-2,

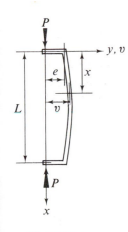

Fig. 13-13. An eccentrically loaded column for derivation of the secant formula

$$\frac{d^2v}{dx^2} = \frac{M}{EI} = -\frac{P}{EI}v$$

where, by again letting $\lambda = \sqrt{P/(EI)}$, the general solution is as before, in Eq. 13-4,

$$v = A \sin \lambda x + B \cos \lambda x$$

However, the remainder of the problem is not the same, since *the boundary conditions are now different.* At the upper end, v is equal to the eccentricity of the applied load, i.e., $v(0) = e$. Hence $B = e$, and Eq. 13-4 becomes

$$v = A \sin \lambda x + e \cos \lambda x \qquad (13\text{-}26)$$

Next, noting that, by virtue of symmetry, the elastic curve has a vertical tangent at the mid-height of the column,

$$v'(L/2) = 0$$

Therefore, by setting the derivative of Eq. 13-26 equal to zero at $x = L/2$, it is found that

$$A = e \, \frac{\sin \lambda L/2}{\cos \lambda L/2}$$

Hence the equation for the elastic curve is

$$v = e \left(\frac{\sin \lambda L/2}{\cos \lambda L/2} \sin \lambda x + \cos \lambda x \right) \qquad (13\text{-}27)$$

No indeterminacy of any constants appears in this equation, and the maximum deflection v_{max} can be found from it. This maximum deflection occurs at $L/2$, since at this point the derivative of Eq. 13-27 is equal to zero. Hence

$$v(L/2) = v_{max} = e \left(\frac{\sin^2 \lambda L/2}{\cos \lambda L/2} + \cos \frac{\lambda L}{2} \right) = e \sec \frac{\lambda L}{2}$$

In the column shown in Fig. 13-13, the largest bending moment M is developed at the point of maximum deflection and numerically is equal to Pv_{max}. Therefore, since the direct force and the largest bending moment are now known, the *maximum* compressive stress occurring in the column (contrast this with the *average* stress P/A acting on the column) can be computed by the usual formula, as

$$\sigma_{max} = \frac{P}{A} + \frac{Mc}{I} = \frac{P}{A} + \frac{Pv_{max}c}{Ar^2} = \frac{P}{A} \left(1 + \frac{ec}{r^2} \sec \frac{\lambda L}{2} \right)$$

But $\lambda = \sqrt{P/(EI)} = \sqrt{P/(EAr^2)}$, hence

$$\sigma_{max} = \frac{P}{A} \left(1 + \frac{ec}{r^2} \sec \frac{L}{r} \sqrt{\frac{P}{4EA}} \right) \qquad (13\text{-}28)$$

This equation, since it contains a secant term, is known as *the secant formula for columns,* and it applies to columns of any length provided the maximum stress does not exceed the elastic limit. A condition of equal eccentricities of the applied forces in the same direction causes the largest deflection.

Note that in Eq. 13-28 the radius of gyration r may not be minimum, since it is obtained from the value of I associated with the axis around which bending occurs. In some cases a more critical condition for buckling can exist in the direction of no definite eccentricity. Also note that in Eq. 13-28 *the relation between* σ_{max} *and P is not linear;* σ_{max} *increases faster than P. Therefore the solutions for maximum stresses in columns caused by different axial forces cannot be superposed; instead the forces must be superposed first,* and then the stresses can be calculated.

For an allowable force P_a on a column, where n is the factor of safety, nP_a, must be substituted for P in Eq. 13-28, while σ_{max} must be set at the yield point of a material, i.e.,

$$\sigma_{max} = \sigma_{yp} = \frac{nP_a}{A}\left(1 + \frac{ec}{r^2}\sec\frac{L}{r}\sqrt{\frac{nP_a}{4EA}}\right) \tag{13-29}$$

This procedure assures a correct factor of safety for the applied force, since such a force can be increased n times before a critical stress is reached. Note the term nP_a appearing under the radical.

Equations 13-28 and 13-29 having been established, mathematically the problem is solved. However, the application of these equations to design is very cumbersome. They can be solved by trial-and-error procedures, or they can be studied graphically. Such a study is shown in Fig. 13-14.* In this

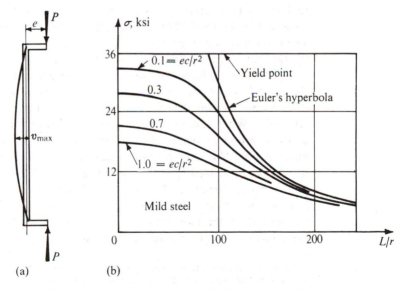

Fig. 13-14. Results of analysis for different columns by the secant formula

*This figure is taken from a paper by D. H. Young, on "Rational Design of Steel Columns," *Trans., ASCE*, 1936, vol. 101, p. 431.

plot, note the large effect that the load eccentricity has on short columns and the negligible one on very slender columns. Graphs of this kind form a suitable aid in practical design. The secant equation covers the whole range of column lengths. The greatest handicap in using this formula is that some eccentricity e must be assumed even for supposedly straight columns, and this is a difficult task.*

The secant formula for *short* columns reverts to a familiar expression when L/r approaches zero. For this case, the value of the secant approaches unity, hence, in the limit, Eq. 13-28 becomes

$$\sigma_{\max} = \frac{P}{A} + \frac{Pec}{Ar^2} = \frac{P}{A} + \frac{Mc}{I}$$

a relation normally used for short blocks.

13-10. DESIGN OF COLUMNS

For economy, the cross-sectional areas of columns, other than short blocks, should possess the largest possible least radius of gyration. This gives a smaller L/r ratio, which permits the use of a higher stress. Tubes form excellent columns. Wide-flange sections (which are sometimes called H sections) are superior to I sections. In columns built up from rolled or extruded shapes, the individual pieces are spread out to obtain the desired effect. Cross sections for typical bridge compression members are shown in Figs. 13-15(a) and (b), for a boom in Fig. 13-15(c), and for an ordinary truss in Fig. 13-15(d). The angles in Fig. 13-15(d) are separated by spacers. The main shapes of Figs. 13-15(a), (b), and (c) are laced (or latticed) together by light bars, as shown in Figs. 13-15(e) and (f).

Obtaining a large r by placing a given amount of material away from the centroid of an area, as illustrated above, can reach a limit. The material can become so thin that it crumples locally. This behavior is termed *local instability*. When failure caused by local instability takes place in the flanges or the component plates of a member, the compression member becomes unserviceable. An illustration of local buckling is shown in Fig. 13-16. It is usually characterized by a change in the shape of a cross section. The equations derived earlier are for the instability of a column as a whole, or for primary instability. Discussion of the possibility of *torsional instability*, exemplified by the twisting of a whole section (which is a form of primary instability), is beyond the scope of this text.

*Moreover, there is some question as to the philosophical correctness of the secant formula. The fact that the stress reaches a certain value does not mean that the column buckles, i.e. stress is not a measure of buckling load in every case. It can be shown that an additional axial load can be resisted beyond the point where the maximum stress at the critical section is reached. See F. Bleich, *Buckling Strength of Metal Structures*, New York: McGraw-Hill, 1952, Chapter 1.

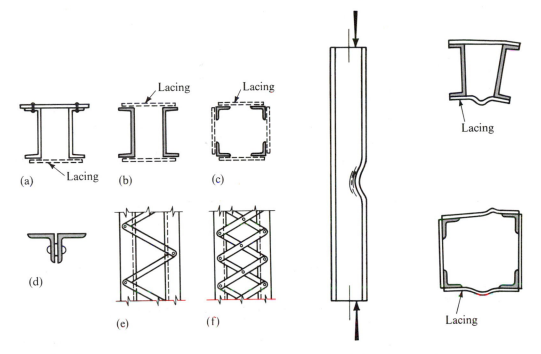

Fig. 13-15. Typical built-up cross sections of columns

Fig. 13-16. Examples of local instability in columns

After the chaotic situation that existed for many years with regard to the column-design formulas, now that the column-buckling phenomenon is more clearly understood only a few formulas are in common use. In most widely used specifications, a pair of formulas is given. One formula is used for small and intermediate values of L/r; the other is used for slender columns with large values of L/r. In the range of small and intermediate values of L/r, either a parabola or a straight line with a stipulated maximum is employed to define the critical stresses. For large values of L/r, Euler's hyperbola for the elastic response is used: See Fig. 13-17. Sometimes the equations of the two complementary formulas have a common tangent at a selected value of L/r. A few specifications make use of the secant-formula approach with an assumed eccentricity based on the manufacturing tolerances.

In applying the design formulas it is important to observe the following items:

1. The material for which the formula is written.
2. Whether the formula gives the working load (or stress) or whether it estimates the ultimate carrying capacity of a member. If the formula is of the latter type, a safety factor must be introduced.
3. The range of the applicability of the formula. Some empirical formulas can lead to unsafe design if used beyond the specified range. (See Fig. 13-17(b).)

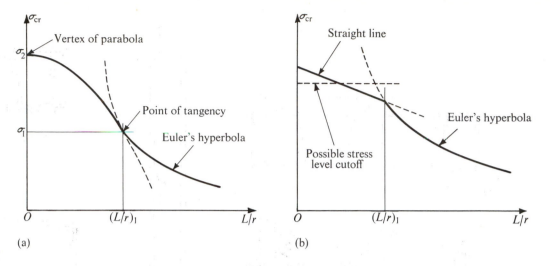

Fig. 13-17. Typical column-buckling curves for design

13-11. COLUMN FORMULAS FOR CONCENTRIC LOADING

As examples of column-design formulas for nominally concentric loading, representative formulas for structural steel, aluminum alloy, and wood are given below. Formulas for eccentrically loaded columns are considered in the next article.

Column Formulas for Structural Steel

The American Institute of Steel Construction* recommends the use of column formulas patterned on the scheme illustrated in Fig. 13-17(a). Since steels of many different yield strengths are manufactured, the formulas are stated in terms of σ_{yp}, which varies for different steels. The elastic modulus E for all steels is approximately the same. Euler's elastic buckling formula is specified for the slender columns beginning with $(L/r)_1 = C_c$ occurring at the slenderness ratio corresponding to one-half of the yield stress σ_{yp} of the steel. In order to fulfill this assumption, from Eq. 13-24, the slenderness ratio $C_c = (L/r)_1 = \sqrt{2\pi^2 E/\sigma_{yp}}$. By using this equation, with $E = 29 \times 10^3$ ksi, the allowable stress for columns having a slenderness ratio larger than C_c becomes

$$\sigma_{allow} = 149{,}000/(L_e/r)^2 \quad \text{[ksi]} \quad (13\text{-}30)$$

where L_e is the effective column length. A safety factor of 1.92 with respect to buckling is incorporated in Eq. 13-30. No column is permitted to exceed an L_e/r of 200.

*See *AISC Steel Construction Manual* (7th ed.), New York: AISC, 1970.

For an L_e/r ratio less than C_c, AISC specifies a parabolic formula:

$$\sigma_{\text{allow}} = \frac{[1 - (L_e/r)^2/(2C_c^2)]\sigma_{\text{yp}}}{\text{F.S.}} \quad [\text{ksi}] \quad (13\text{-}31)$$

where F.S., the factor of safety, is defined as

$$\text{F.S.} = \tfrac{5}{3} + 3(L_e/r)/(8C_c) - (L_e/r)/(8C_c^3)$$

It is interesting to note that F.S. varies, being more conservative for the larger ratios of L_e/r. The equation chosen for F.S. approximates a quarter sine curve with the value of 1.67 at zero L_e/r and 1.92 at C_c. An allowable stress versus slenderness ratio for axially loaded columns of several kinds of structural steels is shown in Fig. 13-18.

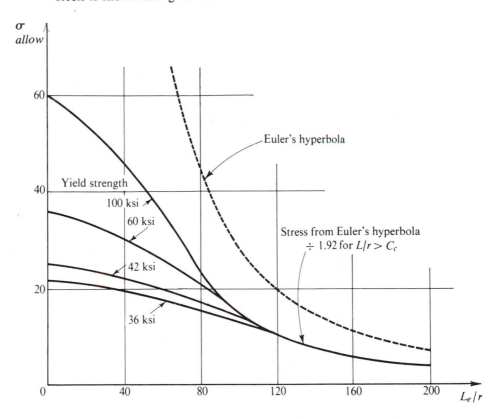

Fig. 13-18. Allowable stress for concentrically loaded columns per AISC specifications

Column Formulas for Aluminum Alloys

A large number of aluminum alloys are available for engineering applications. The yield and the ultimate strengths of such alloys vary over a considerable range. The elastic modulus for the alloys, however, is reasonably constant, and the Aluminum Company of America recommends the use of a factor of 102,000 ksi to represent the quantity $\pi^2 E$.* Hence, using Eq. 13-24, the ultimate strength formula for long columns with effective length L_e is

$$\sigma_{\mathrm{cr}} = 102,000/(L_e/r)^2 \quad \text{[ksi]} \tag{13-32}$$

An inclined straight line, as in Fig. 13-17(b), is used for the column-strength curve for low and intermediate values of L_e/r. For some alloys this straight line is so chosen as to be tangent to Euler's hyperbola; for others, an angle is formed as in the figure. As an example, drawn from a very large number of cases given in the ALCOA handbook, for an extruded 2024-T4 alloy,

$$\sigma_{\mathrm{cr}} = 44.8 - 0.313(L_e/r) \quad \text{[ksi]} \quad (0 \le (L_e/r) \le 64) \tag{13-33}$$

In applying this particular formula the L_e/r ratio must not exceed 64. Extrapolations of such formulas beyond their range of applicability is impermissible as they may give values of higher stress than that which is possible at Euler's buckling load. *A factor of safety must be applied with Eqs. 13-32 and 13-33.*

Column Formulas for Wood

The National Lumber Manufacturers Association† recommends the use of the Euler buckling-load formula for solid wood columns. According to the recommendation the allowable stress is

$$\sigma_{\mathrm{allow}} = \pi^2 E/[2.727(L/r)^2] = 3.619E/(L/r)^2 \tag{13-34}$$

Here the allowable stress cannot exceed the value for compression of short blocks parallel to the grain for the particular species of wood. These stresses are increased for short-duration loading and are decreased for sustained loading. This formula is applicable for pin- and "square"-ended conditions.

For columns of square or rectangular cross section Eq. 13-34 becomes

$$\sigma_{\mathrm{allow}} = 0.30E/(L/d)^2 \tag{13-35}$$

where d is the smallest side dimension of a member.

EXAMPLE 13-2

Compare the allowable axial compressive loads for a 3 in. by 2 in by $\frac{1}{4}$ in. aluminum-alloy angle 43.5 in. long (a) if it acts as a pin-ended column, and (b) if it is so restrained that its effective length L_e is 0.9L. Assume a factor of safety of 2.5 and that Eqs. 13-32 and 13-33 apply.

*See *ALCOA Structural Handbook* (8th ed.), Pittsburgh, Pa.: Aluminum Company of America, 1960, p. 110.

†*NLMA National Design Specification*, Washington, D.C.: National Lumber Manufacturers Association, 1962.

SOLUTION

The proportions of aluminum-alloy angles are the same as those of steel. Hence, from Table 7 of the Appendix, for this size angle, the least radius of gyration $r = 0.435$ in., and its area $A = 1.19$ in.^{2}. The solution follows by applying Eq. 13-32 and using the specified factor of safety. Case (a): $L_e = L = 43.5$ in., $L_e/r = 43.5/(0.43.5) = 100$.

$$\sigma_{cr} = \frac{102,000,000}{(L_e/r)^2} = 10,200 \text{ psi}$$

$$P_{cr} = A\sigma_{cr} = (1.19)10,200 = 12,100 \text{ lb}$$

$$P_{allow} = \frac{P_{cr}}{F.S.} = \frac{12,100}{2.5} = 4,830 \text{ lb}$$

Case (b): $L_e = 0.9L = 39.5$ in., $L_e/r = (39.5)/(0.435) = 90$, and $\sigma_{cr} = 12,600$ psi.

$$P_{cr} = A\sigma_{cr} = (1.19)12,600 = 15,000 \text{ lb}$$

$$P_{allow} = \frac{P_{cr}}{F.S.} = \frac{15,000}{2.5} = 6,000 \text{ lb}$$

EXAMPLE 13-3

Using AISC column formulas, select a 15 ft long pin-ended column to carry a concentric load of 200 kips. The structural steel is to be A441, having $\sigma_{yp} = 50$ ksi.

SOLUTION

The required size of the column can be found directly from the tables in the AISC Steel Construction Manual. However, this example provides an opportunity to demonstrate the trial-and-error procedure that is so often necessary in design, and the solution presented follows from using this method.

First try: Let $L/r = 0$ (a poor assumption for a column 15 ft long). Then, from Eq. 13-31, since F.S. $= \frac{5}{3}$, $\sigma_{allow} = 50/(F.S.) = 30$ ksi and $A = P/\sigma_{allow} = 200/30 = 6.67$ in.^{2} From Table 4 in the Appendix, this requires a W 8×24 section, whose $r_{min} = 1.61$ in. Hence $L/r = 15(12)/(1.61) = 112$. With this L/r, the allowable stress is found using Eq. 13-30 or 13-31, whichever is applicable depending on C_c:

$$C_c = \sqrt{2\pi^2 E/\sigma_{yp}} = \sqrt{2\pi^2 \times 29 \times 10^3/50} = 107 < L/r = 112$$

hence $\qquad \sigma_{allow} = 149,000/(112)^2 = 11.9$ ksi

This is much smaller than the initially assumed stress of 30 ksi, and another section must be selected.

Second try: Let $\sigma_{allow} = 11.9$ ksi as found above. Then $A = 200/11.9 = 16.8$ in.^{2} requiring a W 8×58 section having $r_{min} = 2.10$ in. Now $L/r = 15(12)/(2.10) = 85.7$, which is less than C_c found above. Therefore, Eq. 13-31 applies, and

$$\text{F.S.} = \tfrac{5}{3} + 3(85.7)/(8 \times 107) - (85.7)^3/(8 \times 107^3) = 1.90$$

and $\qquad \sigma_{allow} = [1 - (85.7)^2/(2 \times 107^2)]50/(1.90) = 17.9$ ksi

This stress requires $A = 200/17.9 = 11.2$ in.^{2}, which is met by a W 8 × 40 section with $r_{min} = 2.04$ in. A calculation of the capacity for this section shows that the allowable axial load for it is 204 kips, which meets the requirements of the problem.

*13-12. COLUMN FORMULAS FOR ECCENTRIC LOADING

In practice, columns are designed for load that may be eccentrically applied either intentionally or due to accidental reasons. The secant type formulas discussed in Art. 13-9 have been used in the past as a rational approach to the design of such columns. However, the results are unduly conservative and, in addition, such formulas are rather cumbersome to work with. Alternatively, two other methods, which have found acceptance in practice, are discussed in this article.

Allowable Stress Approach

The *maximum compressive stress* in an eccentrically loaded column may be computed approximately as is done for short blocks, i.e.,

$$\sigma_{max} = \frac{P}{A} + \frac{Mc}{I} = \frac{P}{A} + \frac{Pec}{I} \qquad (13\text{-}36)$$

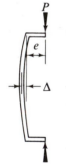

Fig. 13-19. An eccentrically loaded column

where Pe is the bending moment caused by the applied loads. The column can then be designed simply by setting this maximum stress to the *allowable column stress* as given by any one of the formulas for axially loaded columns. The resulting design procedure is usually rather conservative, but has the merit of being very simple to apply. For example, using Eq. 13-33 for 2024-T4 aluminum alloy with an appropriate factor of safety (F.S.) and restricting L/r to be less than or equal to 64,

$$\sigma_{max} = \frac{\sigma_{cr}}{\text{F.S.}} = \frac{44.8 - 0.313 L/r}{\text{F.S.}} = \frac{P}{A} + \frac{Pec}{I}$$

This method is applicable only for columns with moderate L/r, since the secondary (additional) bending moments due to the elastic deflection are not negligible for long, slender members, Fig. 13-19. The theory of beam-columns, which was only briefly discussed in Art. 13-6 (see Fig. 13-9), is needed to deal with such slender members.

Interaction Approach

In an eccentrically loaded column, much of the total stress may result from the applied moment. However, *the allowable stress in flexure is usually higher than the allowable axial stress.* Hence for a particular column, it is desirable to accomplish some balance between the two stresses, depending on the relative magnitudes of the bending moment and the axial force. Thus, since in bending, $\sigma = Mc/I = Mc/Ar_1^2$ where r_1 is the radius of gyration *in the plane of bending*, in effect the area A_b required by the bending moment M is

$$A_b = \frac{Mc}{\sigma_{ab} \, r_1^2}$$

where σ_{ab} is the allowable *maximum stress in bending*. (Also see Art. 13-13.) Similarly, the area A_a required for the axial force P is

$$A_a = \frac{P}{\sigma_{aa}}$$

where σ_{aa} is the *allowable axial average stress for the member acting as a column*, and which depends on the L/r ratio. Therefore the *total* area A required for a column subjected to an axial force and a bending moment is

$$A = A_a + A_b = \frac{P}{\sigma_{aa}} + \frac{Mc}{\sigma_{ab} \, r_1^2} \qquad (13\text{-}37)$$

Whence, dividing by A,

$$\frac{\dfrac{P}{A}}{\sigma_{aa}} + \frac{\dfrac{Mc}{Ar_1^2}}{\sigma_{ab}} = 1 \qquad \text{or} \qquad \frac{\sigma_a}{\sigma_{aa}} + \frac{\sigma_b}{\sigma_{ab}} = 1 \qquad (13\text{-}38)$$

where σ_a is the axial stress caused by the applied vertical loads, and σ_b is the bending stress caused by the applied moment. If a column is carrying only an axial load and the applied moment is zero, the formula above indicates that the column is designed for the stress σ_{aa}. On the other hand, the allowable stress becomes the flexural stress σ_{ab} if there is no direct compressive force acting on the column. Between these two extreme cases, Eq. 13-38 measures the relative importance of the two kinds of action and specifies the nature of their interaction. Hence, it is often referred to as an *interaction formula* and serves as the basis for the specifications in the AISC manual, where it is stated that the sum of these two stress ratios must not exceed unity. The same philosophy has found favor in applications other than those pertaining to structural steel. The Aluminum Company of America suggests a similar relation. The Forest Products Laboratory developed a series of formulas to serve the same purpose.

In terms of the notations used by the AISC, Eq. 13-38 is rewritten as

$$\frac{f_a}{F_a} + \frac{f_b}{F_b} \le 1.0 \qquad (13\text{-}39)$$

In practice, the eccentricity of the load on a column may be such as to cause bending moments about both axes of the cross section. Equation 13-39 is then modified to

$$\frac{f_a}{F_a} + \frac{f_{bx}}{F_{bx}} + \frac{f_{by}}{F_{by}} \leq 1.0 \qquad (13\text{-}40)$$

The subscripts x and y combined with the subscript b indicate the axis of bending about which a particular stress applies, and

F_a = allowable axial stress if axial force alone existed

F_b = allowable compressive bending stress if bending moment alone existed

f_a = computed axial stress

f_b = computed bending stress

At points that are braced in the plane of bending, F_a is equal to 60 per cent of F_y, the yield stress of the material, and

$$\frac{f_a}{0.6F_y} + \frac{f_{bx}}{F_{bx}} + \frac{f_{by}}{F_{by}} \leq 1.0 \qquad (13\text{-}41)$$

At intermediate points in the length of a compression member, the secondary bending moments due to deflection (see Fig. 13-19) can contribute significantly to the combined stress. Following the AISC specifications, this contribution is neglected in cases where f_a/F_a is less than 0.15, i.e., the axial stress is small in relation to the allowable axial stress, and Eq. 13-41 can still be used. When f_a/F_a is greater than 0.15, the effect of the additional secondary bending moments may be approximated by multiplying both f_{bx} and f_{by} by an *amplification factor*, $C_m/[1 - (f_a/F'_e)]$, which takes into account the slenderness ratio in the plane of bending and also the nature of the end moments. The term in the denominator of the amplification factor brings in the effect of the slenderness ratio through the use of F'_e, the Euler buckling stress (using L_e/r in the plane of bending) divided by 23/12, or 1.92, which is the AISC factor of safety for a very long column with L_e/r greater than C_c. (See Art. 13-11 for a definition of C_c.) It can be noted that the amplification factor increases as f_a increases and *blows up* as f_a approaches F'_e. The term C_m, in the numerator, is a correction factor that takes into account the ratio of the end moments as well as their relative sense of direction. The term C_m is larger if the end moments are such that they cause a single curvature of the member, and smaller if they cause a reverse curvature. The formula for $f_a/F_a > 0.15$ then becomes

$$\frac{f_a}{F_a} + \frac{C_{mx}f_{bx}}{(1 - (f_a/F'_{ex}))F_{bx}} + \frac{C_{my}f_{by}}{(1 - (f_a/F'_{ey}))F_{by}} \leq 1.0 \qquad (13\text{-}42)$$

According to the AISC specifications,* the value of C_m shall be taken as follows:

*_AISC Steel Construction Manual_ (7th ed.), New York: AISC, 1970, pp. 5–23.

1. For compression members in frames subject to joint translation (sidesway), $C_m = 0.85$.

2. For restrained compression members in frames braced against joint translation and not subject to transverse loading between their supports in the plane of bending,

$$C_m = 0.6 - 0.4(M_1/M_2)$$

(but not less than 0.4) where M_1/M_2 is the ratio of the smaller to larger moments at the ends of that portion of the member unbraced in the plane of bending under consideration. M_1/M_2 is positive when the member is bent in reverse curvature and negative when it is bent in single curvature.

3. For compression members in frames braced against joint translation in the plane of loading and subjected to transverse loading between their supports, the value of C_m can be determined by rational analysis. However, in lieu of such analysis, the following values may be used: (a) for members whose ends are restrained, $C_m = 0.85$; (b) for members whose ends are unrestrained, $C_m = 1.0$.

EXAMPLE 13-4

Select an aluminum column for the loading shown in Fig. 13-20 by the allowable stress method, using Eq. 13-33 and a factor of safety of 2.5.

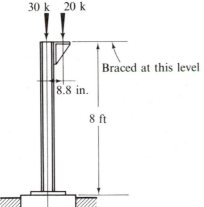

30 k   20 k

Braced at this level

8.8 in.

8 ft

Fig. 13-20

SOLUTION

In this problem, the following relation must be satisfied:

$$\frac{30 + 20}{A} + \frac{20(8.8)}{S} \leq \frac{\sigma_{cr}}{F.S.}$$

or $\quad \dfrac{50}{A} + \dfrac{176}{S} \leq \dfrac{44.8 - 0.313 L_e/r}{2.5} = 17.9 - 0.125\dfrac{L_e}{r}$

where, r, A, and S depend on the column selected. A trial-and-error procedure is used to solve the problem.

First Try: Let the maximum stress due to the loads be equal to the allowable stress corresponding to $L_e/r = 0$ (although it is a poor assumption for a column 8 ft. long). Then

$$\frac{50}{A} + \frac{176}{S} = 17.9$$

or, defining $A/S = B$, where B is termed the *bending factor,**

$$\frac{50}{A} + \frac{176B}{A} = 17.9$$

*The first try can also be made by completely ignoring the bending moment but will result in a poorer estimate. Bending factors B_x around the x-axis, and B_y around the y-axis are listed in aluminum and steel construction manuals.

Whence $A = (50 + 176B)/17.9 = 6.7$ in.² if B is assumed to be 0.4. Try an 8 WF 8.32 section* with a $B_x = 0.337$, $r_{min} = 1.61$ in., and $A = 7.08$ in.².

Hence
$$L_e/r = 8(12)/1.61 = 60$$
$$\sigma_{allow} = 17.9 - 0.125(60) = 10.4 \text{ ksi}$$
$$\sigma_{due\ to\ loads} = 50/(7.08) + 176(0.337)/(7.08) = 15.4 \text{ ksi}$$

These stresses disagree greatly; hence a larger section must be tried.

Second Try: 8 WF 11.24 section† with an average $B_x = 0.337$, $r_{min} = 1.88$, and $A = 9.55$ in.². Hence $L_e/r = 8(12)/1.88 = 51.1$, and

$$\sigma_{allow} = 17.9 - 0.125(51.1) = 11.5 \text{ ksi}$$
$$\sigma_{due\ to\ loads} = 50/(9.55) + 176(0.337)/(9.55) = 11.4 \text{ ksi}$$

Therefore this 8 WF 11.24 section is satisfactory.

EXAMPLE 13-5

Using the AISC specifications, select a steel column for the loading shown in Fig. 13-21, assuming that it is braced at the ends against sidesway bending in both of its bending planes. Assume $E = 29 \times 10^3$ ksi, $F_y = 50$ ksi, and $F_b = 30$ ksi.

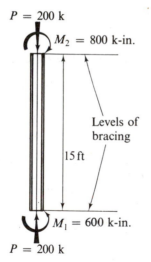

$P = 200$ k

$M_2 = 800$ k-in.

Levels of bracing

15 ft

$M_1 = 600$ k-in.

$P = 200$ k

Fig. 13-21

SOLUTION

In this problem, the interaction formulas Eq. 13-40 or Eq. 13-42 must be satisfied, depending upon whether f_a/F_a is less than or greater than 0.15. The solution is obtained by trial-and-error process as outlined below.

First Try: Let $L_e/r = 0$, although it is a poor assumption for a 15 ft. column. Corresponding to this value of the slenderness ratio, F_a can be calculated, using Eq. 13-31, as $F_a = 50/(5/3) = 30$ ksi. The required area of the section can then be computed using Eq. 13-40:

$$1.0 \geq \frac{f_a}{F_a} + \frac{f_b}{F_b} \quad \text{or} \quad A \geq \frac{Af_a}{F_a} + \frac{Af_b}{F_b}$$

Since
$$f_b = \frac{M}{S_x} = \frac{M}{A}\frac{A}{S_x} = \frac{M}{A}B_x \quad \text{and} \quad f_a = \frac{P}{A}$$

$$A \geq \frac{P}{F_a} + \frac{M}{F_b}B_x$$

---

*See *Aluminum Construction Manual*, 1959, p. 186. Aluminum sections are designated by first giving the depth in inches, followed by WF which means wide-flange beam, and lastly by giving the weight in pounds per linear foot. The proportions of steel and aluminum sections often are very nearly the same. The geometrical properties of 8 WF 8.32 closely correspond to a W 8 × 24 steel section listed in Table 4 of the Appendix.

†No corresponding steel section is listed in Table 4 of the Appendix.

For any one depth of section, the value of B_x does not vary a good deal. Therefore, if a W 10 section is to be chosen, a typical value of B_x is about 0.264 (check a few values of A/S_x in Table 4 of the Appendix). Then

$$A = \frac{200}{30} + \frac{(800)(0.264)}{30} = 13.7 \text{ in.}^2$$

Select a W 10 × 49 section with $A = 14.4$ in.^{2}, $r_{min} = 2.54$ in., $r_x = 4.35$ in., and $B_x = 0.264$, and carry out the calculations necessary to verify if the appropriate interaction formula (Eq. 13-40 or Eq. 13-42) is satisfied.

$$f_a = \frac{P}{A} = \frac{200}{14.4} = 13.9 \text{ ksi}; \qquad f_b = \frac{MB_x}{A} = \frac{(800)(0.264)}{14.4} = 14.7 \text{ ksi}$$

$$\frac{L_e}{r_{min}} = \frac{15 \times 12}{2.54} = 70.9 < C_c \qquad (C_c = \sqrt{2\pi^2 E/F_y} = 107)$$

Using Eq. 13-31, $F_a = 19.3$ ksi, $f_a/F_a = 13.9/19.3 = 0.72 > 0.15$; hence the interaction formula of Eq. 13-42 must be checked. For this purpose, using L_e/r_x in the plane of bending, one determines

$$F_e' = \frac{12\pi^2 E}{23(L_e/r_x)^2} = \frac{149 \times 10^3}{(15 \times 12/4.35)^2} = \frac{149 \times 10^3}{(41.4)^2} = 86.9 \text{ ksi}$$

Then, since the end moments subject the column to a single curvature, $M_1/M_2 = -600/800 = -0.75$, and

$$C_m = 0.6 - 0.4M_1/M_2 = 0.6 - (0.4)(-0.75) = 0.9$$

With bending taking place in one plane only, Eq. 13-42 reduces to

$$\frac{f_a}{F_a} + \frac{C_m f_b}{(1 - f_a/F_e)F_b} \leq 1.0$$

On substituting the appropriate quantities into this relation,

$$\frac{13.9}{19.3} + \frac{(0.9)(14.7)}{(1 - 13.9/86.9)30} = 0.72 + 0.52 = 1.24 > 1.0$$

Since Eq. 13-42 is violated, this is not acceptable, and a larger section must be used.

Second Try: As an aid in choosing a larger section, assume $F_a = 19.3$ ksi, which is the value computed for the section in the previous trial. Also, using $B_x = 0.264$ for W 10 sections,

$$A \geq \frac{P}{F_a} + \frac{MB_x}{F_b} = \frac{200}{19.3} + \frac{(800)(0.264)}{30} = 17.4 \text{ in.}^2$$

Select a W 10 × 60 section with $A = 17.7$ in.^{2}, $r_{min} = 2.57$ in., $r_x = 4.41$ in., and $B_x = 0.264$, and proceed as in the first trial to check the interaction formula.

$$f_a = \frac{P}{A} = \frac{200}{17.7} = 11.3 \text{ ksi}; \qquad f_b = \frac{MB_x}{A} = \frac{(800)(0.264)}{17.7} = 11.9 \text{ ksi}$$

$$\frac{L_e}{r_{\min}} = \frac{15 \times 12}{2.57} = 70.0 < C_c, \text{ and using Eq. 13-31}, F_a = 19.4 \text{ ksi}$$

$f_a/F_a = 11.3/19.4 = 0.58 > 0.15$; hence Eq. 13-42 must be checked.

$$F'_e = \frac{149 \times 10^3}{(L_e/r_x)^2} = \frac{149 \times 10^3}{(15 \times 12/4.41)^2} = 89.4 \text{ ksi}$$

Again using Eq. 13-42 for bending in one plane, and substituting into it the appropriate quantities, one has

$$\frac{11.3}{19.4} + \frac{(0.9)(11.9)}{(1 - 11.3/89.4)30} = 0.58 + 0.41 = 0.99 < 1.0$$

Since this relation satisfies Eq. 13-42, the W 10 $\times$ 60 section is satisfactory.

*13-13. BEAMS WITHOUT LATERAL SUPPORTS

The strength and deflection theory of beams developed in this text applies only if such beams are in *stable equilibrium*. Narrow beams that do not have occasional lateral supports may buckle sideways and thus become unstable. A physical illustration of this situation was cited in Art. 5-2 with reference to a sheet of paper on edge acting as a beam. Actual beams can collapse similarly. The tendency for lateral buckling is developed in the *compression zone* of a beam, which, in a sense, is a continuous column loaded longitudinally by the bending stresses. The theoretical analysis of this problem is beyond the scope of this text. The results of these analyses are, however, very important, and have been so simplified for structural steel beams that they are easily applicable.*

According to the AISC specifications (1970), steel beams whose compression flanges are not laterally braced must be designed by means of the usual flexure formula with a *reduced stress* in the extreme fibers. The formula for the reduced bending stress is

$$F_b = \frac{12 \times 10^3 C_b}{\ell d/A_f} \leq 0.60 F_y \qquad (13\text{-}43)$$

where ℓ is the laterally unsupported length of a beam, d is its depth, A_f is the area of the compression flange, and C_b is defined as follows: $C_b = 1.75 + 1.05(M_1/M_2) + 0.3(M_1/M_2)^2$, but not more than 2.3 (conservatively, C_b can be taken as unity), where M_1 is the smaller and M_2 the larger bending moment

---

*See Karl de Vries, "Strength of Beams as Determined by Lateral Buckling," *Trans. ASCE*, 1947, vol. 112, p. 1245.

at the end of the unbraced length. The ratio M_1/M_2 is positive if the beam is bent in reverse curvature and negative in the case of single curvature bending. Further, C_b is taken as unity if the moment at any point within the unbraced length is larger than that at both the ends.

The formula given by Eq. 13-43 is applicable only when the compression flange is solid and approximately rectangular in cross section, and its area is not less than that of the tension flange. For sections whose compression flanges do not satisfy the above requirement, the AISC specifies a set of alternative formulas* which depend on an ℓ/r_T ratio, where r_T is the radius of gyration of a section comprising the compression flange plus one-third of the compression web area, taken about an axis in the plane of the web. The value of F_b is taken as the larger of Eq. 13-43 or of either Eq. 1.5-6a or 1.5-6b of the AISC Manual as applicable, but not to exceed $0.60F_y$.

These AISC formulas attempt to approximate the results of rational analyses by simple expressions. Similar formulas are also in use for other materials.

PROBLEMS FOR SOLUTION

13-1. A rigid bar is held in a vertical position as shown in the figure by a spring having a spring constant k_1 lb per inch and a torsional spring having stiffness k_2 lb-in. per radian. Find the critical load P_{cr} for this system.

13-2. Two rigid bars of equal length are connected by a frictionless hinge at A and are supported at B and D as shown in the figure. Determine the magnitude of the critical force P_{cr}. The spring attached to the lower bar at C is linearly elastic having a spring constant k. *Ans: $ka/3$.*

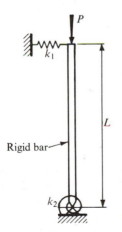

PROB. 13 – 1

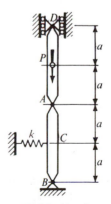

PROB. 13 – 2

*Eqs. 1.5-6a and 1.5-6b in the AISC Manual (7th ed.), 1970.

13-3. The pin-connected aluminum-alloy frame shown carries a concentrated load F. Assuming buckling can only occur in the plane of the frame, determine the value of F that will cause instability. Use the Euler formula as a criterion for member buckling. Take $E = 70 \times 10^6$ kN/m² for the alloy. Both members have 50 mm by 50 mm cross sections.

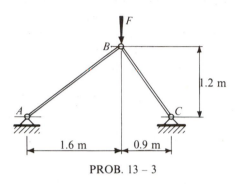

PROB. 13 – 3

13-4. A bracket consists of a curved member AB and a straight member BC. If member BC is a wooden post 40 mm by 80 mm, what vertical force may be supported at the joint B? Assume all joints pin-ended. Neglect the weight of the structure. Use the Euler column formula with a factor of safety of 3.0. Let $E = 11 \times 10^6$ kN/m².

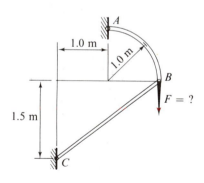

PROB. 13 – 4

13-5. A pin-ended wooden member AB with an effective length of 3 m acts in compression in the planar arrangement shown in the figure. This member has a rectangular cross-section of 0.060

m by 0.100 m. If a factor of safety against buckling of 2 is to be maintained, what can the maximum value of the applied force F be? Use the basic Euler's formula for member AB. Let $E = 1.2 \times 10^{10}$ Pa. *Ans:* 4.14 kN.

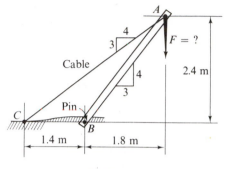

PROB. 13 – 5

13-6. A 1-in. round steel bar 4 ft long acts as a spreader bar in the arrangement shown in the figure. If cables and connections are properly designed, what pull F may be applied to the assembly? Use Euler's formula and assume a factor of safety of 3. Let $E = 29 \times 10^6$ psi. *Ans:* 9.01 k.

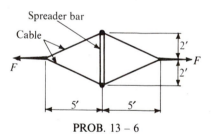

PROB. 13 – 6

13-7. The mast of a derrick is made of a standard 4 in. by 4 in. by $\frac{1}{4}$ in. steel angle. If this derrick is assembled as indicated in the figure, what vertical force F, governed by the size of the angle used for the mast, may be applied at A? Assume that all joints are pin-connected, and that the connection details are so made that the mast is loaded concentrically. The top of the mast is braced to prevent sideways displacement. Use

Euler's formula with a factor of safety of 3.5. Let $E = 29 \times 10^6$ psi. *Ans: 9.28 k.*

tension rod AB are adequate. Let $E = 29 \times 10^3$ ksi. *Ans: 4.86 k.*

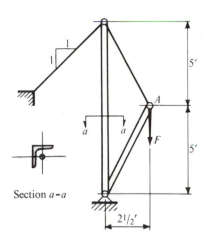

Section a-a

PROB. 13 – 7

13-8. If the capacity of the jib crane, the dimensions of which are shown in the figure, is to be 4 kips, what size standard steel pipe AB should be used? Use the Euler formula with a factor of safety of 3.5. Let $E = 29 \times 10^6$ psi. Neglect the weight of construction.

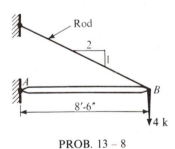

PROB. 13 – 8

13-9. A tripod made from 2 in. standard steel pipes is to be used for lifting loads vertically with a pulley at A as shown in the figure. What load rating may be assigned to this structure? Use the Euler formula and a factor of safety of 3. All joints may be considered pinned, and assume that the connection details, anchorage, and the

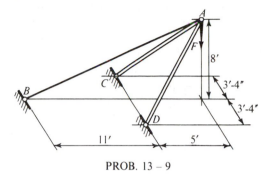

PROB. 13 – 9

13-10. A tripod is made from three 3-in.-by-3-in.-by-$\frac{1}{4}$ in. steel angles each 10 ft long. In a plan view these angles are placed 120° apart, and the tripod is 8 ft high, see figure. Assuming that the ends of these angles are pin-ended, what is the magnitude of the allowable vertical downward force F that may be applied to the tripod? Use the Euler formula with a factor of safety equal to 3. Neglect the weight of the angles. Let $E = 30 \times 10^6$ psi. *Ans: 8.25 kips.*

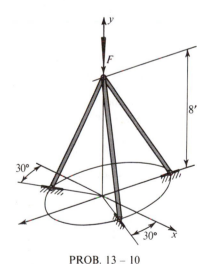

PROB. 13 – 10

13-11. A thin bar of stainless steel is axially precompressed 100 N between two plates which are fixed at a constant distance of 150 mm apart, see figure. This assembly is made at 20°C. How high may the temperature of the bar rise, so as to have a factor of safety of 2 with respect to buckling? Assume $E = 200 \times 10^6$ kN/m², and $\alpha = 15 \times 10^{-6}$ per °C. *Ans: 29.1 °C.*

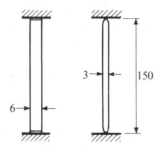

PROB. 13 – 11

*13-12. Derive Eq. 13-14, and show that a typical eigenfunction for an *n*-th mode in this case is $v_n = C_n(1 - \cos \lambda_n x)$ where $\lambda_n = (2n + 1)\pi/(2L)$ with *n* an integer. (For $n = 0$ see Fig. 13-8(b).)

*13-13. Derive Eq. 13-12. (Here the transcendental equation for determining the critical roots is $\tan \lambda L = \lambda L$, which is satisfied when $\lambda L = 4.493$.)

13-14. An allowable axial load for a 4 m long pin-ended column of a certain linearly elastic material is 20 kN. Five different columns made of the same material and having the same cross section have the supporting conditions shown in

the figure. Using the column capacity for the 4 m column as the criterion, what are the allowable loads for the five columns shown?

13-15. A piece of mechanical equipment is to be supported at the top of a 5-in. nominal diameter standard steel pipe as shown in the sketch. The equipment and its supporting platform weigh 5,500 lb. The base of the pipe will be anchored in a concrete pad, the top end will be unsupported. If the factor of safety required against buckling is 2.5, what is the maximum height of the column on which the equipment can be supported? Let $E = 30 \times 10^6$ psi. *Ans: 23.8 ft.*

PROB. 13 – 15

13-16. A bar is initially curved so that its axis has the shape of a one-half sine wave, $v_0 = a \sin \pi x/L$. If this bar is subjected to an axial compression as shown in the figure, show that the total deflection

$$v = v_0 + v_1 = [1/(1 - P/P_{cr})]a \sin \pi x/L$$

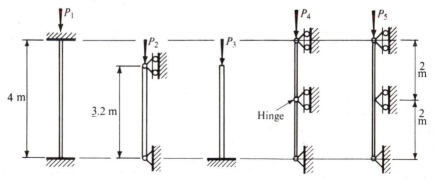

PROB. 13 – 14

where $P_{cr} = \pi^2 EI/L^2$ and the expression in the brackets is the magnification factor.

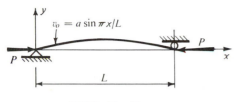

$$v_0 = a \sin \pi x/L$$

PROB. 13 – 16

13-17. If the proportional limit of steel is 250 000 kN/m² and $E = 200 \times 10^6$ kN/m², in what range of slenderness ratio is the use of the Euler formula for fixed end columns not permissible? *Ans:* Less than 178.

13-18. (a) If a pin-ended solid circular shaft is 1.5 m long and its diameter is 50 mm, what is the shaft's slenderness ratio? (b) If the same amount of material as above is reshaped into a square bar of the same length, what is the slenderness ratio of the bar? *Ans:* (a) 120.

13-19. The cross-section of a compression member for a small bridge is made as shown in Fig. 13-15(a). The top cover plate is $\frac{1}{2}$ in. by 18 in. and the two C 12 × 20.7 channels are placed 10 in. from back to back. If this member is 20 ft long, what is its slenderness ratio? (Check L/r in two directions.) *Ans:* 51.5.

13-20. What size W column is required to carry a concentric load of 200 kips on a pin-ended column if it is 12 ft long? Use the AISC formulas and assume that A36 steel with $\sigma_{yp} = 36$ ksi will be used.

13-21. Two 10 in.-15.3 lb channels form a 24 ft long square compression member; the flanges are turned in, and the channels thoroughly laced together. What is the allowable axial force on this member according to either Eq. 13-30 or Eq. 13-31 whichever is appropriate? (Check L/r in both directions.) The channels are made of A36 steel, i.e. $\sigma_{yp} = 36$ ksi.

13-22. A compression member made of two C 8 × 11.5 channels is arranged as shown in Fig. 13-15(b). (a) Determine the distance back to back of the channels so that the moments of inertia about the two principal axes are equal. (b) If the member is pin-ended and is 32 ft long, what axial load may be applied according to the AISC code? Assume that A441 steel with $\sigma_{yp} = 50$ ksi will be used, and that the lacing is adequate. *Ans:* (a) 4.94 in.

13-23. A boom for an excavating machine is made of four $2\frac{1}{2}$ in. by $2\frac{1}{2}$ in. by $\frac{1}{2}$ in. steel angles arranged as shown in Fig. 13-15(c). Out to out dimensions of the square column (exclusive of the dimensions of the lacing bars) are 14 in. What axial load may be applied to this member if it is 52 ft long? Use the Euler formula for pinned columns and a factor of safety of 4.

13-24. A compression chord of a small truss consists of two 4 in. by 4 in. by $\frac{3}{8}$ in. steel angles arranged as shown in Fig. 13-15(d). The vertical legs of the angles are separated $\frac{1}{2}$ in. apart by spacers. If the length of this member between braced points is 8 ft, what axial load may be applied according to the AISC code? $\sigma_{yp} = 36$ ksi.

13-25. A W 14 × 320 core section has two 24-in.-by-3-in. cover plates as shown in the figure. If this member is 20 ft long and is assumed to be pin-ended, what axial compressive force may be applied according to the AISC code? Assume that A242 steel with $\sigma_{yp} = 42$ ksi will be used.

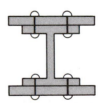

PROB. 13 – 25

13-26. An 8 WF 8.32 aluminum section (assume same geometric properties as that of

W 8 × 24 steel section; see footnote to Example 13-4) is used as a column with an effective length of 15 ft. Determine the carrying capacity of this column, with a factor of safety of 2. Use Eq. 13-32 or 13-33, whichever is applicable.

13-27. Determine the allowable axial loads for two 100 mm by 100 mm (actual size) Douglas Fir posts, one 3 m and the other 5 m long. $E = 11 \times 10^6$ kN/m².

13-28. A round platform 6 ft in diameter, to be used by a "flag-pole sitter," is attached to a standard 6 in. pipe 20 ft long. If the "sitter" weighs 150 lb, what weight of equipment for his comfort may be allowed on the platform? Locate the weights in the most critical position, i.e., at 3 ft from the vertical centerline of the pipe. The answer is to be based on an allowable stress approach: $P/A + Mc/I \leq$ allow stress for columns by Euler formula with FS = 3. (*Hint:* see Fig. 13-8(b).) $E = 29 \times 10^3$ ksi. *Ans: 325 lb.*

13-29. A W 12 × 85 column 20 ft long is subjected to an eccentric load of 180 kips located as shown in the figure. Using the AISC interaction formula, determine whether this column is adequate. Use the same allowable stresses as in Example 13-5.

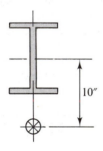

PROB. 13 – 29

13-30. A W 14 × 68 column made of A36 grade steel ($\sigma_{yp} = 36$ ksi) is 20 ft long and is loaded eccentrically as shown in the figure. Determine the allowable load P using the AISC formulas. Assume pin-ended conditions. Let $F_b = 27$ ksi.

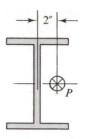

PROB. 13 – 30

13-31. A W 12 × 40 column has an effective length of 20 ft. Using the AISC formulas, determine the magnitude of an eccentric load that may be applied to this column at A, as shown in the figure, in addition to a concentric load of 20 kips. The column is braced at top and bottom. The allowable bending stress is given by Eq. 13-43. *Ans: 68.3 k.*

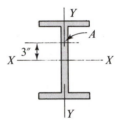

PROB. 13 – 31

13-32. What may the magnitude be of the maximum beam reaction that can be carried by a W 10 × 49 column having an effective length of 14 ft, according to the AISC interaction formula? Assume that the beam delivers the reaction at the outside flange of the column as shown in the figure and is concentric with respect to the

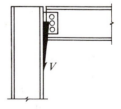

PROB. 13 – 32

minor axis. The top and bottom of the column are held laterally. Assume $F_y = 36$ ksi, and $F_b = 22$ ksi.

13-33. A W 24 × 100 simple beam 24 ft long is laterally unsupported. What uniformly distributed load, including the weight of the beam, may be applied according to the AISC specifications?

13-34.* Using the AISC code, select a W shape column to carry a concentric load of 60 kips and an eccentric load of 25 kips applied on the $Y\text{-}Y$ axis at a distance of 6 in. from the $X\text{-}X$ axis. The column is braced top and bottom, and is 14 ft long. The allowable bending stress is 22 ksi. Let $\sigma_{yp} = 36$ ksi.

*14 Structural Connections

14-1. INTRODUCTION

The design of various members based on strength, stiffness, and stability considerations was discussed in the preceding chapters. In this chapter, the methods of analysis and design of connections for these members will be treated. The analysis of connections cannot be made on as rigorous a basis as used in much of the preceding work. The design of connections is largely empirical and is based on available experience as well as sound interpretation of experimental research.

Riveted connections played a dominant role in past engineering construction. However, the increasing use of welding and high-strength bolting due to the economic advantages they offer has caused a rapid decline in the use of rivets and ordinary bolts as fasteners in the fabrication of structural products. The manner in which riveted and bolted joints can fail will be discussed, followed by methods for selecting the proper size and number of rivets or bolts (ordinary or high strength) for transmitting a given force through a joint. The remainder of the chapter will be concerned with welded connections.

14-2. RIVETED AND BOLTED CONNECTIONS

Most members are made up of platelike parts such as actual plates, webs and flanges of beams, legs of angles, etc. The design of a riveted or bolted connection is mainly concerned with the transfer of forces through these plates. The transfer of forces by means of single and double shear was discussed in Art. 1-5. It is also possible for more than two shear planes to occur in complex joints involving interaction between several plate elements.

For a riveted assembly of ordinary parts, matching holes are punched in the plates for the insertion of hot rivets.* For more accurate work, as well

*Ordinary steel rivets are driven at a temperature between 1000°F (blood red) and 1950°F (light yellow). Small steel rivets and some aluminum rivets are driven without preheating, although hot (1040°F ±) and cold (32°F) rivets are frequently employed in aluminum work. Several patented rivet types are available for nonferrous metals.

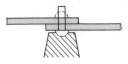

Fig. 14-1. One of the rivet heads formed from the original shank

as for the larger size of rivets (1 in. to $1\frac{1}{4}$ in. diameter), the holes are drilled or first punched and then reamed (enlarged) to size. One end of a rivet has a head, while the other head is formed from the shank by a pneumatic hammer as a back-up tool is held against the prefabricated head, Fig. 14-1. It is desirable to have the clearance of the rivet in a prepared hole as small as possible. For rivets of $\frac{1}{2}$ in. diameter and larger, an increase of $\frac{1}{32}$ to $\frac{1}{16}$ in. in the diameter of the hole over the nominal rivet diameter is customarily made. On proper driving, a hot rivet expands in the prepared hole, and, while hot, it presses the surrounding material outward. Upon cooling, the rivet diameter diminishes somewhat, but the elastic return of the surrounding material helps to reduce this tendency for undesirable looseness. Moreover, *a rivet also shrinks lengthwise upon cooling*. This effect causes a permanent tensile stress in the shank of a rivet and a compression stress in the assembled plates (see Example 12-6). The compressive force so set up between the plates may be rather large, and this force is capable of developing a large frictional resistance normal to the axis of the rivet. This action may be sufficient to carry the applied working load. However, in determining the required number of rivets, it is customary to *disregard* the frictional resistance of the joint. The capacity of a joint is thought of in terms of its ultimate capacity after the frictional resistance between the plates is "broken."

Ingenuity must be exercised in arranging the rivets at a joint. Accessibility for riveting and proper tool clearances are important considerations. Several typical arrangements of rivets, with self-explanatory titles, are shown in Fig. 14-2. Riveted joints are normally designed to transmit shear rather than tension through the rivets. The design of a connection is more involved than a check of stresses in an existing joint.

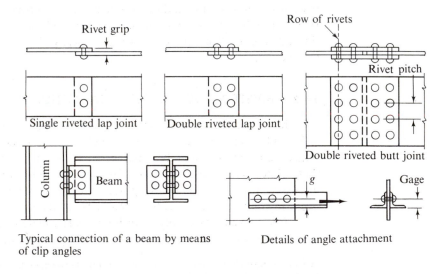

Fig. 14-2. Typical arrangements of riveted connections

Ordinary bolts can also be used to transmit forces in a connection. The frictional resistance of the joint is disregarded as the initial tension developed by such bolts (also known as common or unfinished bolts) is uncertain. Hence, the design of connections using ordinary bolts follows the same general procedure as that for rivets, except that lower values of allowable stresses are specified to account for the possibility of the threaded portion of the bolt extending into the shear plane.

The use of high strength bolts, Fig. 14-3(a), is a relatively new development that has gained acceptance as the preferred fastener in field work.

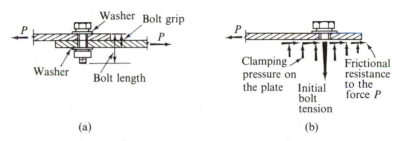

Fig. 14-3

The bolt is tightened, using the calibrated wrench method or the turn-of-nut method, to obtain a specified minimum initial tension of approximately 70 percent of its tensile strength.* A reliable clamping force is thus developed, and the load transfer is achieved through frictional resistance between the plates, Fig. 14-3(b). Since the bolt does not slip at allowable loads due to the clamping pressure, there is no shearing stress on the bolt itself, nor does the bolt bear against the plates through which it passes. However, for the purposes of a simplified analysis, an allowable shear stress is specified, which takes into account the load transfer due to the frictional resistance developed. This enables the design of a connection using high strength bolts to be carried out in the same manner as that using rivets and ordinary bolts. If a direct load applied on a bolt causes a reduction of the initial clamping force, the frictional resistance is reduced, and the AISC Manual specifies certain formulas to reduce the allowable shear stresses correspondingly.

14-3. METHODS OF FAILURE OF A RIVETED OR BOLTED JOINT

Riveted or ordinary bolted joints may fail in a number of different ways (also see Art. 1-5). These are as follows:

*AISC Manual (7th ed.), 1970. Also, the interested reader may consult modern books on steel design for more details, for example, B. G. Johnston and F. J. Lin, *Basic Steel Design*, Englewood Cliffs, N.J.: Prentice-Hall, 1974, Y. Y. Lin, *Elementary Steel Structures*, Englewood Cliffs, N.J.: Prentice-Hall, 1973, and B. Bresler, T. Y. Lin, and J. B. Scalzi, *Design of Steel Structures*, John Wiley, 2nd ed., 1968.

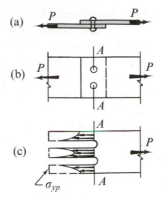

Fig. 14-4. Tensile stresses in a plate of a riveted joint

Failure in Tension

A riveted joint transmitting a tensile force may fail in a plate weakened by the rivet holes. For example, in a single-riveted lap joint, Fig. 14-4(a), the *net area* in either plate across the section *A-A* in Fig. 14-4(b) is the least area, and a tear would occur there. At working loads, before failure occurs, large stress concentrations exist at the rivet holes in a plate, Fig. 14-4(c), since the holes interrupt the continuity of the plate. However, as the applied force is increased, a nearly uniform stress distribution will prevail at the yield point for ductile materials (Art. 2-11). Since riveting is employed only for ductile materials, it is customary to base the capacity of a joint in tension on the assumption of a *uniform stress distribution across the net section of a plate.* Because of the fatigue properties of materials, this assumption should not be used for riveted joints subjected to severe fluctuating loadings, since stress concentrations are important for such cases. Usually no deduction for the rivet holes in a compression joint is made.

Failure in Shear

In a riveted joint, the rivets themselves may fail in shear. This type of failure

Fig. 14-5. (a) Single and (b) double shear of a rivet

is shown in Figs. 14-5(a) and (b). In analyzing this possible manner of failure, one must always note whether a rivet acts in single or double shear. In the latter case, *two* cross-sectional areas of the same rivet resist the applied force. In practical calculations, the shearing stress is assumed to be *uniformly distributed over the cross section of a rivet.* This assumption is justified for ductile materials after the elastic limit has been exceeded.

Failure in Bearing

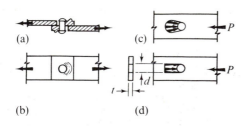

Fig. 14-6. A bearing failure of a riveted joint and bearing stress distributions, (c) probably, (d) assumed

A riveted joint may fail if a rivet crushes the material of the plate against which it bears, Figs. 14-6(a) and (b), or if the rivet itself is deformed by the plate acting on it. The stress distribution is very complicated in this type of failure and is somewhat like that shown in Fig. 14-6(c). In practice, this stress distribution is approximated on the basis of an *average bearing stress acting over the projected area of the rivet's shank onto a plate,* i.e., on area td in

Fig. 14-6(d). It is difficult to justify this procedure theoretically. However, the allowable bearing stress is determined from experiments and is interpreted on the basis of this *average stress* acting on the projected area of a rivet. Therefore the inverse process used in design is satisfactory. Tests show that plates *confined* between other plates, as the middle plate in Fig. 14-5(b), can withstand a greater average bearing pressure than unconfined plates, as in Fig. 14-6(a) or the outer plates in Fig. 14-5(b). For this reason, several specifications allow a greater bearing stress for rivets in double shear acting on inside plates than for rivets in single shear acting on outside plates.

Other Methods of Failure

In a multiple-riveted joint with several rows of rivets, the net area to be used in calculations is sometimes difficult to determine. For example, if, as in Fig. 14-7(a), the rivet rows are very close together, a *zig-zag* tear is more likely to occur than a tear across the normal section *B-B*. However, the standard specifications for shop practice are so written that this type of failure is not common. Therefore the possibility of a zig-zag tear will not be considered here, as it will be assumed that in all cases the rows of rivets are sufficiently far apart.* It will also be assumed that *e*, the *edge distance*, is large enough to prevent failure in shear across the *a-a* planes in Fig. 14-7(b). Likewise, in conformity with usual practice, local bending of plates in lap joints, Fig. 14-8, will be neglected.

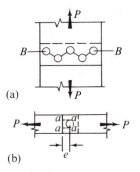

(a)

(b)

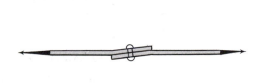

Fig. 14-7. Possible methods of failure of a riveted joint, (a) zig-zag tear, and (b) "tear out" due to insufficient edge (end) distance

Fig. 14-8. Bending of the plates in lap joints, commonly neglected

Some Further Remarks

In this text the capacity of a riveted or ordinary bolted joint will be based only on the probable tear, shear, and bearing capacities. This assumed action

*For details see AISC Manual (7th ed.), 1970.

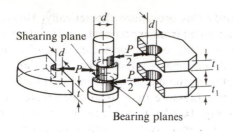

Shearing plane

Bearing planes

Fig. 14-9. The assumed action of a riveted joint

of a riveted joint is illustrated in Fig. 14-9.* The frictional resistance between the plates is neglected.† The *smallest* of the three resistances is the strength of a joint. The ratio of this strength divided by the strength of a solid plate or member, expressed in percent, is called the *efficiency of a joint*, i.e.,

$$efficiency = \frac{strength\ of\ the\ joint}{strength\ of\ the\ solid\ member} \times 100$$

(14-1)

The total force acting concentrically on a joint is assumed to be equally distributed between rivets of equal size. In many cases this cannot be justified by elastic analyses, but the ductile deformation of the rivets permits an equal redistribution of the applied force before the ultimate capacity of a connection is reached. This redistribution is possible as the rivets continue to carry a full load during yielding. Therefore, the practice of assigning equal forces to the rivets of a concentrically loaded joint is close to being correct.‡ The same assumptions are also commonly used in the analysis of bolted joints.

Rivets and bolts are usually specified to conform to the American Society for Testing Materials (ASTM) standards. Following the AISC Manual, the allowable loads in shear are given in the table that follows for $\frac{3}{4}$ in. and $\frac{7}{8}$ in. diameter rivets and bolts, which are widely used in practice. The allowable load in bearing, for riveted and ordinary bolted joints, is obtained by multiplying the projected area of the rivet or bolt by the allowable bearing stress $F_p = 1.35\ F_y$, where F_y is the yield stress of the connected part.

In general, AISC nomenclature will be largely adhered to in this chapter; namely, depending on the subscript, F corresponds to the allowable or ultimate (fiber) stress, and f, to the calculated (fiber) stress due to the applied loads.

EXAMPLE 14-1

Find the capacity of the tension member AB of the Fink truss shown in Fig. 14-10(a) if it is made from two 3 in. by 2 in. by $\frac{5}{16}$ in. angles attached to a $\frac{3}{8}$ in. thick gusset plate by four $\frac{3}{4}$ in. A502-1 rivets in $\frac{13}{16}$ in. diameter holes. The allowable stresses (AISC) are 22 ksi in tension, 15 ksi in shear, and 48.6 ksi in bearing on the angles as well as the gusset.

---

*From G. Dreyer, *Festigkeitslehre und Elastizitätslehre*, Leipzig: Jänecke, 1938, p. 34.

†Note, on the other hand, that *only* the frictional resistance is considered when connections are designed using high-strength bolts, with the capacity of each such bolt being the allowable shear stress times the nominal bolt area.

‡A conclusive experimental vertification of this assumption may be found in the paper by R. E. Davis, G. B. Woodruff, and H. E. Davis, "Tension Tests of Large Riveted Joints," *Trans. ASCE*, 1940, vol. 105, p. 1193.

Description	Diameter, in.	ASTM Designation or Yield Stress, ksi	Shear, F_v, ksi	Single Shear, kips	Double Shear, kips
Power driven field and shop rivets	$\frac{3}{4}$	A502–1 A502–2	15.0 20.0	6.63 8.84	13.25 17.67
	$\frac{7}{8}$	A502–1 A502–2	15.0 20.0	9.02 12.03	18.04 24.05
Unfinished bolts, ASTM 307, and threaded parts, $F_y = 36$ ksi[a]	$\frac{3}{4}$	A307 $F_y = 36$	10.0 10.8[a]	4.42 4.77	8.84 9.54
	$\frac{7}{8}$	A307 $F_y = 36$	10.0 10.8[a]	6.01 6.49	12.03 12.99
High strength bolts, friction-type	$\frac{3}{4}$	A325–F A490–F	15.0 20.0	6.63 8.84	13.25 17.67
	$\frac{7}{8}$	A325–F A490–F	15.0 20.0	9.02 12.03	18.04 24.05

*From AISC Manual, 7th edition, 1970.
[a]For threaded parts of material other than $F_y = 36$ ksi, use $F_v = 0.30F_y$.

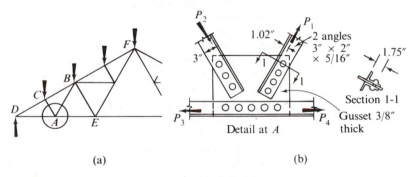

(a) (b)

Fig. 14-10

SOLUTION

The net cross-sectional area of the angles is obtained by deducting the width of the rivet holes (rivet diameter plus $\frac{1}{8}$ in.) multiplied by the thickness of the angle from the gross area of the angles (Table 7 of the Appendix). The net area multiplied by the allowable tensile stress gives the allowable force controlled by tensile stresses. Similar calculations are made for the other modes of failure, based on the following observations: There are *eight* bearing surfaces with single shear between the rivets and the angles, four rivets provide *eight* shearing surfaces, and *four* bearing surfaces with double shear are provided for the rivets by the gusset plate. In following through the solution of this problem, the reader is urged to note the sequence of load transfer, from angles to rivets, and from rivets to the gusset plate.

Tear at Section 1-1:

$$A_{net} = 2[1.46 - \tfrac{5}{16}(\tfrac{3}{4} + \tfrac{1}{8})] = 2.37 \text{ in.}^2$$

$$P_{allow} = A_{net}F_t = (2.37)(22) = 52.1 \text{ kips} \qquad \text{(governs)}$$

Bearing on angles:

$$A = 8td = (\tfrac{5}{16})(\tfrac{3}{4}) = \tfrac{15}{8} \text{ in.}^2$$

$$P_{allow} = AF_p = \tfrac{15}{8}(48.6) = 91 \text{ kips}$$

Bearing on gusset:

$$A = 4t_1d = 4(\tfrac{3}{8})(\tfrac{3}{4}) = \tfrac{9}{8} \text{ in.}^2$$

$$P_{allow} = AF_p = \tfrac{9}{8}(48.6) = 54.6 \text{ kips}$$

Shear of rivets:

$$A_{one \ rivet} = \tfrac{1}{4}\pi (\tfrac{3}{4})^2 = 0.4418 \text{ in.}^2$$

$$P_{allow} = 8A_{one \ rivet}F_v = 8(0.4418)15 = 53 \text{ kips}$$

The tensile stress at section 1-1 governs the capacity of the connection.

Note that standard gage lines are generally used for the rivet lines (see AISC Manual), and the rivets are usually spaced 3 in. apart. The standard gage line for the angles of this example is 1.75 in. On the other hand, the centroid of the cross-sectional area for the same angle is 1.02 in. from the corner (Table 7 of the Appendix). Hence there is an eccentricity of $1.75 - 1.02 = 0.73$ in. of the applied force with respect to the rivet line. For small trusses, it is customary to neglect the effect of this eccentricity, as presumably the allowable stresses are set low enough to provide for this contingency. No less than two rivets are used at all main connections. An excessively large number of rivets in one line is avoided.

The compression members of small trusses are often connected to a gusset plate in the same manner as shown for the tension member. The compression members of a truss are designed as columns. Effective column lengths are determined from a centerline diagram such as shown in Fig. 14-10(a) and a study of bracing in the direction perpendicular to the plane of the truss. The *gross areas* of columns are used to determine their load-carrying capacities, while the connection is governed either by bearing or shearing stresses. In some cases, members of a truss may act in tension, then, under a different loading condition, in compression. Such members must be designed accordingly.

EXAMPLE 14-2

Rework Example 14-1 if A307 unfinished or ordinary bolts of $\tfrac{3}{4}$ in. diameter are used instead of rivets.

SOLUTION

The allowable stresses in tension and bearing are the same as in Example 14-1. The allowable shear stress F_v in A307 bolts is 10.0 ksi. Hence, with $A_{\text{one bolt}}$ = 0.4418 in.^{2},

Shear of bolts:

$$P_{\text{allow}} = (8)(0.4418)(10) = 35.4 \text{ kips}$$

which is less than the P_{allow} calculated earlier for tearing at section 1-1, bearing on angles and the gusset plate. Thus the capacity of the connection is governed by the shear of the bolts in this case.

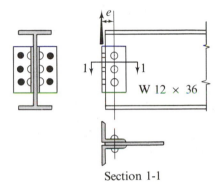

Section 1-1

Fig.14-11

EXAMPLE 14-3

Find the capacity of the standard AISC connection for the W 12 × 36 beam shown in Fig. 14-11. The connection consists of two 4 in. by $3\frac{1}{2}$ in. by $\frac{3}{8}$ in. angles, each $8\frac{1}{2}$ in. long; $\frac{7}{8}$ in. A502-1 rivets spaced 3 in. apart are used in $\frac{15}{16}$ in. holes. Use the AISC allowable stresses given in Example 14-1.

SOLUTION

A tension tear cannot occur in this connection, so only shearing and bearing capacities need be investigated. *Bearing on the web of W 12 × 36 beam:* (Thickness of the web is found from Table 4 of the Appendix.)

$$P_b = 3(0.305)\tfrac{7}{8}(48.6) = 38.9 \text{ kips} \qquad \text{(governs)}$$

Shear of rivets: There are six cross-sectional areas.

$$A_{\text{one rivet}} = 0.601 \text{ in.}^2$$
$$P_s = (6)(0.601)(15) = 54.1 \text{ kips}$$

Bearing of rivets on angles:

$$P_b = 6(\tfrac{7}{8})\tfrac{3}{8}(48.6) = 95.7 \text{ kips}$$

Hence the capacity of this connection is 38.9 kips.

The legs of the angles perpendicular to the web of the beam are called the *outstanding legs*. They are used to attach the connection to a column or a plate. Adequate shear resistance across the additional rivets and bearing in the outstanding legs is always available. However, the thickness of the plate to which the connection is made must be investigated. If this plate is thin, the capacity of the joint may be lower than computed above. Note that the effect of eccentricity e, as well as bending and shear in the connection angles, is usually not investigated. The capacities of standard connections are tabulated in the AISC Manual.

EXAMPLE 14-4

Rework Example 14-3 if A490-F friction-type high strength bolts of $\frac{7}{8}$ in. diameter are used instead of the rivets.

SOLUTION

The allowable shear stress for the A490-F bolts is 20 ksi.
Shear of high strength bolts:

$$P_s = (6)(0.601)(20) = 72.1 \text{ kips}$$

Since no slip occurs in a friction-type connection, no bearing of the bolt with connected parts takes place, and the capacity of this connection is given as 72.1 kips.

With the wide use of welding and the high capacities of high strength bolts, instead of two connection angles, only a single plate to one side of the beam web is often used. Such a plate is shop-welded to the column and bolted to the beam in the field.

EXAMPLE 14-5

Find the allowable tensile force that the multiple-riveted structural joint shown in Figs. 14-12(a) and (b) can transmit. Also find the efficiency of this joint. All rivets are nominally $\frac{7}{8}$ in. in $\frac{15}{16}$ in. diameter holes. Use the AISC allowable stresses given in Example 14-1.

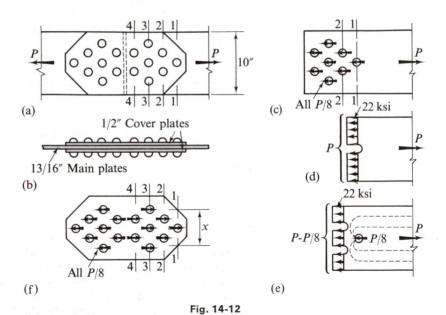

Fig. 14-12

SOLUTION

The analysis of a multiple-riveted structural joint is based on the assumption that the *concentrically applied force is distributed equally between rivets of equal size.* Therefore the free body for one of the main plates is as shown in Fig. 14-12(c), while that for the *two* cover plates is as shown in Fig. 14-12(f). Tensile stresses must be investigated at several sections. This is done by considering the free bodies as shown in Figs. 14-12(d) and (e) and assuming uniform stress distribution in the plates. Nominal rivet diameter is used for shear and bearing calculations.

Capacity P_s of the joint in shear: 16 cross-sectional areas of rivets are available. $F_v = 15$ ksi.

$$A_{\text{one rivet}} = \tfrac{1}{4}\pi(\tfrac{7}{8})^2 = 0.601 \text{ in.}^2$$

$$P_s = 16(0.601)15 = 144 \text{ kips} \qquad \text{(governs)}$$

Capacity P_B of the joint in bearing on the main plate: 8 bearing surfaces in double shear are available. $F_p = 48.6$ ksi.

$$P_B = 8(\tfrac{7}{8})(\tfrac{13}{16})(48.6) = 276 \text{ kips}$$

Capacity P_b of the joint in bearing on the cover plates: 16 bearing surfaces in single shear. $F_p = 48.6$ ksi.

$$P_b = 16(\tfrac{7}{8})(\tfrac{1}{2})(48.6) = 340 \text{ kips}$$

Capacity of the main plate in tension:

$$F_t = 22 \text{ ksi.}$$

Without holes:

$$P_t = (\tfrac{13}{16})(10)22 = 179 \text{ kips.}$$

Section 1-1, free body, Fig. 14-12(d):

$$P_{1-1} = (\tfrac{13}{16})[10 - (\tfrac{7}{8} + \tfrac{1}{8})]22 = 161 \text{ kips.}$$

Section 2-2, free body, Fig. 14-12(e):

$$P_{2-2} = \tfrac{1}{8}P_{2-2} + \tfrac{13}{16}[10 - 2(\tfrac{7}{8} + \tfrac{1}{8})]22 \qquad \text{or} \qquad \tfrac{7}{8}P_{2-2} = 143 \text{ kips}$$

hence $P_{2-2} = 163$ kips.

This result means that, if a tensile force $P_{2-2} = 163$ kips were applied to the joint, only seven-eighths of this force need be resisted by section 2-2 at a 22 ksi stress, since one-eighth of this force is resisted by the *outer rivet.* Similarly,

$$\tfrac{5}{8}P_{3-3} = \tfrac{13}{16}(10 - 3)22 \qquad \text{or} \qquad P_{3-3} = 200 \text{ kips}$$

and $\qquad \tfrac{2}{8}P_{4-4} = \tfrac{13}{16}(10 - 2)22 \qquad \text{or} \qquad P_{4-4} = 572 \text{ kips}$

Capacity of the two cover plates in tension: $F_t = 22$ ksi. Across section 4-4, Fig. 14-12(f), the whole force P is transmitted, hence,

$$P_{4-4} = 2(\tfrac{1}{2})[10 - 2(\tfrac{7}{8} + \tfrac{1}{8})]22 = 176 \text{ kips}$$

Then, by considering free-body diagrams of the cover plates to one side of a section, as was done for the main plate,

$$\tfrac{6}{8}P_{3-3} = 1(10-3)22 \quad \text{or} \quad P_{3-3} = 205 \text{ kips}$$

and $\quad\tfrac{3}{8}P_{2-2} = 1(10-2)22 \quad \text{or} \quad P_{2-2} = 469 \text{ kips}$

Similarly, section 1-1 is not critical. Therefore the width of the *cover plates* at section 1-1 may be reduced to a width x to provide, at an allowable stress, a net area for only one-eighth of the applied force.

The capacity of this joint is limited by the allowable shearing stress in the rivets. Hence the capacity of this joint is 144 kips. The efficiency of this joint is $(144/179)100 = 80.4\%$, which is good.

The same calculation procedure would be used for a compression splice if the ends of the main plates were not machined. However, if the ends are milled, the transfer of a compression force is direct. Nevertheless, even in such cases, it is customary to provide a riveted connection calculated at a certain fraction, say 50%, of the total actual force.

14-4. ECCENTRIC RIVETED AND BOLTED CONNECTIONS

The foregoing discussion of riveted and bolted connections applies to situations where the line of action of the applied force passes through the centroid of a rivet group. More precisely, it is assumed that each rivet or bolt possesses a certain allowable shear or bearing resistance in *every* direction. Thus each rivet in a group of rivets has a certain resistance, say in the horizontal direction, indicated by $P_1, P_2, \ldots, P_5$ in Fig. 14-13. The best place to apply horizontal force to this group of rivets is in the direction of E_1, which is an equilibrant of the allowable rivet forces. A force applied along this line of action distributes itself among the various rivets without any tendency to twist the attached plate. By similar reasoning, the line of action of E_2 can be established in the vertical direction. Then, if the point of intersection of E_1 and E_2 is denoted by B, it is seen that an inclined force E_3 must pass through B to cause no twisting of the plate, since E_3 may be resolved into horizontal and vertical components. Point B is synonymous with the *centroid of rivet areas*, since in most practical cases the resistance of a rivet depends on its cross-sectional area. A force acting in any direction, but passing through the centroid of rivet areas, is the former problem of a concentrically loaded connection. However, since forces are not always applied through this point, eccentrically loaded connections result.

Consider a joint, such as shown in Fig. 14-14(a), with a force P applied eccentrically with respect to the centroid B of the rivet areas. This problem

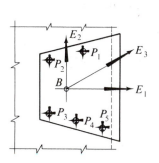

Fig. 14-13. The applied force must act through the centroid of rivet areas to prevent twisting of a connection

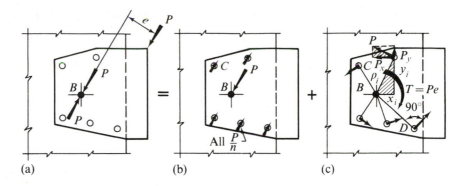

Fig. 14-14. Resolution of a problem of an eccentrically loaded riveted connection into two problems.

is not changed by introducing at B two equal and opposite forces P parallel to the applied force P. The new problem can be conveniently resolved into the two problems shown in Figs. 14-14(b) and (c). The first problem, Fig. 14-14(b), was discussed earlier, and the average *direct* shearing stress f_d in *all* rivets is

$$f_d = \frac{P}{\sum A_i} \tag{14-2}$$

where $\sum A_i$ is the sum of all the cross-sectional areas of the rivets. If all n rivets are of equal size, $\sum A_i = nA$, where A is the cross-sectional area of one rivet, and

$$f_d = \frac{P}{nA} \quad \text{or} \quad P_d = Af_d = \frac{P}{n} \tag{14-2a}$$

where P_d is the direct force in each rivet. For equilibrium, f_d and P_d act in a direction opposite to the force P applied at B.

The second problem, Fig. 14-14(c), is a problem in torsion, in which the applied torque T is equal to Pe. In this problem, if it is assumed that the plate is *rigid* and that it twists around the point B, the shearing strains in rivets vary linearly from B. Further, if the rivets are assumed to be elastic, the average shearing stress in each rivet also varies linearly from B. Therefore this problem resembles the torsion problem of a circular shaft (or a bolted coupling, Art. 3-12), and the torsion formula $\tau = T\rho/J$ may be adapted for its solution. By means of this formula, if each rivet is assumed to be concentrated at a point, the distance ρ (c to the farthest rivet) from the *centers* of the various rivets to B can be easily determined, and only the quantity J needs further comment. Thus, since by definition $J = \int \rho^2 dA$, and the "area of the torsion member" in this case is a discrete number of rivet areas, with sufficient accuracy,

$$J \approx \sum \rho_i^2 A_i$$

where the summation includes the product of the cross-sectional area A_i of *every* rivet by the square of *its* distance ρ_i from the centroid of all rivet areas. Moreover, since from Fig. 14-14(c), each $\rho_i^2 = x_i^2 + y_i^2$, where x_i and y_i are the co-ordinates of a particular rivet's center from the centroid of all rivet areas, $J \approx \sum (x_i^2 + y_i^2)A_i$.

By using the approximation for J established above, the *torsional* shearing stress f_t on any one rivet at a distance ρ_i from the centroid of all rivet areas becomes

$$(f_t)_i = \frac{T\rho_i}{J} \approx \frac{T\rho_i}{\sum \rho_i^2 A_i} = \frac{T\rho_i}{\sum(x_i^2 + y_i^2)A_i} \qquad (14\text{-}3)$$

whereas, if all rivets are of equal size,

$$(f_t)_i = \frac{T\rho_i}{A\sum(x_i^2 + y_i^2)} \quad \text{or} \quad (P_t)_i = A(f_t)_i = \frac{T\rho_i}{\sum(x_i^2 + y_i^2)} \qquad (14\text{-}3a)$$

where $(P_t)_i$ is a force due to a torque T acting on a rivet at a distance ρ from the centroid of all rivet areas. *Either $(f_t)_i$ or $(P_t)_i$ acts perpendicular to the direction of ρ_i.* Further, by noting the similarity of the triangle with sides x_i, y_i, and ρ_i to the triangle of force at a rivet and its components, shown shaded in Fig. 14-14(c), it follows that the *components* of the torsional stresses in the x- and y-directions, respectively, are

$$(f_{tx})_i = \frac{Ty_i}{\sum(x_i^2 + y_i^2)A_i} \quad \text{and} \quad (f_{ty})_i = \frac{Tx_i}{\sum(x_i^2 + y_i^2)A_i} \qquad (14\text{-}3b)$$

A *vectorial superposition* of the direct and torsional shearing stresses (or forces) given by the above equations gives the *total* shearing stress (or force) acting on any one rivet. The highest stressed rivet can usually be found by inspection. For example, in Fig. 14-14 the rivet C does not appear to be highly stressed, since the force P/n and the force at C due to torque act approximately in the opposite directions. On the other hand, the corresponding forces at the rivet D are nearly colinear and of the same sense.

The same procedure of analysis is also applicable for both ordinary and high strength bolted connections. It should be noted, however, that, if the load eccentricity causes both shear and *tension* of the fasteners, the AISC specifies a reduction of the allowable tensile stress for rivets and ordinary bolts, and a reduction of the allowable shear stress in the friction-type high strength bolts.

Note that the solution for an eccentrically loaded connection is based on the superposition of two entirely different solutions. The first, for the direct stresses, is based on a nonelastic hypothesis; the second, for the torsional stresses, is based on the elastic concepts. The superposition of these two solutions is not completely consistent. The potential capacity of the joint in torsion is not accurately appraised, since rivets situated close to the centroid of the whole rivet group can resist higher forces at ultimate loads.

The solution obtained is conservative, however; hence its use in practice is justified.*

EXAMPLE 14-6

Find the maximum shearing stress caused by an inclined force $P = 12$ kips in the rivets of the connection shown in Fig. 14-15(a). The rivets are of 1 in. diameter ($A = 0.785$ in.²).

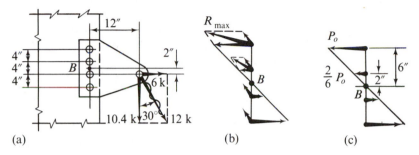

Fig. 14-15

SOLUTION

The applied force is first resolved into the horizontal and vertical components, which simplifies the determination of the components of the direct shearing stress, as well as the torque T. The centroid of all rivet areas is between the top and bottom rivets at B. Inspection of Fig. 14-15(b), where the anticipated direct and torsional stresses are shown acting simultaneously, shows that the top rivet is the highest stressed rivet. This rivet is $c = 6$ in. from B. Note that the rivet pattern is symmetrical, and that the x's in the $\sum (x^2 + y^2)$ term are zero.

$$P_x = 12 \sin 30° = 6 \text{ kips} \rightarrow, \quad P_y = 12 \cos 30° = 10.4 \text{ kips} \downarrow$$

$$T = 10.4(12) - 6(2) = 113 \text{ kips-in.} \circlearrowleft$$

$$f_{dy} = \frac{P_y}{nA} = \frac{10.4}{4(0.785)} = 3.31 \text{ ksi, resisting} \uparrow$$

$$f_{dx} = \frac{P_x}{nA} = \frac{6}{4(0.785)} = 1.91 \text{ ksi, resisting} \leftarrow$$

$$f_{tx} = f_t = \frac{Tc}{A \sum (x^2 + y^2)} = \frac{113(6)}{0.785(6^2 + 2^2)2} = 10.8 \text{ ksi} \leftarrow$$

$$f_{max} = \sqrt{(1.91 + 10.8)^2 + 3.31^2} = 13.1 \text{ ksi}$$

The torsional part of an eccentrically loaded connection problem can always be solved from the first principles without the use of the torsion formula. Thus, assuming, as before, that the plate is *rigid*, also assuming a linear variation of strain in the rivets from the center of twist, and using Hooke's law, it is found that the stresses in the rivets vary linearly from the center of

*Also see the last paragraph in Art. 3-12.

twist. Hence, for the above problem with equal cross-sectional areas of all rivets, in terms of the force P_o at the outer rivets, from Fig. 14-15(c),

$$P_o(12) + \tfrac{2}{6}P_o(4) = T = 113 \qquad \text{and} \qquad P_o = 8.48 \text{ kips}$$

where, as above, $f_t = 8.48/0.785 = 10.8$ ksi. This procedure amounts to a rederivation of the torsion formula.

14-5. WELDED CONNECTIONS

The connection of members by means of welding is very widely used in industry. *Butt welds* have already been mentioned in Art. 9-6 for use in the manufacture of pressure vessels. The strength of these welds is simply found by multiplying the cross-sectional area of the thinnest plate being connected by the allowable tensile or compressive stress for a weld. The AISC specifications, based on the recommendations of the American Welding Society (AWS), allow the same tensile stress in the weld as in the base metal in the case of butt welds subjected to static loads in steel buildings.

Another common type of weld, which is designed on a semi-empirical basis, is the so-called *fillet weld* shown in Fig. 14-16. These welds are desig-

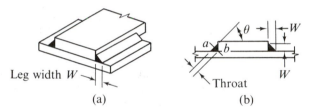

Leg width W Throat

W

(a)   (b)

Fig. 14-16. An example of a fillet weld

nated by the size of the legs, Fig. 14-16(b), which are usually made of equal width W. The smallest dimension across a weld is called its *throat*. For example, a standard $\frac{1}{2}$ in. weld has both legs $\frac{1}{2}$ in. wide and a throat equal to $(0.5) \sin \theta = (0.5) \sin 45° = 0.707(0.5)$ in. The strength of a fillet weld, *regardless of the direction of the applied force*,* is based on the cross-sectional area at the *throat* multiplied by the allowable *shearing stress* for the weld metal. The AWS allowable shear stress is 0.3 times the electrode tensile strength. For example, E70 electrodes (i.e., tensile strength of 70 ksi) used as weld metal has an allowable shear stress of $0.3 \times 70 = 21$ ksi. The allowable force q per inch of the weld is then given as

$$q = (21)(0.707)\,W = 14.85\,W \qquad [\text{k/in}] \qquad (14\text{-}4)$$

where W is the width of the legs. For a $\frac{1}{4}$ in. fillet weld this reduces to 3.71 kips per in.; for a $\frac{3}{8}$ in. fillet weld to 5.56 kips per in., etc.

*This is a considerable simplification of the real problem.

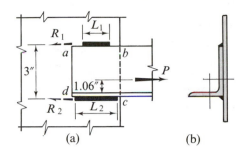

(a)   (b)

Fig. 14-17

EXAMPLE 14-7

Determine the proper lengths of welds for the connection of a 3 in. by 2 in. by $\frac{7}{16}$ in. steel angle to a steel plate, Fig. 14-17. The connection is to develop the full strength in the angle uniformly stressed to 20 ksi. Use $\frac{3}{8}$ in. fillet welds, whose strength per AWS specification is 5.56 kips per linear inch.

SOLUTION

Many arrangements of welds are possible. If two welds of length L_1 and L_2 are to be used, their strength must be such as to maintain the applied force P in equilibrium without any tendency to twist the connection. This requires the resultant of the forces R_1 and R_2 developed by the welds to be equal and opposite to P. For the optimum performance of the angle, the force P must act through the centroid of the cross-sectional area (see Table 7 of the Appendix). For the purposes of computation, the welds are assumed to have only linear dimensions.

$$A_{\text{angle}} = 2.00 \text{ in.}^2 \qquad \sigma_{\text{allow}} = 20 \text{ ksi}$$

$$P = A\sigma_{\text{allow}} = 2(20) = 40 \text{ kips}$$

$$\Sigma M_d = 0 \circlearrowright +, \quad R_1(3) - 40(1.06) = 0, \qquad R_1 = 14.1 \text{ kips}$$

$$\Sigma M_a = 0 \circlearrowleft +, \quad R_2(3) - 40(3 - 1.06) = 0, \quad R_2 = 25.9 \text{ kips}$$

Check: $\qquad R_1 + R_2 = 14.1 + 25.9 = 40 \text{ kips} = P$

Hence, using the specified value for the strength of the $\frac{3}{8}$ in. weld, note that $L_1 = 14.1/5.56 = 2.54$ in. and $L_2 = 25.9/5.56 = 4.66$ in. The actual length of welds is usually increased a small amount over the lengths computed to account for craters at the beginning and end of the welds. The eccentricity of the force P with respect to the plane of the welds is neglected.

To minimize the length of the connection, end fillet welds are often used. Thus, in the above example, a weld along the line ad could be added. The centroid of the resistance for this weld is midway between a and d. For this arrangement, the lengths L_1 and L_2 are so reduced that the resultant force for all three welds coincides with the resultant of R_1 and R_2 of the former case. To accomplish the same purpose, slots and notches in the attached member are also used.

14-6. ECCENTRIC WELDED CONNECTIONS

A fillet weld, for purposes of calculation, is concentrated into a line and is assumed to resist an equal force per lineal inch in any direction. Therefore the most advantageous line of action for a force applied to a welded connection passes through the *centroid of weld lines*. If this does not occur, an eccentrically loaded connection results. The analysis of these connections is analogous to that of the eccentrically loaded bolted connections. The problem is resolved into two problems as shown in Fig. 14-18. The first problem,

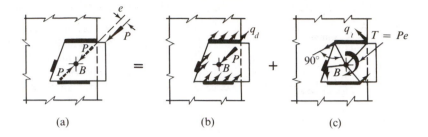

(a) (b) (c)

Fig. 14-18. Resolution of a problem of an eccentrically loaded welded connection into two separate problems

Fig. 14-18(b), is a concentrically loaded connection where the applied force P acts through the centroid B of all welds. For this case, the *direct* force q_d per inch length of weld acting in the direction opposite to P is

$$q_d = \frac{P}{\sum L_i} \qquad \text{[lb/in.]} \qquad (14\text{-}5)$$

where $\sum L_i$ is the *total length* of all welds. If the force P is resolved into the horizontal and vertical components P_x and P_y, respectively, the components of the direct force per inch of weld are

$$q_{dx} = \frac{P_x}{\sum L_i} \qquad \text{and} \qquad q_{dy} = \frac{P_y}{\sum L_i} \qquad (14\text{-}5a)$$

The second problem, Fig. 14-18(c), is analogous to the torsion problem, if the plate is assumed to be *rigid* and to twist around the point B. Then, by further assuming elastic action of the welds, the torsion formula, with a modified value of J, may be applied. A particular value of J for a group of *straight* welds can be found as outlined below.

The polar moment of inertia J is equal to the sum of two rectangular moments of inertia with respect to any two mutually perpendicular axes, i.e., $J = I_{xx} + I_{yy}$. Likewise, bearing in mind the parallel axis theorem, Eq. 5-2, the contribution to J with respect to the center of twist B of *any individual straight weld* such as *ab* in Fig. 14-19 is*

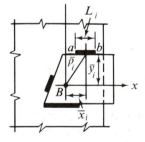

Fig. 14-19

$$(I_{xx})_i + (I_{yy})_i = 0 + L_i \bar{y}_i^2 + \tfrac{1}{12} L_i^3 + L_i \bar{x}_i^2 \qquad \text{[in.}^3\text{]}$$

where L_i is the length of a weld, and $\bar{x}_i$ and $\bar{y}_i$ are the respective coordinate distances from the centroid B of all welds *to the center of the weld considered*. This expression remains the same for any position or inclination of straight weld, since $\bar{x}_i^2 + \bar{y}_i^2 = \bar{p}_i^2 = \bar{x}_i^2 + \bar{y}_i^2$. Therefore the total equivalent J for several

*Note that I of a line around itself equals zero.

straight welds is

$$J = \sum J_i = \sum \left(\frac{L_i^3}{12} + L_i \bar{x}_i^2 + L_i \bar{y}_i^2 \right)$$

where the expression in parentheses is computed for *every* weld in the connection whose individual length is L_i, and the results are added.

Using the above equivalent value of J in the torsion formula, the *torsional* force q_t per inch of weld is

$$q_t = \frac{T\rho}{\sum \left(\frac{L_i^3}{12} + L_i \bar{x}_i^2 + L_i \bar{y}_i^2 \right)} \qquad [\text{lb/in.}] \qquad (14\text{-}6)$$

where ρ is the distance from the centroid of all welds to *a particular point on any weld*. The torsional force q_t acts perpendicular to the radius vector ρ. The horizontal and vertical *components* of q_t, in a manner analogous to that employed in the analysis of riveted connections, may be shown to be, respectively,

$$q_{tx} = \frac{Ty}{\sum \left(\frac{L_i^3}{12} + L_i \bar{x}_i^2 + L_i \bar{y}_i^2 \right)}$$

and

$$q_{ty} = \frac{Tx}{\sum \left(\frac{L_i^3}{12} + L_i \bar{x}_i^2 + L_i \bar{y}_i^2 \right)} \qquad (14\text{-}6a)$$

where as above, T is the *total torque* Pe on the connection, and x and y are the co-ordinate distances to a selected point on the weld.

The vectorial superposition of the direct and torsional forces per inch of weld gives the *intensity of the total force per inch of weld*. Inspection of the diagram for the joint on which the anticipated q_d and q_t are estimated generally reveals the highest stressed point on the welds. Usually the size of all welds is based on the highest stressed point. Like the eccentrically loaded riveted connection, this solution is obtained by superposition of the inelastic and elastic solutions.

In conclusion it should be noted that vector summation of forces per inch is used for fillet welds only. In the case of butt welds, the direct and torsional stresses are combined by using Mohr's circle.

EXAMPLE 14-8

Find the size of the two welds required to attach a plate to a machine as shown in Fig. 14-20(a) if the plate carries an inclined force $P = 10$ kips. Use the stresses allowed by the AWS.

SOLUTION

The applied force is resolved into horizontal and vertical components. The centroid of the two welds is seen to be at B. In computing J, symmetry and the

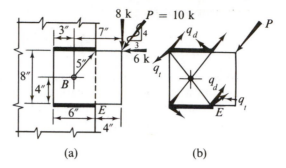

(a)   (b)

Fig. 14-20

fact that $\bar{x}$'s are zero are noted. Inspection of the anticipated direct and torsional forces on the welds in Fig. 14-20(b) shows that point E has the highest stress. Application of Eqs. 14-5(a) and 14-6(a) yields the components of q_d and q_t, which are then used to find q_{max}.

$$P_x = \tfrac{3}{5}(10) = 6 \text{ kips} \qquad P_y = \tfrac{4}{5}(10) = 8 \text{ kips}$$

$$T = 8(7) - 6(4) = 32 \text{ kips-in.}$$

$$J = 2\left(\frac{L^3}{12} + L\bar{y}^2\right) = 2\left[\frac{6^3}{12} + 6(4)^2\right] = 228 \text{ in.}^3$$

$$(q_{dx})_E = \frac{P_x}{\sum L_i} = \frac{6,000}{6+6} = 500 \text{ lb/in.} \rightarrow$$

$$(q_{dy})_E = \frac{P_y}{\sum L_i} = \frac{8,000}{12} = 667 \text{ lb/in.} \uparrow$$

$$(q_{tx})_E = \frac{Ty_E}{J} = \frac{32,000(4)}{228} = 562 \text{ lb/in.} \rightarrow$$

$$(q_{ty})_E = \frac{Tx_E}{J} = \frac{32,000(3)}{228} = 421 \text{ lb/in.} \uparrow$$

$$q_{max} = q_E = \sqrt{(500 + 562)^2 + (667 + 421)^2} = 1,520 \text{ lb/in.}$$

Finally, since by Eq. 14-4 the allowable force per inch of weld, regardless of the direction of the applied force, is 14,850 W, where W is the width of the leg,

$$q = 14\,850\,W = 1,520 \text{ lb/in.} \qquad \text{or} \qquad W = 0.103 \text{ in.}$$

Hence a uniform size of $\tfrac{1}{8}$ in. fillet weld should be used throughout.

PROBLEMS FOR SOLUTION

14-1.   A member of a truss to be made of two 4 in. by 4 in. by $\tfrac{1}{2}$ in. angles must transmit an axial force of 70 kips to a $\tfrac{1}{2}$ in. gusset plate. The angles are to be arranged as shown in Fig. 14-10.

How many $\tfrac{7}{8}$ in. rivets are required? How many rivets should be used to develop the full tensile capacity of the angles? Use the allowable stresses given in Example 14-1.   *Ans: 4, 7.*

14-2. A framed beam shear connection for an S 20 × 65.4 beam consists of two 4 in. by $3\frac{1}{2}$ in. by $\frac{1}{4}$ in. angles with four $\frac{7}{8}$ in. A502-1 rivets through the web and eight rivets in the outstanding legs. (A similar connection is shown in Fig. 14-11.) Determine the capacity of the connection. Use the allowable stresses given in Example 14-1.

14-3. A W 18 × 50 beam is attached to two W 12 × 65 columns by means of connections, each of which consists of two 4 in. by $3\frac{1}{2}$ in. by $\frac{3}{8}$ in. angles. Four $\frac{3}{4}$ in. rivets go through the web of the beam and eight $\frac{3}{4}$ in. rivets are used at each column. What force P, governed by allowable bearing and shear on rivets, may be applied to the beam as shown in the figure? Use the *old* AASHO (American Association of State Highway Officials) stresses: shear, 13.5 ksi; single *or* double bearing, 27 ksi. *Ans:* 58 kips.

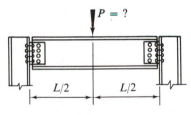

PROB. 14 – 3

14-4. If the beam to column connections described in Prob. 14-3 were made with $\frac{3}{4}$ in. A325-F friction type high strength bolts instead of rivets, what would the allowable force P be?

14-5. What direct tensile force may be applied to the multiple-riveted, structural butt joint shown in the figure? What is the efficiency of the joint? The main plates are 13 mm thick by 250 mm wide; the *two* cover plates are each 7 mm thick. The rivets are 22 mm in diameter. The allowable stresses are 140 MN/m² in tension, 100 MN/m² in shear, and 340 MN/m² in bearing.

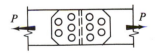

PROB. 14 – 5

14-6. A multiple-riveted, structural butt joint has the same arrangement of rivets as the joint of the preceding problem. The main plates are 10 mm thick by 200 mm wide; the *two* cover plates are each 6 mm thick by 200 mm wide. The 25 mm rivets are used in 27 mm diameter holes. If the joint trnasmits a force of 120 kN: (a) what is the tensile stress in the cover plates at the inner row of rivets; (b) what is the tensile stress in the main plate at the inner and outer rows of rivets?

14-7. A 1 in. by 8 in. plate transmitting a pull of 105 kips is reinforced with two side plates at a 2.00 in. diameter pin for bearing requirements as shown in the figure. (a) Determine the required thickness of the side plates if the allowable bearing stress in single or double shear is 27 ksi. State the answer to the nearest $\frac{1}{16}$th of an inch. (b) Determine the required number of $\frac{7}{8}$ in. rivets for attaching the side plates to the main plate if the allowable shearing stress is 13.5 ksi. *Ans:* (a) $\frac{1}{2}$ in., (b) 4.

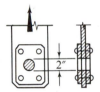

PROB. 14 – 7

14-8. Two 13 mm by 300 mm plates are lapped and riveted with 25 mm rivets as shown in the figure. Determine the allowable load and efficiency of the joint. Use the allowavle stresses given in Prob. 14-5. *Ans:* 80.9%.

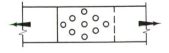

PROB. 14 – 8

14-9. A structural, multiple-riveted lap joint, such as is shown in the figure, is designed for a 42 kip load. The plates are $\frac{3}{8}$ in. thick by 10 in. wide. The rivets are $\frac{3}{4}$ in. (a) What is the shearing stress in the middle rivet? (b) What are the tensile

stresses in the *upper* plate in row 1-1 and row 2-2?
Ans: (a) 13.6 ksi, (b) 6.5 ksi, 7.02 ksi.

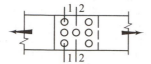

PROB. 14 – 9

14-10.   Rework Example 14-5 if, instead of rivets, $\frac{7}{8}$ in. A325-F friction type high strength bolts are used.

14-11.   A structural butt joint is to transmit a tension of 110 kips. The main plate is 8 in. by 1 in. Use $\frac{1}{2}$ in. cover plates and $\frac{7}{8}$ in. A325-F high strength bolts. Design the connection, i.e., determine the number of bolts and the necessary dimensions.

14-12.   An S 20 × 85 beam has two 1 in. by 10 in. cover plates as shown in the figure. This composite beam carries a total vertical shear of 150 kips at a particular section. Using the stresses allowed by the AISC code (Example 14-1), specify the required spacing of $\frac{3}{4}$ in. rivets between the beam and the cover plates.

PROB. 14 – 12

14-13.   A steel plate is attached to a machine with five $\frac{3}{4}$ in. bolts as shown in the figure. Deter-

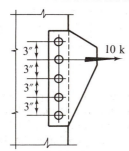

PROB. 14 – 13

mine the stress caused by the applied force in the highest stressed bolt.

14-14.   If the connection of Prob. 14-13 were made with $\frac{3}{4}$ in. A439-F friction type high strength bolts, what would be the maximum allowable horizontal pull which could be applied as shown in the figure?

14-15.   If the shearing stress in the rivets governs the allowable load P which may be applied to the connection shown in the figure, what is the allowable force P? The rivets are 19 mm and the allowable shearing stress is 100 MN/m². All dimensions shown in the figure are in mm.

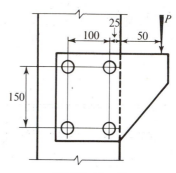

PROB. 14 – 15

14-16.   For the riveted connection shown in the figure, and used in an aeroplane, (a) determine the maximum stress on the highest stressed rivet. All rivets are $\frac{1}{8}$ in. in diameter. (b) If rivet A were knocked out, what would be the stress on the highest stressed rivet? *Ans:* (a) 1,920 psi, (b) 1,680 psi.

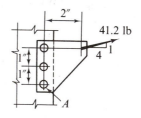

PROB. 14 – 16

14-17.   Calculate the maximum shearing stress in the rivet group shown in the figure if all rivets are 19 mm. The plates form a lap joint. All dimensions shown in the figure are in mm.

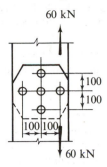

60 kN

100
100
100 100

60 kN

PROB. 14 – 17

14-18. Calculate the shearing stress caused by the applied force in each of the 1 in. diameter rivets for the connection shown in the figure.

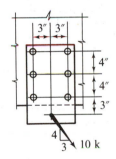

3″ 3″
4″
4″
3″
4
3 ↘ 10 k

PROB. 14 – 18

14-19. Determine the maximum shearing stress in the rivets in the bracket loaded as shown in the figure. All rivets are 1 in. in diameter. *Ans:* 9,140 psi.

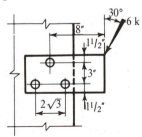

30°
6 k
8″
1¹/₂″
3″
2√3 1¹/₂″

PROB. 14 – 19

14-20. In order to obtain a wood beam of sufficient length, two short pieces of 2 in. by 10 in. timbers were spliced to the end of a 4 in. by 10 in. timber by six ¾ in. bolts as shown in the figure. If the working load of a ¾ in. bolt in double shear in any direction is limited to 1,000 lb, what is the maximum load P that the beam can support? Neglect the weight of the beam. *Ans:* 1,210 lb.

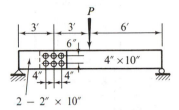

P
3′   3′   6′
6″
4″ × 10″
4″ 4″
2 – 2″ × 10″

PROB. 14 – 20

14-21. Determine the allowable force P that may be applied to a riveted connection having the dimensions shown in the figure. Rivets A and B are 0.2 in.², and rivet C is 0.6 in.² in cross-sectional area. The allowable shearing stress in the rivets is 20 ksi. *Ans:* 8.37 kips.

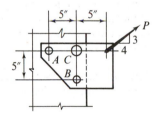

5″ 5″
P
3
4
5″
A C
B

PROB. 14 – 21

14-22. Calculate the maximum shearing stress in the rivets for the connection shown in the figure. Rivets A and B have cross-sections of 1 in.² each, and rivet C has a cross-section of 2 in.² *Ans:* 10 ksi (max).

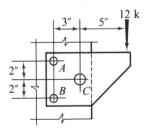

3″ 5″   12 k
2″ A
2″ B C

PROB. 14 – 22

14-23. A column bracket is made by attaching $\frac{1}{2}$ in. gusset plates to the flanges of a column with $\frac{7}{8}$ in. A325-F high strength bolts. The dimensions of the connection are shown in the figure. Determine the safe load P for the bracket.

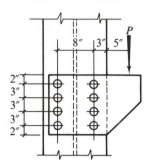

PROB. 14 – 23

14-24. Revise the design of Prob. 14-7 by welding the reinforcing plates to the main plate. Investigate two alternative welding schemes: (a) continuous welds all around the reinforcing plates, and (b) intermittent welds of sufficient size to develop the required forces. Which solution do you favor? Use AWS stresses.

14-25. Rework Example 14-7 for an 8 in. by 6 in. by $\frac{3}{4}$ in. angle using $\frac{1}{2}$ in. fillet welds.

14-26. Determine the proper lengths of welds L_1 and L_2 in Example 14-7 if, to minimize the length of connection, a weld is made along ad.

14-27. A bracket is loaded as shown in the

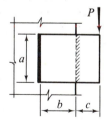

PROB. 14 – 27

figure. If the allowable shearing stress of the weld is 145 MN/m², what size of fillet welds to the nearest millimeter, should be used? Let $a = 200$ mm, $b = 150$ mm, $c = 100$ mm, and $P = 50$ kN.

14-28. Rework the preceding problem by letting $a = 8$ in., $b = 6$ in., $c = 4$ in., and $P = 12$ kips.

14-29. A bracket is to be attached to a body of a machine by means of three welds as shown in the figure. If the applied inclined force is 10 kips, what size fillet weld is required?   *Ans:* $\frac{1}{4}$ in.

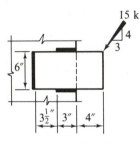

PROB. 14 – 29

14-30.* Determine the required size of fillet welds for the connection loaded as shown in the figure. *Ans:* $\frac{3}{8}$ in.

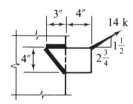

PROB. 14 – 30

14-31. If in Prob. 14-12 the cover plates are to be attached to the beam by means of welds, specify the size of fillet welds required. (*Hint:* determine shear flow per foot of beam's length, then decide on the size and length of the weld needed in such a distance. Use intermittent welds.)

15 The Energy Methods

15-1. INTRODUCTION

In the preceding chapters of the text, static equilibrium equations were always employed to solve the basic problems in mechanics of materials. An equally fundamental concept for the solution of these problems is based on the principle of conservation of energy. Methods of problem analysis which are generally applicable result by using this concept, and the study of these methods will be the object of this chapter.

In mechanics, energy is defined as the capacity to do work, and work is the product of a force by the distance in the direction the force moves. In solid deformable bodies, stresses multiplied by their respective areas are forces, and deformations are distances. The product of these two quantities is the *internal work* done in a body by externally applied forces. This internal work is stored in a body as the *internal elastic energy of deformation*, or *the elastic strain energy*. Methods of computing this internal energy will be discussed first. Then, by using the principle of the conservation of energy and equating the internal work to the *external work*, one can obtain deflections of axially loaded members, torsional members, and beams. This procedure will permit the investigation of stresses and deflections not only of members subjected to steadily applied forces but also of members subjected to *energy* or *impact loads*.

The direct solution of problems by equating the external to the internal work is limited to cases where only one force is applied to a member. Therefore a generalized procedure will be treated under the caption of *virtual* work, and will be used in discussing deflection problems of a very general nature, such as those caused by any loading of trusses and curved bars. In all cases, the members will be assumed to be in equilibrium, as the solution of instability problems by the energy methods is beyond the scope of this text.

15-2. ELASTIC STRAIN ENERGY FOR UNIAXIAL STRESS

Consider an infinitesimal element, such as shown in Fig. 15-1(a), subjected to a normal stress σ_x. The force acting on the right or the left

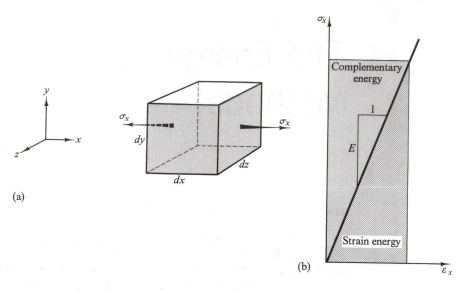

Fig. 15-1. (a) An element in tension and (b) a stress-strain diagram

face of this element is $\sigma_x \, dy \, dz$, where $dy \, dz$ is an infinitesimal area of the element. Because of this force, the element elongates an amount $\varepsilon_x \, dx$, where ε_x is strain in the x direction. If the element is made of a linearly elastic material, stress is proportional to strain, Fig. 15-1(b). Therefore, if the element is initially free of stress, the force which finally acts on the element increases linearly from zero until it attains its full value. The average force acting on the element while deformation is taking place is $\sigma_x \, dy \, dz/2$. This average force multiplied by the distance through which it acts is the work done on the element. For a perfectly elastic body no energy is dissipated, and the work done on the element is stored as recoverable internal strain energy. Thus, the internal elastic strain energy U for an infinitesimal element subjected to uniaxial stress is

$$dU = \tfrac{1}{2} \underbrace{\sigma_x \, dy \, dz}_{\substack{\text{average} \\ \text{force}}} \times \underbrace{\varepsilon_x \, dx}_{\text{distance}} = \tfrac{1}{2} \sigma_x \varepsilon_x \, dx \, dy \, dz = \tfrac{1}{2} \sigma_x \varepsilon_x dV \qquad (15\text{-}1)$$

where dV is the volume of the element.

By recasting Eq. 15-1, one obtains the strain energy stored in an elastic body per unit volume of the material, or its *strain-energy density* U_o. Thus

$$U_o = \frac{dU}{dV} = \frac{\sigma_x \varepsilon_x}{2} \qquad (15\text{-}1a)$$

This expression may be graphically interpreted as an area under the inclined line on the stress-strain diagram, Fig. 15-1(b). The corresponding area enclosed by the inclined line and the vertical axis is called the *complementary energy*. For linearly elastic materials the two areas are equal. Expressions analogous to Eq. 15-1a apply to the normal stresses σ_y and σ_z and to the corresponding extensional strains ε_y and ε_z.

Since in the elastic range Hooke's law applies, $\sigma_x = E\varepsilon_x$, Eq. 15-1a may be written as

$$U_o = \frac{dU}{dV} = \frac{E\varepsilon_x^2}{2} = \frac{\sigma_x^2}{2E} \qquad (15\text{-}1b)$$

or

$$U = \int_{\text{vol}} \frac{\sigma_x^2}{2E}\, dV \qquad (15\text{-}1c)$$

These forms of the equation for the elastic strain energy are convenient in applications, although they mask the dependence of the energy expression on the force and distance.

For a particular material, substitution into Eq. 15-1b of the value of the stress at the proportional limit gives an index of the material's ability to store or absorb energy without permanent deformation. The quantity so found is termed the *modulus of resilience* and is used to differentiate materials for applications where energy must be absorbed by members. For example, a steel with a proportional limit of 30,000 psi and an E of 30×10^6 psi has a modulus of resilience of $\sigma^2/(2E) = (30,000)^2/2(30)10^6 = 15$ in.-lb per cu in., whereas a good grade of Douglas fir, having a proportional limit of 6,450 psi and an E of 1,920,000 psi. has a modulus of resilience of $(6,450)^2/2(1,920,000) = 10.8$ in.-lb per cu in.

By reasoning analogous to the above, the area under a complete stress-strain diagram, Fig. 15-2, gives a measure of a material's ability to resist energy load up to fracture and is called its *toughness*. The larger the total area under the stress-strain diagram, the tougher the material. In the inelastic range, only a small part of the energy absorbed by a material is recoverable. Most of the energy is *dissipated* in permanently deforming the material and is lost in heat. The energy which may be recovered when a specimen has been stressed to some such point as A in Fig. 15-2 is represented by the triangle ABC. The line AB of this triangle is parallel to the line OD, since all materials behave elastically upon the release of stress.

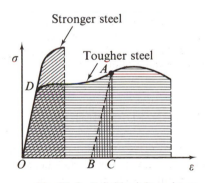

Fig. 15-2. Definition of toughness

EXAMPLE 15-1

Two elastic bars, whose proportions are shown in Fig. 15-3, are to absorb the same amount of energy delivered by axial forces. Neglecting stress concentrations, compare the stresses in the two bars.

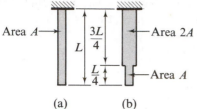

Area A →   ← Area $2A$

L ⟨ $\dfrac{3L}{4}$

$\dfrac{L}{4}$ ⟩ ← Area A

(a)   (b)

Fig. 15-3

SOLUTION

The bar shown in Fig. 15-3(a) is of uniform cross-sectional area, therefore the normal stress σ_1 is constant throughout. Using Eq. 15-1c and integrating over the volume V of the bar, one can write the total energy for the bar as

$$U_1 = \int_V \frac{\sigma_1^2}{2E}\, dV = \frac{\sigma_1^2}{2E} \int_V dV = \frac{\sigma_1^2}{2E}(AL)$$

where A is the cross-sectional area of the bar, and L is its length.

The bar shown in Fig. 15-3(b) is of variable cross section. Therefore, if the stress σ_2 acts in the lower part of the bar, the stress in the upper part is $\frac{1}{2}\sigma_2$. Again using Eq. 15-1c and integrating over the volume of the bar, it is found that the total energy that this bar will absorb in terms of the stress σ_2 is

$$U_2 = \int_V \frac{\sigma^2}{2E}\, dV = \frac{\sigma_2^2}{2E} \int_{\text{lower part}} dV + \frac{(\sigma_2/2)^2}{2E} \int_{\text{upper part}} dV$$

$$= \frac{\sigma_2^2}{2E}\left(\frac{AL}{4}\right) + \frac{(\sigma_2/2)^2}{2E}\left(2A\frac{3L}{4}\right) = \frac{\sigma_2^2}{2E}\left(\frac{5}{8}AL\right)$$

If both bars are to absorb the same amount of energy, $U_1 = U_2$ and

$$\frac{\sigma_1^2}{2E}(AL) = \frac{\sigma_2^2}{2E}\left(\frac{5}{8}AL\right) \qquad \text{or} \qquad \sigma_2 = 1.265\sigma_1$$

The enlargement of the cross-sectional area over a part of the bar in the second case is actually detrimental. For the same energy load, the stress in the "reinforced" bar is 26.5% higher than in the first bar. This situation is not found in the design of members for static loads.

15-3.   ELASTIC STRAIN ENERGY IN PURE BENDING

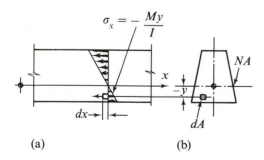

$\sigma_x = -\dfrac{My}{I}$

NA

x

$-y$

dx →

dA

(a)   (b)

Fig. 15-4. A segment of a beam used in deriving the expression for strain energy in bending

The elastic strain energy for an infinitesimal element in uniaxial stress having been established, one can obtain the elastic strain energy for beams in pure bending. For this special case, the normal stress is known to vary linearly from the neutral axis as is shown in Fig. 15-4(a), and the stress acting on an arbitrary element is $\sigma = -My/I$, Eq. 5-1a. The volume of this element is $dx\, dA$, where dx is the element's length and dA is its cross-sectional area, Fig. 15-4(b). Hence, using Eq. 15-1c and integrating over the volume V of the beam, the expression for the internal elastic strain energy for a beam is

$$U = \int_V \frac{\sigma_x^2}{2E} dV = \int_V \frac{1}{2E} \left(-\frac{My}{I}\right)^2 dx\, dA$$

Rearranging terms and remembering that M at a section of a beam is constant and that the order of performing the integration may be chosen arbitrarily, one obtains

$$U = \int_{\text{length}} \frac{M^2}{2EI^2} dx \int_{\text{area}} y^2\, dA = \int_0^L \frac{M^2\, dx}{2EI} \qquad (15\text{-}2)$$

where the last simplification is possible since by definition $I = \int y^2\, dA$. Equation 15-2 reduces the volume integral for the elastic energy of prismatic beams in pure flexure to a single integral to be taken over the length L of a beam.

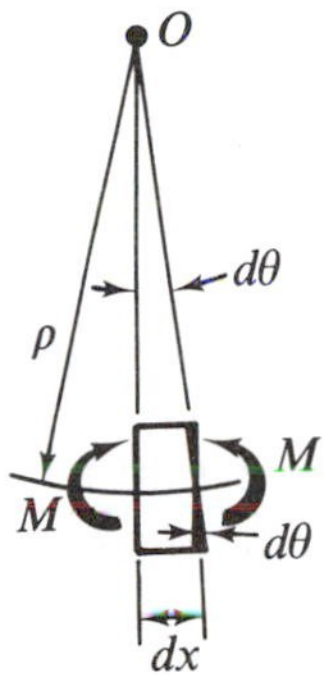

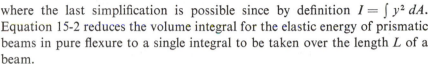

For further emphasis, Eq. 15-2 will be rederived from a different point of view, by considering an elementary segment of a beam dx long, as is shown in Fig. 15-5. Before the application of the bending moments M, the two planes perpendicular to the axis of the beam are parallel. After the application of the bending moments, extensions of the same two planes, which remain planes, intersect at O, and the angle included between these two planes is $d\theta$. Moreover, since the full value of the moment M is attained *gradually*, the *average* moment acting through an angle $d\theta$ is $\frac{1}{2}M$. Hence the external work U_e done on a segment of a beam is $dW_e = \frac{1}{2} M\, d\theta$. Further, since for small deflections $dx \approx \rho\, d\theta$, where ρ is the radius of curvature of the elastic curve and $1/\rho = M/EI$, from the principle of conservation of energy the internal strain energy of an element of a beam is

Fig. 15-5. A segment of a beam used in an alternate derivation of the strain energy in bending

$$dU = dW_e = \frac{1}{2} M\, d\theta = \frac{1}{2} M \frac{dx}{\rho} = \frac{M^2\, dx}{2EI}$$

which has the same meaning as Eq. 15-2.

EXAMPLE 15-2

Find the elastic strain energy stored in a rectangular cantilever beam subjected to a bending moment M applied at the end, Fig. 15-6.

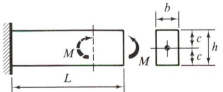

Fig. 15-6

SOLUTION

The bending moment at every section of this beam, as well as the flexural rigidity EI, is constant. By direct application of Eq. 15-2,

$$U = \int_0^L \frac{M^2\, dx}{2EI} = \frac{M^2}{2EI} \int_0^L dx = \frac{M^2 L}{2EI}$$

It is instructive to write this result in another form. Thus, since $\sigma_{\max} = Mc/I$,

$$M = \sigma_{max} I/c = 2\sigma_{max} I/h, \text{ and } I = bh^3/12,$$

$$U = \frac{\left(\dfrac{2\sigma_{max}I}{h}\right)^2 L}{2EI} = \frac{\sigma_{max}^2}{2E}\left(\frac{bhL}{3}\right) = \frac{\sigma_{max}^2}{2E}\left(\frac{1}{3}\text{vol}\right)$$

For a given maximum stress, the volume of the material in this beam is only a third as effective for absorbing energy as it would be in a uniformly stressed bar where $U = (\sigma^2/2E)(\text{vol})$. This results from the presence of *variable* stresses in a beam. If the bending moment also varies along a prismatic beam, the volume of the material becomes even less effective.

15-4. ELASTIC STRAIN ENERGY FOR SHEARING STRESSES

An expression for the elastic strain energy for an infinitesimal element in pure shear may be established in a manner analogous to that for one in uniaxial stress. Thus consider an element in a state of shear as shown in Fig. 15-7(a). The deformed shape of this element is shown in Fig. 15-7(b), where it is assumed that the bottom plane of the element is fixed in position.*

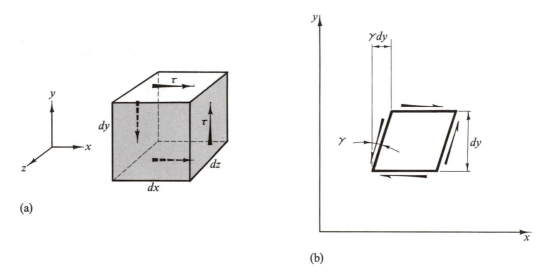

(a)

(b)

Fig. 15-7. An element for deriving the expression of strain energy due to shearing stresses

As this element is deformed, the force on the top plane reaches a final value of $\tau\,dx\,dz$. The total displacement of this force for small deformation of the element is $\gamma\,dy$, Fig. 15-7(b). Therefore, since the external work done on the

*This assumption does not make the expression less general.

element is equal to the internal, recoverable, elastic strain energy,

$$dU_{\text{shear}} = \underbrace{\tfrac{1}{2}\tau\,dx\,dz}_{\text{average force}} \times \underbrace{\gamma\,dy}_{\text{distance}} = \tfrac{1}{2}\tau\gamma\,dx\,dy\,dz = \tfrac{1}{2}\tau\gamma\,dV \qquad (15\text{-}3)$$

where dV is the volume of the infinitesimal element.

By recasting Eq. 15-3, the strain-energy density for shear becomes

$$(U_o)_{\text{shear}} = \left(\frac{dU}{dV}\right)_{\text{shear}} = \frac{\tau\gamma}{2} \qquad (15\text{-}3a)$$

Using Hooke's law for shearing stresses, $\tau = G\gamma$, Eq. 15-3a may be transformed as

$$(U_o)_{\text{shear}} = \left(\frac{dU}{dV}\right)_{\text{shear}} = \frac{\tau^2}{2G} \qquad (15\text{-}3b)$$

or

$$U_{\text{shear}} = \int_{\text{vol}} \frac{\tau^2}{2G}\,dV \qquad (15\text{-}3c)$$

Note the similarity of Eqs. 15-3, 15-3a, 15-3b and 15-3c to Eqs. 15-1, 15-1a, 15-1b, and 15-1c for elements in a state of uniaxial stress.

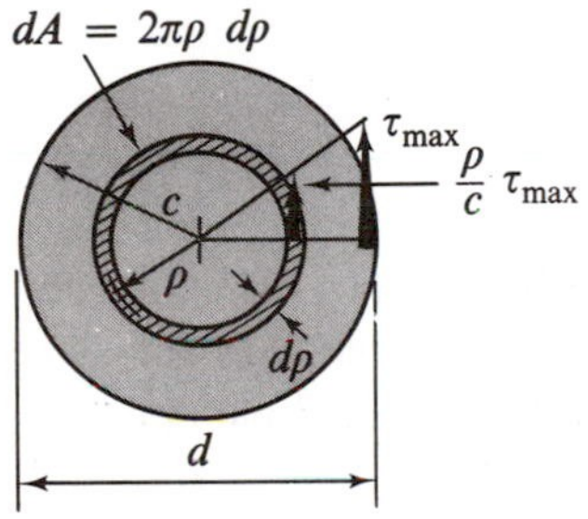

EXAMPLE 15-3

Find the energy absorbed by an elastic circular rod subjected to a constant torque in terms of the maximum shearing stress and the volume of the material, Fig. 15-8.

SOLUTION

$dA = 2\pi\rho\,d\rho$

Fig. 15-8

The shearing stress in an elastic circular rod subjected to a torque varies linearly from the longitudinal axis. Hence the shearing stress acting on an element at a distance ρ from the center of the cross section is $\tau_{\max}\rho/c$. Then, using Eq. 15-3(c) and integrating over the volume V of the rod L inches long, one obtains

$$U = \int_V \frac{\tau^2}{2G}\,dV = \int_V \frac{\tau_{\max}^2\rho^2}{2Gc^2}\,2\pi\rho\,d\rho\,L$$

$$= \frac{\tau_{\max}^2}{2G}\frac{2\pi L}{c^2}\int_0^c \rho^3\,d\rho = \frac{\tau_{\max}^2}{2G}\frac{2\pi L}{c^2}\frac{c^4}{4}$$

$$= \frac{\tau_{\max}^2}{2G}\left(\frac{1}{2}\text{vol}\right)$$

If there were uniform shearing stress throughout the member, a more efficient arrangement for absorbing energy would be obtained. In practice this condition is nearly attained in sandwich-type rubber mountings for machinery.

*15-5. STRAIN ENERGY FOR MULTIAXIAL STATES OF STRESS

As pointed out in Art. 15-2, analogous equations to that of Eq. 15-1a can be written for normal stresses σ_y and σ_z and the corresponding extensional strains ε_y and ε_z. To generalize the form of Eq. 15-3a to include all possible shearing stress components (see Fig. 1-3) and the corresponding shearing strains, it must be recognized that Eq. 15-3a was written for the xy coordinate axes. Therefore, it can be rewritten as $(U_o)_{\text{shear}} = \tau_{xy}\gamma_{xy}/2$. Then, by simply permuting the subscripts, the shear strain energy associated with τ_{yz} and τ_{zx} can also be written. On this basis the strain energy expression for a three-dimensional state of stress follows directly by superposition of the energy associated with each stress component, i.e.,

$$U_o = \frac{dU}{dV} = \frac{1}{2}\sigma_x\varepsilon_x + \frac{1}{2}\sigma_y\varepsilon_y + \frac{1}{2}\sigma_z\varepsilon_z$$

$$+ \frac{1}{2}\tau_{xy}\gamma_{xy} + \frac{1}{2}\tau_{yz}\gamma_{yz} + \frac{1}{2}\tau_{zx}\gamma_{zx} \qquad (15\text{-}4)$$

Upon substituting into this equation the relations for strains as given by the generalized Hooke's law, Eqs. 2-6 and 2-9a, and after some simple algebraic manipulations, one obtains

$$V_o = \frac{dU}{dV} = \frac{1}{2E}(\sigma_x^2 + \sigma_y^2 + \sigma_z^2) - \frac{v}{E}(\sigma_x\sigma_y + \sigma_y\sigma_z + \sigma_z\sigma_x)$$

$$+ \frac{1}{2G}(\tau_{xy}^2 + \tau_{yz}^2 + \tau_{zx}^2) \qquad (15\text{-}4a)$$

as the expression for the elastic strain energy per unit volume for *isotropic* materials. The last term in the equation vanishes for situations where no shearing stresses exist. For the two-dimensional plane stress case with $\sigma_z = 0$ and $\tau_{yz} = \tau_{zx} = 0$, Eq. 15-4a is greatly simplified to

$$U_o = \frac{dU}{dV} = \frac{\sigma_x^2}{2E} + \frac{\sigma_y^2}{2E} - \frac{v}{E}\sigma_x\sigma_y + \frac{\tau_{xy}^2}{2G} \qquad (15\text{-}4b)$$

*15-6. DESIGN OF MEMBERS FOR ENERGY LOADS

Examination of the results in the examples solved shows certain common characteristics which must be considered in the design of members to resist energy loads. Thus, whether an axially loaded member, a beam, or a torsion member is considered, the maximum stress for a given energy load U absorbed by a member may be expressed as

$$\sigma_{\text{max}} = \sqrt{\frac{2EU}{fV}} \qquad \text{or} \qquad \tau_{\text{max}} = \sqrt{\frac{2GU}{fV}}$$

where f is a fraction by which the total volume V of the member must be multiplied, depending on the type of stress distribution.

It may be seen from these expressions that, for a given U, the smallest stresses will be obtained:

(1) By selecting a material with a low E or G;
(2) By making the total *volume* of the member large;
(3) By stressing the material uniformly.

When constant stress exists throughout a body, $f = 1$. Solid circular members in torsion, as commonly found in springs, are reasonably good since $f = 0.5$. For the same stress, the volume of the material in a beam is at best only one-third as effective as it is in an axially loaded rod. This situation is found only in constant-strength beams. For example, it can be shown* that if a cantilever with a rectangular section is used to absorb energy delivered by a concentrated force at the end, $f = \frac{1}{9}$.

Several cases can be cited as illustrations of practical situations where the above principles are used. Wood is used in railroad ties since its E is low, and the cost per unit volume of the material is small. In pneumatic cylinders and jackhammers, Fig. 15-9, very long bolts are used to attach the ends to the tube. Long bolts provide a large volume of material, which, in operation, is uniformly stressed in tension. In the early stages of the development of this equipment, short bolts were used, and frequent failures occurred. A practical approximation to a constant-strength beam is found in leaf springs, Fig. 15-10(b) and (c). The various leaves of the spring, when spread out, Fig. 15-10(a), are approximately equivalent to a beam of constant strength† (see Fig. 10-18(c)).

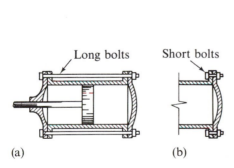

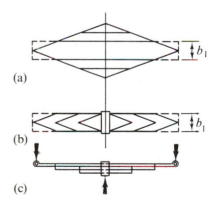

Long bolts   Short bolts

(a)   (b)

Fig. 15-9. (a) Good and (b) bad design of a pneumatic cylinder

(a)

(b)

(c)

Fig. 15-10. Leaf spring (a) composite leaves of a spring in one plane approximate a beam of constant strength

*Prob. 15-6.

†In operation, some energy is dissipated through friction between the leaves of the spring.

15-7. DEFLECTIONS BY THE ENERGY METHOD

As stated in the introduction, the principle of conservation of energy may be used to find the deflection of a loaded member. For this purpose, the internal strain energy U for a member is determined by using the equations derived above. Then, by equating this energy to the external work W_e done by an applied force, one can establish a relation from which the deflection in the direction of the applied force is found. This direct procedure will be used only in situations where one force is applied to a member. A general method for finding the deflection of any point on a member caused by any loading will be discussed in Art. 15-9.

For the present, it will be assumed that an external force is *gradually applied* to a body. This means that, as a force or a torque is being applied, its full effect on the material is reached gradually from zero. Therefore the external work U_e is equal to one-half of the total force multiplied by the deflection in the direction of its action.

Note that since $W_e = U$, unlike the procedure used in the following examples, the strain energy may often be easily computed by using expressions for the deflection of members derived earlier in the text.

EXAMPLE 15-4

Find the deflection of the free end of an elastic rod of constant cross-sectional area A and of length L due to an axial force P applied at the free end.

SOLUTION

If the force P is gradually applied to the rod, the external work $W_e = \frac{1}{2}P\Delta$, where Δ is the deflection of the end of the rod. The expression for the internal strain energy U of the rod was found in Example 15-1, and since $\sigma_1 = P/A$, it is

$$U = \frac{\sigma_1^2}{2E}AL = \frac{P^2L}{2AE}$$

Then, from $W_e = U$

$$\frac{P\Delta}{2} = \frac{P^2L}{2AE} \quad \text{and} \quad \Delta = \frac{PL}{AE}$$

which is the same as Eq. 2-4.

EXAMPLE 15-5

Find the rotation of the end of an elastic circular shaft with respect to the built-in end when a torque T is applied at the free end.

SOLUTION

If the torque T is gradually applied to the shaft, the external work $W_e = \frac{1}{2}T\phi$, where ϕ is the angular rotation of the free end in radians. The expression for the internal strain energy U for the circular shaft was found in Example 15-3. This relation may be written in a more convenient form by noting that $\tau_{\max} = Tc/J$, the volume of the rod is $\pi c^2 L$, and $J = \pi c^4/2$. Thus

$$U = \frac{\tau_{\max}^2}{2G}\left(\frac{1}{2}\text{vol}\right) = \frac{T^2 c^2}{2J^2 G}\frac{1}{2}\pi c^2 L = \frac{T^2 L}{2JG}$$

Then, from $W_e = U$,

$$\frac{T\phi}{2} = \frac{T^2 L}{2JG} \quad \text{and} \quad \phi = \frac{TL}{JG}$$

which is the same as Eq. 3-9.

EXAMPLE 15-6

Find the maximum deflection due to a force P applied at the end of a cantilever having a rectangular cross section, Fig. 15-11. Consider the effect of the flexural and shearing deformations.

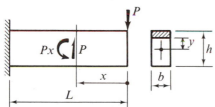

Fig. 15-11

SOLUTION

If the force P is gradually applied to the beam, the external work $W_e = \frac{1}{2}P\Delta$, where Δ is the total deflection of the end of the beam. The internal strain energy consists of two parts. One part is due to the bending stresses, the other is caused by the shearing stresses. According to Art. 15-5, these strain energies may be directly superposed.

The strain energy in pure bending is obtained from Eq. 15-2, $U = \int M^2\, dx/(2EI)$, by noting that $M = -Px$. The strain energy in shear is found from Eq. 15-3b, $dU_{\text{shear}} = [\tau^2/(2G)]\, dV$. In this particular case, the shear at every section is equal to the applied force P, while the shearing stress τ, according to Example 6-3, is distributed parabolically, as

$$\tau = [P/(2I)][(h/2)^2 - y^2].$$

At any one level y, this shearing stress does not vary across the breadth b and the length L of the beam. Therefore the infinitesimal volume dV in the shear energy expression is taken as $Lb\, dy$. By equating the sum of these two internal strain energies to the external work, the total deflection is obtained:

$$U_{\text{bending}} = \int_0^L \frac{M^2\, dx}{2EI} = \int_0^L \frac{(-Px)^2\, dx}{2EI} = \frac{P^2 L^3}{6EI}$$

$$U_{\text{shear}} = \int_{\text{vol}} \frac{\tau^2}{2G}\, dV = \frac{1}{2G}\int_{-h/2}^{+h/2} \left\{\frac{P}{2I}\left[\left(\frac{h}{2}\right)^2 - y^2\right]\right\}^2 Lb\, dy$$

$$= \frac{P^2 Lb}{8GI^2}\frac{h^5}{30} = \frac{P^2 Lbh^5}{240G}\left(\frac{12}{bh^3}\right)^2 = \frac{3P^2 L}{5AG}$$

where $A = bh$ is the cross section of the beam. Then

$$W_e = U = U_{\text{bending}} + U_{\text{shear}}$$

$$\frac{P\Delta}{2} = \frac{P^2 L^3}{6EI} + \frac{3P^2 L}{5AG} \qquad \text{or} \qquad \Delta = \frac{PL^3}{3EI} + \frac{6PL}{5AG}$$

The first term in this answer, $PL^3/(3EI)$, is the deflection of the beam due to the flexure. The second term is the deflection due to shear. The factor, such as $\frac{6}{5}$ in this term, varies for different shapes of the cross section, since it depends on the nature of the shearing-stress distribution.

It is instructive to recast the expression for the total deflection Δ as

$$\Delta = \frac{PL^3}{3EI}\left(1 + \frac{3E}{10G}\frac{h^2}{L^2}\right)$$

where as before, the last term gives the deflection due to shear.

To gain further insight into this problem, replace in the last expression the ratio E/G by 2.5, a typical value for steels. Then

$$\Delta = (1 + 0.75h^2/L^2)\Delta_{\text{bending}}$$

From this equation it can be seen that for a short beam (for example, one with $L = h$) the total deflection is 1.75 times that due to bending. Hence shear deflection is very important in comparable cases. On the other hand, if $L = 10h$, the deflection due to shear is less than 1 per cent. Small deflections due to shear are typical for ordinary, slender beams. This fact can be noted further from the original equation for Δ. There, the deflection due to bending increases as the cube of the span length, whereas the deflection due to shear increases directly. Hence, as beam length increases, the bending deflection quickly becomes dominant. For this reason it is usually possible to neglect the deflection due to shear;

*15-8. IMPACT LOADS

A freely falling weight, or a moving body, that strikes a structure delivers what is called a *dynamic* or *impact* load or force. Problems involving such forces may be analyzed rather simply on the basis of the following idealizing assumptions:

1. Materials behave elastically, and no dissipation of energy takes place at the point of impact or at the supports owing to local inelastic deformation of materials.
2. The inertia of a system resisting an impact may be neglected.
3. The deflection of a system is directly proportional to the magnitude of the applied force whether a force is dynamically or statically applied.

Then, on the basis of the principle of conservation of energy, it may be further assumed that at the *instant* a moving body is stopped, its kinetic energy is completely transformed into the internal strain energy of the resisting system. At this instant, the maximum deflection of a resisting system occurs and vibrations begin. However, since only maximum stresses and deflections are of primary interest in this text, the latter subject will not be pursued.

As an example of a dynamic force applied to an elastic system, consider a falling weight striking a spring. This situation is illustrated in Fig. 15-12(a), where a weight W falls from a height h above the free length of a spring. *This system represents a very general case, since, in a broad sense, every elastic system may be treated as an equivalent spring.* The spring constant k is defined as the force required to deflect the "spring" (such as a beam or an actual helical spring) a unit distance. In terms of the spring constant, the static deflection Δ_{st} of the spring due to the weight W is $\Delta_{st} = W/k$.

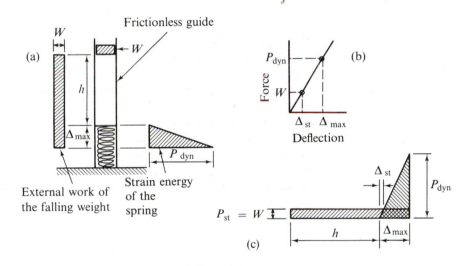

Fig. 15-12. Behavior of an elastic system under an impact force

Similarly, the maximum dynamic deflection $\Delta_{max} = P_{dyn}/k$, where P_{dyn} is the maximum dynamic force experienced by the spring. Therefore the dynamic force in terms of the weight W and the deflections of the spring is

$$P_{dyn} = \frac{\Delta_{max}}{\Delta_{st}} W \qquad (15\text{-}5)$$

This relationship is shown in Fig. 15-12(b).

At the instant the spring deflects its maximum amount, all energy of the falling weight is transformed into the strain energy of the spring. Therefore an equation representing the equality of external work to internal strain energy may be written as

$$W(h + \Delta_{max}) = \frac{1}{2} P_{dyn}\Delta_{max}$$

A graphical interpretation of this equation is shown in Fig. 15-12(c). Note that a factor of one-half appears in front of the strain energy expression, since the spring takes on the load *gradually*. Then, from Eq. 15-5,

$$W(h + \Delta_{max}) = \frac{1}{2} \frac{(\Delta_{max})^2}{\Delta_{st}} W$$

or

$$(\Delta_{max})^2 - 2\Delta_{st}\Delta_{max} - 2h\Delta_{st} = 0$$

hence

$$\Delta_{max} = \Delta_{st} + \sqrt{(\Delta_{st})^2 + 2h\Delta_{st}}$$

or

$$\Delta_{max} = \Delta_{st}\left(1 + \sqrt{1 + \frac{2h}{\Delta_{st}}}\right) \tag{15-6}$$

and again using Eq. 15-5,

$$P_{dyn} = W\left(1 + \sqrt{1 + \frac{2h}{\Delta_{st}}}\right) \tag{15-7}$$

Equation 15-6 gives the maximum deflection occurring in a spring struck by a weight W falling from a height h, while Eq. 15-7 gives the maximum force experienced by the spring for the same condition. To apply these equations, the static deflection Δ_{st} caused by the gradually applied known weight W is computed by the formulas derived earlier in the text.

After the effective dynamic force P_{dyn} is found, it may be used in computations as a static force. The magnification effect of a static force when dynamically applied is termed the *impact factor* and is given by the expression in parentheses appearing in Eqs. 15-6 and 15-7. The impact factor is surprisingly large in most cases. For example, if a force is applied to an elastic system *suddenly*, i.e., $h = 0$, it is equivalent to *twice* the same force *gradually* applied. If h is large compared to Δ_{st}, the impact factor is approximately equal to $\sqrt{2h/\Delta_{st}}$.

Similar equations may be derived for the case where a weight W is moving horizontally with a velocity v and is suddenly stopped by an elastic body. For this purpose, it is necessary to replace the external work done by the falling weight in the preceding derivation by the kinetic energy of a moving body, *using a consistent system of units*. Therefore, since the kinetic energy of a moving body is $Wv^2/2g$, where g is the acceleration of gravity, it can be shown that

$$P_{dyn} = W\sqrt{\frac{v^2}{g\Delta_{st}}} \quad \text{and} \quad \Delta_{max} = \Delta_{st}\sqrt{\frac{v^2}{g\Delta_{st}}}$$

where Δ_{st} is the static deflection caused by W assumed acting in the horizontal direction.

EXAMPLE 15-7

Find the maximum stress in the steel rod shown in Fig. 15-13 caused by a 30 N weight falling freely through 0.5 m. The steel helical spring of 35 mm outside

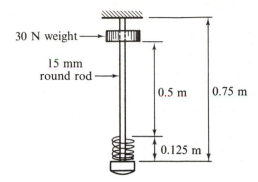

30 N weight

15 mm round rod

0.5 m 0.75 m

0.125 m

Fig. 15-13

diameter inserted into the system is made of 5 mm round wire and has 10 live coils. Let $E = 200 \times 10^9$ N/m², and $G = 80 \times 10^9$ N/m².

SOLUTION

The static deflection of this system due to the 30 N weight is computed first. It consists of two parts: the deflection of the rod given by Eq. 2-4, and the deflection of the spring given by Eq. 7-6a. For use in Eq. 7-6a, $\bar{r} = 15$ mm. Then, from Eq. 15-7, the dynamic force experienced by the spring and the rod is found. This force is used as a static force to find the stress in the rod.

$$\Delta_{st} = \Delta_{rod} + \Delta_{spr} = \frac{PL}{AE} + \frac{64F\bar{r}^3 N}{Gd^4}$$

$$= \frac{(30)(0.75)}{(177)(10^{-6})(200)(10^9)} + \frac{(64)(30)(15 \times 10^{-3})^3(10)}{(80)(10^9)(5 \times 10^{-3})^4} = 1\,296 \times 10^{-6} \text{ m}$$

$$P_{dyn} = W\left(1 + \sqrt{1 + \frac{2h}{\Delta_{st}}}\right) = 30\left(1 + \sqrt{1 + \frac{2(0.5)}{(1\,296) \times 10^{-6}}}\right) = 865 \text{ N}$$

$$\sigma_{dyn} = \frac{P_{dyn}}{A} = \frac{865}{177 \times 10^{-6}} = 4.9 \times 10^6 \text{ Pa} = 4.9 \text{ MPa}$$

EXAMPLE 15-8

Find the instantaneous maximum deflections and bending stresses for the 50 mm × 50 mm steel beam shown in Fig. 15-14 when struck by a 150 N weight falling from a height 75 mm above the top of the beam, if (a) the beam is on rigid supports, and (b) the beam is supported at each end on springs. The constant k for each spring is 300 000 N/m. Let $E = 200 \times 10^9$ N/m².

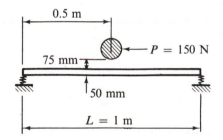

0.5 m

75 mm

$P = 150$ N

50 mm

$L = 1$ m

Fig. 15-14

SOLUTION

The deflection of the system due to the 150 N weight statically applied is computed for each case. In the first case, this deflection is that of the beam only; see Table 11 of the Appendix. In the second case, the static deflection of the beam is augmented by sagging of the springs subjected to a 75 N force each. The impact factors from Eqs. 15-6 and 15-7 are then computed. Static deflections and stresses are multiplied by the impact factors to obtain the answers.

Case (a):

$$\Delta_{st} = \frac{PL^3}{48EI} = \frac{(150)(1)^3}{(48)(200)(10^9)(50 \times 10^{-3})^4/12} = 30 \times 10^{-6} \text{ m}$$

$$\text{impact factor} = 1 + \sqrt{1 + \frac{2h}{\Delta_{st}}}$$

$$= 1 + \sqrt{1 + \frac{2(75 \times 10^{-3})}{30 \times 10^{-6}}} = 71.7$$

Case (b):

$$\Delta_{st} = \Delta_{beam} + \Delta_{spr} = 30 \times 10^{-6} + \frac{75}{300 \times 10^3} = 280 \times 10^{-6} \text{ m}$$

$$\text{impact factor} = 1 + \sqrt{1 + \frac{2(75 \times 10^{-3})}{280 \times 10^{-6}}} = 24.2$$

For either case, the maximum bending stress in the beam due to a static application of P is

$$(\sigma_{max})_{st} = \frac{M}{S} = \frac{PL}{4S} = \frac{(150)(1)}{(4)(50 \times 10^{-3})^3/6} = 1.8 \times 10^6 \text{ Pa} = 1.8 \text{ MPa}$$

Multiplying the static deflections and stress by the respective impact factors gives the answers.

	Static		Dynamic	
	With Springs	No Springs	With Springs	No Springs
Δ_{max}, m $\times$ 10^6	280	30	6 780	2 150
σ_{max}, MPa	1.8	1.8	43.6	129

It is apparent from this table that large deflections and stresses are caused by a dynamically applied load. The stress for the condition with no springs is particularly large; however, owing to the flexibility of the beam, it is not excessive. The results for the dynamic load are probably somewhat high, since in both cases the ratio of h/Δ_{st} is large, and, in such cases, the equations used are only approximately true.

*15-9. VIRTUAL WORK METHOD FOR DEFLECTIONS

The method of obtaining deflections by directly equating the external and internal work, as discussed in Art. 15-7, has the disadvantage that usually only the deflection caused by one force can be found. The concept of conservation of energy enables one to devise methods which overcome this difficulty. The virtual work method is one such technique, and will be discussed in the remainder of this chapter.

It is possible to imagine that a real mechanical or a structural system in static equilibrium is arbitrarily displaced consistent with its boundary conditions or constraints. During this process the real forces acting on the system move through imaginary or virtual displacements. Alternatively, imaginary or virtual forces in equilibrium with the given system can be given real, kinematically admissible displacements. In either case one can formulate the imaginary or virtual work done. Here the discussion will be limited to the consideration of virtual forces undergoing real displacements.

For forces and displacements occurring in the above manner, the principle of conservation of energy remains valid and the virtual change in the external work must be equal to the virtual change in the external work on the internal elements of a body, i.e.,

$$\delta W_e = \delta W_{ei} \qquad (15\text{-}8)$$

where the notation δW is used instead of dW to emphasize that the change in work is virtual. Eq. 15-8 expresses mathematically the virtual work principle. The term on the right side of this equation vanishes for rigid body systems whereas it equals the virtual change in the internal strain energy, δU, for elastic systems. The restriction of the principle to elastic response, however, is not implied in Eq. 15-8 which is equally applicable to cases involving inelastic material behavior, deformations due to temperature, movement of the supports, etc. It is the complete generality of the virtual work equation that makes it such a particularly valuable tool of analysis.

For determining the deflection of any point of a body due to any deformations occurring within a body, Eq. 15-8 can be put into a more suitable form. For example, consider a body such as shown in Fig. 15-15, for which the deflection of some point A in the direction A-B caused by deformation of the body is sought. For this, the virtual work equation can be formulated by employing the following sequence of reasoning:

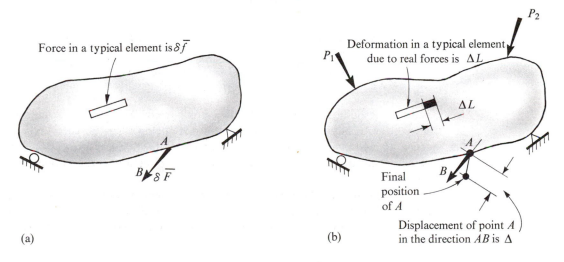

Fig. 15-15. Derivation of deflection formula by virtual work

First, apply to the unloaded body an imaginary or virtual force $\delta \bar{F}$ acting in the direction A-B. This force causes internal forces throughout the body. These internal forces, designated as $\delta \bar{f}$, Fig. 15-15(a), can be found in statically determinate systems.

Next, with the virtual force remaining on the body, apply the actual or real forces, Fig. 15-15(b), or introduce the specified deformations, such as those due to a change in temperature. This causes real internal deformations ΔL, which can be computed. Owing to these deformations, the virtual force system does work.

Therefore, since the external work done by the virtual force $\delta \bar{F}$ moving a real amount Δ in the direction of this force is equal to the total work done on the internal elements by the virtual forces $\delta \bar{f}$'s moving their respective real amounts ΔL, the special form of the virtual work equation becomes*

$$\underbrace{\delta \bar{F} \cdot \Delta}_{\text{virtual}} = \sum \overbrace{\delta \bar{f} \cdot \Delta L}^{\text{real}} \qquad (15\text{-}9)$$

Since all virtual forces attain their full values before real deformations are imposed, no factor of one-half appears anywhere in the equation. The summation, or, in general, an integration sign, is necessary on the right side of Eq. 15-9 to indicate that all internal work must be included.

Note that $\delta \bar{F}$ and $\delta \bar{f}$ need not be infinitesimal quantities. In Eq. 15-9 only their ratio is of consequence. Therefore, it is particularly convenient in applications to choose $\delta \bar{F}$ equal to unity, and to restate Eq. 15-9 as

$$\underbrace{\mathbf{1} \cdot \Delta}_{\text{virtual}} = \sum \overbrace{\bar{f} \cdot \Delta L}^{\text{real}} \qquad (15\text{-}10)$$

where Δ = real deflection of a point in the direction of the applied virtual unit force
$\bar{f}$ = internal forces caused by the virtual unit force
ΔL = real internal deformations of a body

Symbols designating virtual quantities are barred. The real deformations can be due to any cause with the elastic ones being a special case. Tensile forces and elongations of members are taken positive. A positive result indicates that the deflection occurs in the same direction as the applied virtual force.

In determining the angular rotations of a member a unit couple is used instead of the unit force. In practice, the procedure of using the unit force or the unit couple in conjunction with virtual work is referred to as the *unit-dummy-load method*.

*15-10. VIRTUAL WORK EQUATIONS FOR ELASTIC SYSTEMS

For linearly elastic systems Eq. 15-10 can be specialized to facilitate the solution of problems. This is done here for axially loaded and for flexural members. Applications are illustrated by examples.

*This equation represents the scalar (dot) product of the vectors.

Trusses

A virtual unit force must be applied at a point in the direction of the deflection to be determined.

If the real deformations are linearly elastic and are due only to axial deformations, $\Delta L = PL/(AE)$, and Eq. 15-10 becomes

$$\bar{1} \times \Delta = \sum_{i=1}^{n} \frac{\bar{p}_i P_i L_i}{A_i E_i} \tag{15-11}$$

where $\bar{p}_i$ is the axial force in a member due to the virtual unit force and P_i is the force in the same member due to the real loads. The summation extends over all members of a truss.

Beams

If the deflection of a point on an elastic beam is wanted by the virtual work method, a virtual unit force must be applied first in the direction in which the deflection is sought. This virtual force will set up internal bending moments at various sections of the beam designated by $\bar{m}$, as is in Fig. 15-16(a).

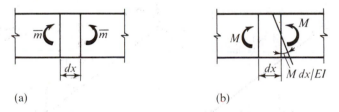

(a)   (b)

Fig. 15-16. Elements of a beam (a) virtual bending moments m, and (b) real bending moment M and the rotation of sections they cause

Next, as the real forces are applied to the beam, bending moments M rotate the "plane sections" of the beam $M\,dx/(EI)$ radians (Eq. 11-27). Hence the work done on an element of a beam by the virtual moments $\bar{m}$ is $\bar{m}M\,dx/(EI)$. Integrating this over the length of the beam gives the external work on the internal elements. Hence the special form of Eq. 15-10 for beams becomes

$$\bar{1} \times \Delta = \int_0^L \frac{\bar{m}M\,dx}{EI} \tag{15-12}$$

An analogous expression may be used to find the angular rotation of a particular section in a beam. For this case, instead of applying a virtual unit force, a virtual unit couple is applied to the beam at the section being investigated. This virtual couple sets up internal moments $\bar{m}$ along the beam.

Then, as the real forces are applied, they cause rotations $M\,dx/(EI)$ of the cross sections. Hence the same integral expression as in Eq. 15-12 applies here. The external work by the virtual unit couple is obtained by multiplying it by the real rotation θ of the beam at this couple. Hence

$$\bar{1} \times \theta = \int_0^L \frac{\bar{m}M\,dx}{EI} \tag{15-13}$$

In Eqs. 15-12 and 15-13, $\bar{m}$ is the bending moment due to the virtual loading, and M is the bending moment due to the real loads. Since both $\bar{m}$ and M usually vary along the length of the beam, both must be expressed by appropriate functions.

EXAMPLE 15-9

Find the vertical deflection of point B in the pin-jointed steel truss shown in Fig. 15-17(a) due to the following causes: (a) the elastic deformation of the members, (b) a shortening by 0.125 in. of the member AB by means of a turnbuckle, and (c) a drop in temperature of 120°F occurring in the member BC. The coefficient of thermal expansion of steel is 0.0000065 in. per inch per degree Fahrenheit. Neglect the possibility of lateral buckling of the compression member. Let $E = 30 \times 10^6$ psi.

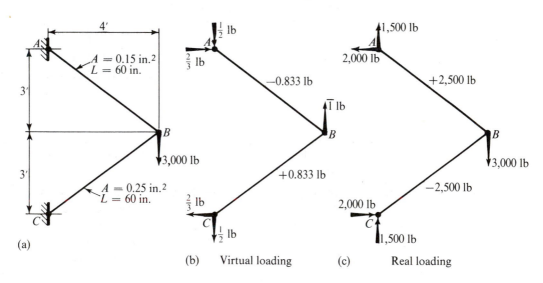

Fig. 15-17

SOLUTION

Case (a). A virtual unit force is applied in the vertical direction as shown in Fig. 15-17(b), and the resulting forces $\bar{p}$ are determined and recorded on the same diagram (check these). Then the forces in each member due to the real force are also determined and recorded, Fig. 15-17(c). The solution follows by means of Eq. 15-11. The work is carried out in tabular form.

Member	$\bar{p}$, lb	P, lb	L, in.	A, in.^{2}	$\bar{p}PL/A$
AB	−0.833	+2,500	60	0.15	−833,000
BC	+0.833	−2,500	60	0.25	−500,000

From this table $\sum \bar{p}PL/A = -1{,}333{,}000$. Hence

$$\bar{1} \times \Delta = \sum \frac{\bar{p}PL}{AE} = \frac{-1{,}333{,}000}{(30)10^6} = -0.0444 \text{ lb-in.}$$

and

$$\Delta = -0.0444 \text{ in.}$$

The negative sign means that point B deflects down. In this case, "negative work" is done by the virtual force acting upward when it is displaced in a downward direction. Note particularly the units and the signs of all quantities. Tensile forces in members are taken positive, and vice versa.

Case (b). Equation 15-10 is used to find the vertical deflection of point B due to the shortening of the member AB by 0.125 in. The forces set up in the bars by the virtual force acting in the direction of the deflection sought are shown in Fig. 15-17(b). Then, since ΔL is −0.125 in. (shortening) for the member AB and is zero for the member BC,

$$\bar{1} \times \Delta = (-0.833)(-0.125) + (+0.833)(0) = +0.1042 \text{ lb-in.}$$

and

$$\Delta = +0.1042 \text{ in. up}$$

Case (c). Again using Eq. 15-10 and noting that due to the drop in temperature, $\Delta L = -0.0000065(120)60 = -0.0468$ in. in the member BC,

$$\bar{1} \times \Delta = (+0.833)(-0.0468)$$

$$= -0.0390 \text{ lb-in.}$$

and

$$\Delta = -0.0390 \text{ in. down}$$

By superposition, the net deflection of point B due to all three causes is $-0.0444 + 0.1042 - 0.0390 = +0.0208$ in. up. To find this quantity, all three effects could have been considered simultaneously in the virtual work equation.

EXAMPLE 15-10

Find the deflection at the midspan of a cantilever beam loaded as in Fig. 15-18(a). The EI of the beam is constant.

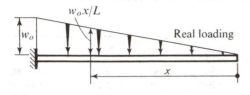

(a)

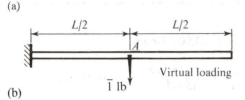

(b)

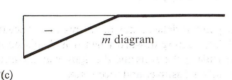

$\bar{m}$ diagram

(c)

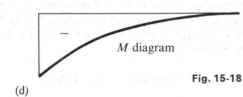

M diagram

(d)

Fig. 15-18

SOLUTION

The virtual force is applied at point A, whose deflection is sought, Fig. 15-18(b). The $\bar{m}$ diagram and the M diagram are shown in Figs. 15-8(c) and 15-8(d), respectively. For these functions, the same origin of x is taken at the free end of the cantilever. After these moments are determined, Eq. 13-25 is applied to find the deflection.

$$M = -\frac{x}{2}\frac{w_o x}{L}\frac{x}{3} = -\frac{w_o x^3}{6L} \qquad (0 \le x \le L)$$

$$\bar{m} = 0 \qquad (0 \le x \le L/2)$$

$$\bar{m} = -1(x - L/2) \qquad (L/2 \le x \le L)$$

$$\bar{1} \times \Delta = \int_0^L \frac{\bar{m}M\,dx}{EI}$$

$$= \frac{1}{EI}\int_0^{L/2} (0)\left(-\frac{w_o x^3}{6L}\right) dx$$

$$+ \frac{1}{EI}\int_{L/2}^L \left(-x + \frac{L}{2}\right)\left(-\frac{w_o x^3}{6L}\right) dx$$

$$= \frac{49 w_o L^4}{3{,}480 EI} \text{ N·m}$$

The deflection of point A is numerically equal to this quantity. The deflection due to shear has been neglected.

EXAMPLE 15-11

Find the downward deflection of the end C caused by the applied force of 2 kN in the structure shown in Fig. 15-19(a). Neglect deflection caused by shear. Let $E = 7 \times 10^7$ kN/m².

SOLUTION

A unit virtual force of 1 kN is applied vertically at C. This force causes an axial force in member DB and in the part AB of the beam, Fig. 15-19(b). Owing to this force, bending moments are also caused in the beam AC, Fig. 15-19(c). Similar computations are made and are shown in Figs. 15-19(d) and (e) for the applied real force. The deflection of point C depends on the deformations caused by the axial forces, as well as flexure, hence the virtual work equation is

$$\bar{1} \times \Delta = \sum \frac{\bar{p}PL}{AE} + \int_0^L \frac{\bar{m}M\,dx}{EI}$$

The first term on the right side of this equation is computed in the table. Then the integral for the internal virtual work due to bending is found. For the

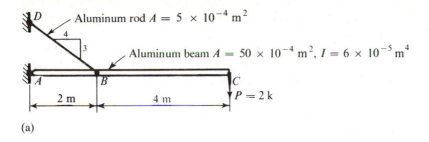

(a)

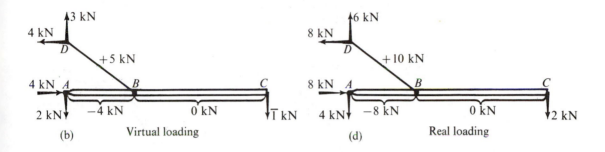

(b) Virtual loading

(d) Real loading

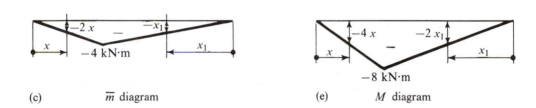

(c) $\bar{m}$ diagram

(e) M diagram

Fig. 15-19

different parts of the beam, two origins of x's are used in writing the expressions for $\bar{m}$ and M, Figs. 15-19(c) and (e).

Member	$\bar{p}$, kN	P, kN	L, m	A, m²	$\bar{p}PL/A$
DB	+5	+10	2.5	5 × 10⁻⁴	+250 000
AB	−4	−8	2.0	50 × 10⁻⁴	+12 800

From the table, $\qquad \Sigma \bar{p}PL/A = +262\,800,$

or $\qquad\qquad \Sigma \bar{p}PL/(AE) = 3.75 \times 10^{-3} \text{ kN} \cdot \text{m}$

$$\int_0^L \frac{\bar{m}M\,dx}{EI} = \int_0^2 \frac{(-2x)(-4x)\,dx}{EI} + \int_0^4 \frac{(-x_1)(-2x_1)\,dx_1}{EI}$$

$$= +15.25 \times 10^{-3} \text{ kN} \cdot \text{m}$$

Therefore $\bar{1} \times \Delta = (3.75 + 15.25)10^{-3} = 19 \times 10^{-3}$ kN·m and point C deflects 19×10^{-3} m $= 19$ mm down.

Note that the work due to the two types of action was superposed. Also note that the origins for the coordinate system for moments may be chosen as convenient; however, the same origin must be used for the corresponding $\bar{m}$ and M.

EXAMPLE 15-12

Find the horizontal deflection, caused by the concentrated force P, of the end of the curved bar shown in Fig. 15-20(a). The flexural rigidity EI of the bar is constant. Neglect the effect of shear on the deflection.

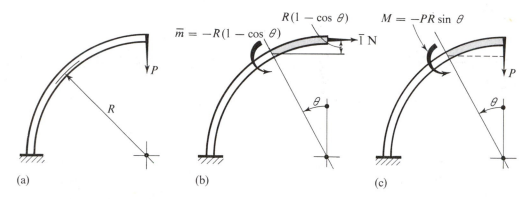

Fig. 15-20

SOLUTION

If the radius of curvature of a bar is large in comparison with the cross-sectional dimensions (Art. 5-11), ordinary beam deflection formulas may be used replacing dx by ds. In this case, $ds = R\,d\theta$.

Applying a horizontal virtual force at the end in the direction of the deflection wanted, Fig. 15-20(b), it is seen that $\bar{m} = -R(1 - \cos\theta)$. Similarly, for the real load, from Fig. 15-20(c), $M = -PR\sin\theta$. Therefore

$$\bar{1} \times \Delta = \int_0^L \frac{\bar{m}M\,ds}{EI}$$

$$= \int_0^{\pi/2} \frac{-R(1 - \cos\theta)(-PR\sin\theta)R\,d\theta}{EI} = +\frac{PR^3}{2EI}\,\text{N·m}$$

The deflection of the end to the right is numerically equal to this expression.

*15-11. STATICALLY INDETERMINATE PROBLEMS

Statically indeterminate problems can be solved with the aid of the virtual work method. For the linearly elastic systems, the force method discussed in Art. 12-5 can be used to particular advantage. In applying this approach the virtual work method merely provides the means for determining deflections of structures artificially reduced to statical determinacy. Much confusion is avoided if this is clearly kept in mind.

EXAMPLE 15-13

Find the forces in the pin-jointed bars of the steel structure shown in Fig. 15-21(a) if a force of 3,000 lb is applied at B.

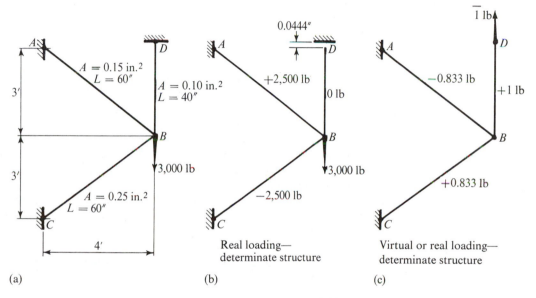

(a)   (b)   (c)

Real loading—
determinate structure

Virtual or real loading—
determinate structure

Fig. 15-21

SOLUTION

The structure can be rendered statically determinate by cutting the bar DB at D. Then the forces in the members are as shown in Fig. 15-21(b). In this determinate structure, the movement of point D must be found. This can be done by applying a vertical virtual force at D, Fig. 15-21(c), and using the virtual-work method. However, since the $\bar{p}PL/(AE)$ term for the member BD is zero, the vertical movement of point D is the same as that of B. In Example 15-9 the latter quantity was found to be 0.0444 in. down and is so shown in Fig. 15-21(b).

The movement of point D, shown in Fig. 15-21(b), violates the conditions of the problem, and a force must be applied to bring it back where it belongs. According to Eq. 12-11 this can be stated as

$$\Delta_D = f_{DD}X_D + \Delta_{DP} = 0$$

where the gap $\Delta_{DP} = -0.0444$ in.

To determine f_{DD}, a 1-lb real force is applied at D and the virtual-work method is used to find the deflection due to this force. The forces set up in the determinate structure by the virtual and the real forces are numerically the same, Fig. 15-21(c). To differentiate between the two, forces in members caused by a virtual force are designated by $\bar{p}$, and the real force by p. The solution is carried out in tabular form.

Member	$\bar{p}$, lb	p, lb	L, in.	A, in.^{2}	$\bar{p}p\,L/A$
AB	-0.833	-0.833	60	0.15	$+278$
BC	$+0.833$	$+0.833$	60	0.25	$+167$
BD	$+1.000$	$+1.000$	40	0.10	$+400$

From the table, $\sum \bar{p}p\,L/A = +845$. Therefore, since

$$\bar{1} \times \Delta = \sum \frac{\bar{p}pL}{AE} = \frac{+845}{30(10)^6} = 0.0000281 \text{ lb-in.}$$

$$f_{DD} = 0.0000281 \text{ in.} \quad \text{and} \quad 0.0000281X_D - 0.0444 = 0$$

To close the gap of 0.0444 in., the 1-lb real force at D must be increased $X_D = 0.0444/0.0000281 = 1{,}580$ times. Therefore the actual force in the member DB is 1,580 lb. The forces in the other two members may now be determined from statics or by superposition of the forces shown in Fig. 15-21(b) with X_D times the p forces shown in Fig. 15-21(c). By either method, the force in AB is found to be $+1{,}180$ lb (tension), and in BC, $-1{,}180$ lb (compression).

In any given case to make certain that the elastic analysis (such as the one in the above example) is applicable, maximum stresses must be determined. For the solution to be correct, these must be in the linearly elastic range for the material used.

PROBLEMS FOR SOLUTION

15-1. What is the modulus of resilience for an aluminum alloy if its proportional limit at $70°F$ is 28,000 psi and $E = 10.3 \times 10^6$ psi?

15-2. A 1 m long steel rod of 40 mm diameter is subjected to an axial energy load of 4 N·m that causes a tensile stress in the rod. (a) Determine the maximum tensile stress. $E = 200\,000$ MN/m^2. (b) If the same rod is machined down to a 20 mm diameter in the middle half of the bar, i.e., for a distance of 0.5 m, will the maximum stress increase or decrease and by how much? *Ans:* (b) 90.3 MN/m^2.

15-3. A 50 mm square alloy-steel bar 1 m long is a part of a machine and must resist an axial energy load of 100 N·m. What must the proportional limit of the steel be to safely resist the energy load elastically with a factor of safety of 4? $E = 200$ GPa. *Ans:* 253 MPa.

15-4. Show that, in an axially loaded rod, when an initial stress σ_i changes by an amount σ_c to a final stress $\sigma_f = \sigma_i + \sigma_c$, the change in the elastic strain energy per unit volume of the material is $(\sigma_c^2 + 2\,\sigma_i\sigma_c)/(2E)$. Interpret the result on a diagram similar to Fig. 15-1(b).

15-5. Show that the elastic strain energy due to bending for a simple uniformly loaded beam of rectangular cross-section is $(\sigma_{max}^2/2E)(\frac{8}{45}AL)$ where σ_{max} is the maximum bending stress, A is the cross-sectional area, and L is the length of the beam.

15-6. Show that $(U)_{bending} = (\sigma_{max}^2/2E)$ (Vol/9) for a cantilever of rectangular cross-section supporting a concentrated load P at the end.

15-7. Show that $(U)_{bending} = (\sigma_{max}^2/2E)$ (Vol/3) for a cantilever of constant strength having a parabolic profile (Fig. 10-18d) supporting a concentrated load P at the end.

15-8. Determine the maximum amount of strain energy which a helical spring can absorb under a tensile load if it is $8\frac{3}{4}$ in. in outside diameter and is made of $\frac{3}{4}$ in. diameter steel wire. There are 10 active coils and the allowable shearing stress is 80,000 psi. Neglect correction for stress concentration and the effect of direct shear. $G = 12 \times 10^6$ psi.

15-9. Consider a small steel cantilever with a rectangular cross-section of 2 in. by 6 in. deep ($E = 30 \times 10^6$ psi, and $G = 12 \times 10^6$ psi) with a concentrated force P applied at the end. Neglecting the weight of the beam, (a) determine the deflection due to flexure, and that due to shear, if $L = 6$ in. and $P = 3$ kips; (b) what must L be such that the deflection due to bending is equal to that caused by shear? (*Hint:* See Example 15-6.) *Ans:* (a) 0.00020 in., 0.00015 in., (b) 5.20 in.

15-10. A simple beam of rectangular cross section and span L is loaded with a concentrated force P at the middle of the span. Neglecting the weight of the beam and equating internal to external energy, (a) determine the maximum deflection caused by bending; (b) determine the maximum deflection caused by the shearing deformations. *Ans:* (a) $PL^3/(48EI)$.

15-11. Using an energy method, determine the vertical deflection of the free end of the cantilever beam shown in the figure due to the application of a force $P = 100$ lb. Consider only

flexural effects, i.e., neglect shear deformations. Let $E = 30 \times 10^6$ psi. *Ans:* 0.112 in.

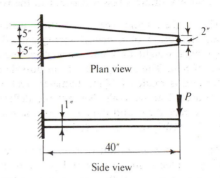

PROB. 15 – 11

15-12. (a) In terms of P, L, and EI, calculate the amount of elastic strain energy stored in the beam shown in the figure, caused by the applied loads. (b) By equating the work done by the external forces to the change in the elastic strain energy, determine the deflection at the loads. (*Hint:* Due to symmetry, deflections at both loads are equal.) *Ans:* (b) $PL^3/(48EI)$.

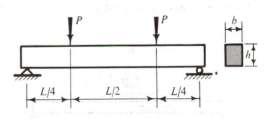

PROB. 15 – 12

15-13. For the beam shown in the figure, using an energy method, determine the deflection of the beam at the points of application of the loads. The moment of inertia of the cross section in the middle half of the beam is I_o. *Ans:* 0.029 $PL^3/(EI_o)$.

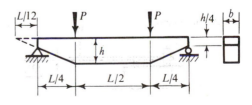

PROB. 15 – 13

15-14. If the free fall of the 30 N weight in Example 15-7 is 0.50 m when the spring is removed, what maximum stress will occur in the rod?

15-15. Determine the maximum instantaneous deflection of a helical spring caused by dropping a weight of 100 lb through a free distance of 8 in. The spring is 16 in. in outside diameter and is made of $1\frac{7}{8}$ in. diameter steel wire. There are 12 active coils. Neglect the deflection due to direct shear and the inertia of the spring. $G = 12 \times 10^6$ psi. *Ans:* 1.90 in.

15-16. An 8 in. round wooden cantilever 12 ft long is placed in a horizontal position. If its end is struck by a 200 lb weight dropped from a height of 6 in., what will the maximum instantaneous deflection be? Neglect the inertia of the beam. $E = 1.2 \times 10^6$ psi. *Ans:* 4.08 in.

15-17. Find the instantaneous maximum deflections and bending stresses for the square steel beam shown in the figure when struck by a 150 N weight falling from a height 60 mm above the top of the beam, if (a) the beam is on rigid supports, and (b) the beam is supported at each end on springs. The constant k for each spring is 300 kN/m. *Ans:* (a) 116 MPa.

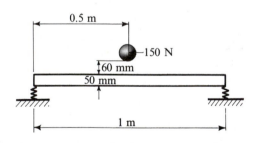

PROB. 15 – 17

15-18. A man weighing 180 lb jumps onto a diving board from a height of 2 ft. If the board

is of the dimensions shown in the figure, what is the maximum bending stress? Let $E = 1.6 \times 10^6$ psi. Use any method to establish the deflection characteristics of the board. *Ans:* 8,560 psi.

15-19. In Example 15-9 determine the horizontal movement of point B for the three cases enumerated.

15-20. For the mast and boom arrangement shown in the figure, (a) determine the vertical movement of the load W caused by lengthening the rod AB a distance of $\frac{1}{2}$ in. (b) By how much must the rod BC be shortened to bring the weight W to its original position? *Ans:* (a) 0.167 in., (b) 0.347 in.

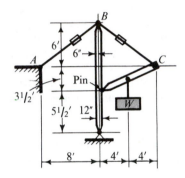

PROB. 15 – 20

15-21. A pin-joined system of three bars, each having a cross-sectional area A, is loaded as shown in the figure. (a) Determine the vertical and horizontal displacements of the joint B caused by the load P. (b) If by means of a turnbuckle the length of the member AC is shortened by $\frac{1}{2}$ in., what is the movement of the joint B? *Ans:* (a) $-9PL/(4AE)$, $-\sqrt{3}\,PL/(12AE)$; (b) $\frac{1}{4}$ in., $\sqrt{3}/12$ in.

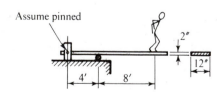

PROB. 15 – 18

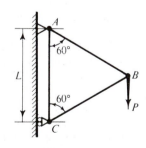

PROB. 15 – 21

15-22. Using the method of virtual work, determine the vertical and the horizontal displacements of joint C for the truss shown in the figure due to an applied force $P = 10$ kN. For simplicity, assume $AE = 1$ for all members.

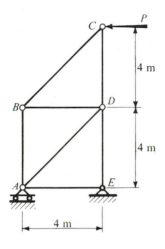

PROB. 15 – 22

15-23. Using the method of virtual work, determine the maximum deflection for a uniformly loaded simple beam having a constant EI in terms of w, L, and EI. *Ans: $5\,wL^4/(384EI)$.*

15-24. Using the method of virtual work, determine the maximum deflection of a simple beam of span L caused by two equal loads P applied at the third points. The EI is constant. *Ans: $23PL^3/(648EI)$.*

15-25. Using the method of virtual work, find the deflection of the point of application of the force P on the beam of variable cross section shown in the figure. *Ans: $13PL^3/(1,458EI)$.*

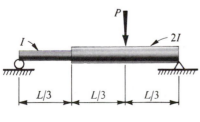

PROB. 15 – 25

15-26. For the beam shown in the figure, using the method of virtual work, determine (a) the deflection at the center of the beam, and (b) the tip deflection of the beam. Consider only flexural deformations. The EI is constant. *Ans: (a) $PL^3/(3EI)$, (b) $5PL^3/(6EI)$.*

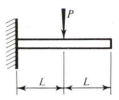

PROB. 15 – 26

15-27. Using the method of virtual work, find the deflection of the point of application of the force P. The flexural rigidity EI is constant over the entire length. Consider only bending deformations. *Ans: $54P/(EI)$.*

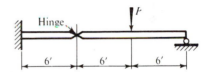

PROB. 15 – 27

15-28. Using the method of virtual work, determine the deflection at the center of the beam loaded as shown in the figure. The EI is constant. *Ans: $5w_oL^4/(768EI)$.*

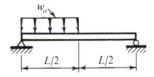

PROB. 15 – 28

15-29. An overhanging beam is loaded with a couple M_o at the end as shown in the figure. Using the method of virtual work, deter-

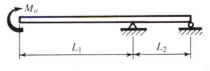

PROB. 15 – 29

mine the deflection and rotation of the over-hanging end due to M_o. Ans: $M_oL_1(3L_1 + 2L_2)/(6EI)$, $M_o[L_1 + \frac{1}{3}L_2]/(EI)$.

15-30.   By applying the method of virtual work, determine the horizontal deflection of point A caused by the force F applied to the frame as shown in the figure. Consider only the deflection due to bending. The EI is constant. Ans: $2Fa^3/(EI)$.

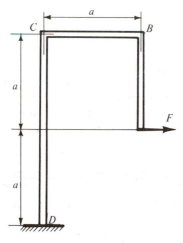

PROB. 15 – 30

15-31.   A planar elastic member is loaded as shown in the figure. Determine (a) the horizontal deflection of end A and (b) the rotations at points A and B caused by the application of the force P. Consider only flexural deformations. The EI is constant. Ans: (a) $(\frac{3}{2} + \sqrt{2}/3)$ [$Pa^3/(EI)$].

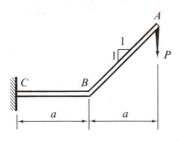

PROB. 15 – 31

15-32.   An L-shaped bar is loaded by a horizontal force F as shown in the figure. Determine

the horizontal and the vertical deflections of the end caused by F. Consider the bending effect only. EI is constant. Ans: $v_y = 0.145\ FL^3/(EI)$.

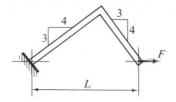

PROB. 15 – 32

15-33.   A bent planar bar of constant EI has the dimensions shown in the figure. Find the vertical deflection of the tip due to the application of the force P. Consider only flexural deflections. Comment on the virtual work method compared to that based on the solution of differential equations as discussed in Chapter 11.

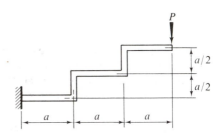

PROB. 15 – 33

15-34.   Determine the vertical deflection of point D, and the slope at A caused by the applied moment $M_1 = 900$ k-ft at end C. For all members EI is constant. Consider only flexural deformations. Ans: $27,000/(EI)$, $2,700/(EI)$.

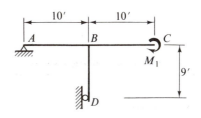

PROB. 15 – 34

15-35. For the planar structural system shown in the figure, made of aluminum, determine the vertical deflection of D due to both bending and direct (axial) stress. Consider only the effect of the applied load of 12 k. Use the method of virtual work. For the rod $A = 0.5$ in.^{2}; for the beam $A = 4$ in.^{2} and $I = 15$ in.^{4} Let $E = 10 \times 10^3$ ksi. *Ans:* 1.57 in.

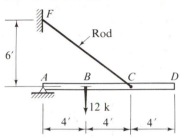

PROB. 15 – 35

15-36. In the preceding problem it was determined that due to the applied load of 12 k the end D moves 1.57 in. up. If, without removing this load, it is necessary to return point D to its initial position to make a connection, what is the required change in the length of rod CF? This change in length can be accomplished by means of a turnbuckle. *Ans:* 0.628 in.

15-37. An inclined steel bar 2 m long having a cross-section of 4 000 mm^2 and an I of 8.53×10^6 mm^4 is supported as shown in the figure. The inclined steel hanger DB has a cross-section of 600 mm.^{2} Determine the downward deflection of point C due to the application of the vertical load of $2\sqrt{2}$ kN. Let $E = 200\ 000$ MPa.

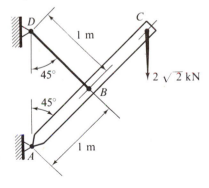

PROB. 15 – 37

15-38. A U-shaped member of constant EI has the dimensions shown in the figure. Determine the deflection of the applied forces away from each other. Consider only flexural effects. (*Hint:* Take advantage of symmetry.) *Ans:* $(2PL^3/3 + PL^2R\pi + PR^3\pi/2 + 4PLR^2)/(EI)$.

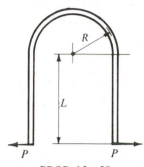

PROB. 15 – 38

15-39. A bent bar having a circular cross section is built in at one end and is loaded with a force F at the other end as shown in the figure. The force F acts normal to the plane of the bent bar. Determine (a) the translations of the free end along the coordinate axes, and (b) the rotation of the free end about the same axes. The constants E, G, I, and J are given.

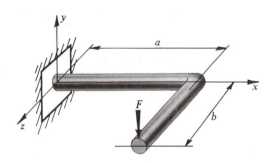

PROB. 15 – 39

15-40. For the beam shown in the figure, using the method of virtual work, determine the

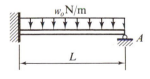

PROB. 15 – 40

reaction at A, treating this reaction as the redundant. *Ans:* $\frac{3}{8} wL$.

15-41. For the beam shown in the figure, using the method of virtual work, (a) determine the reaction at A, treating it as the redundant. (b) Determine the moment at B, treating it as a redundant. *Ans:* (a) $\frac{2}{3}P$, (b) $-\frac{1}{3} PL$.

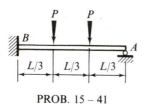

PROB. 15 – 41

15-42. A planar frame with hinges at A, B, C, and D has the dimensions shown in the figure.

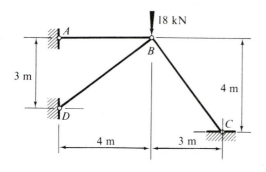

PROB. 15 – 42

By the method of virtual work, find the forces in members AB, DB, and CB caused by the application of the force of 18 kN. Let the values of L/A be as follows: 1 for AB, 2 for DB, and 3 for CB. Neglect buckling of members. Consider member BC to be redundant. *Ans:* $+2.57$ kN for BA, -12.86 kN for BC and BD.

15-43. Determine the reaction at A, treating it as a redundant, for the truss loaded as shown in the figure. The material of the truss is linearly elastic. The value of L/A for all members is 1. *Ans:* 18.75 kips.

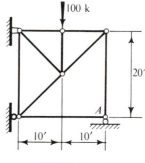

PROB. 15 – 43

*16 Thick-Walled Cylinders

16-1. INTRODUCTION

The characteristic method of mechanics of materials for deriving stress analysis formulas depends on assumptions regarding deformations. In the torsion of circular rods, an assumption that shearing strains vary linearly from the axis is made; in bending, it is assumed that plane sections through the beam remain plane. In more complicated problems, it is usually impossible to make analogous assumptions. Therefore the analysis begins with considerations of a general infinitesimal element; Hooke's law is postulated, and the solution is said to be found after stresses acting on any element and its deformations are known. At the boundaries of a body, the equilibrium of known forces or prescribed displacements must be satisfied by the corresponding infinitesimal elements. This is the technique of the *mathematical theory of elasticity*. Therefore it seems fitting in this last chapter of the book to solve a technically significant problem by these methods. This will be the problem of a thick-walled cylinder under pressure. Mathematically, it is a simple problem, yet the solution will display the characteristic method used in elasticity.

The inelastic behavior of a thick-walled cylinder under internal pressure is also briefly studied at the end of the chapter. Both the elastic-plastic and the fully plastic states are examined, using the Tresca maximum shear theory (Art. 9-8) as the yield criterion for ideally plastic material.

16-2. SOLUTION OF THE GENERAL PROBLEM

Consider a long cylinder with axially restrained ends whose cross section has the dimensions shown in Fig. 16-1(a).* The inside radius of this cylinder is r_i; the outside radius is r_o. Let the internal pressure in the cylinder be p_i; the outside or external pressure be p_o. Stresses in the wall of the cylinder caused by these pressures are sought.

*This problem was originally solved by Lamé, a French engineer, in 1833 and is sometimes referred to as the Lamé problem.

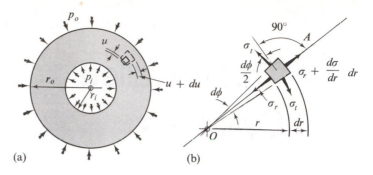

Fig. 16-1. Thick-walled cylinder

This problem can be conveniently solved by using cylindrical coordinates. Since the cylinder is long, every ring of unit thickness measured perpendicular to the plane of the paper is stressed alike. A typical infinitesimal element of unit thickness is defined by two radii, r and $r + dr$, and an angle $d\phi$, as shown in Fig. 16-1(b).

If the normal *radial* stress acting on the infinitesimal element at a distance r from the center of the cylinder is σ_r, this variable stress at a distance $r + dr$ will be $\sigma_r + (d\sigma_r/dr)dr$. (Recall an analogous situation in beams where, in a distance dx along the beam, M changes by $(dM/dx)dx$.) Both normal *tangential* stresses acting on the other two faces of the element are σ_t. These stresses, analogous to the hoop stresses in a thin cylinder, are equal. Moreover, since from the condition of symmetry every element at the same radial distance from the center must be stressed alike, no shearing stresses act on the element shown. Further, the axial stresses σ_z on the two faces of the element are equal and opposite normal to the plane of the paper.

The nature of the stresses acting on an infinitesimal element having been formulated, a characteristic elasticity solution proceeds along the following pattern of reasoning.

Static Equilibrium

The element chosen must be in static equilibrium. To express this mathematically requires the evaluation of *forces* acting on the element. These forces are obtained by multiplying stresses by their respective areas. The area on which σ_r acts is $(1)(r\,d\phi)$; that on which $\sigma_r + d\sigma_r$ acts is $(1)(r + dr)\,d\phi$; and each area on which σ_t acts is $(1)\,dr$. The weight of the element itself is neglected. Since the angle included between the sides of the element is $d\phi$, both tangential stresses are inclined $\frac{1}{2}\,d\phi$ to the line perpendicular to OA. Then, summing the forces along a radial line, $\sum F_r = 0$,

$$\sigma_r r\,d\phi + 2\sigma_t\,dr\left(\frac{d\phi}{2}\right) - \left(\sigma_r + \frac{d\sigma_r}{dr}\,dr\right)(r + dr)\,d\phi = 0$$

Simplifying, and neglecting the infinitesimals of higher order,

$$\sigma_t - \sigma_r - r\frac{d\sigma_r}{dr} = 0 \qquad \text{or} \qquad \frac{d\sigma_r}{dr} + \frac{\sigma_r - \sigma_t}{r} = 0 \qquad (16\text{-}1)$$

This one equation has two unknown stresses, σ_t and σ_r. Intermediate steps are required to express this equation in terms of one unknown so that it can be solved. This is done by introducing into the problem the geometry of deformations and properties of materials.

Geometric Compatibility

The deformation of an element is described by its strains in the radial and tangential directions. If u represents the *radial displacement* or *movement* of a cylindrical surface of radius r, Fig. 16-1(a), $u + (du/dr)dr$ is the radial displacement or movement of the adjacent surface of radius $r + dr$. Hence, the strain ε_r of an element in the radial direction is

$$\varepsilon_r = \frac{\left(u + \dfrac{du}{dr}\,dr\right) - u}{dr} = \frac{du}{dr} \qquad (16\text{-}2)$$

The strain ε_t in the tangential direction follows by subtracting from the length of the circumference of the deformed cylindrical surface of radius $r + u$ the circumference of the unstrained cylinder of radius r and dividing the difference by the latter length. Hence

$$\varepsilon_t = \frac{2\pi(r + u) - 2\pi r}{2\pi r} = \frac{u}{r} \qquad (16\text{-}3)$$

Since the stresses acting on the element are principal stresses,* Eqs. 16-2 and 16-3 give the principal strains expressed in terms of *one* unknown variable u.

Properties of Material

The generalized Hooke's law relating strains to stresses is given by Eq. 2-6, and can be restated here in the form

$$\varepsilon_r = \frac{1}{E}(\sigma_r - \nu\sigma_t - \nu\sigma_z) \qquad (16\text{-}4a)$$

$$\varepsilon_t = \frac{1}{E}(-\nu\sigma_r + \sigma_t - \nu\sigma_z) \qquad (16\text{-}4b)$$

$$\varepsilon_z = \frac{1}{E}(-\nu\sigma_r - \nu\sigma_t + \sigma_z) \qquad (16\text{-}4c)$$

---

*Since an infinitesimal cylindrical element includes an *infinitesimal* angle between two of its sides, it can be treated as if it were an element in a Cartesian coordinate system.

However, in the case of the thick-walled cylinder with axially restrained deformation, the problem is one of *plane strain*, i.e., $\varepsilon_z = 0$. The last equation then leads to a relation for the axial stress as

$$\sigma_z = v(\sigma_r + \sigma_t) \tag{16-5}$$

Introducing this result into Eqs. 16-4a and b and solving them simultaneously gives rise to expressions for the stresses σ_r and σ_t in terms of the principal strains:

$$\sigma_r = \frac{E}{(1+v)(1-2v)}[(1-v)\varepsilon_r + v\varepsilon_t] \tag{16-6a}$$

$$\sigma_t = \frac{E}{(1+v)(1-2v)}[v\varepsilon_r + (1-v)\varepsilon_t] \tag{16-6b}$$

These equations bring the plane strain condition into the problem for elastic material.

Formation of the Differential Equation

Now the equilibrium equation, Eq. 16-1, can be expressed in terms of one variable u. Thus, one eliminates the strains ε_r and ε_t from Eqs. 16-6a and 16-6b by expressing them in terms of the displacement u, as given by Eqs. 16-2 and 16-3; then the radial and tangential stresses are

$$\sigma_r = \frac{E}{(1+v)(1-2v)}\left[(1-v)\frac{du}{dr} + v\frac{u}{r}\right]$$

and
$$\sigma_t = \frac{E}{(1+v)(1-2v)}\left[v\frac{du}{dr} + (1-v)\frac{u}{r}\right] \tag{16-7}$$

and, by substituting these values into Eq. 16-1 and simplifying, the desired differential equation is obtained,

$$\frac{d^2u}{dr^2} + \frac{1}{r}\frac{du}{dr} - \frac{u}{r^2} = 0 \tag{16-8}$$

Solution of the Differential Equation

As can be verified by substitution, the general solution of Eq. 16-8, which gives the radial displacement u of any point on the cylinder, is

$$u = A_1 r + A_2/r \tag{16-9}$$

where the constants A_1 and A_2 must be determined from the conditions at the *boundaries* of the body.

Unfortunately, for the determination of the constants A_1 and A_2, the displacement u is not known at either the inner or the outer boundary of the cylinder's wall. However, the known pressures are equal to the radial stresses acting on the elements at the respective radii. Hence

$$\sigma_r(r_i) = -p_i \qquad \text{and} \qquad \sigma_r(r_o) = -p_o \qquad (16\text{-}10)$$

where the minus signs are used to indicate compressive stresses. Moreover, since u as given by Eq. 16-9 and $du/dr = A_1 - A_2/r^2$ can be substituted into the expression for σ_r given by Eq. 16-7, the boundary conditions given by Eqs. 16-10 become

$$\sigma_r(r_i) = -p_i = \frac{E}{(1+v)(1-2v)}\left[A_1 - (1-2v)\frac{A_2}{r_i^2}\right]$$
$$\sigma_r(r_o) = -p_o = \frac{E}{(1+v)(1-2v)}\left[A_1 - (1-2v)\frac{A_2}{r_o^2}\right] \qquad (16\text{-}11)$$

Solving these equations simultaneously for A_1 and A_2 yields

$$A_1 = \frac{(1+v)(1-2v)}{E}\frac{p_i r_i^2 - p_o r_o^2}{r_o^2 - r_i^2}$$
$$A_2 = \frac{1+v}{E}\frac{(p_i - p_o)r_i^2 r_o^2}{r_o^2 - r_i^2} \qquad (16\text{-}12)$$

These constants, when used in Eq. 16-9, permit the determination of the radial displacements of any point on the elastic cylinder subjected to the specified pressures. Thus, displacements of the inner and outer boundaries of the cylinder can be computed.

If Eq. 16-9 and its derivative (together with the constants given by Eqs. 16-12) are substituted into Eqs. 16-7, and the results are simplified, general equations for the radial and tangential stresses at any point of an elastic cylinder are obtained. These are

$$\sigma_r = C_1 - \frac{C_2}{r^2} \qquad \text{and} \qquad \sigma_t = C_1 + \frac{C_2}{r^2} \qquad (16\text{-}13)$$

where $\qquad C_1 = \dfrac{p_i r_i^2 - p_o r_o^2}{r_o^2 - r_i^2} \qquad$ and $\qquad C_2 = \dfrac{(p_i - p_o)r_i^2 r_o^2}{r_o^2 - r_i^2}$

Note that $(\sigma_r + \sigma_t)$ is constant over the whole cross-sectional area of the cylinder. This means that the axial stress σ_z as given by Eq. 16-5 is also constant over the entire cross-sectional area of the thick-walled cylinder.

Remarks on the Thin Disk Problem

The stress-strain relations used above for a thick-walled cylinder corresponded to a *plane strain* condition. If, on the other hand, an annular thin disk were to be considered, the *plane stress* condition (i.e., $\sigma_z = 0$ and

$\varepsilon_z = -v(\sigma_x + \sigma_y)/E)$ governs. Then the stress-strain relations as given by Eq. 8-20 must be used in the solution process instead of Eqs. 16-6a and b. However, the resulting differential equation remains the same as Eq. 16-8, and the radial and tangential stresses are also identical to those in the thick-walled cylinder and are given by Eq. 16-13. The only difference is that a different constant A_1 must be used in Eq. 16-9 for determining the radial displacement u. The constant A_2 remains the same as in Eq. 16-12, whereas A_1 becomes

$$A_1 = \frac{1-v}{E} \frac{p_i r_i^2 - p_o r_o^2}{r_o^2 - r_i^2}$$

16-3. SPECIAL CASES

Internal pressure only, i.e., $p_i \neq 0$ and $p_o = 0$. For this case, Eqs. 16-13 simplify to

$$\sigma_r = \frac{p_i r_i^2}{r_o^2 - r_i^2}\left(1 - \frac{r_o^2}{r^2}\right)$$

$$\sigma_t = \frac{p_i r_i^2}{r_o^2 - r_i^2}\left(1 + \frac{r_o^2}{r^2}\right)$$

(16-13a)

Since $r_o^2/r^2 \geq 1$, σ_r is always a compressive stress and is maximum at $r = r_i$. Similarly, σ_t is always a tensile stress, and its maximum also occurs at $r = r_i$.

For materials such as mild steel, which fail in shear rather than in direct tension, the maximum shear theory of failure (Art. 9-8) should be used in design. For internal pressure only, the maximum shearing stress occurs on the inner surface of the cylinder, Fig. 16-2. At this surface, the tensile stress σ_t and the compressive stress σ_r reach their respective maximum values. Further, from Eq. 16-5 it can be concluded that the axial stress σ_z is the intermediate principal stress for $0 < v \leq 0.5$. Substituting the maximum and minimum principal stresses into Eq. 8-8, and using Eq. 9-4 ($\tau_{max} = \sigma_{yp}/2$), gives

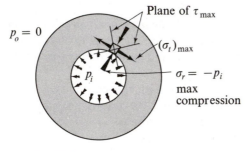

$P_o = 0$

Plane of τ_{max}

$(\sigma_t)_{max}$

P_i

$\sigma_r = -p_i$
max
compression

Fig. 16-2. An element in which τ_{max} occurs

$$\tau_{max} = \frac{\sigma_1 - \sigma_2}{2} = \frac{(\sigma_t)_{max} - (\sigma_r)_{max}}{2} = \frac{p_i r_o^2}{r_o^2 - r_i^2} = \frac{\sigma_{yp}}{2} \quad (16\text{-}13b)$$

and

$$p_{yp} = \sigma_{yp}(r_o^2 - r_i^2)/(2\,r_o^2) \quad (16\text{-}13c)$$

External pressure only, i.e., $p_i = 0$ and $p_o \neq 0$. For this case, Eqs. 16-12 simplify to

$$\sigma_r = -\frac{p_o r_o^2}{r_o^2 - r_i^2}\left(1 - \frac{r_i^2}{r^2}\right)$$

$$\sigma_t = -\frac{p_o r_o^2}{r_o^2 - r_i^2}\left(1 + \frac{r_i^2}{r^2}\right)$$

(16-13d)

Since $r_i^2/r^2 \leq 1$, both stresses are always compressive. The maximum compressive stress is σ_t and occurs at $r = r_i$.

Equations 16-13d must not be used for very thin-walled cylinders. Buckling of the walls may occur and strength formulas give misleading results.

EXAMPLE 16-1

Make a comparison of the tangential stress distribution caused by the internal pressure p_i as given by the exact formula of this chapter with the distribution given by the approximate formula for thin-walled cylinders of Chapter 9 if (a) $r_o = 1.1r_i$, and if (b) $r_o = 4r_i$, Fig. 16-3.

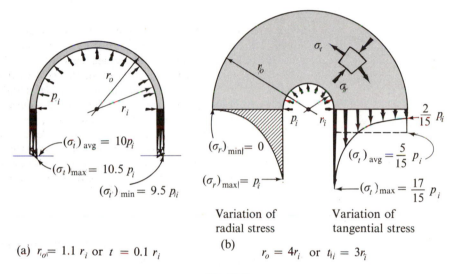

(a) $r_o = 1.1\,r_i$ or $t = 0.1\,r_i$

(b) $r_o = 4r_i$ or $t_{li} = 3r_i$

Fig. 16-3

SOLUTION

Case (a). Using Eq. 16-13b for σ_t,

$$(\sigma_t)_{r=r_i} = (\sigma_t)_{max} = \frac{p_i r_i^2}{(1.1r_i)^2 - r_i^2}\left[1 + \frac{(1.1r_i)^2}{r_i^2}\right] = 10.5p_i$$

$$(\sigma_t)_{r=r_o} = (\sigma_t)_{min} = \frac{p_i r_i^2}{(1.1r_i)^2 - r_i^2}\left[1 + \left(\frac{1.1r_i}{1.1r_i}\right)^2\right] = 9.5p_i$$

while, since the wall thickness $t = 0.1r_i$, the *average* "hoop" stress given by Eq. 9-2 is

$$(\sigma_t)_{avg} = \frac{p_i r_i}{t} = \frac{p_i r_i}{0.1r_i} = 10\,p_i$$

These results are shown in Fig. 16-3(a). Note particularly that in using Eq. 9-2 no appreciable error is involved.

Case (b). By using Eq. 16-13b for σ_t, the exact tangential stresses are obtained as above. These are

$$(\sigma_t)_{r=r_i} = (\sigma_t)_{\max} = \frac{p_i r_i^2}{(4r_i)^2 - r_i^2}\left[1 + \frac{(4r_i)^2}{r_i^2}\right] = \frac{17}{15}p_i$$

$$(\sigma_t)_{r=r_o} = (\sigma_t)_{\min} = \frac{p_i r_i^2}{(4r_i)^2 - r_i^2}\left[1 + \left(\frac{4r_i}{4r_i}\right)^2\right] = \frac{2}{15}p_i$$

The values of the tangential stress computed for a few intermediate points are plotted in Fig. 16-3(b). A striking variation of the tangential stress can be observed from this figure. The average tangential stress given by Eq. 9-2, using $t = 3r_i$, is

$$(\sigma_t)_{\mathrm{av}} = \frac{p_i r_i}{t} = \frac{5}{15}p_i = \frac{1}{3}p_i$$

This stress is nowhere near the true maximum stress.

The radial stresses were also computed by using Eq. 16-12(a) for σ_r, and the results are shown by the shaded area in Fig. 16-3(b).

It is interesting to note that no matter how thick a cylinder is made resisting internal pressure, the maximum tangential stress* will not be smaller than p_i. In practice, this necessitates special techniques to reduce the maximum stress. For example, in gun manufacture, instead of using a single cylinder, another cylinder is shrunk onto the smaller one, which sets up initial *compressive stresses* in the inner cylinder and tensile stresses in the outer one. In operation, the compressive stress in the inner cylinder is released first, and only then does this cylinder begin to act in tension. A greater range of operating pressures is obtained thereby.

16-4. BEHAVIOR OF IDEALLY PLASTIC THICK-WALLED CYLINDERS

The case of a thick-walled cylinder under internal pressure alone was considered in the previous article, and Eq. 16-13(c) was derived for the onset of yield at the inner surface of the cylinder according to the maximum shear criterion of Eq. 8-8. Upon subsequent increase in the internal pressure, the yielding progresses towards the outer surface, and an elastic-plastic state prevails in the cylinder with a limiting radius c beyond which the cross section remains elastic. As the pressure increases, the radius c also increases until, eventually, the entire cross section becomes fully plastic at the ultimate collapse load.

In the following discussion, the maximum shear criterion for ideally plastic material, Eq. 8-8,

$$\tau_{\max} = \frac{\sigma_1 - \sigma_2}{2} = \frac{\sigma_{\mathrm{yp}}}{2}$$

*See Prob. 16-3.

is used with the maximum and minimum principal stresses being σ_t and σ_r, respectively. This is based on the assumption that σ_z, the axial stress, is the intermediate principal stress. An inspection of Eqs. 16-5 and 16-13a shows this to be true in the elastic range, provided that $0 < v < 0.5$, but in the plastic range this applies only if the ratio of outer to inner radius r_o/r_i is less than a certain value.* For $v = 0.3$, this ratio can be established to be 5.75; hence, the solutions to be obtained in this article will be valid only so long as $r_o < 5.75\, r_i$ (with $v = 0.3$). The task of finding the stress distribution is more complicated when this condition is not satisfied and is beyond the scope of this book.

Plastic Behavior of Thick-Walled Cylinders

The equations of static equilibrium are applicable, regardless of whether the elastic or plastic state is considered. Hence, Eq. 16-1 is applicable but must be supplemented by a yield condition.

Static equilibrium (Eq. 16-1):

$$\frac{d\sigma_r}{dr} + \frac{\sigma_r - \sigma_t}{r} = 0$$

Yield condition (Eq. 8-8):

$$\frac{\sigma_t - \sigma_r}{2} = \frac{\sigma_{yp}}{2}$$

By combining these two equations, the basic differential equation becomes

$$\frac{d\sigma_r}{dr} - \frac{\sigma_{yp}}{r} = 0 \qquad \text{or} \qquad d\sigma_r = \frac{\sigma_{yp}}{r}\, dr \tag{16-14}$$

The solution of this can be written as

$$\sigma_r = \sigma_{yp} \ln r + C \tag{16-15}$$

For a cylinder with inner radius a and outer radius b, the boundary condition (zero external pressure) can be expressed as

$$\sigma_r(b) = 0 = \sigma_{yp} \ln b + C \tag{16-16}$$

Hence, the integration constant C is given as

$$C = -\sigma_{yp} \ln b$$

The radial and tangential stresses are then obtained, using Eqs. 16-15 and 8-8,

---

*See W. T. Koiter, "On partially plastic thick-walled tubes," *Biezeno Anniversary Volume on Applied Mechanics*, Haarlem, Holland: 1953, pp. 233–251.

respectively, Thus,

$$\sigma_r = \sigma_{yp}(\ln r - \ln b) = \sigma_{yp} \ln r/b \qquad (16\text{-}17a)$$

$$\sigma_t = \sigma_{yp} + \sigma_r = \sigma_{yp}(1 + \ln r/b) \qquad (16\text{-}17b)$$

The stress distributions given by Eqs. 16-17a and b are shown in Fig. 16-4(c), whereas Fig. 16-4(b) shows the elastic stress distributions. Since the fully plastic state represents the ultimate collapse of the thick-walled cylinder, the ultimate internal pressure, using Eq. 16-17a, is given as

$$p_{ult} = \sigma_r(a) = \sigma_{yp} \ln a/b \qquad (16\text{-}18)$$

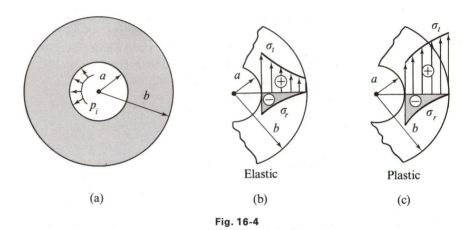

(a)   Elastic   Plastic

                        (b)   (c)

Fig. 16-4

Elastic-Plastic Behavior of Thick-Walled Cylinders

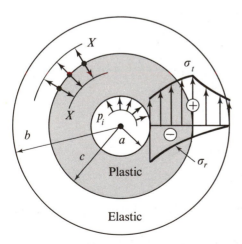

Fig. 16-5

For any value of p_i that is intermediate to the yield and ultimate values given by Eq. 16-13c and Eq. 16-18, respectively, i.e., $p_{yp} < p_i < p_{ult}$, the cross section of the cylinder between the inner radius a and an intermediate radius c is fully plastic, while that between c and the outer radius b is in the elastic domain, Fig. 16-5. At the elastic-plastic interface, the yield condition is just satisfied, and the corresponding radial stress X can be computed using Eq. 16-13c with $r_i = c$ and $r_o = b$; hence

$$X = \frac{\sigma_{yp}}{2} \frac{b^2 - c^2}{b^2} \qquad (16\text{-}19)$$

This stress becomes the boundary condition to be used in conjunction with Eq. 16-15 for a fully plastic segment with inner radius a and outer radius c. Hence,

$$\sigma_r(c) = -X = -\frac{\sigma_{yp}}{2}\frac{b^2 - c^2}{b^2} = \sigma_{yp}\ln c + C \qquad (16\text{-}20)$$

and

$$C = -\frac{\sigma_{yp}}{2}\frac{b^2 - c^2}{b^2} - \sigma_{yp}\ln c \qquad (16\text{-}21)$$

Substituting this value of C into Eq. 16-15, the radial stress in the plastic region is obtained as

$$\sigma_r = \sigma_{yp}\ln\frac{r}{c} - \frac{\sigma_{yp}}{2}\frac{b^2 - c^2}{b^2} \qquad (16\text{-}22a)$$

and, using Eq. 8-8, the tangential stress in the plastic zone becomes

$$\sigma_t = \sigma_{yp} + \sigma_r = \sigma_{yp}\left(1 + \ln\frac{r}{c}\right) - \frac{\sigma_{yp}}{2}\frac{b^2 - c^2}{b^2} \qquad (16\text{-}22b)$$

The internal pressure p_i at which the plastic zone extends from a to c can be obtained, using Eq. 16-22a, simply as $p_i = \sigma_r(a)$. Equation 16-13a, with $r_i = c$ and $r_o = b$, provide the necessary relations for calculating the stress distributions in the elastic zone.

PROBLEMS FOR SOLUTION

16-1. Verify the solution of Eq. 16-8.

16-2. Show that the ratio of the maximum tangential stress to the average tangential stress for a thick-walled cylinder subjected only to internal pressure is $(1 + \beta^2)/(1 + \beta)$, where $\beta = r_o/r_i$.

16-3. Show that no matter how large the outside diameter of a cylinder, subjected only to internal pressure, is made, the maximum tangential stress is not less than p_i (*Hint:* let $r_o \rightarrow \infty$.)

16-4. An alloy-steel cylinder is 6 in. *I.D.* (inside diameter) and 18 in. O.D. If it is subjected to an internal pressure of $p_i = 24{,}000$ psi ($p_o = 0$), (a) determine the radial and tangential stress distribution and show the results on a plot. (b) Determine the maximum (principal) shearing stress. (c) Determine the change in external and internal diameters. $E = 30 \times 10^6$ psi, $v = 0.3$.

Ans: (b) 27 ksi, (c) 1.64×10^{-3} in. and 3.67×10^{-3} in.

16-5. An alloy-steel cylinder is 0.15 m I.D. (inside diameter) and 0.45 m O.D. (outside diameter). If it is subjected to an internal pressure of $p_i = 160$ MPa (and $p_o = 0$): (a) determine the radial and tangential stress distributions and show the results on a plot, (b) determine the maximum (principal) shearing stress, and (c) determine the changes in external and internal diameters. Use $E = 200 \times 10^3$ MPa and $v = 0.3$.

16-6. Rework Prob. 16-5 with $p_i = 0$ and $p_o = 80$ MPa.

16-7. Rework Prob. 16-5 with $p_i = 160$ MPa and $p_o = 80$ MPa.

16-8. Isolate one-half of the cylinder of Prob. 16-7 by passing a plane through the axis of the cylinder. Then, by integrating the tangential stresses over the respective areas, show that the isolated free-body is in equilibrium.

16-9. Design a thick-walled cylinder of a 4 in. internal diameter for an internal pressure of 8,000 psi such as to provide: (a) a factor of safety of 2 against any yielding in the cylinder, and (b) a factor of safety of 3 against ultimate collapse. The yield stress of steel in tension is 36 ksi. *Ans:* (a) 12 in., (b) 7.8 in.

16-10. A 16 in. O.D. steel cylinder with approximately a 10 in. bore (I.D.) is shrunk onto another steel cylinder of 10 in., O.D. with a 6 in. I.D. Initially the internal diameter of the outer cylinder was 0.01 in. smaller than the external diameter of the inner cylinder. The assembly was accomplished by heating the larger cylinder in oil. For both cylinders $E = 30 \times 10^6$ psi and $v = 0.3$. (a) Determine the pressure at the boundaries between the two cylinders. (*Hint:* the elastic increase in the diameter of the outer cylinder with the elastic decrease in the diameter of the inner cylinder accommodates the initial interference between the two cylinders.) (b) Determine the tangential and radial stresses caused by the pressure found in (a). Show the results on a plot.

(c) Determine the internal pressure to which the composite cylinder may be subjected without exceeding a tangential stress of 20,000 psi in the inner cylinder. (*Hint:* after assembly, the cylinders act as one unit. The initial compressive stress in the inner cylinder is released first.) (d) Superpose the tangential stresses found in (b) with the tangential stresses resulting from the internal pressure found in (c). Show the results on a plot. *Ans:* (a) 7,480 psi.

16-11. Set up the differential equation for a thin disk rotating with an angular velocity of ω rad per sec. The unit weight of the material is γ. (*Hint:* consider an element as in Fig. 16-1(b) and add an inertia term.) *Ans:* add a term $(1 - v^2) \gamma \omega^2 r/gE$ to Eq. 16-8.

16-12. For a thick-walled cylinder of inner radius a and outer radius $b = 2a$, (a) calculate the internal pressure at which the elastic-plastic boundary is at $r = 1.5\ a$, (b) determine the radial and tangential stress distributions due to the internal pressure found in (a) and show them on a plot, and (c) calculate the ultimate collapse load. Assume the material to be elastic-perfectly plastic with a yield stress of 250 MPa.

Appendix

Tables

Acknowledgement: Data for Tables 3 through 10 are taken from *AISC Manual of Steel Construction* and are reproduced by permission of the American Institute of Steel Construction, Inc.

Table 1   TYPICAL PHYSICAL PROPERTIES OF AND ALLOWABLE STRESSES FOR SOME COMMON MATERIALS[a]

Material	Unit Weight, lb/in.³	Ultimate Strength, ksi			Yield Strength[g], ksi		Allow. Stresses[i], psi		Elastic Moduli × 10⁶ psi		Coef. of Thermal Expans. × 10⁻⁶ per °F
		Tens.	Comp.[c]	Shear	Tens.[h]	Shear	Tens. or Comp.	Shear	Tens. or Comp.	Shear	
Aluminum alloy (extruded) 2024–T4	0.100	60	...	32	44	25			10.6	4.00	12.9
6061–T6		38	...	24	35	20			10.0	3.75	13.0
Cast iron Gray	0.276	30	120	...[e]	...	...			13	6	5.8
Malleable		54	...	48	36	24			25	12	6.7
Concrete[b] 8 gal/sack	0.087	...	3	...[e]	...	...	−1,350[f]	66	3	...	6.0
6 gal/sack		...	5	...	...	...	−2,250[f]	86	5	...	
Magnesium alloy, AM100A	0.065	40	...	21	22	...	±24,000	14,500	6.5	2.4	14.0
Steel 0.2% Carbon (hot-rolled)	0.283	65	...	48	36	24			30[k]	12	6.5
0.6% Carbon (hot-rolled)		100	...	80	60	36					
0.6% Carbon (quenched)		120	...	100	75	45					
3½% Ni, 0.4% C		200	...	150	150	90					
Wood Douglas Fir (coast)	0.018	...	7.4[d]	1.1[f]	...	...	±1,900[j]	120[f]	1.76	...	...
Southern Pine (longleaf)	0.021	...	8.4[d]	1.5[f]	...	...	±2,250[j]	135[f]	1.76	...	...

[a] Mechanical properties of metals depend not only on composition but also on heat treatment, previous cold working, etc. Data for wood are for clear 2-in.-by-2-in. specimens at 12 per cent moisture content. True values vary.
[b] 8 gal/sack means 8 gallons of water per 94-lb sack of Portland cement. Values for 28-day-old concrete.
[c] For short blocks only. For ductile materials the ultimate strength in compression is indefinite; may be assumed to be the same as that in tension.
[d] Compression parallel to grain on short blocks. Compression perpendicular to grain at proportional limit 950 psi, 1,190 psi, respectively. Values from *Wood Handbook*, U.S. Dept. of Agriculture.
[e] Fails in diagonal tension.   [f] Parallel to grain.   [g] For ductile materials compressive yield strength may be assumed the same.
[h] For most materials at 0.2 per cent set.   [i] For static loads only. Much lower stresses required in machine design because of fatigue properties and dynamic loadings.
[j] In bending only. No tensile stress is allowed in concrete. Timber stresses are for select or dense grade.   [k] AISC recommends the value of 29 × 10⁶ psi.

Table 2   USEFUL PROPERTIES OF AREAS

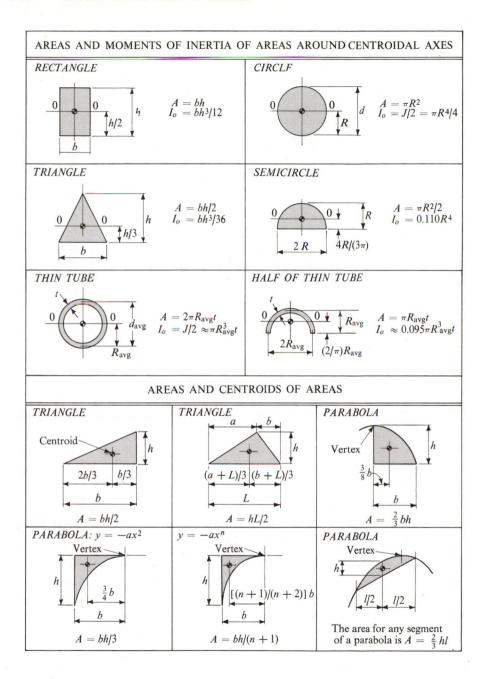

AREAS AND MOMENTS OF INERTIA OF AREAS AROUND CENTROIDAL AXES

RECTANGLE

$A = bh$
$I_o = bh^3/12$

CIRCLF

$A = \pi R^2$
$I_o = J/2 = \pi R^4/4$

TRIANGLE

$A = bh/2$
$I_o = bh^3/36$

SEMICIRCLE

$A = \pi R^2/2$
$I_o = 0.110 R^4$

$4R/(3\pi)$

THIN TUBE

$A = 2\pi R_{avg} t$
$I_o = J/2 \approx \pi R_{avg}^3 t$

HALF OF THIN TUBE

$A = \pi R_{avg} t$
$I_o \approx 0.095 \pi R_{avg}^3 t$

$(2/\pi) R_{avg}$

AREAS AND CENTROIDS OF AREAS

TRIANGLE

Centroid

$2b/3$ $b/3$

b

$A = bh/2$

TRIANGLE

a b

$(a + L)/3$ $(b + L)/3$

L

$A = hL/2$

PARABOLA

Vertex

$\frac{3}{8} b$

b

$A = \frac{2}{3} bh$

PARABOLA: $y = -ax^2$

Vertex

$\frac{3}{4} b$

b

$A = bh/3$

$y = -ax^n$

Vertex

$[(n + 1)/(n + 2)]\, b$

b

$A = bh/(n + 1)$

PARABOLA

Vertex

$l/2$ $l/2$

The area for any segment of a parabola is $A = \frac{2}{3} hl$

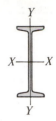

Table 3 AMERICAN STANDARD STEEL BEAMS, S SHAPES, PROPERTIES FOR DESIGNING

Designation*	Area	Depth	Flange Width	Flange Thickness	Web Thickness	Axis X-X I	Axis X-X $\frac{I}{c}$	Axis X-X r	Axis Y-Y I	Axis Y-Y $\frac{I}{c}$	Axis Y-Y r
	in.²	in.	in.	in.	in.	in.⁴	in.³	in.	in.⁴	in.³	in.
S 24 × 120	35.3	24.00	8.048	1.102	0.798	3030	252	9.26	84.2	20.9	1.54
× 105.9	31.1	24.00	7.875	1.102	0.625	2830	236	9.53	78.2	19.8	1.58
S 24 × 100	29.4	24.00	7.247	0.871	0.747	2390	199	9.01	47.8	13.2	1.27
× 90	26.5	24.00	7.124	0.871	0.624	2250	187	9.22	44.9	12.6	1.30
× 79.9	23.5	24.00	7.001	0.871	0.501	2110	175	9.47	42.3	12.1	1.34
S 20 × 95	27.9	20.00	7.200	0.916	0.800	1610	161	7.60	49.7	13.8	1.33
× 85	25.0	20.00	7.053	0.916	0.653	1520	152	7.79	46.2	13.1	1.36
S 20 × 75	22.1	20.00	6.391	0.789	0.641	1280	128	7.60	29.6	9.28	1.16
× 65.4	19.2	20.00	6.250	0.789	0.500	1180	118	7.84	27.4	8.77	1.19
S 18 × 70	20.6	18.00	6.251	0.691	0.711	926	103	6.71	24.1	7.72	1.08
× 54.7	16.1	18.00	6.001	0.691	0.461	804	89.4	7.07	20.8	6.94	1.14
S 15 × 50	14.7	15.00	5.640	0.622	0.550	486	64.8	5.75	15.7	5.57	1.03
× 42.9	12.6	15.00	5.501	0.622	0.411	447	59.6	5.95	14.4	5.23	1.07
S 12 × 50	14.7	12.00	5.477	0.659	0.687	305	50.8	4.55	15.7	5.74	1.03
× 40.8	12.0	12.00	5.252	0.659	0.472	272	45.4	4.77	13.6	5.16	1.06
S 12 × 35	10.3	12.00	5.078	0.544	0.428	229	38.2	4.72	9.87	3.89	0.980
× 31.8	9.35	12.00	5.000	0.544	0.350	218	36.4	4.83	9.36	3.74	1.00
S 10 × 35	10.3	10.00	4.944	0.491	0.594	147	29.4	3.78	8.36	3.38	0.901
× 25.4	7.46	10.00	4.661	0.491	0.311	124	24.7	4.07	6.79	2.91	0.954
S 8 × 23	6.77	8.00	4.171	0.425	0.441	64.9	16.2	3.10	4.31	2.07	0.798
× 18.4	5.41	8.00	4.001	0.425	0.271	57.6	14.4	3.26	3.73	1.86	0.831
S 7 × 20	5.88	7.00	3.860	0.392	0.450	42.4	12.1	2.69	3.17	1.64	0.734
× 15.3	4.50	7.00	3.662	0.392	0.252	36.7	10.5	2.86	2.64	1.44	0.766
S 6 × 17.25	5.07	6.00	3.565	0.359	0.465	26.3	8.77	2.28	2.31	1.30	0.675
× 12.5	3.67	6.00	3.332	0.359	0.232	22.1	7.37	2.45	1.82	1.09	0.705
S 5 × 14.75	4.34	5.00	3.284	0.326	0.494	15.2	6.09	1.87	1.67	1.01	0.620
× 10	2.94	5.00	3.004	0.326	0.214	12.3	4.92	2.05	1.22	0.809	0.643
S 4 × 9.5	2.79	4.00	2.796	0.293	0.326	6.79	3.39	1.56	0.903	0.646	0.569
× 7.7	2.26	4.00	2.663	0.293	0.193	6.08	3.04	1.64	0.764	0.574	0.581
S 3 × 7.5	2.21	3.00	2.509	0.260	0.349	2.93	1.95	1.15	0.586	0.468	0.516
× 5.7	1.67	3.00	2.330	0.260	0.170	2.52	1.68	1.23	0.455	0.390	0.522

*American Standard I-shaped beams are referred to as S shapes, and are designated by the letter S followed by their depth in inches, with their weight in pounds per linear foot given last. For example, S 24 × 120 means that this S-shape is 24 in. deep and weighs 120 lb/ft.

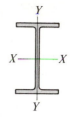

Table 4   AMERICAN WIDE-FLANGE STEEL BEAMS, W SHAPES,
PROPERTIES FOR DESIGNING (Abridged List)

Designation*	Area	Depth	Flange Width	Flange Thick-ness	Web Thick-ness	Axis X-X I	Axis X-X $\frac{I}{c}$	Axis X-X r	Axis Y-Y I	Axis Y-Y $\frac{I}{c}$	Axis Y-Y r
	in.²	in.	in.	in.	in.	in.⁴	in.³	in.	in.⁴	in.³	in.
W 36 × 230	67.7	35.88	16.471	1.260	0.761	15000	837	14.9	940	114	3.73
× 150	44.2	35.84	11.972	0.940	0.625	9030	504	14.3	270	45.0	2.47
W 33 × 200	58.9	33.00	15.750	1.150	0.715	11100	671	13.7	750	95.2	3.57
× 130	38.3	33.10	11.510	0.855	0.580	6710	406	13.2	218	37.9	2.38
W 30 × 172	50.7	29.88	14.985	1.065	0.655	7910	530	12.5	598	79.8	3.43
× 108	31.8	29.82	10.484	0.760	0.548	4470	300	11.9	146	27.9	2.15
W 27 × 145	42.7	26.88	13.965	0.975	0.600	5430	404	11.3	443	63.5	3.22
× 94	27.7	26.91	9.990	0.747	0.490	3270	243	10.9	124	24.9	2.12
W 24 × 130	38.3	24.25	14.000	0.900	0.565	4020	332	10.2	412	58.9	3.28
× 100	29.5	24.00	12.000	0.775	0.468	3000	250	10.1	223	37.2	2.75
× 76	22.4	23.91	8.985	0.682	0.440	2100	176	9.69	82.6	18.4	1.92
W 21 × 112	33.0	21.00	13.000	0.865	0.527	2620	250	8.92	317	48.8	3.10
× 82	24.2	20.86	8.962	0.795	0.499	1760	169	8.53	95.6	21.3	1.99
× 62	18.3	20.99	8.240	0.615	0.400	1330	127	8.54	57.5	13.9	1.77
× 55	16.2	20.80	8.215	0.522	0.375	1140	110	8.40	48.3	11.8	1.73
W 18 × 96	28.2	18.16	11.750	0.831	0.512	1680	185	7.70	225	38.3	2.82
× 64	18.9	17.87	8.715	0.686	0.403	1050	118	7.46	75.8	17.4	2.00
× 50	14.7	18.00	7.500	0.570	0.358	802	89.1	7.38	40.2	10.7	1.65
× 45	13.2	17.86	7.477	0.499	0.335	706	79.0	7.30	34.8	9.32	1.62
× 35	10.3	17.71	6.000	0.429	0.298	513	57.9	7.05	15.5	5.16	1.23
W 16 × 96	28.2	16.32	11.533	0.875	0.535	1360	166	6.93	224	38.8	2.82
× 88	25.9	16.16	11.502	0.795	0.504	1220	151	6.87	202	35.1	2.79
× 58	17.1	15.86	8.464	0.645	0.407	748	94.4	6.62	65.3	15.4	1.96
× 50	14.7	16.25	7.073	0.628	0.380	657	80.8	6.68	37.1	10.5	1.59
× 36	10.6	15.85	6.992	0.428	0.299	447	56.5	6.50	24.4	6.99	1.52
× 26	7.67	15.65	5.500	0.345	0.250	300	38.3	6.25	9.59	3.49	1.12
W 14 × 320	94.1	16.81	16.710	2.093	1.890	4140	493	6.63	1640	196	4.17
× 136	40.0	14.75	14.740	1.063	0.660	1590	216	6.31	568	77.0	3.77
× 87	25.6	14.00	14.500	0.688	0.420	967	138	6.15	350	48.2	3.70
× 84	24.7	14.18	12.023	0.778	0.451	928	131	6.13	225	37.5	3.02
× 78	22.9	14.06	12.000	0.718	0.428	851	121	6.09	207	34.5	3.00

*American wide-flange I- or H-shaped steel beams are referred to as W shapes, and are designated by the letter W followed by their *nominal* depth in inches, with their weight in pounds per linear foot given last. For example, W 21 × 62 means that this W shape is 21 in. deep and weighs 62 lb/ft. This list is abridged.

Table 4 (*Continued*)

Designation	Area	Depth	Flange Width	Flange Thickness	Web Thickness	Axis X-X I	Axis X-X $\frac{I}{c}$	Axis X-X r	Axis Y-Y I	Axis Y-Y $\frac{I}{c}$	Axis Y-Y r
	in.²	in.	in.	in.	in.	in.⁴	in.³	in.	in.⁴	in.³	in.
W 14 × 74	21.8	14.19	10.072	0.783	0.450	797	112	6.05	133	26.5	2.48
× 68	20.0	14.06	10.040	0.718	0.418	724	103	6.02	121	24.1	2.46
× 61	17.9	13.91	10.000	0.643	0.378	641	92.2	5.98	107	21.5	2.45
× 53	15.6	13.94	8.062	0.658	0.370	542	77.8	5.90	57.5	14.3	1.92
× 43	12.6	13.68	8.000	0.528	0.308	429	62.7	5.82	45.1	11.3	1.89
W 14 × 38	11.2	14.12	6.776	0.513	0.313	386	54.7	5.88	26.6	7.86	1.54
× 34	10.0	14.00	6.750	0.453	0.287	340	48.6	5.83	23.3	6.89	1.52
× 30	8.83	13.86	6.733	0.383	0.270	290	41.9	5.74	19.5	5.80	1.49
W 12 × 85	25.0	12.50	12.105	0.796	0.495	723	116	5.38	235	38.9	3.07
× 65	19.1	12.12	12.000	0.606	0.390	533	88.0	5.28	175	29.1	3.02
× 53	15.6	12.06	10.000	0.576	0.345	426	70.7	5.23	96.1	19.2	2.48
× 40	11.8	11.94	8.000	0.516	0.294	310	51.9	5.13	44.1	11.0	1.94
W 12 × 36	10.6	12.24	6.565	0.540	0.305	281	46.0	5.15	25.5	7.77	1.55
× 31	9.13	12.09	6.525	0.465	0.265	239	39.5	5.12	21.6	6.61	1.54
× 27	7.95	11.96	6.497	0.400	0.237	204	34.2	5.07	18.3	5.63	1.52
W 10 × 112	32.9	11.38	10.415	1.248	0.755	719	126	4.67	235	45.2	2.67
× 100	29.4	11.12	10.345	1.118	0.685	625	112	4.61	207	39.9	2.65
× 89	26.2	10.88	10.275	0.998	0.615	542	99.7	4.55	181	35.2	2.63
× 77	22.7	10.62	10.195	0.868	0.535	457	86.1	4.49	153	30.1	2.60
× 60	17.7	10.25	10.075	0.683	0.415	344	67.1	4.41	116	23.1	2.57
× 49	14.4	10.00	10.000	0.558	0.340	273	54.6	4.35	93.0	18.6	2.54
W 10 × 45	13.2	10.12	8.022	0.618	0.350	249	49.1	4.33	53.2	13.3	2.00
× 39	11.5	9.94	7.990	0.528	0.318	210	42.2	4.27	44.9	11.2	1.98
× 33	9.71	9.75	7.964	0.433	0.292	171	35.0	4.20	36.5	9.16	1.94
W 10 × 29	8.54	10.22	5.799	0.500	0.289	158	30.8	4.30	16.3	5.61	1.38
× 21	6.20	9.90	5.750	0.340	0.240	107	21.5	4.15	10.8	3.75	1.32
W 8 × 67	19.7	9.00	8.287	0.933	0.575	272	60.4	3.71	88.6	21.4	2.12
× 58	17.1	8.75	8.222	0.808	0.510	227	52.0	3.65	74.9	18.2	2.10
× 48	14.1	8.50	8.117	0.683	0.405	184	43.2	3.61	60.9	15.0	2.08
× 40	11.8	8.25	8.077	0.558	0.365	146	35.5	3.53	49.0	12.1	2.04
× 35	10.3	8.12	8.027	0.493	0.315	126	31.1	3.50	42.5	10.6	2.03
× 31	9.12	8.00	8.000	0.433	0.288	110	27.4	3.47	37.0	9.24	2.01
W 8 × 28	8.23	8.06	6.540	0.463	0.285	97.8	24.3	3.45	21.6	6.61	1.62
× 24	7.06	7.93	6.500	0.398	0.245	82.5	20.8	3.42	18.2	5.61	1.61
W 8 × 20	5.89	8.14	5.268	0.378	0.248	69.4	17.0	3.43	9.22	3.50	1.25
× 17	5.01	8.00	5.250	0.308	0.230	56.6	14.1	3.36	7.44	2.83	1.22

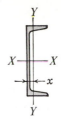

Table 5 AMERICAN STANDARD STEEL CHANNELS, PROPERTIES FOR DESIGNING

Designation*	Area	Depth	Flange Width	Average Thickness	Web Thickness	Axis X-X I	$\dfrac{I}{c}$	r	Axis Y-Y I	$\dfrac{I}{c}$	r	x
	in.²	in.	in.	in.	in.	in.⁴	in.³	in.	in.⁴	in.³	in.	in.
C 15 × 50	14.7	15.00	3.716	0.650	0.716	404	53.8	5.24	11.0	3.78	0.867	0.799
× 40	11.8	15.00	3.520	0.650	0.520	349	46.5	5.44	9.23	3.36	0.886	0.778
× 33.9	9.96	15.00	3.400	0.650	0.400	315	42.0	5.62	8.13	3.11	0.904	0.787
C 12 × 30	8.82	12.00	3.170	0.501	0.510	162	27.0	4.29	5.14	2.06	0.763	0.674
× 25	7.35	12.00	3.047	0.501	0.387	144	24.1	4.43	4.47	1.88	0.780	0.674
× 20.7	6.09	12.00	2.942	0.501	0.282	129	21.5	4.61	3.88	1.73	0.799	0.698
C 10 × 30	8.82	10.00	3.033	0.436	0.673	103	20.7	3.42	3.94	1.65	0.669	0.649
× 25	7.35	10.00	2.886	0.436	0.526	91.2	18.2	3.52	3.36	1.48	0.676	0.617
× 20	5.88	10.00	2.739	0.436	0.379	78.9	15.8	3.66	2.81	1.32	0.691	0.606
× 15.3	4.49	10.00	2.600	0.436	0.240	67.4	13.5	3.87	2.28	1.16	0.713	0.634
C 9 × 20	5.88	9.00	2.648	0.413	0.448	60.9	13.5	3.22	2.42	1.17	0.642	0.583
× 15	4.41	9.00	2.485	0.413	0.285	51.0	11.3	3.40	1.93	1.01	0.661	0.586
× 13.4	3.94	9.00	2.433	0.413	0.233	47.9	10.6	3.48	1.76	0.962	0.668	0.601
C 8 × 18.75	5.51	8.00	2.527	0.390	0.487	44.0	11.0	2.82	1.98	1.01	0.599	0.565
× 13.75	4.04	8.00	2.343	0.390	0.303	36.1	9.03	2.99	1.53	0.853	0.615	0.553
× 11.5	3.38	8.00	2.260	0.390	0.220	32.6	8.14	3.11	1.32	0.781	0.625	0.571
C 7 × 14.75	4.33	7.00	2.299	0.366	0.419	27.2	7.78	2.51	1.38	0.779	0.564	0.532
× 12.25	3.60	7.00	2.194	0.366	0.314	24.2	6.93	2.60	1.17	0.702	0.571	0.525
× 9.8	2.87	7.00	2.090	0.366	0.210	21.3	6.08	2.72	0.968	0.625	0.581	0.541
C 6 × 13	3.83	6.00	2.157	0.343	0.437	17.4	5.80	2.13	1.05	0.642	0.525	0.514
× 10.5	3.09	6.00	2.034	0.343	0.314	15.2	5.06	2.22	0.865	0.564	0.529	0.500
× 8.2	2.40	6.00	1.920	0.343	0.200	13.1	4.38	2.34	0.692	0.492	0.537	0.512
C 5 × 9	2.64	5.00	1.885	0.320	0.325	8.90	3.56	1.83	0.632	0.449	0.489	0.478
× 6.7	1.97	5.00	1.750	0.320	0.190	7.49	3.00	1.95	0.478	0.378	0.493	0.484
C 4 × 7.25	2.13	4.00	1.721	0.296	0.321	4.59	2.29	1.47	0.432	0.343	0.450	0.459
× 5.4	1.59	4.00	1.584	0.296	0.184	3.85	1.93	1.56	0.319	0.283	0.449	0.458
C 3 × 6	1.76	3.00	1.596	0.273	0.356	2.07	1.38	1.08	0.305	0.268	0.416	0.455
× 5	1.47	3.00	1.498	0.273	0.258	1.85	1.24	1.12	0.247	0.233	0.410	0.438
× 4.1	1.21	3.00	1.410	0.273	0.170	1.66	1.10	1.17	0.197	0.202	0.404	0.437

*American Standard Steel Channels are designated by the letter C followed by their depth in inches, with their weight per linear foot given last. For example, C 10 × 15.3 means that this channel is 10 in. deep and weighs 15.3 lb/ft.

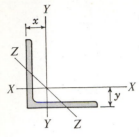

Table 6 STEEL ANGLES WITH EQUAL LEGS, PROPERTIES FOR DESIGNING

Size and Thickness	Weight per Foot	Area	Axis X-X and Axis Y-Y				Axis Z-Z
			I	$\dfrac{I}{c}$	r	x or y	r
in.	lb	in.^{2}	in.^{4}	in.^{3}	in.	in.	in.
L 8 × 8 × 1$\frac{1}{8}$	56.9	16.7	98.0	17.5	2.42	2.41	1.56
1	51.0	15.0	89.0	15.8	2.44	2.37	1.56
$\frac{7}{8}$	45.0	13.2	79.6	14.0	2.45	2.32	1.57
$\frac{3}{4}$	38.9	11.4	69.7	12.2	2.47	2.28	1.58
$\frac{5}{8}$	32.7	9.61	59.4	10.3	2.49	2.23	1.58
$\frac{9}{16}$	29.6	8.68	54.1	9.34	2.50	2.21	1.59
$\frac{1}{2}$	26.4	7.75	48.6	8.36	2.50	2.19	1.59
L 6 × 6 × 1	37.4	11.0	35.5	8.57	1.80	1.86	1.17
$\frac{7}{8}$	33.1	9.73	31.9	7.63	1.81	1.82	1.17
$\frac{3}{4}$	28.7	8.44	28.2	6.66	1.83	1.78	1.17
$\frac{5}{8}$	24.2	7.11	24.2	5.66	1.84	1.73	1.18
$\frac{9}{16}$	21.9	6.43	22.1	5.14	1.85	1.71	1.18
$\frac{1}{2}$	19.6	5.75	19.9	4.61	1.86	1.68	1.18
$\frac{7}{16}$	17.2	5.06	17.7	4.08	1.87	1.66	1.19
$\frac{3}{8}$	14.9	4.36	15.4	3.53	1.88	1.64	1.19
$\frac{5}{16}$	12.4	3.65	13.0	2.97	1.89	1.62	1.20
L 5 × 5 × $\frac{7}{8}$	27.2	7.98	17.8	5.17	1.49	1.57	.973
$\frac{3}{4}$	23.6	6.94	15.7	4.53	1.51	1.52	.975
$\frac{5}{8}$	20.0	5.86	13.6	3.86	1.52	1.48	.978
$\frac{1}{2}$	16.2	4.75	11.3	3.16	1.54	1.43	.983
$\frac{7}{16}$	14.3	4.18	10.0	2.79	1.55	1.41	.986
$\frac{3}{8}$	12.3	3.61	8.74	2.42	1.56	1.39	.990
$\frac{5}{16}$	10.3	3.03	7.42	2.04	1.57	1.37	.994
L 4 × 4 × $\frac{3}{4}$	18.5	5.44	7.67	2.81	1.19	1.27	.778
$\frac{5}{8}$	15.7	4.61	6.66	2.40	1.20	1.23	.779
$\frac{1}{2}$	12.8	3.75	5.56	1.97	1.22	1.18	.782
$\frac{7}{16}$	11.3	3.31	4.97	1.75	1.23	1.16	.785
$\frac{3}{8}$	9.8	2.86	4.36	1.52	1.23	1.14	.788
$\frac{5}{16}$	8.2	2.40	3.71	1.29	1.24	1.12	.791
$\frac{1}{4}$	6.6	1.94	3.04	1.05	1.25	1.09	.795
L 3$\frac{1}{2}$ × 3$\frac{1}{2}$ × $\frac{1}{2}$	11.1	3.25	3.64	1.49	1.06	1.06	.683
$\frac{7}{16}$	9.8	2.87	3.26	1.32	1.07	1.04	.684
$\frac{3}{8}$	8.5	2.48	2.87	1.15	1.07	1.01	.687
$\frac{5}{16}$	7.2	2.09	2.45	.976	1.08	.990	.690
$\frac{1}{4}$	5.8	1.69	2.01	.794	1.09	.968	.694
L 3 × 3 × $\frac{1}{2}$	9.4	2.75	2.22	1.07	.898	.932	.584
$\frac{7}{16}$	8.3	2.43	1.99	.954	.905	.910	.585
$\frac{3}{8}$	7.2	2.11	1.76	.833	.913	.888	.587
$\frac{5}{16}$	6.1	1.78	1.51	.707	.922	.869	.589
$\frac{1}{4}$	4.9	1.44	1.24	.577	.930	.842	.592
$\frac{3}{16}$	3.71	1.09	.962	.441	.939	.820	.596
L 2$\frac{1}{2}$ × 2$\frac{1}{2}$ × $\frac{1}{2}$	7.7	2.25	1.23	.724	.739	.806	.487
$\frac{3}{8}$	5.9	1.73	.984	.566	.753	.762	.487
$\frac{5}{16}$	5.0	1.46	.849	.482	.761	.740	.489
$\frac{1}{4}$	4.1	1.19	.703	.394	.769	.717	.491
$\frac{3}{16}$	3.07	0.92	.547	.303	.778	.694	.495

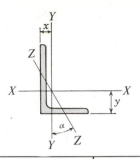

Table 7   STEEL ANGLES WITH UNEQUAL LEGS, PROPERTIES FOR DESIGNING *This list is abridged.

Size and Thickness*	Weight per Foot	Area	Axis X-X				Axis Y-Y				Axis Z-Z	
			I	$\frac{I}{c}$	r	y	I	$\frac{I}{c}$	r	x	r	Tan α
in.	lb	in.²	in.⁴	in.³	in.	in.	in.⁴	in.³	in.	in.	in.	
L 8 × 6 × 1	44.2	13.0	80.8	15.1	2.49	2.65	38.8	8.92	1.73	1.65	1.28	.543
¾	33.8	9.94	63.4	11.7	2.53	2.56	30.7	6.92	1.76	1.56	1.29	.551
½	23.0	6.75	44.3	8.02	2.56	2.47	21.7	4.79	1.79	1.47	1.30	.558
L 8 × 4 × 1	37.4	11.0	69.6	14.1	2.52	3.05	11.6	3.94	1.03	1.05	.846	.247
¾	28.7	8.44	54.9	10.9	2.55	2.95	9.36	3.07	1.05	.953	.852	.258
½	19.6	5.75	38.5	7.49	2.59	2.86	6.74	2.15	1.08	.859	.865	.267
L 6 × 4 × ¾	23.6	6.94	24.5	6.25	1.88	2.08	8.68	2.97	1.12	1.08	.860	.428
½	16.2	4.75	17.4	4.33	1.91	1.99	6.27	2.08	1.15	.987	.870	.440
L 5 × 3 × ½	12.8	3.75	9.45	2.91	1.59	1.75	2.58	1.15	.829	.750	.648	.357
⅜	9.8	2.86	7.37	2.24	1.61	1.70	2.04	.888	.845	.704	.654	.364
¼	6.6	1.94	5.11	1.53	1.62	1.66	1.44	.614	.861	.657	.663	.371
L 4 × 3½ × ½	11.9	3.50	5.32	1.94	1.23	1.25	3.79	1.52	1.04	1.00	.722	.750
⅜	9.1	2.67	4.18	1.49	1.25	1.21	2.95	1.17	1.06	.955	.727	.755
¼	6.2	1.81	2.91	1.03	1.27	1.16	2.09	.808	1.07	.909	.734	.759
L 4 × 3 × ½	11.1	3.25	5.05	1.89	1.25	1.33	2.42	1.12	.827	.864	.639	.543
⅜	8.5	2.48	3.96	1.46	1.26	1.28	1.92	.866	.879	.782	.644	.551
¼	5.8	1.69	2.77	1.00	1.28	1.24	1.36	.599	.896	.736	.651	.558
L 3½ × 2½ × ½	9.4	2.75	3.24	1.41	1.09	1.20	1.36	.760	.704	.705	.534	.486
7/16	8.3	2.43	2.91	1.26	1.09	1.18	1.23	.677	.711	.682	.535	.491
⅜	7.2	2.11	2.56	1.09	1.10	1.16	1.09	.592	.719	.660	.537	.496
5/16	6.1	1.78	2.19	.927	1.11	1.14	.939	.504	.727	.637	.540	.501
¼	4.9	1.44	1.80	.755	1.12	1.11	.777	.412	.735	.614	.544	.506
L 3 × 2½ × ½	8.5	2.50	2.08	1.04	.913	1.00	1.30	.744	.722	.750	.520	.667
7/16	7.6	2.21	1.88	.928	.920	.978	1.18	.664	.729	.728	.521	.672
⅜	6.6	1.92	1.66	.810	.928	.956	1.04	.581	.736	.706	.522	.676
5/16	5.6	1.62	1.42	.688	.937	.933	.898	.494	.744	.683	.525	.680
¼	4.5	1.31	1.17	.561	.945	.911	.743	.404	.753	.661	.528	.684
3/16	3.39	.996	.907	.430	.954	.888	.577	.310	.761	.638	.533	.688
L 3 × 2 × ½	7.7	2.25	1.92	1.00	.924	1.08	.672	.474	.546	.583	.428	.414
7/16	6.8	2.00	1.73	.894	.932	1.06	.609	.424	.553	.561	.429	.421
⅜	5.9	1.73	1.53	.781	.940	1.04	.543	.371	.559	.539	.430	.428
5/16	5.0	1.46	1.32	.664	.948	1.02	.470	.317	.567	.516	.432	.435
¼	4.1	1.19	1.09	.542	.957	.993	.392	.260	.574	.493	.435	.440
3/16	3.07	.902	.842	.415	.966	.970	.307	.200	.583	.470	.439	.446
L 2½ × 2 × ⅜	5.3	1.55	.912	.547	.768	.831	.514	.363	.577	.581	.420	.614
5/16	4.5	1.31	.788	.466	.776	.809	.446	.310	.584	.559	.422	.620
¼	3.62	1.06	.654	.381	.784	.787	.372	.254	.592	.537	.424	.626
3/16	2.75	.809	.509	.293	.793	.764	.291	.196	.600	.514	.427	.631

Table 8   STANDARD STEEL PIPE

	Dimensions					Properties		
				Weight per Foot				
Nom. Diam.	Outside Diam.	Inside Diam.	Thick-ness	Plain Ends	Thread & Cplg.	I	A	r
in.	in.	in.	in.	lb	lb	in.^{4}	in.^{2}	in.
$\frac{1}{8}$	.405	.269	.068	.24	.25	.001	.072	.12
$\frac{1}{4}$	.540	.364	.088	.42	.43	.003	.125	.16
$\frac{3}{8}$	.675	.493	.091	.57	.57	.007	.167	.21
$\frac{1}{2}$	.840	.622	.109	.85	.85	.017	.250	.26
$\frac{3}{4}$	1.050	.824	.113	1.13	1.13	.037	.333	.33
1	1.315	1.049	.133	1.68	1.68	.087	.494	.42
$1\frac{1}{4}$	1.660	1.380	.140	2.27	2.28	.195	.669	.54
$1\frac{1}{2}$	1.900	1.610	.145	2.72	2.73	.310	.799	.62
2	2.375	2.067	.154	3.65	3.68	.666	1.075	.79
$2\frac{1}{2}$	2.875	2.469	.203	5.79	5.82	1.530	1.704	.95
3	3.500	3.068	.216	7.58	7.62	3.017	2.228	1.16
$3\frac{1}{2}$	4.000	3.548	.226	9.11	9.20	4.788	2.680	1.34
4	4.500	4.026	.237	10.79	10.89	7.233	3.174	1.51
5	5.563	5.047	.258	14.62	14.81	15.16	4.300	1.88
6	6.625	6.065	.280	18.97	19.19	28.14	5.581	2.25
8	8.625	7.981	.322	28.55	28.81	72.49	8.399	2.94
10	10.750	10.020	.365	40.48	41.13	160.7	11.91	3.67
12	12.750	12.000	.375	49.56	50.71	279.3	14.58	4.38

Table 9   PLASTIC SECTION MODULI AROUND THE X-X AXIS

Shape	Plastic Modulus Z in.^{3}	Shape	Plastic Modulus Z in.^{3}
W 36 × 230	943	W 24 × 68	176
W 33 × 220	838	W 21 × 68	160
W 36 × 194	768	W 24 × 61	152
W 36 × 182	718	W 24 × 55	134
W 36 × 170	668	W 21 × 55	126
W 36 × 160	625	W 18 × 55	112
W 36 × 150	581	W 21 × 49	108
W 33 × 141	514	W 21 × 44	95.3
W 36 × 135	510	W 18 × 40	78.4
W 33 × 130	467	W 16 × 40	72.8
W 33 × 118	415	W 18 × 35	66.8
W 30 × 116	378	S 12 × 50	61.2
W 30 × 108	346	W 16 × 31	54.0
W 30 × 99	313	W 14 × 26	40.0
W 27 × 94	278	W 14 × 22	33.1
W 24 × 94	253	S 10 × 25.4	28.4
W 27 × 84	244	W 8 × 20	19.1
S 24 × 90	222	W 8 × 17	15.9
S 24 × 79.9	205	S 7 × 20	14.5

Table 10 AMERICAN STANDARD TIMBER SIZES, PROPERTIES FOR DESIGNING

Nominal Size	American Standard Dressed Size	Area of Section	Weight per Foot	Moment of Inertia	Section Modulus	Nominal Size	American Standard Dressed Size	Area of Section	Weight per Foot	Moment of Inertia	Section Modulus
in.	in.	in.²	lb	in.⁴	in.³	in.	in.	in.²	lb	in.⁴	in.³
2 × 4	1⅝ × 3⅝	5.89	1.64	6.45	3.56	10 × 10	9½ × 9½	90.3	25.0	679	143
6	5⅝	9.14	2.54	24.1	8.57	12	11½	109	30.3	1204	209
8	7½	12.2	3.39	57.1	15.3	14	13½	128	35.6	1948	289
10	9½	15.4	4.29	116	24.4	16	15½	147	40.9	2948	380
12	11½	18.7	5.19	206	35.8	18	17½	166	46.1	4243	485
14	13½	21.9	6.09	333	49.4	20	19½	185	51.4	5870	602
16	15½	25.2	6.99	504	65.1	22	21½	204	56.7	7868	732
18	17½	28.4	7.90	726	82.9	24	23½	223	62.0	10274	874
3 × 4	2⅝ × 3⅝	9.52	2.64	10.4	5.75	12 × 12	11½ × 11½	132	36.7	1458	253
6	5⅝	14.8	4.10	38.9	13.8	14	13½	155	43.1	2358	349
8	7½	19.7	5.47	92.3	24.6	16	15½	178	49.5	3569	460
10	9½	24.9	6.93	188	39.5	18	17½	201	55.9	5136	587
12	11½	30.2	8.39	333	57.9	20	19½	224	62.3	7106	729
14	13½	35.4	9.84	538	79.7	22	21½	247	68.7	9524	886
16	15½	40.7	11.3	815	105	24	23½	270	75.0	12437	1058
18	17½	45.9	12.8	1172	134	14 × 14	13½ × 13½	182	50.6	2768	410
4 × 4	3⅝ × 3⅝	13.1	3.65	14.4	7.94	16	15½	209	58.1	4189	541
6	5⅝	20.4	5.66	53.8	19.1	18	17½	236	65.6	6029	689
8	7½	27.2	7.55	127	34.0	20	19½	263	73.1	8342	856
10	9½	34.4	9.57	259	54.5	22	21½	290	80.6	11181	1040
12	11½	41.7	11.6	459	79.9	24	23½	317	88.1	14600	1243
14	13½	48.9	13.6	743	110	16 × 16	15½ × 15½	240	66.7	4810	621
16	15½	56.2	15.6	1125	145	18	17½	271	75.3	6923	791
18	17½	63.4	17.6	1619	185	20	19½	302	83.9	9578	982
						22	21½	333	92.5	12837	1194
6 × 6	5½ × 5½	30.3	8.40	76.3	27.7	24	23½	364	101	16763	1427
8	7½	41.3	11.4	193	51.6						
10	9½	52.3	14.5	393	82.7	18 × 18	17½ × 17½	306	85.0	7816	893
12	11½	63.3	17.5	697	121	20	19½	341	94.8	10813	1109
14	13½	74.3	20.6	1128	167	22	21½	376	105	14493	1348
16	15½	85.3	23.6	1707	220	24	23½	411	114	18926	1611
18	17½	96.3	26.7	2456	281	26	25½	446	124	24181	1897
20	19½	107.3	29.8	3398	349	20 × 20	19½ × 19½	380	106	12049	1236
8 × 8	7½ × 7½	56.3	15.6	264	70.3	22	21½	419	116	16150	1502
10	9½	71.3	19.8	536	113	24	23½	458	127	21089	1795
12	11½	86.3	23.9	951	165	26	25½	497	138	26945	2113
14	13½	101.3	28.0	1538	228	28	27½	536	149	33795	2458
16	15½	116.3	32.0	2327	300	24 × 24	23½ × 23½	552	153	25415	2163
18	17½	131.3	36.4	3350	383	26	25½	599	166	32472	2547
20	19½	146.3	40.6	4634	475	28	27½	646	180	40727	2962
22	21½	161.3	44.8	6211	578	30	29½	693	193	50275	3408

All properties and weights given are for dressed size only. The weights given above are based on assumed average weight of 40 lb per cubic foot. Table compiled by the National Lumber Manufacturers Association.

Table 11 DEFLECTIONS AND SLOPES OF ELASTIC CURVES FOR VARIOUSLY LOADED BEAMS

Loading	Equation of Elastic Curve	
	Maximum Deflection	Slope at End
	$v = \dfrac{P}{6EI}(2L^3 - 3L^2x + x^3)$ $v_{max} = v(0) = \dfrac{PL^3}{3EI}$	$\theta(0) = -\dfrac{PL^2}{2EI}$
	$v = \dfrac{q_o}{24EI}(x^4 - 4L^3x + 3L^4)$ $v_{max} = v(0) = \dfrac{q_oL^4}{8EI}$	$\theta(0) = -\dfrac{q_oL^3}{6EI}$
	$v = \dfrac{q_ox}{24EI}(L^3 - 2Lx^2 + x^3)$ $v_{max} = v(L/2) = \dfrac{5q_oL^4}{384EI}$	$\theta(0) = -\theta(L) = \dfrac{q_oL^3}{24EI}$
	When $0 \le x \le a$, then $v = \dfrac{Pb}{6EIL}[(L^2 - b^2)x - x^3]$ When $a = b = \dfrac{L}{2}$, then $v = \dfrac{Px}{48EI}(3L^2 - 4x^2)$ $v_{max} = v(L/2) = \dfrac{PL^3}{48EI}$	See Example 11-4. $\left(0 \le x \le \dfrac{L}{2}\right)$ $\theta(0) = -\theta(L) = \dfrac{PL^2}{16EI}$
	$v = -\dfrac{M_ox}{6EIL}(L^2 - x^2)$ $v_{max} = v(L/\sqrt{3}) = -\dfrac{M_oL^2}{9\sqrt{3}\,EI}$	$\theta(0) = -\dfrac{\theta(L)}{2} = -\dfrac{M_oL}{6EI}$
	$v_a = v(a) = \dfrac{Pa^2}{6EI}(3L - 4a)$ $v_{max} = v(L/2) = \dfrac{Pa}{24EI}(3L^2 - 4a^2)$	$\theta(0) = \dfrac{Pa}{2EI}(L - a)$

Table 12   FIXED-END ACTIONS OF PRISMATIC BEAMS*

Loading	Moments*	Reactions*
	$M_A = -M_B = -\dfrac{q_o L^2}{12}$	$R_A = R_B = -\dfrac{q_o L}{2}$
	$M_A = -\dfrac{Pab^2}{L^2}$ $M_B = \dfrac{Pba^2}{L^2}$	$R_A = -\dfrac{Pb^2}{L^3}(3a + b)$ $R_B = -\dfrac{Pa^2}{L^3}(a + 3b)$
	$M_A = -\dfrac{q_o L^2}{30}$ $M_B = \dfrac{q_o L^2}{20}$	$R_A = -\dfrac{3q_o L}{20}$ $R_B = -\dfrac{7q_o L}{20}$
	$M_A = \dfrac{2EI}{L}\theta_B$ $M_B = \dfrac{4EI}{L}\theta_B$	$R_A = \dfrac{6EI}{L^2}\theta_B$ $R_B = -\dfrac{6EI}{L^2}\theta_B$
	$M_A = -\dfrac{6EI}{L^2}\Delta$ $M_B = -\dfrac{6EI}{L^2}\Delta$	$R_A = -\dfrac{12EI}{L^3}\Delta$ $R_B = \dfrac{12EI}{L^3}\Delta$
	$M_A = M_o\left(-1 + 4\dfrac{a}{L} - \dfrac{3a^2}{L^2}\right)$ $M_B = \dfrac{M_o a}{L}\left(2 - 3\dfrac{a}{L}\right)$	$R_A = \dfrac{6M_o a}{L^2}\left(1 - \dfrac{a}{L}\right)$ $R_B = -\dfrac{6M_o a}{L^2}\left(1 - \dfrac{a}{L}\right)$

*For all the cases tabulated, the positive senses of the end moments and reactions are the same as those shown in the first diagram for uniformly distributed loading. *The special sign convention used here is that adopted for the displacement method in Art. 12-6.*

Index